AF411171

Natural Fiber Based Composites

Natural Fiber Based Composites

Editor

Philippe Evon

MDPI • Basel • Beijing • Wuhan • Barcelona • Belgrade • Manchester • Tokyo • Cluj • Tianjin

Editor
Philippe Evon
Université de Toulouse
France

Editorial Office
MDPI
St. Alban-Anlage 66
4052 Basel, Switzerland

This is a reprint of articles from the Special Issue published online in the open access journal *Coatings* (ISSN 2079-6412) (available at: https://www.mdpi.com/journal/coatings/special_issues/fiber_based).

For citation purposes, cite each article independently as indicated on the article page online and as indicated below:

LastName, A.A.; LastName, B.B.; LastName, C.C. Article Title. *Journal Name* **Year**, *Volume Number*, Page Range.

ISBN 978-3-0365-2002-5 (Hbk)
ISBN 978-3-0365-2003-2 (PDF)

Contents

About the Editor

Philippe Evon

Dr. EVON is Research Engineer at the Laboratoire de Chimie Agro-industrielle (LCA). He specializes in the valorization of wastes from biomass to produce extracts and to design agromaterials. He is mainly developing studies for using biomass as raw material for:

- Producing bioactive extracts through fractionation processes using "green" solvents and the twin-screw extrusion technology as continuous extraction technique.

- The manufacture of agromaterials by combining single- or twin-screw extrusion technologies with molding processes (e.g. injection-molding or thermopressing).

He is the Manager of the LCA's Industrial Technological Hall "AGROMAT" dedicated to agromaterial's (https://www6.toulouse.inra.fr/lca/AGROMAT), which is located in Tarbes (South-West of France).

Editorial

Special Issue "Natural Fiber Based Composites"

Philippe Evon

Laboratoire de Chimie Agro-Industrielle, Université de Toulouse, INP, ENSIACET, 4 Allée Emile Monso, 31030 Toulouse, France; Philippe.Evon@toulouse-inp.fr

Citation: Evon, P. Special Issue "Natural Fiber Based Composites". *Coatings* **2021**, *11*, 1031. https://doi.org/10.3390/coatings11091031

Received: 25 August 2021
Accepted: 26 August 2021
Published: 27 August 2021

Publisher's Note: MDPI stays neutral with regard to jurisdictional claims in published maps and institutional affiliations.

In the last twenty years, the use of cellulosic and lignocellulosic agricultural by-products for composite applications has been of great interest, especially for reinforcing matrices. Fibers of renewable origin have many advantages. They are abundant and cheap, they have a reduced impact on the environment, and they are also independent from fossil resources. Their ability to mechanically reinforce thermoplastic matrices is well known, as is their natural heat insulation ability. The matrices can themselves be of renewable origin (e.g., proteins, starch, polylactic acid, polyhydroxyalkanoates, polyamides, etc.), thus contributing to the development of 100% bio-based composites with a controlled end of life.

This Special Issue's objective is to give an inventory of the latest research in this area of composites reinforced with natural fibers, focusing in particular on the preparation and molding processes of such materials (e.g., extrusion, injection-molding, hot pressing, etc.) and their characterization. It contains one review and nineteen research reports authored by researchers from four continents and sixteen countries, namely, Brazil, China, France, Italy, Japan, Malaysia, Mexico, Pakistan, Poland, Qatar, Serbia, Slovenia, Spain, Sweden, Tunisia, and Vietnam.

An overview of the chemical treatments, manufacturing techniques, and potential lightweight engineering applications of natural fiber based biodegradable composites is provided in a comprehensive review paper [1]. Many natural fibers are presented, including jute and sisal. These have been utilized for ages in several applications, such as ropes, building materials, particle boards, automotive industry, etc., and the chemical treatments and surface modifications are presented as solutions to improve the quality of the natural fibers.

Some of the research articles in this Special Issue suggest the use of very different raw materials to obtain biobased materials that may find future applications in many fields, e.g., agriculture, ecological engineering, construction, load-bearing composites, and packaging.

- Especially, once denaturated in a twin-screw extruder, sunflower proteins can be used for their thermoplastic behavior to produce slow release fertilizers through injection-molding with urea and/or municipal biowastes acting as additional sources of nutrients for plants [2].
- Fibers from linseed flax straw may be mechanically extracted in a continuous mode before their use in geotextile applications [3].
- Constituting an interesting source of vegetable squalene in its seeds, the stems of the amaranth plant are presented as a new perspective for building applications with low-density insulation blocks obtained from their pith fraction and hardboards from the bark one [4].
- After their mechanical mixing, flax and polylactide fibers can be transformed into green sound-absorbing composite materials through hot pressing, and various structure and profiling can be obtained by using different multilayer structures of nonwovens and adjusting the pressing conditions [5].
- The same sound-absorption ability can be obtained from hot pressed composites made from polycaprolactone (PCL) and kapok fibers as thermoplastic matrix and mechanical

1

reinforcement, respectively, their sound-absorption performance depending on their volume density, the mass fraction of kapok fibers inside the materials, and their thickness [6].

- A thermoplastic biocomposite based on hemp fibers and polyamide 11 (PA11) has also been developed through a three-step process, i.e., a wrapping operation to obtain 100% biosourced commingled yarns followed by their weaving to produce two different fabrics and then the molding through thermocompression, and the resulting composites, with stacking of two cross-plies, were characterized regarding mechanical strength [7].
- Electrospun films developed from wastewater treatment plant sludge (WTPS) are also presented as new, local, circular, renewable, and environmentally friendly packaging materials, their characteristics (i.e., tensile properties, contact angle, and surface properties) being largely influenced by both purification and treatment of WTPS to make it suitable for the electrospinning process [8].

Some other research articles propose various pre-treatments on natural fibers with the objectives to favor both the manufacturing and use properties of the resulting natural fiber-based composites or cellulosic textiles. Pre-treatments for plant fibers are very popular for increasing the fineness of bundles, promoting individualization of fibers, modifying the fiber-matrix interface inside composites, or even reducing their water uptake. For subsequent textile applications, the pre-treatments can also be proposed to make the cellulosic fibers more resistant to fungi, bacteria, fire, etc.

- In particular, flax-polylactic acid (PLA) non-woven load-bearing composites were obtained after two different physical pre-treatments of flax tows (i.e., ultrasound or gamma irradiation) rather than the addition of chemicals, and this resulted in effective fiber individualization and a significant increase in the stress at break after the ultrasound pre-treatment [9].
- Hot water, green liquor, and sodium chlorite pre-treatments were also suggested to improve the thermal stability and UV resistance performance of films made from cellulose nanofibers (CNFs) from sugarcane bagasse and spruce [10].
- Another pre-treatment example consists in the surface modification on the nanoscale of cotton fabrics using electrospun sericin/PNIPAM, and this pre-treatment allowed the fast forming of cotton fabrics that also revealed promising antimicrobial activity for subsequent textile applications [11].
- Antibacterial efficiency of textile cellulosic fibers can also be enhanced by means of microencapsulation and green grafting strategies using the bactericidal activity of chemicals from natural essential oils [12].
- In the same way, the grafting of chitosan-essential oil microcapsules onto cellulosic fibers would make it possible to consider the manufacture of protective textile substrates such as antimicrobial masks, bacteriostatic fabrics and healthcare textiles [13].
- It is also possible to enhance the flame resistance of cellulosic fibers using an eco-friendly coating consisting in the grafting of acrylic acid onto the surface of cotton using plasma technology for improved attachment of acrylate phosphate monomer, and this opens up the possibility of producing flame-retardant cellulosic textiles [14].

Additionally, last but not least, olive pomace (OP) is presented as a potential fibrous reinforcement in polypropylene (PP) and polyethylene (PE) biocomposite materials [15]. More specifically, this contribution proposes a life cycle assessment (LCA) of these new construction materials (e.g., decking), and the latter confirmed their environmental friendliness. This constitutes a new perspective for the valorization of OP, an agricultural by-product widely present in southern Europe.

The plant world is also the source of natural substances other than fibers, which can also find applications in the field of composite materials, e.g., in the building industry. By way of example, one research article in this Special Issue evidences the interest to add mucilages extracted from five specific plants to improve the quality of lime-based mortars, making these natural extracts promising additives for restoring and protecting historical buildings [16]. Vegetable raw materials can also be the source of natural dyes, possibly useful in the construction sector (case of the exterior wall paints [17]), but also for textiles, paper, etc. In particular, using natural dyes on textiles has many advantages, as they are healthier and more environmentally friendly than the synthetic ones. A natural dye extracted from the purple petals of an invasive plant has thus been successfully used for the screen printing of both cotton and polyester woven fabrics just as various papers [18].

Plants can also be the source of natural adsorbents. Another research report in this Special Issue thus presents the obtaining of efficient biobased carbon adsorbents [19]. Prepared from spruce bark residues, these will find application for efficient removal of reactive dyes and colors from synthetic effluents, and may be used for treating contaminated wastewater to remove pollutants.

In addition to the vegetable fibers and other chemicals from natural resources, the animal world is also the source of promising renewable biopolymers for material applications. In particular, chitin is one of the most abundant polysaccharides, known to possibly act as a structural material in biological systems. In [20], nanochitin/polystyrene composite particles are prepared by Pickering emulsion polymerization using scaled-down chitin nanofibers. Owing to both their biocompatibility and hydrophobicity, these are promising candidates for future medical uses, e.g., by being used as carriers to control the release of hydrophobic drugs over time.

This Special Issue thus presents a very wide range of topics. It provides an update on current research in the field of natural fiber based composite materials. I am convinced that all these contributions will be a source of inspiration for the development of new composites. Generally speaking, these new materials are environmentally friendly and will undoubtedly find numerous applications in the years to come in many sectors.

Acknowledgments: I would like to thank all the authors for their valuable contributions to this Special Issue, the reviewers for their reviews and useful comments allowing the improvement of the submitted papers, and the journal editors for their kind support throughout the production of this Special Issue.

Conflicts of Interest: The authors declare no conflict of interest.

References

1. Khalid, M.; Imran, R.; Arif, Z.; Akram, N.; Arshad, H.; Al Rashid, A.; García Márquez, F. Developments in Chemical Treatments, Manufacturing Techniques and Potential Applications of Natural-Fibers-Based Biodegradable Composites. *Coatings* **2021**, *11*, 293. [CrossRef]
2. Evon, P.; Labonne, L.; Padoan, E.; Vaca-Garcia, C.; Montoneri, E.; Boero, V.; Negre, M. A New Composite Biomaterial Made from Sunflower Proteins, Urea, and Soluble Polymers Obtained from Industrial and Municipal Biowastes to Perform as Slow Release Fertiliser. *Coatings* **2021**, *11*, 43. [CrossRef]
3. Khan, S.; Labonne, L.; Ouagne, P.; Evon, P. Continuous Mechanical Extraction of Fibres from Linseed Flax Straw for Subsequent Geotextile Applications. *Coatings* **2021**, *11*, 852. [CrossRef]
4. Evon, P.; de Langalerie, G.; Labonne, L.; Merah, O.; Talou, T.; Ballas, S.; Véronèse, T. Low-Density Insulation Blocks and Hardboards from Amaranth (Amaranthus cruentus) Stems, a New Perspective for Building Applications. *Coatings* **2021**, *11*, 349. [CrossRef]
5. Gliscinska, E.; Perez de Amezaga, J.; Michalak, M.; Krucinska, I. Green Sound-Absorbing Composite Materials of Various Structure and Profiling. *Coatings* **2021**, *11*, 407. [CrossRef]
6. Lyu, L.; Zhang, D.; Tian, Y.; Zhou, X. Sound-Absorption Performance and Fractal Dimension Feature of Kapok Fibre/Polycaprolactone Composites. *Coatings* **2021**, *11*, 1000. [CrossRef]
7. Laqraa, C.; Ferreira, M.; Rashed Labanieh, A.; Soulat, D. Elaboration by Wrapping Process and Multiscale Characterisation of Thermoplastic Bio-Composite Based on Hemp/PA11 Constituents. *Coatings* **2021**, *11*, 770. [CrossRef]
8. Lavrič, G.; Miletić, A.; Pilić, B.; Medvešček, D.; Nastran, S.; Vrabič-Brodnjak, U. Development of Electrospun Films from Wastewater Treatment Plant Sludge. *Coatings* **2021**, *11*, 733. [CrossRef]
9. Gautreau, M.; Kervoelen, A.; Barteau, G.; Delattre, F.; Colinart, T.; Pierre, F.; Hauguel, M.; Le Moigne, N.; Guillon, F.; Bourmaud, A.; et al. Fibre Individualisation and Mechanical Properties of a Flax-PLA Non-Woven Composite Following Physical Pre-Treatments. *Coatings* **2021**, *11*, 846. [CrossRef]
10. Luo, L.; Wang, X.; Zhang, S.; Yuan, X.; Li, M.; Wang, S. Contribution of Different Pretreatments to the Thermal Stability and UV Resistance Performance of Cellulose Nanofiber Films. *Coatings* **2021**, *11*, 247. [CrossRef]

11. Li, J.; Wang, B.; Cheng, D.; Liu, Z.; Lv, L.; Guo, J.; Lu, Y. Electrospun Sericin/PNIPAM-Based Nano-Modified Cotton Fabric with Multi-Function Responsiveness. *Coatings* **2021**, *11*, 632. [CrossRef]
12. Dridi, D.; Bouaziz, A.; Gargoubi, S.; Zouari, A.; B'chir, F.; Bartegi, A.; Majdoub, H.; Boudokhane, C. Enhanced Antibacterial Efficiency of Cellulosic Fibers: Microencapsulation and Green Grafting Strategies. *Coatings* **2021**, *11*, 980. [CrossRef]
13. Bouaziz, A.; Dridi, D.; Gargoubi, S.; Zouari, A.; Majdoub, H.; Boudokhane, C.; Bartegi, A. Study on the Grafting of Chitosan-Essential Oil Microcapsules onto Cellulosic Fibers to Obtain Bio Functional Material. *Coatings* **2021**, *11*, 637. [CrossRef]
14. Zouari, R.; Gargoubi, S. Enhancing Flame Resistance of Cellulosic Fibers Using an Ecofriendly Coating. *Coatings* **2021**, *11*, 179. [CrossRef]
15. Espadas-Aldana, G.; Guaygua-Amaguaña, P.; Vialle, C.; Belaud, J.; Evon, P.; Sablayrolles, C. Life Cycle Assessment of Olive Pomace as a Reinforcement in Polypropylene and Polyethylene Biocomposite Materials: A New Perspective for the Valorization of This Agricultural By-Product. *Coatings* **2021**, *11*, 525. [CrossRef]
16. Alisi, C.; Bacchetta, L.; Bojorquez, E.; Falconieri, M.; Gagliardi, S.; Insaurralde, M.; Martinez, M.; Orozco, A.; Persia, F.; Sprocati, A.; et al. Mucilages from Different Plant Species Affect the Characteristics of Bio-Mortars for Restoration. *Coatings* **2021**, *11*, 75. [CrossRef]
17. Huang, X.; Wang, C.; Zhu, D. An Experimental Study for the Improvement of the Stain Resistance for Exterior Wall Paints in a Western City in China. *Coatings* **2021**, *11*, 220. [CrossRef]
18. Klančnik, M. Printing with Natural Dye Extracted from Impatiens glandulifera Royle. *Coatings* **2021**, *11*, 445. [CrossRef]
19. dos Reis, G.; Larsson, S.; Thyrel, M.; Pham, T.; Claudio Lima, E.; de Oliveira, H.; Dotto, G. Preparation and Application of Efficient Biobased Carbon Adsorbents Prepared from Spruce Bark Residues for Efficient Removal of Reactive Dyes and Colors from Synthetic Effluents. *Coatings* **2021**, *11*, 772. [CrossRef]
20. Watanabe, R.; Izaki, K.; Yamamoto, K.; Kadokawa, J. Preparation of Nanochitin/Polystyrene Composite Particles by Pickering Emulsion Polymerization Using Scaled-Down Chitin Nanofibers. *Coatings* **2021**, *11*, 672. [CrossRef]

Article

Coffee Wastes as Sustainable Flame Retardants for Polymer Materials

Henri Vahabi [1,*], Maryam Jouyandeh [1], Thibault Parpaite [2], Mohammad Reza Saeb [3] and Seeram Ramakrishna [4,*]

1   CentraleSupélec, Université de Lorraine, LMOPS, 57000 Metz, France; maryam.jouyande@gmail.com
2   Precision Macromolecular Chemistry Group, Institut Charles Sadron, CNRS-UPR 22, 23 Rue du Loess, CEDEX 2, 67034 Strasbourg, France; thibault.parpaite@ics-cnrs.unistra.fr
3   Department of Polymer Technology, Faculty of Chemistry, Gdańsk University of Technology, G. Narutowicza 11/12, 80-233 Gdańsk, Poland; mrsaeb2008@gmail.com
4   Center for Nanofibers & Nanotechnology, National University of Singapore, Singapore 119260, Singapore
*   Correspondence: henri.vahabi@univ-lorraine.fr (H.V.); seeram@nus.edu.sg (S.R.); Tel.: +33-(0)3-72-74-98-66 (H.V.)

Abstract: Development of green flame retardants has become a core part of the attention of material scientists and technologists in a paradigm shift from general purpose to specific sustainable products. This work is the first report on the use of coffee biowastes as sustainable flame retardants for epoxy, as a typical highly flammable polymer. We used spent coffee grounds (SCG) as well as SCG chemically modified with phosphorus (P-SCG) to develop a sustainable highly efficient flame retardant. A considerable reduction in the peak of heat release rate (pHRR) by 40% was observed in the pyrolysis combustion flow calorimeter analysis (PCFC), which proved the merit of the used coffee biowastes for being used as sustainable flame retardants for polymers. This work would open new opportunities to investigate the impact of other sorts of coffee wastes rather than SCG from different sectors of the coffee industry on polymers of different family.

Keywords: epoxy; sustainability; epoxy; flame retardancy; flame retardant; coffee wastes; biowaste

Citation: Vahabi, H.; Jouyandeh, M.; Parpaite, T.; Saeb, M.R.; Ramakrishna, S. Coffee Wastes as Sustainable Flame Retardants for Polymer Materials. *Coatings* **2021**, *11*, 1021. https://doi.org/10.3390/coatings11091021

Academic Editor: Philippe Evon

Received: 4 August 2021
Accepted: 24 August 2021
Published: 26 August 2021

Publisher's Note: MDPI stays neutral with regard to jurisdictional claims in published maps and institutional affiliations.

1. Introduction

Polymer materials scientists together with environmental engineers today are at a critical juncture to find practical solutions to the huge amount of plastic wastes. In this regard, exploring and formulating green materials, developing green and clean technologies, and unifying low-cost recycling strategies have become the subjects of heated debates. In line with such attempts, several guidelines and legislations have been proposed and launched by the policy makers and managers for the treatment of biowastes in a value-added pathway rather than making biowastes landfilled or burnt [1–3]. The polymer industry these days makes good use of biowastes to develop sustainable polymer composites. Nevertheless, biowastes in their raw form may lead to poor properties mainly arising from inadequate interfacial adhesion. Therefore, it seems necessary to modify biowastes to attain high-performance polymer/biowastes composites [4,5].

Flame retardancy is a signature of performance of a polymer. Different sorts of flame retardants have been examined in developing flame-retardant polymer systems. In recent years, there has been a worldwide shift in the use of flame retardant additives from conventional additives to bio-based flame retardants. Several sorts of bio-based flame retardants have been used in polymer matrices, Figure 1 [6]. The majority of these bio-based molecules being used as a flame retardant, or as a part of a flame retardant system, are extracted from the raw biomaterials, which itself necessitates applying several steps of pre- and/or post-treatment. Moreover, in some cases like lignin, the botanical origin of molecule as well as the method of extraction may to a large extent govern the final properties of product [7]. Thus, modification of bio-based flame retardants seems crucial.

5

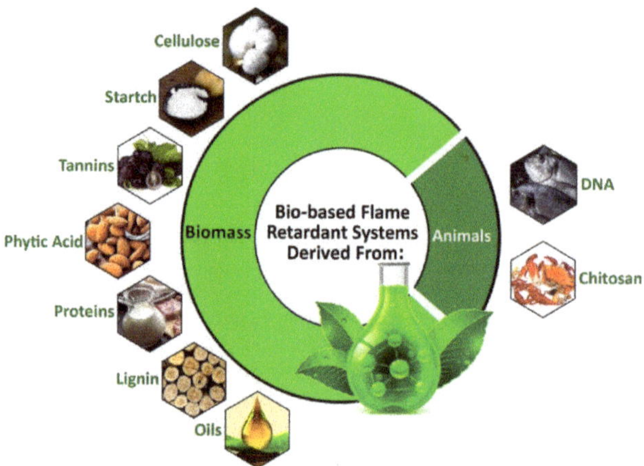

Figure 1. Schematic representation of bio-based flame retardant systems derived from animals (DNA and chitosan) and biomass (cellulose, starch, tannins, phytic acid, proteins, lignin, and oils) [6]. Reproduced/Adapted with permission from [6]. Copyright (2021, Elsevier).

From the sustainability point of view, the use of biowaste materials in flame retardant systems can be considered as a promising solution to plastic disposal and biowaste landfill, mainly due to the availability and low cost of biowastes. So far, several biowastes have been incorporated as flame retardants into several polymers, including "oyster shell powder" in polypropylene [8], polyurethane [9], polyethylene [10], and acrylonitrile-butadiene-styrene copolymer (ABS) [11], "eggshell" in ethylene-vinyl acetate copolymer (EVA) [12] and for intumescent coating system [13], and "rice husk" in epoxy [14]. Sustainable flame retardants are at the core of attention in view of providing environmental safety, cost-effectiveness, and health safety considerations (Figure 2). Nevertheless, finding industrial biowastes in large quantities that guarantee all aspects of sustainability remains as a debate.

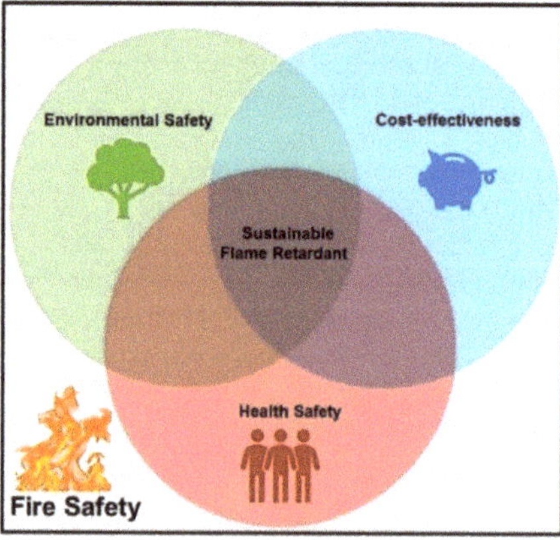

Figure 2. Schematic representation of some important parameters centered to development of sustainable flame retardants.

The byproducts of the coffee industry as well as the spent coffee grounds are huge in amount [15,16]. Some research works already reported the use of spent coffee grounds to produce biodiesel [17,18], biosorbent [19], biogas [20], biocomposites [21], and biofoams [22,23]. Coffee biowastes have also been used in water treatment [24], and gas storage [25] (Figure 3). Recycled coffee grounds contain proteins, minerals, carbohydrates, and especially lignin, cellulose, and hemicellulose [26–28]. Therefore, such biowastes could potentially be useful in flame retardancy because they possess lignin, cellulose, and hemicellulose, which have been found to be promising green flame retardants for different polymers [29]. Their mode of action in flame retardancy is generally in the condensed phase where they generate and increase the char residue. The char formed thereof acts as a physical barrier against heat and mass transfer during the combustion. It is demonstrated that the combination of these materials with phosphorus-based flame retardants improve the quantity and quality of the formed char [29,30].

Figure 3. Applications of spent coffee grounds including agriculture, food industry deodorization, water treatment, construction materials, and bioenergy.

This work aims to unravel, for the first time, the potential of the spent coffee grounds as a flame retardant for polymer materials. In a comprehensive review, we already discussed flame retardancy of epoxy and classified epoxy composites in terms of Flame Retardancy Index (FRI) [31]. Thus, epoxy resin was selected herein as a host polymer due to its versatility in a wide range of application from structural composites to organic coatings and also because of high flammability of epoxy as a model polymer. Two types of the biowastes including spent coffee grounds and phosphorus-modified spent coffee grounds were used varying the loading level. Several composites containing the aforementioned coffee grounds were prepared and evaluated for thermal degradation and micro-calorimetry flame retardancy behaviors.

2. Materials and Methods

Diglycidyl ether of bisphenol a epoxy resin (EPON™ 828) with epoxide equivalent weight of 450–550 g/eq. was purchased from Hexion (Shanghai, China). The amine-based curing agent of Epikure™ F205 with hydrogen equivalent weight of 105 g/eq. and viscosity of 500–700 mPa.s was provided by Hexion (China). Dimethyl phosphite (98%) was manufactured by Sigma-Aldrich (Saint Louis, MO, USA) and used as received. The coffee waste was obtained from the spent coffee grounds collected from a coffee machine. The brand of coffee beans was Lavazza, Arabica 40%/Robusta 60%.

For the preparation of phosphite-treated coffee biowaste (P-coffee), 15 g of coffee waste was added to 150 mL dimethyl phosphite under stirring at 30 °C for 24 h. The resulting phosphite-treated coffee biowaste was collected utilizing centrifugation (2000× *g*, 5 min) and dried at room temperature for seven days.

Epoxy containing coffee biowaste and phosphite-treated coffee biowaste (P-coffee) were prepared varying the amount of biowaste as 5, 15, and 30 wt.% (Table 1). To prepare composites, the fixed amounts of coffee biowaste ground and epoxy resin were mixed and sonicated two times for 5 min. Then, F205 curing agent was added to the resulting mixtures at 2:1 resin to curing agent stoichiometric ratio and thoroughly mixed for 2 min. The resulting composite was casted on a glass substrate and then put in an oven for 3 h at 70 °C for curing. Then, the post-curing was performed for two weeks at ambient temperature. For peeling off films from glass substrate, they were immersed in water for 24 h.

Table 1. Formulations and sample codes of epoxy/coffee biowaste composites (Ep.: epoxy, Coffee: untreated coffee waste, P-coffee: treated coffee waste with dimethyl phosphite).

Sample Code	Epoxy (wt.%)	Coffee (wt.%)	P-Coffee (wt.%)
Ep.	100	0	0
Ep./Coffee5	95	5	0
Ep./Coffee15	85	15	0
Ep./Coffee30	70	30	0
Ep./P-coffee5	95	0	5
Ep./P-coffee15	85	0	15
Ep./P-coffee30	70	0	30

Microstructure and chemical constituents were analyzed using a scanning electron microscope (SEM) Zeiss Gemini SEM 500 (Jena, Germany) with accelerating voltage of 6 keV, coupled with an energy-dispersive X-ray spectroscopy (EDAX) Element SDD Silicon Drift Detector X-Ray Spectrometer 30 mm^2.

Thermal decomposition of samples was investigated using a Setaram Labsys Evo thermogravimetric analyzer (France) in the temperature range of 25–900 °C, at a heating rate of 10 °C·min^{-1} under nitrogen atmosphere with gas flow rate fixed at 100 cm^3·min^{-1}.

Fourier-transform infrared (FTIR) spectroscopy was also conducted on pure coffee and treated coffee wastes in form of powder using a Bruker FTIR Alpha with a resolution of 4 cm^{-1} and 64 scans.

Pyrolysis combustion flow calorimeter analysis (PCFC) was performed to evaluate the flammability behavior of samples according to ASTM D7309 [32]. This test was conducted on a Fire Testing Technology (FTT) Company machine, East Grinstead, UK. Small quantity of samples between 2 and 4 mg was pyrolyzed at a heating rate of 1 °C/s. The gases obtained were conducted to another chamber and combustion took place in presence of oxygen (20%) at 900 °C. The Huggett's relation [33] (1 kg of consumed oxygen corresponds to 13.1 MJ of released energy) was used to calculate the flammability parameters including the peak of heat release rate (pHRR), temperature at pHRR (T_{pHRR}), and the total heat release (THR). A non-conventional horizontal burning test was performed on samples with dimensions of 30 × 100 × 4 mm^3. A Bunsen burner flame was placed on the bottom of sample for 3 s and after the apparition of flame it moved up to continue burning to the end of test. The digital videos were recorded, and selected images were also extracted.

3. Results and Discussions

3.1. Analysis of Phosphorus Modification of Coffee

Pure coffee biowastes (hereafter referred to as Coffee) and treated-coffee (P-coffee) were imaged on SEM. Their chemical composition was also investigated using EDX. Figure 4 shows micrographs obtained from Coffee and P-coffee, and Table 2 summarizes their chemical compositions. The size of particles was between 10 μm and 100 μm. Some more SEM micrographs are available in the Appendix A.

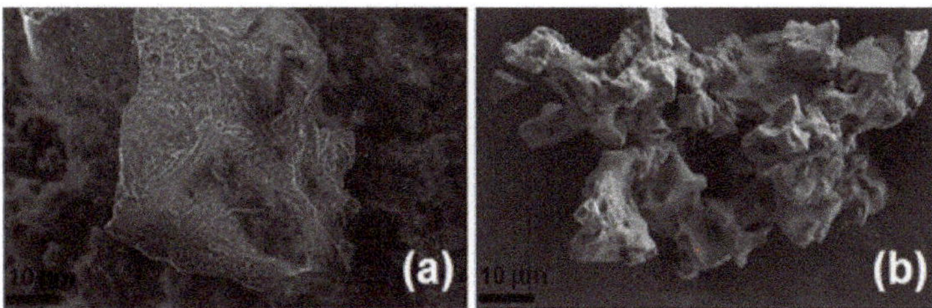

Figure 4. SEM micrographs of unmodified coffee biowaste (Coffee) (**a**) and coffee biowaste modified with phosphorus (P-coffee) (**b**).

Table 2. Chemical composition of unmodified coffee biowaste (Coffee) and coffee biowaste modified with phosphorus (P-coffee) obtained in EDX (the mean of the obtained values at three different points). The elemental analysis proves the presence of phosphorus in P-coffee. The percentage of phosphorus was estimated to be 20 wt.% for P-coffee. Moreover, carbon and oxygen are increased after modification of Coffee.

Sample Code	Carbon (wt.%)	Oxygen (wt.%)	Phosphorus (wt.%)
Coffee	82.6	13.8	-
P-coffee	41.0	39.0	20

Fourier-transform infrared (FTIR) spectroscopy was also conducted on the treated and untreated coffee. Figure 5 displays the FTIR spectra of coffee and P-coffee. The spent coffee grounds generally contain several components such as cellulose, lignin, fatty acids, and hemicellulose [18]. The broad band between 3100 cm^{-1} and 3500 cm^{-1} was attributed to the stretching of −OH groups related to the inter- and intra-molecular hydrogen bonding of some polymeric compounds, such as cellulose and lignin. Two bands at 2921 cm^{-1} and 2851 cm^{-1} can be assigned to symmetric or asymmetric C-H stretching vibration of aliphatic acids, respectively. The band at 1742 cm^{-1} was associated to the vibration of C=O groups in aliphatic esters. For the P-coffee sample, the absorption peak of P=O was situated in the range of 1350–1250 cm^{-1}. Moreover, two bands at 994 cm^{-1} and 925 cm^{-1} corresponds to P-O-C stretching. Although it is not possible to confirm chemical reaction between coffee biowastes and phosphite, the intensity of peaks in the case of P-coffee are indicative of appropriate interaction between components.

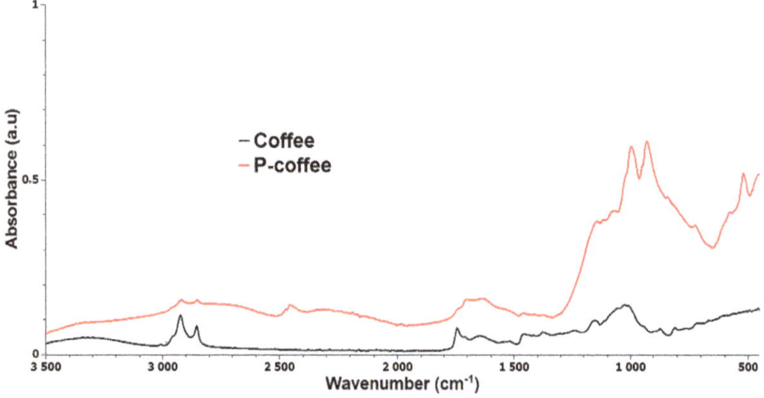

Figure 5. FTIR spectra of untreated (Coffee) and phosphorus treated coffee (P-coffee) samples.

3.2. Analysis of Thermal Degradation

TGA thermograms of Coffee and P-coffee biowastes are compared in Figure 6. Increase in the residual mass is generally a signature of flame retardancy improvement, which reflects the contribution of condensed phase action in controlling the flammability. On one hand, the formation of more char residue is indicative of the decrease in the amount of gases in the vapor phase. Although the content of phosphorus ingredient was 20%, the difference in the TGA thermograms is almost close to 30%, a further evidence for strong interaction between phosphorus and coffee wastes leading to formation of more char residue. On the other hand, the presence of char during the combustion process on the surface of the polymer reduces the release rate of gases to the vapor phase and also protects the underlying polymer from the flame [6].

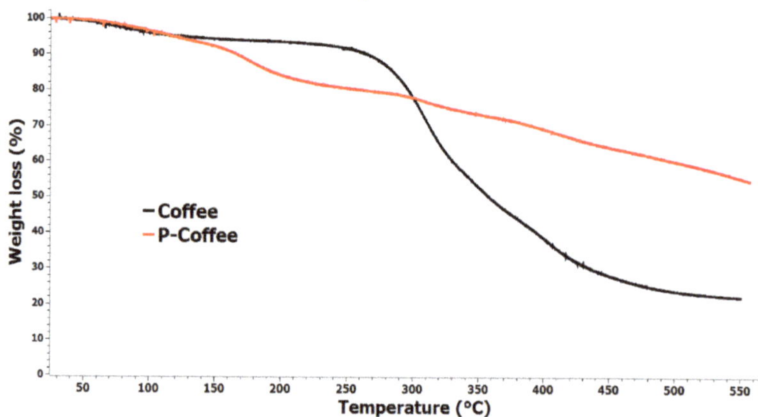

Figure 6. TGA thermograms of unmodified coffee biowaste (Coffee) and coffee biowaste modified with phosphorus (P-coffee) under nitrogen at 10 °C·min^{-1}.

The graph confirmed successful modification of coffee biowastes with phosphorus, on the bedrock of lower thermal stability of P-coffee between 150 °C and 300 °C compared with Coffee. Surprisingly, a significant improvement in thermal stability was observed for P-coffee above 300 °C. At 350 °C, P-coffee lost only 27% of its initial masse, while for Coffee it was ca. 48%. More prominently, degradation of P-coffee is single-stage, indicative of strong interaction or formation of a complex between coffee biowastes and phosphite. The final residues of Coffee and P-coffee coffee were 22 and 55%, respectively. Therefore, the difference of remaining residue for Coffee and P-coffee at the end of the test is more than a theoretical addition of dimethyl phosphite on coffee waste. This difference showed that some chemical interactions occurred during the decomposition, mainly consisting of the formation of double bonds (C=C), cyclization, aromatization, fusion of aromatics, and graphitization leading to formation of char residue.

TGA thermograms of epoxy and epoxy composites containing Coffee and P-coffee biowastes are compared in Figure 7a,b, and some important parameters extracted from the figures are listed in Table 3. Overall, lower thermal stability of composites at the early stage of TGA can be explained by P-coffee and the release of some chemicals. In fact, the presence of P-coffee stabilized and increased the char yield.

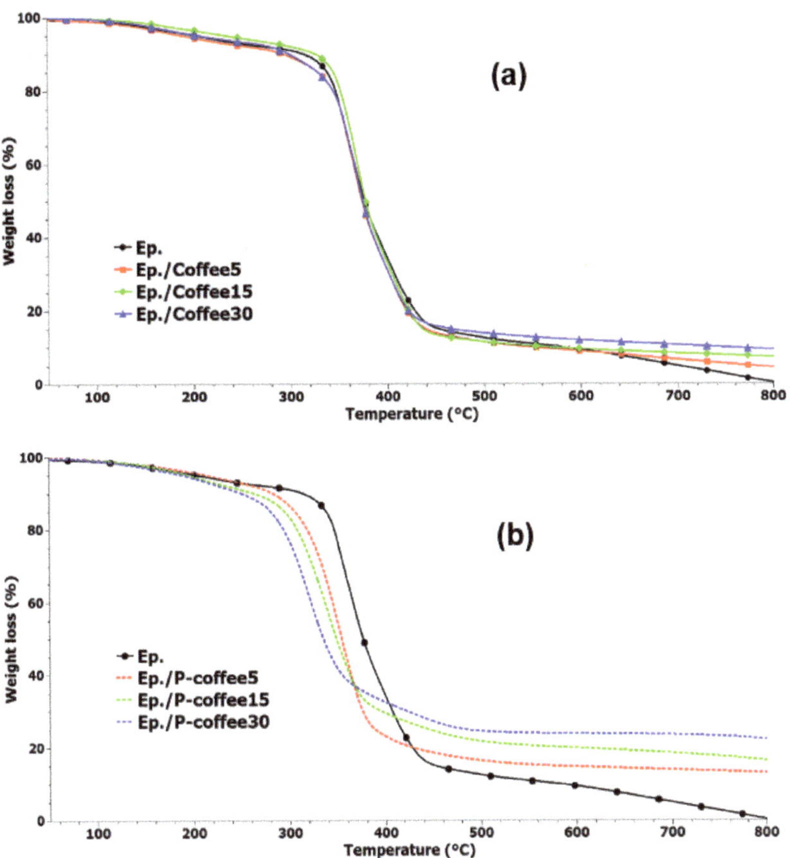

Figure 7. TGA thermograms of epoxy and epoxy composites containing various amounts of unmodified coffee biowaste (Coffee); (**a**) those of coffee biowaste modified with phosphorus (P-coffee) (**b**).

Table 3. Thermal decomposition characteristics of neat epoxy and epoxy composites extracted from TGA diagrams.

Sample Code	T_5 (°C)	T_{10} (°C)	Residue at 800 °C (%)
Coffee	113	264	22.0
P-coffee	113	166	55.0
Ep.	208	314	0.0
Ep./Coffee5	187	292	4.5
Ep./Coffee15	234	326	7.5
Ep./Coffee30	208	299	9.5
Ep./P-coffee5	213	284	13.0
Ep./P-coffee15	194	263	16.5
Ep./P-coffee30	188	250	22.5

For samples containing Coffee (Figure 7a), the thermal decomposition mechanism of epoxy remained almost unaffected, while above 500 °C, composite samples show improved thermal stability. For samples containing P-coffee (Figure 7b), however, decomposition started at a lower temperature with respect to the neat epoxy; the more the P-coffee amount the higher the rate of thermal decomposition. At higher temperatures (above 450 °C), composites obviously have higher thermal stability compared to epoxy and systems containing Coffee, seen in Figure 7a. In Table 3, T_5 and T_{10} are temperatures at which 5% and 10% weight loss takes place, respectively. T_{10} vividly increased from 314 °C for neat epoxy to 326 °C for Ep./Coffee15 sample. Overall, composites were not thermally stable and roughly degraded up to 800 °C. The residue at 800 °C were 4.5, 7.5, and 9.5% for epoxy composites containing 5, 15, and 30 wt.% Coffee, respectively. For instance, the value of T_{10} for Ep./P-coffee30 sample was 250 °C, while it was 314 °C for the neat epoxy. It can be concluded that stabilization of char was observed after 428 °C and all composites showed improved thermal stability compared with the neat epoxy. The remaining residue at 800 °C was 13, 16.5, and 22.5% for epoxy composites containing 5, 15, and 30 wt.% P-coffee, respectively. This suggests the effectiveness of phosphorus modification of coffee biowastes. Derivative thermogravimetric (DTG) curves of epoxy and epoxy composites containing Coffee and P-coffee are also shown in Figure 8a,b. It demonstrates that the main degradation steps of all samples containing unmodified coffee biowaste (Figure 8a) occurs between 280 °C and 470 °C. The presence of modified coffee waste at 15 wt.% and 30 wt.% of loading led to decrease of T_{max} (temperature, at which sample lost maximum of its weight). T_{max} of pure epoxy and Ep./P-coffee5 was 353 °C, while it was 341 °C and 318 °C for Ep./P-coffee15 and Ep./P-coffee30, respectively.

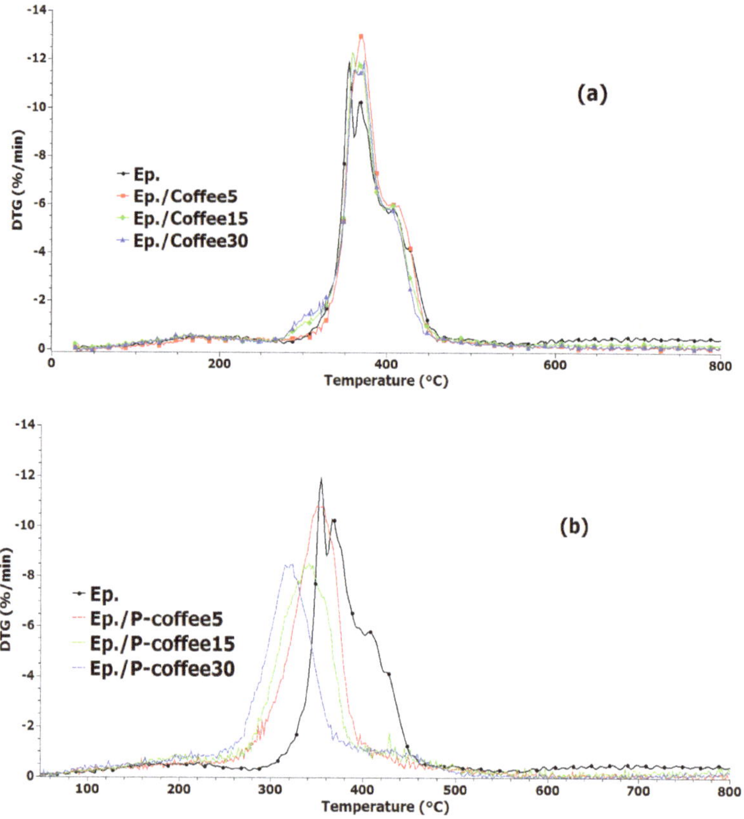

Figure 8. DTG curves of epoxy and epoxy composites containing unmodified coffee biowaste (Coffee) (**a**) and coffee biowaste modified with phosphorus (P-coffee) (**b**).

3.3. Analysis of Flammability

Figure 9 compares the heat release rate (HRR) curves of samples as a function of temperature. The main characteristics of PCFC including peak of heat release rate (pHRR), total heat release (THR), and temperature at pHRR (T_{pHRR}) are extracted from curves and summarized in Table 4.

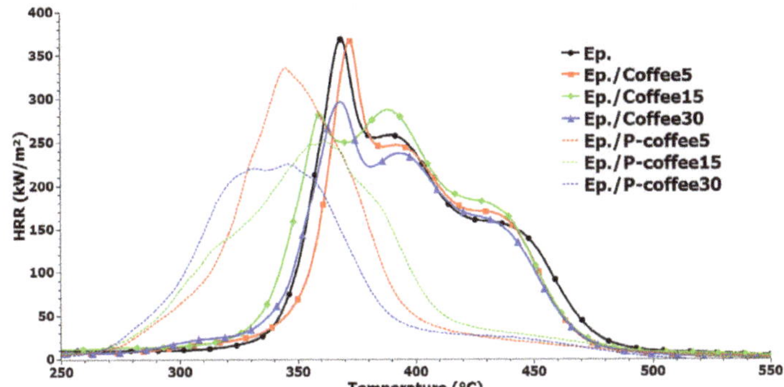

Figure 9. Heat release rate (HRR) curves for the neat epoxy and epoxy composites containing both Coffee and P-coffee obtained in PCFC test.

Table 4. Summary results obtained in PCFC tests for the neat epoxy and epoxy containing Coffee and P-coffee.

Sample Code	pHRR (W/g)	T_{pHRR} (°C)	THR (kJ/g)	Reduction in pHRR (%)
Ep.	370	368	24.5	-
Ep./Coffee5	365	372	25.4	1.4
Ep./Coffee15	282	359	22.2	24
Ep./Coffee30	296	368	24	20
Ep./P-coffee5	334	345	24	10
Ep./P-coffee15	250	360	23.4	32
Ep./P-coffee30	225	346	21.6	40

From Figure 9, two different flammability behaviors are observed for composite samples containing Coffee or P-coffee. For all composites containing Coffee, the curve of HRR initiated around 325 °C, which was more similar to that of neat epoxy. Then, a sharp increase in HRR was observed, before reaching the pHRR. A completely different behavior was observed for samples containing P-coffee, where HRR curves followed a moderate ascending trend in the vicinity of 275 °C. A moderate slope compared with those of the composites containing Coffee was also seen, which ended in pHRR rise and then fall around 420 °C. Such early-stage variation in HRR curves can be explained on account of phosphorous chemical used in treatment of coffee biowaste.

The values of pHRR were almost similar for neat epoxy and epoxy/Coffee5 (370 $W \cdot g^{-1}$ and 365 $W \cdot g^{-1}$, respectively). At higher loadings of 15 and 30 wt.%, however, Coffee more substantially affected pHRR compared to P-coffee (with a reduction of 24% and 20% in pHRR compared with that of neat epoxy). The temperature at which pHRR occurred also remained in 359–372 °C for neat epoxy and epoxy composites containing Coffee. Interestingly, reduction in pHRR was more salient for the samples containing P-coffee, which demonstrated the efficiency of chemical treatment of coffee biowastes in improvement of flammability of polymers. For samples containing 30 wt.% of P-coffee, pHRR was significantly increased by 40%, while for 5 wt.% and 15 wt.% loadings the values of reduction in pHRR were 10% and 32%, respectively. Moreover, temperature at pHRR was significantly decreased for samples containing P-coffee. Since PCFC is not a real flame test, a horizontal flame test was performed to evaluate the flame behavior of samples. Figure 9 displays the extracted photos from recorded videos during the horizontal flame test. All videos are presented as Supporting Information section for experts. After the ignition, a big flame instantaneously appeared in the case of neat epoxy sample leading to a complete and rapidly burning of this sample. At flameout around 30 s, no remaining residue was observed for neat epoxy (Figure 10a). In the presence of Coffee, the inflammation rate was decreased, leading to an increase in time to inflammation up to 90 s. A small quantity of remaining residue was obtained at the end of tests, which is essentially ascribed to the presence of coffee biowaste. Surprisingly, the flame behavior of epoxy was significantly changed in the presence of P-coffee. The rate of burning decreased even for Ep./P-coffee5 sample. An self-extinguishable behavior was observed for samples containing 15 wt.% and 30 wt.% P-coffee. For Ep./P-coffee15, the self-extinguishability was appeared after 40 s of ignition. The self-extinguishability character was more salient for the Ep./P-coffee30 sample even at the beginning of the test. Therefore, the ignition was repeated to initiate the inflammation, but the self-extinguishability character came back after 40 s. Thus, these samples were not consumed due to the aforementioned behavior and a compact char formed, on the zone of inflammation (Figure 10f,g). These flame tests clearly demonstrated the efficiency of treatment of coffee biowaste with phosphorus in flame retardancy of epoxy resins.

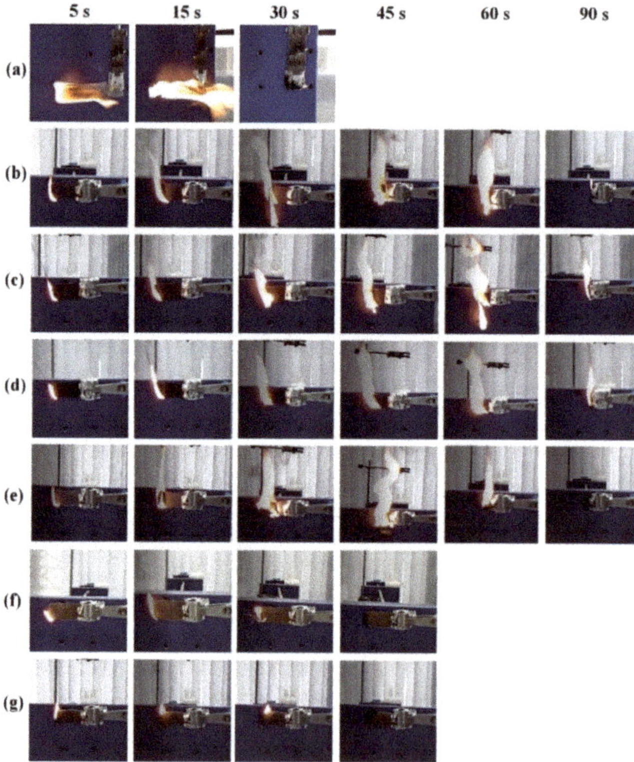

Figure 10. Snapshots taken from videos showing the evolution of flame propagation as a function of time in a horizontal flame test configuration; (**a**) Ep., (**b**) Ep./Coffee5, (**c**) Ep./Coffee15, (**d**) Ep./Coffee30, (**e**) Ep./P-coffee5, (**f**) Ep./P-coffee15, (**g**) Ep./P-coffee30.

3.4. Possible Flame Retardancy Mechanism

The sustainable coffee biowastes used as flame retardant for epoxy as a typical highly flammable polymer can be explained mechanistically. Possible mechanisms of action of P-coffee in epoxy are both actions in the gas and condensed phases. The simultaneous presence of a carbon source, coffee biowaste, and phosphorus in epoxy facilitated and accelerated the formation of a carbonaceous residue playing the role of a barrier against heat and mass diffusion of gases to the gas phase. In the meanwhile, the released phosphorus chemicals captured free radicals in the gas phase and avoided the propagation of flame (Figure 11).

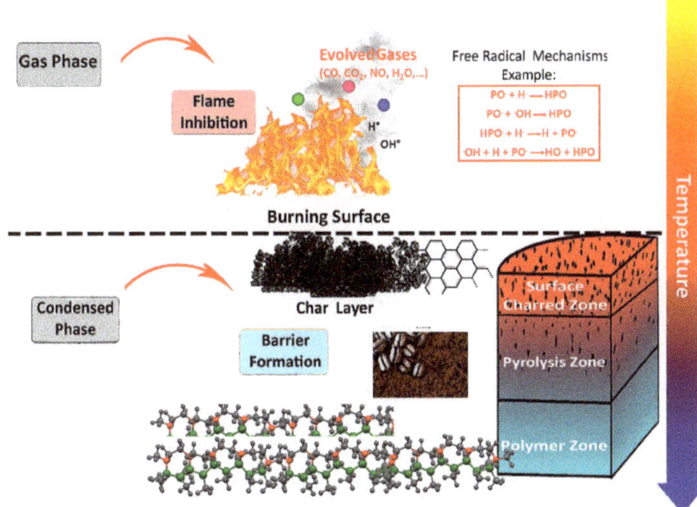

Figure 11. Possible mechanism of flame retardancy of sustainable coffee biowastes in epoxy.

4. Conclusions and Future Perspectives

In this work, spent ground coffee was treated with dimethyl phosphite using a simple method and then incorporated into epoxy resin as flame retardant at 5, 15, and 30 wt.% of loading. The analysis of modified coffee biowaste showed that it contained a high amount of phosphorus (20 wt.%). The remaining residue at the end of TGA test for epoxy containing 30 wt.% modified coffee (P-coffee) was about 22.5% against 0% for the blank epoxy and 9.5% for the epoxy containing 30 wt.% unmodified coffee biowaste (Coffee). The flammability analysis in PCFC demonstrated a significant reduction in the pHRR (40%) for epoxy containing 30 wt.% P-coffee. Moreover, the burning tests revealed an self-extinguishable character for this sample. From these results, it is clear that coffee biowaste has a huge potential to be used as flame retardant. A recent survey highlighted the importance of visionary management of food industry biowastes such as agricultural, food industry, and spent coffee ground biowastes [34]. The need for improving the rate of recycling and composting as well as sustainable strategies for future developments have also been discussed. Nevertheless, there is a global need to enrich the cultural standpoints in a paradigm shift towards minimizing household biowastes. The idea of circular economy is grounded on prevention of waste production as well as management of the natural resources and investment on green and clean productions rather than recycling or reuse of waste materials. There have been some attempts in recent years in the quest for implementing circular economy in polymer industry. For example, life cycle assessment (LCA) is applied as a criterion to assess environmental impact of recycling polyethylene terephthalate (PET) wastes in relation to additives used in processing and also economical features [35]. LCA is a measure of the possibility of management of wastes having eyes at the life cycle of plastics. The use of sustainable biowastes makes it possible to significantly shorten LCA and circular economy interval. The outcome of this short communication unveiled the potential of coffee biowaste as a flame retardant for polymers. Considering the huge amount of coffee industry byproducts as well as collection of coffee biowastes all around the world, future horizon seems promising. The point is that using coffee biowastes not merely allows for taking big steps in line with circular economy requirements, but auspiciously gives rise to developing high-performance polymer materials, herein highly flame-retardant polymer composites.

Author Contributions: Conceptualization, H.V., M.R.S.; methodology, H.V., M.R.S. and S.R.; validation, all authors; formal analysis, H.V., M.J. and T.P.; investigation, H.V., M.J. and T.P.; data curation, H.V., M.J.; writing—original draft preparation, H.V., M.J.; writing—review and editing, all authors; visualization, H.V., M.R.S.; supervision, H.V. and M.R.S. All authors have read and agreed to the published version of the manuscript.

Funding: This research received no external funding.

Institutional Review Board Statement: Not applicable.

Informed Consent Statement: Not applicable.

Data Availability Statement: The data presented in this study are available on request from the corresponding author.

Acknowledgments: H.V. would like to thank IUT de Moselle Est-"Plastinnov" for the PCFC tests.

Conflicts of Interest: The authors declare no conflict of interest.

Appendix A SEM Photos Provided from Coffee and P-Coffee Powders (First Line: Coffee, Second Line: P-Coffee)

Table A1. SEM photos provided from Coffee and P-coffee powders.

Coffee		
P-coffee	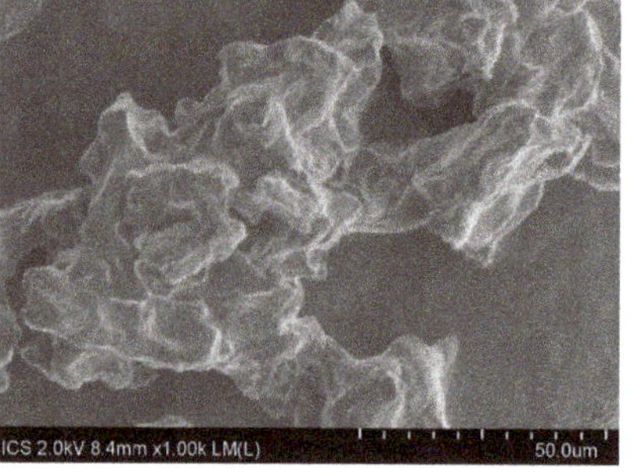	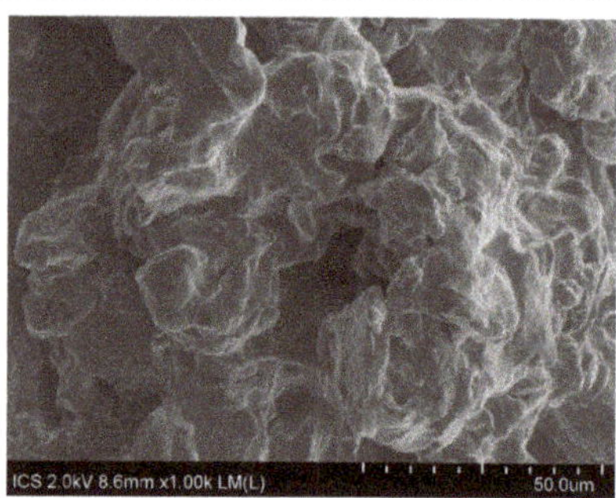

References

1. Lohri, C.R.; Diener, S.; Zabaleta, I.; Mertenat, A.; Zurbrügg, C. Treatment technologies for urban solid biowaste to create value products: A review with focus on low- and mid-dle-income settings. *Rev. Environ. Sci. Bio/Technol.* **2017**, *16*, 81–130. [CrossRef]
2. Lee, P.; Sims, E.; Bertham, O.; Symington, H.; Bell, N.; Pfaltzgraff, L.; O'Brien, M. Towards a Circular Economy: Waste Management in the EU.; Study: Europ. Union. 2017. Available online: https://www.europarl.europa.eu/RegData/etudes/STUD/2017/581913/EPRS_STU%282017%29581913_EN.pdf (accessed on 25 August 2021).
3. Brusselaers, J.; Van Der Linden, A. Bio-waste in Europe—turning challenges into opportunities. *EEA Rep.* **2020**, *4*, 56.
4. Thomas, S.; Mishra, R.K.; Asiri, A.M. *Sustainable Polymer Composites and Nanocomposites*; Springer: Berlin/Heidelberg, Germany, 2019.
5. Shamsi, R.; Sadeghi, G.M.M.; Vahabi, H.; Seyfi, J.; Sheibani, R.; Zarintaj, P.; Laoutid, F.; Saeb, M.R. Hopes beyond PET recycling: Environmentally clean and engineeringly applicable. *J. Polym. Environ.* **2019**, *27*, 2490–2508. [CrossRef]
6. Vahabi, H.; Laoutid, F.; Mehrpouya, M.; Saeb, M.R.; Dubois, P. Flame retardant polymer materials: An update and the future for 3D printing developments. *Mater. Sci. Eng. R Rep.* **2021**, *144*, 100604. [CrossRef]
7. Costes, L.; Laoutid, F.; Brohez, S.; Dubois, P. Bio-based flame retardants: When nature meets fire protection. *Mater. Sci. Eng. R Rep.* **2017**, *117*, 1–25. [CrossRef]
8. Shah, A.U.R.; Prabhakar, M.N.; Saleem, M.; Song, J. IDevelopment of biowaste encapsulated polypropylene composites: Thermal, optical, dielectric, flame retardant, me-chanical, and morphological properties. *Polym. Compos.* **2017**, *38*, 236–243. [CrossRef]
9. Liu, C.-H.; Lee, H.-T.; Tsou, C.-H.; Wang, C.-C.; Gu, J.-H.; Suen, M.-C. Preparation and characterization of biodegradable polyurethane composites containing oyster shell powder. *Polym. Bull.* **2019**, *77*, 3325–3347. [CrossRef]
10. Chong, M.H.; Chun, B.C.; Chung, Y.-C.; Cho, B.G. Fire-retardant plastic material from oyster-shell powder and recycled polyethylene. *J. Appl. Polym. Sci.* **2005**, *99*, 1583–1589. [CrossRef]
11. Moustafa, H.; Youssef, A.M.; Duquesne, S.; Darwish, N. ACharacterization of bio-filler derived from seashell wastes and its effect on the mechanical, thermal, and flame re-tardant properties of ABS composites. *Polym. Compos.* **2017**, *38*, 2788–2797. [CrossRef]
12. Oualha, M.A.; Omri, N.; Oualha, R.; Nouioui, M.A.; Abderrabba, M.; Amdouni, N.; Laoutid, F. Development of metal hydroxide nanoparticles from eggshell waste and seawater and their application as flame retardants for ethylene-vinyl acetate copolymer (EVA). *Int. J. Biol. Macromol.* **2019**, *128*, 994–1001. [CrossRef]
13. Yew, M.C.; Sulong, N.H.R.; Amalina, M.; Johan, M.R. Eggshells: A novel bio-filler for intumescent flame-retardant coatings. *Prog. Org. Coatings* **2015**, *81*, 116–124. [CrossRef]
14. Krishnadevi, K.; Selvaraj, V. Biowaste material reinforced cyanate ester based epoxy composites for flame retardant applications. *High Perform. Polym.* **2016**, *28*, 881–894. [CrossRef]
15. Pulgarin, C.; Schwitzguebel, J.P.; Tabacchi, R. Utilization of wastes from coffee production. *Biofutur* **1991**, *102*, 43–50.
16. Murthy, P.S.; Naidu, M.M. Sustainable management of coffee industry by-products and value addition—A review. *Resour. Conserv. Recycl.* **2012**, *66*, 45–58. [CrossRef]
17. Uddin, M.N.; Techato, K.; Rasul, M.; Hassan, N.; Mofijur, M. Waste coffee oil: A promising source for biodiesel production. *Energy Procedia* **2019**, *160*, 677–682. [CrossRef]
18. Kondamudi, N.; Mohapatra, S.K.; Misra, M. Spent coffee grounds as a versatile source of green energy. *J. Agric. Food Chem.* **2008**, *56*, 11757–11760. [CrossRef] [PubMed]
19. Mariana, M.; Mahidin, M.; Mulana, F.; Aman, F. Utilization of activated carbon prepared from aceh coffee grounds as bio-sorbent for treatment of fertilizer industrial waste water. In *Proceedings of the IOP Conference Series: Materials Science and Engineering*; IOP Publishing: Bristol, UK, 2018; Volume 358, p. 012027.
20. Kim, J.; Kim, H.; Baek, G.; Lee, C. Anaerobic co-digestion of spent coffee grounds with different waste feedstocks for biogas production. *Waste Manag.* **2017**, *60*, 322–328. [CrossRef]
21. Reis, K.C.; Pereira, L.; Melo, I.C.N.A.; Marconcini, J.M.; Trugilho, P.F.; Tonoli, G.H.D. Particles of coffee wastes as reinforcement in polyhydroxybutyrate (phb) based composites. *Mater. Res.* **2015**, *18*, 546–552. [CrossRef]
22. Gama, N.V.; Ferreira, A.; Barros-Timmons, A. Polyurethane foams: Past, present, and future. *Materials* **2018**, *11*, 1841. [CrossRef]
23. Gama, N.; Soares, B.; Freire, C.; Silva, R.; Neto, C.P.; Timmons, A.B.; Ferreira, A. Bio-based polyurethane foams toward applications beyond thermal insulation. *Mater. Des.* **2015**, *76*, 77–85. [CrossRef]
24. Santos, D.C.D.S.; Adebayo, M.; Pereira, S.; Prola, L.; Cataluña, R.; Lima, E.C.; Saucier, C.; Gally, C.R.; Machado, F. New carbon composite adsorbents for the removal of textile dyes from aqueous solutions: Kinetic, equilibrium, and thermodynamic studies. *Korean J. Chem. Eng.* **2014**, *31*, 1470–1479. [CrossRef]
25. Akasaka, H.; Takahata, T.; Toda, I.; Ono, H.; Ohshio, S.; Himeno, S.; Kokubu, T.; Saitoh, H. Hydrogen storage ability of porous carbon material fabricated from coffee bean wastes. *Int. J. Hydrog. Energy* **2011**, *36*, 580–585. [CrossRef]
26. Ballesteros, L.F.; Teixeira, J.A.; Mussatto, S.I. Chemical, functional, and structural properties of spent coffee grounds and coffee silverskin. *Food Bioprocess Technol.* **2014**, *7*, 3493–3503. [CrossRef]
27. Mussatto, S.I.; Carneiro, L.M.; Silva, J.P.A.; Roberto, I.C.; Teixeira, J.A. A study on chemical constituents and sugars extraction from spent coffee grounds. *Carbohydr. Polym.* **2011**, *83*, 368–374. [CrossRef]
28. Pujol, D.; Liu, C.; Gominho, J.; Olivella, M.; Fiol, N.; Villaescusa, I.; Pereira, H. The chemical composition of exhausted coffee waste. *Ind. Crop. Prod.* **2013**, *50*, 423–429. [CrossRef]

29. Mandlekar, N.; Cayla, A.; Rault, F.; Giraud, S.; Salaün, F.; Malucelli, G.; Guan, J.-P. An overview on the use of lignin and its derivatives in fire retardant polymer systems. In *Lignin-Trends and Applications*; Poletto, M., Ed.; IntechOpen: London, UK, 2018.
30. Vahabi, H.; Shabanian, M.; Aryanasab, F.; Mangin, R.; Laoutid, F.; Saeb, M.R. Inclusion of modified lignocellulose and nano-hydroxyapatite in development of new bio-based adjuvant flame retard-ant for poly (lactic acid). *Thermochim. Acta* **2018**, *666*, 51–59. [CrossRef]
31. Movahedifar, E.; Vahabi, H.; Saeb, M.R.; Thomas, S. Flame retardant epoxy composites on the road of innovation: An analysis with flame retardancy index for future development. *Molecules* **2019**, *24*, 3964. [CrossRef]
32. Sonnier, R.; Vahabi, H.; Ferry, L.; Lopez-Cuesta, J.-M. Pyrolysis-combustion flow calorimetry: A powerful tool to evaluate the flame retardancy of polymers. In *Fire and Polymers VI: New Advances in Flame Retardant Chemistry and Science*; ACS Symposium Series; Wilkie, C.A., Morgan, A.B., Nelson, G.L., Eds.; American Chemical Society: Washington, DC, USA, 2012.
33. Huggett, C. Estimation of rate of heat release by means of oxygen consumption measurements. *Fire Mater.* **1980**, *4*, 61–65. [CrossRef]
34. Bigdeloo, M.; Teymourian, T.; Kowsari, E.; Ramakrishna, S.; Ehsani, A. Sustainability and circular economy of food wastes: Waste reduction strategies, higher recycling methods, and improved valorization. *Mater. Circ. Econ.* **2021**, *3*, 1–9. [CrossRef]
35. Sadeghi, B.; Marfavi, Y.; Aliakbari, R.; Kowsari, E.; Ajdari, F.B.; Ramakrishna, S. Recent studies on recycled PET fibers: Production and applications: A review. *Mater. Circ. Econ.* **2021**, *3*, 1–18. [CrossRef]

Article

Sound-Absorption Performance and Fractal Dimension Feature of Kapok Fibre/Polycaprolactone Composites

Lihua Lyu *, Duoduo Zhang, Yuanyuan Tian and Xinghai Zhou *

School of Textile and Material Engineering, Dalian Polytechnic University, Dalian 116034, China; zd19976@163.com (D.Z.); dllvlh1978@163.com (Y.T.)
* Correspondence: lvlh@dlpu.edu.cn (L.L.); zhouxh@dlpu.edu.cn (X.Z.);
  Tel.: +86-138-4283-9565 or +86-130-3223-1622 (L.L.)

Abstract: This article introduces a kind of composite material made of kapok fibre and polycaprolactone by the hot-pressing method. The effects of volume density, mass fraction of kapok fibre, and thickness on the sound-absorption performance of composites were researched using a single-factor experiment. The sound-absorption performance of the composites was investigated by the transfer function method. Under the optimal process parameters, when the density of the composite material was 0.172 g/cm^3, the mass fraction of kapok was 40%, and the thickness was 2 cm, the composite material reached the maximum sound-absorption coefficient of 0.830, and when the sound-absorption frequency was 6300 Hz, the average sound-absorption coefficient was 0.520, and the sound-absorption band was wide. This research used the box dimension method to calculate composites' fractal dimensions by using the Matlab program based on the fractal theory. It analysed the relationships between fractal dimension and volume density, fractal dimension and mass fraction of kapok fibre, and fractal dimension and thickness. The quantitative relations between fractal dimension and maximum sound-absorption coefficient, fractal dimension, and resonant sound-absorption frequency were derived, which provided a theoretical basis for studying sound-absorption performance. The results showed that kapok fibre/polycaprolactone composites had strong fractal characteristics, which had important guiding significance for the sound-absorption performance of kapok fibre composites.

Keywords: kapok fibre; polycaprolactone; composite materials; sound-absorption performance; fractal dimension

Citation: Lyu, L.; Zhang, D.; Tian, Y.; Zhou, X. Sound-Absorption Performance and Fractal Dimension Feature of Kapok Fibre/Polycaprolactone Composites. *Coatings* **2021**, *11*, 1000. https://doi.org/10.3390/coatings11081000

Academic Editor: Philippe Evon

Received: 30 July 2021
Accepted: 19 August 2021
Published: 22 August 2021

1. Introduction

Noise pollution belongs to one kind of environmental pollution, which is regarded as one of four major environmental problems in the world together with water pollution, air pollution, and light pollution. Noise pollution causes people's mood level and sleep quality to decline, and continues to slowly affect the human body system and increase the incidence of various diseases, even to a life-threatening level [1]. Secondly, noise will also accelerate the aging rate of conveyor belts, gears, bolts, and other mechanical structures, thus affecting the accuracy and service life of instruments and equipment, seriously affecting the safety of a building [2]. More and more scholars focus on developing environmentally friendly sound-absorbing materials and striving to improve their sound-absorption performance in recent years.

Lignocellulose insulation materials, natural materials, and recycled materials have been widely used in the field of sound-absorption in recent years. Tudor et al. [3] studied the sound-absorption coefficient of bark insulation board made of cork bark spruce and larch. The results showed that cork bark was an underrated material, and that compared with wood-based composites, engineered spruce bark was even better at absorbing sound than MDF, particleboard, or oriented particleboard. Tudor et al. [4] also analysed the acoustic performance of bark boards and found that the optimal density of bark boards to obtain the best sound-absorption coefficient was about 350 kg/m^3, and that these lightweight panels achieved better sound-absorption performance at a higher thickness (especially at lower frequencies). Smardzewski et al. [5] determined the normal surface impedance and sound-absorption coefficient of several woods from Europe and tropical regions, and the results showed that oak, ash, sapele, and pine had the highest sound-absorption coefficient at the frequency of 2 kHz. Asdrubali et al. [6] studied the acoustic properties of sustainable materials and found that sustainable products made from natural and recycled materials are effective substitutes for traditional synthetic materials. Fouladi et al. [7] used fresh coconut shell fibres and industrially prepared coconut shell fibres with an added binder as raw materials to study the sound-absorption coefficients of the two fibres as porous materials. The results showed that the industrially prepared fibres had poorer sound-absorption performance at low frequencies than fresh coconut shell fibres due to the addition of the binder. For commercial use, however, fibres must be mixed with additives to enhance properties such as hardness, antifungal, and flammability. Therefore, methods such as increasing air gaps or perforating plates should be used to improve the acoustic properties of industrially treated coconut shell fibres.

According to its formation mechanism and structure, porous sound-absorption materials can be divided into fibrous sound-absorption materials, granular sound-absorption materials, and foam plastic plates. When sound waves penetrate the surface of porous materials, the sound waves cause the vibration of the object, and then cause the vibration of the pores and gases inside the material. A large part of the sound energy can be consumed by the motion between the sound waves and the pore walls, and the transmission of sound waves is reduced due to the friction effect and viscous effect. The transmitting sound wave is weakened, thus achieving the purpose of sound insulation and absorption. At the same time, due to the different temperature of the pore wall and the small hole of the material, the gas will flow to realise heat exchange, and then lead to the loss of heat, resulting in the reduction in sound energy, to achieve the purpose of sound absorption. In addition, because of the influence of the frequency and friction of high frequency sound waves, the vibration speed of air particles will become faster, and then improve the speed of heat exchange; therefore, porous materials can achieve good sound-absorption effect in the range of high frequency sound waves.

Sound-absorbing materials have a wide range of application prospects, but the traditional sound-absorbing materials have many shortcomings, such as short service life, secondary pollution, performance instability, and so on. The emergence of new sound-absorbing materials makes up for the deficiency of traditional sound-absorbing materials to some extent [8]. Compared with traditional fibre sound-absorption materials, the sound-absorption coefficient of fibre sound-absorption composite materials is greatly improved, and many of them can reach more than 0.8, which makes up for the limitation of traditional fibre sound-absorption materials in the application range, to a large extent. Moreover, fibrous composites have less impact on the environment than conventional materials, generally have lower overall energy requirements for manufacturing and installing these materials, and are less harmful to human health. Replacing traditional sound-absorption materials with fibre composites has a significant impact on all phases of the building life cycle (construction, operation, end of life) [9].

Kapok fibre is a single cell fibre with a smooth surface, cylindrical shape, and no torsion. As a natural cellulose fibre, it is biodegradable and recyclable [10]. Due to its advantages such as softness, high hollow rate, anti-bacterial, and anti-mite, it has attracted extensive attention and has a good development prospect in the textile field [11]. Kapok is a kind of natural fibre, which has excellent sound-absorption performance due to its high degree of emptiness and small fibre diameter [12]. Because kapok fibre belongs to the short fibre category and is affected by its hollow structure, fibre strength is low, and has weak adhesion and poor elasticity; therefore, the technical difficulty of single spinning kapok yarn is high, and the application of economic benefits in textile and clothing is poor [13]. The large hollow structure of kapok fibre can be well utilised by making it into sound-absorbing composite material, which has low technical requirements and high economic benefits. The research and development of kapok fibre products can not only reduce the over-dependence on oil and other non-renewable resources, but also respond to the national strategic goal of ecological environment protection and sustainable development [14].

Dresden Technical University in Germany [15] focused on the study of kapok fibre and wool fibre. By comparing their differences in heat insulation and sound insulation, it was finally concluded that kapok fibre material had better heat insulation and sound insulation performance than wool fibre material. Liu et al. [16] developed a sound-absorbing nonwoven composite material based on kapok fibre and hollow polyester fibre, and used the impedance tube method to study the sound-absorbing performance of kapok fibre nonwoven composite material in the low-frequency region of 100–500 Hz. Makki et al. [17] studied the sound-absorption properties of nonwoven, layered warp knitting, and layered double-layer fabric made of kapok fibre at frequencies from 100 to 5000 Hz. The results showed that the best sound-absorption effect was obtained by combining kapok nonwovens with double-layer fabric with a thickness of 11.25 mm. Azieyanti et al. [18] prepared reinforced polypropylene composites using jute and kapok fibres as raw materials and compared jute and kapok fibre-reinforced polypropylene sound insulation effect composites; the results showed that the performance of the composite material, along with the increase in the quality percentage of jute and kapok fibre, increased the filling amount to 30% when the performance of the composite material was best. Liu [19] used polyethene film to fabricate the kapok fibre nonwoven fabric. The research on the sound-absorption coefficient and specific surface impedance of composite materials in the frequency range of 100–2500 Hz was carried out using the impedance tube method. The results showed that the multilayer composite had better sound-absorption performance than the single-layer non-woven fabric at low frequency.

A fractal is defined as a mathematical object with a fraction (not an integer) dimension. In the 1990s, Yu proposed a method to determine the effective thermal conductivity of reinforced fibre materials and irregular porous materials, that is, the fractal theory method, but in the research process, the calculation method and process were too complex to be applied in practice. In 2001, Yu conducted another study on fractal calculation methods and published a paper [20] in which it was pointed out that the fractal characteristics of porous media could be described by a general model. Chen's [21] use of fluid science, fractal theory, the study of space structure of the fibre porous metal processing, the fractal model of flow resistance rate, and the effectiveness of the proposed model was verified by experiment, which showed that the fractal model could intuitively reflect the relationship between the geometric parameters and flow resistance rate of material. In order to design fibre-porous metal, sound-absorbing material provides a theoretical support. According to the heat conduction model of down fibre aggregate, Fu [22] calculated the fractal dimension of hollow polyester fibre under different volume fractions by the box-counting method and calculated the heat transfer coefficient of the aggregate. The predicted value calculated by the model was in good agreement with the measured value, which proved that the heat conduction model was effective. According to the box-counting method fractal theory, Lyu [23] calculated the fractal dimension of discarded feather/EVA sound-absorption composite material. By Matlab programming, the fractal dimension between the mass and density of discarded feathers was calculated, and then the relationship between the fractal dimension and the maximum sound-absorption coefficient was derived quantitatively.

Fractal geometric properties mainly included self-similarity, wholeness, fractal dimension, organisational depth, and recursion, which interweaved, reinforced, and supported each other [24]. Kapok fibre is the most hollow fibre, and its hollow structure is self-similar to the porous structure of the whole nonwoven fibres. In kapok fibre, the intercross and stacking of macromolecular chains are similar to the whole, and most macromolecular chains are six membered rings, benzene rings, and other molecular chains. The gap between macromolecular chains is similar to that of the whole porous. Kapok fibre sound absorption of the pore size and distribution of composite materials has obvious fractal characters;

according to the fractal theory and image processing technology, intuitive characterisation of fibre morphology parameters can be directly extracted: one is the characterisation of porosity, a single pore area, fractal dimension, and pore number, pore structure parameters, such as the size and distribution of these parameters can reflect the pore; the second is the parameters that characterise the state of the fibre. These parameters can reflect the orientation distribution of the fibre. The relationship between structure and sound-absorption property, structure, and fractal dimension of kapok fibre sound-absorption composite was established. Finally, the fractal method is used to solve the sound-absorption performance of kapok fibre.

This article introduces a kind of composite material made of kapok fibre and polycaprolactone by the hot-pressing method. The effects of volume density, mass fraction of kapok fibre, and thickness on the sound-absorption performance of composites were researched using a single-factor experiment. The sound-absorption performance of the composites was investigated by the transfer function method. This research used the box dimension method to calculate composites' fractal dimension by using the Matlab program based on the fractal theory. It analysed the relationships between fractal dimension and volume density, fractal dimension and mass fraction of kapok fibre, and fractal dimension and thickness. The quantitative relations between fractal dimension and maximum sound-absorption coefficient, fractal dimension, and resonant sound-absorption frequency were derived, which provided a theoretical basis for studying sound-absorption performance.

2. Materials and Methods

2.1. Materials

Kapok fibre—the micronaire is rated A (Guangzhou CUHK Textile Co., Ltd., Guangzhou, China). Polycaprolactone (PCL)—800 mesh (Solvay Company, Dongguan, China).

2.2. Equipment

The surface structure of the material was observed using a scanning electron microscope (Nippon Electronics Co., Ltd., Beijing, China). FA2004B, (Shanghai Precision Experimental Instrument Co., Ltd., Shanghai, China); Pressure Forming Machine, QLB-50D/QMN (Jiangsu Wuxi Zhongkai Rubber Machinery Co., Ltd., Wuxi, China); SW422/SW477, (Beijing Wangsheng Electronic Technology Co., Ltd., Beijing, China).

2.3. Preparation of Composites

Using kapok fibre and polycaprolactone as raw materials, a kapok fibre/polycaprolactone composite material was prepared by the hot-pressing method. Kapok fibre and polycaprolactone were mixed in a particular proportion and pressed by QLB-50D/QMN press. Yang tested the influence of hot-pressing pressure on the sound-absorption coefficient of composite materials in the frequency range of 2000–6300 Hz, and found that the influence of hot-pressing pressure on the sound-absorption coefficient of composite materials was small. In a certain range, the increase in hot-pressing pressure could accelerate the flow of fibre and matrix in the die, improve the chamber of the material, and result in uniform composite molding and good appearance. At the same temperature, the lower hot-pressing pressure was not enough to make the polycaprolactone powder melt and flow between the fibres, resulting in the pressure pointing to the fibre-extrusion fibres, resulting in a large number of fibre-aggregation phenomena. Suitable hot-pressing pressure could allow the polycaprolactone matrix to overcome the resistance between the fibres and fill the gap between the fibres, so that the fibres in the material were evenly distributed. However, too-high pressure would damage the fibre and reduce the mechanical properties of the material. When the hot-pressing pressure was 10 MPa, the composite material was well formed, the noise reduction coefficient and the average sound-absorption coefficient were relatively high, and the overall sound-absorption performance of the material was excellent. Therefore, the most suitable hot-pressing pressure for the preparation of the composite material was 10 MPa [25]. The ideal hot-pressing processing parameters were as follows:

The pressure was 10 MPa; the temperature was 130 °C and the time was 15 min. The volume density was 0.156, 0.172, 0.188, and 0.205 g/cm^3, the mass fraction of kapok fibre was 30%, 40%, 50%, and 60%, and the thickness was 1.0, 1.5, 2.0, and 2.5 cm. The composite material was cut into circular samples with diameters of 100 and 30 mm according to the mold size for the sound-absorption performance test.

2.4. Testing of the Composites

2.4.1. Sound-Absorption Coefficient Test

The composite material was tested using a stationary standing wave tube. The composite material was placed in a sample bag at room temperature prior to the test. Test specifications were 010534-2:1998 and GB/T1869.2-2002; the experiment used the transfer function method. The SW422/SW477 impedance tube sound-absorption test system was shown in Figure 1.

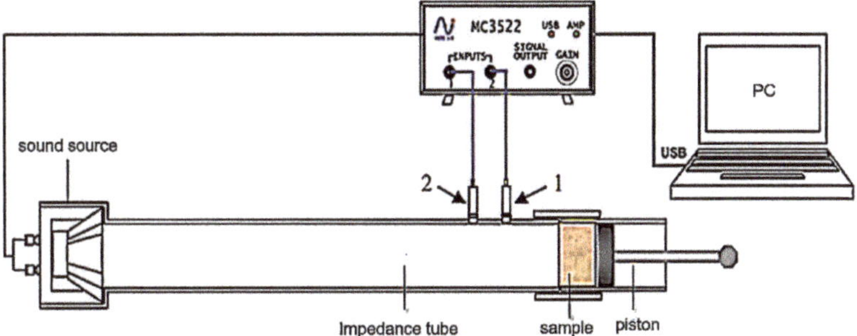

Figure 1. Schematic diagram of the sound-absorption test.

Test environment setting: the temperature was 25 °C, sound speed was 346.11 m/s, atmospheric pressure was 101.325 Pa, relative humidity was 65%.

The sound-absorbing performance could be characterised by the average sound-absorption coefficient. In the experiment, the SW422/SW477 impedance sound-absorption test system was used to measure the sound-absorption coefficient of samples at different frequencies. The measured sound-absorption coefficient was the average of six tests, and then drew the material's sound-absorbing curve at various frequencies.

For materials, when the average sound-absorption coefficient is greater than 0.20, it is called sound-absorption material, and when the average sound-absorption coefficient is greater than 0.56, it is called efficient sound-absorption material [26]. The average sound-absorption coefficient was the arithmetic average of six sound-absorption coefficients at 125, 250, 500, 1000, 2000, and 4000 Hz frequencies. Equation (1) was used to calculate the average sound-absorption coefficient.

$$\alpha = \frac{\alpha_{125} + \alpha_{250} + \alpha_{500} + \alpha_{1000} + \alpha_{2000} + \alpha_{4000}}{6} \tag{1}$$

Generally speaking, materials with a noise reduction coefficient (NRC) greater than or equal to 0.20 were called sound-absorbing materials [27]. NRC was the average sound-absorption coefficient of 250, 500, 1000, and 2000 Hz. Equation (2) represented the calculation method of the noise reduction coefficient.

$$NRC = \frac{\alpha_{250} + \alpha_{500} + \alpha_{1000} + \alpha_{2000}}{4} \tag{2}$$

2.4.2. Calculation of Porosity

The sound-absorption performance of composites is closely related to the porosity, which can be calculated indirectly according to the thickness and surface density of the sample [28]. Equation (3) was used to calculate the porosity.

$$\eta\% = \left(1 - \frac{G}{\rho \times \sigma}\right) \times 100 \tag{3}$$

In the formula: η was the sample's porosity (%), G was the model's surface density (G/cm^2), ρ was the fibre's mixing specific gravity (G/cm^3), and σ was the sample's thickness (m).

2.4.3. Fractal Characterisation

The sample's two-dimensional image was taken by a scanning electron microscope (Beijing Jeol Co. Ltd., Beijing, China) with a pixel size of 1024 × 1024. Additionally, the Matlab software (Matlab 2014) was used for a preprocessing series so that the image background was consistent and the picture was more precise. Then, the grey image was converted into a binary image that the computer could recognise, and Matlab's box-counting method was used to compute the sample's fractal dimension (see Appendix A for the program) [29].

The calculation method used in this experiment was the box-counting method. The principle was to take small cube boxes with a side length of ε and stack the boxes into the shape of a graph curve. Then, it was found that some boxes were empty, and some boxes contained part of the graph curve. Finally, the number of boxes containing the curve was denoted as N(ε) [30]. When $\varepsilon \to 0$, the fractal dimension of the curve could be obtained by using Equation (4):

$$D = -\lim_{\varepsilon \to 0} \frac{\log N(\varepsilon)}{\log \varepsilon} \tag{4}$$

In the calculation process, the value of ε was not equal to 0, and N(ε) was also numerical. A line was obtained by the least square fitting and mapping, and the slope of the line gave the fractal dimension [31]. Matlab provided a wealth of visual graphical representation functions and convenient programming capabilities in the box-counting analysis process of two-dimensional digital images. Therefore, in this research, image processing, numerical analysis, and other operations were all performed using Matlab.

3. Results and Discussion

3.1. Influence of Process Parameters on the Sound-Absorption Coefficient

3.1.1. Influence of Volume Density on the Sound-Absorption Coefficient

The samples' thickness was 1.0 cm when the composite material maintained the hot-pressing temperature at 130 °C, and the hot-pressing time was 15 min. The volume density was designed as 0.156, 0.172, 0.188, and 0.205 g/cm^3. Figure 2 showed the influence of volume density on the sound-absorption coefficient.

The influence of volume density on the sound-absorption coefficient of composite materials; that is, extremely low or extremely high density is not conducive to the sound absorption of composite materials. When the volume density of the composite material was extremely low, the material's gap was large and the material became extremely loose. Therefore, when the sound wave passed through the composite material, there was not enough friction and vibration between the air. The sound wave had less resistance and less sound energy loss and even could pass directly through the material. The lack of friction and vibration airflow between fibre walls led to the reduction in sound loss. At high volume density, if the volume density of the composites increased, the number of fibres per unit volume would increase, and the pore structure between the fibres became more complex than that under the condition of low density. The built-in surface area increased, and the porosity decreased. Under the action of sound waves, the molecular chains of fibres moved and consumed sound energy. When the volume density of composite material was very high, sound waves could not quickly go in the material, and the advantages of the porous sound-absorbing material would be eliminated due to the low porosity. As noted in Figure 1, the sound-absorption coefficient with the lowest volume density of kapok fibre in the low and mid-low range was also the lowest; that was, the sound-absorption coefficient increased with increasing volume density. This was because the denser composites (higher density and less open structure) behaved as reflective materials, absorbing lower frequencies of sound. In the middle and high-frequency part, the composites' sound-absorption performance became better and better with the composite material's decreased surface volume density. This was because the low density and high porosity allowed sound to enter the matrix more easily in order to dissipate, which helped improve the sound-absorption performance of the material. The volume density had extreme value, and the highest sound-absorption performance could be obtained. When the kapok fibre composite material's volume density was 0.172 g/cm^3, the average sound-absorption coefficient reached 0.388, with a maximum range of sound-absorbing. Overall, when the volume density of the composite material was 0.172 g/cm^3, the sound-absorbing performance reached the highest [32].

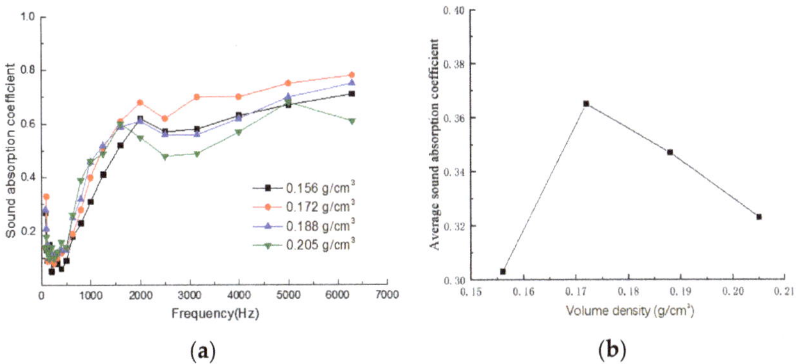

(a)   (b)

Figure 2. Influence of volume density on sound-absorption coefficient: (**a**) sound-absorption coefficient, (**b**) average sound-absorption coefficient.

3.1.2. Influence of Mass Fraction of Kapok Fibre on the Sound-Absorption Coefficient

The influence of mass fraction of kapok fibre on the sound-absorption coefficient was investigated. Under the unchanged conditions of other experimental conditions (hot-pressing temperature 130 °C, time 15 min, volume density 0.172 g/cm^3, thickness 1.0 cm), samples with a mass fraction of kapok fibre of 30%, 40%, 50%, and 60% were prepared, respectively, and their sound-absorption coefficients were measured. Figure 3 showed the influence of mass fraction on sound-absorption coefficients.

In the low and mid-low frequency regions, the sound-absorbing effect was better when the fibre mass fraction was larger. As the mass fraction increased, the sample's sound-absorption coefficient increased in a particular range. In the mid-high frequency and high-frequency regions, the sound-absorption effect became worse when the mass fraction was high. When the mass fraction of kapok fibre was 30%, the sample's maximum sound-absorption coefficient could reach 0.92. However, the sound-absorbing performance of sample with low-medium and low-frequency was not good, and the sound-absorbing results were good at high frequency.

The results showed that the excessive addition of kapok fibre made the composites too fluffy, the strength was low, and the forming effect was not good. However, when kapok fibre was too sparse, the material gap became more extensive, and the sound-absorbing effect was low. Based on the above analysis, the mass fraction of kapok fibre was 40% [32].

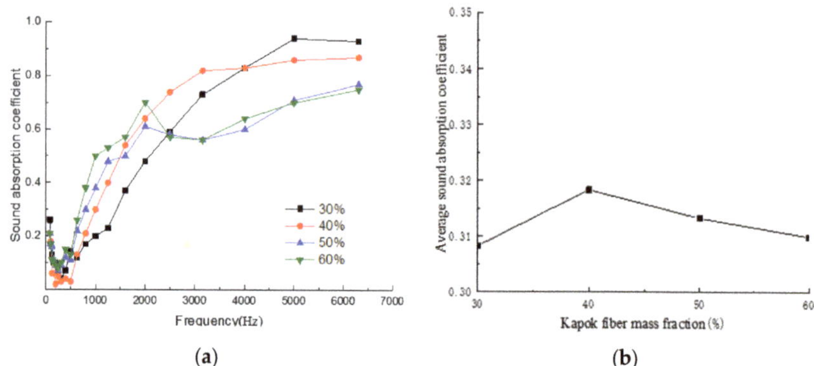

(a)

(b)

Figure 3. Influence of mass fraction on sound-absorption coefficient: (**a**) sound-absorption coefficient and (**b**) average sound-absorption coefficient.

3.1.3. Influence of Thickness on the Sound-Absorption Coefficient

To study the relationship between the sound-absorbing performance of kapok fibre composite materials and the thickness of composites, the sample's volume density was 0.172 g/cm^3, the hot-pressing time was 15 min, the hot-pressing temperature was 130 °C, and the mass fraction of the model was 40%. The samples' thickness was 1.0, 1.5, 2.0, and 2.5 cm, respectively. Figure 4 showed the influence of thickness on sound-absorption coefficient.

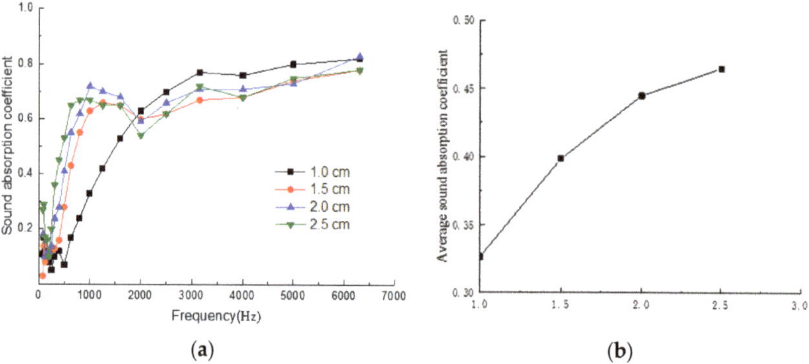

(a)

(b)

Figure 4. Influence of thickness on sound-absorption coefficient: (**a**) sound-absorption coefficient and (**b**) average sound-absorption coefficient.

As shown in Figure 4, as the thickness increased, the average sound-absorption coefficient would also increase. Its increasing trend was evident and showed a linear relationship. When the thickness increased, the peak of its sound-absorption coefficient would move to the low frequency region. When the thickness was 1.5 cm, the sound-absorption coefficient's peak frequency was 1600 Hz, and when the composite material increased the thickness to 2.5 cm, the frequency of the peak sound-absorption coefficient was reduced to 800 Hz. Additionally, the composite's thickness was 2.0 cm [32].

When the thickness of the composites increased, the air permeability of the composites would decrease, and the flow resistance would increase accordingly, thereby enhancing the sound-absorbing effect. When the material's thickness was reduced, the time and distance for the sound to pass through the composite would be shorter, and the sound wave could not be reflected and refracted multiple times, thereby reducing the sound-absorption coefficient of the composites. Consequently, the thickness of the material was generally considered an essential factor in controlling sound-absorption performance. However, when the thickness of the material remained unchanged to a certain extent, the sound-absorption coefficient would be substantially unchanged [33].

3.2. Fractal Characterisation Results

3.2.1. Kapok Fibre Sound-Absorbing Composites Image Acquisition

The EOL-JSM-6460LV scanning electron microscope was used for the image acquisition of composite materials. The magnification of the image was 100 times. Afterimage processing took the basic fibre layer for pore structure analysis (sample 1#~17#). This processing method was based on the fact that the material was composed of multiple layers of randomly oriented fibres. Most of them were considered to be perpendicular to the average propagation direction of sound waves in the composite material. Some performances of a single layer (base plane fibre layer), such as sound-absorption coefficient, could reflect the entire composite material's performance.

The SEM images of composites with volume densities of 0.139, 0.155, 0.172, 0.188, 0.205, and 0.211 g/cm^3 were numbered from 1# to 6#, and were shown in Figure 5.

(a)   (b)
(c)   (d)
(e)   (f)

Figure 5. SEM images of composite materials with different volume densities. (**a**) 1# (0.139 g/cm^3); (**b**) 2# (0.155 g/cm^3); (**c**) 3# (0.172 g/cm^3); (**d**) 4# (0.188 g/cm^3); (**e**) 5# (0.205 g/cm^3); (**f**) 6# (0.205 g/cm^3).

The composites images with mass fraction of kapok fibre 30, 40, 50, and 60% were numbered as 7#–10#, respectively. The SEM images were shown in Figure 6.

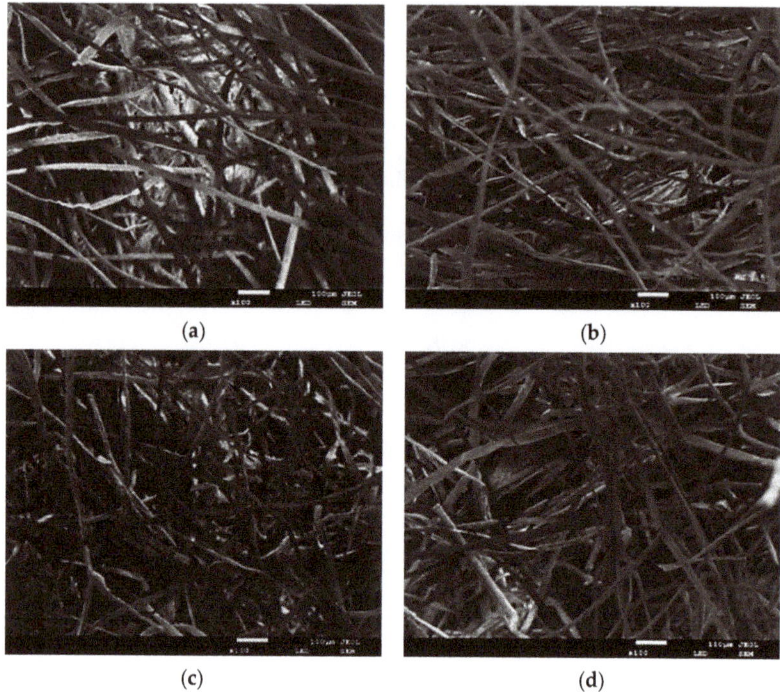

(a)

(b)

(c)

(d)

Figure 6. SEM of composite materials with different kapok fibre mass fractions. (**a**) 7# 30%; (**b**) 8# 40%; (**c**) 9# 50%; (**d**) 10# 60%.

The SEM images with a thickness of 0.5, 1.0, 1.5, 2.0, 2.5, and 3.0 cm were numbered as 11#–16# in Figure 7.

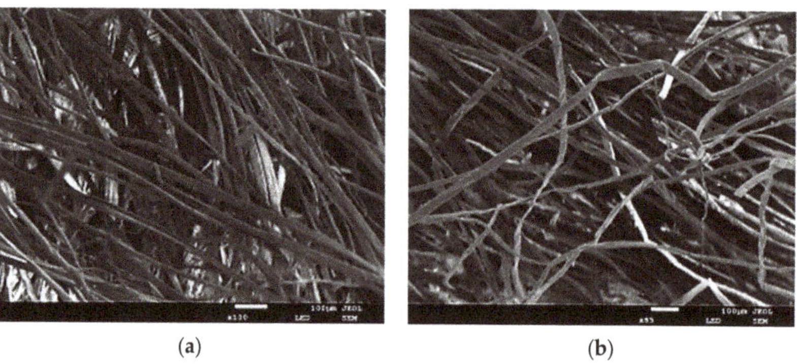

(a)

(b)

Figure 7. *Cont.*

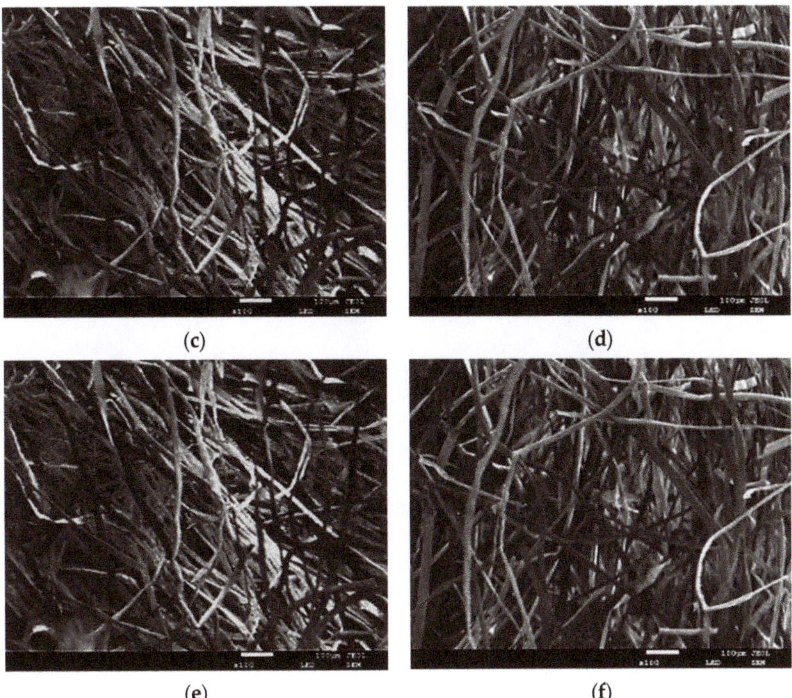

Figure 7. SEM images of composite materials with different thicknesses. (**a**) 11# (0.5 cm); (**b**) 12# (1 cm); (**c**) 13# (1.5 cm); (**d**) 14# (2 cm); (**e**) 15# (2.5 cm); (**f**) 16# (3.0 cm).

3.2.2. Preprocessing of Composites Images

To obtain accurate data, it was necessary to process the SEM images. Matlab software realised the image preprocessing of each part. The following procedures were shown in Appendix A.

3.2.3. Image Binarisation

The process of converting multiple grayscale images into two grayscale images through a specific threshold was called image binarisation; that was, all pixels were set as black or white. The grey level of pores and their surrounding areas were significantly different; therefore, the points whose grey level was more significant than or equal to a threshold could be turned into white points, and those whose grey story was less than a threshold could be turned into black points through binarisation of the image by selecting appropriate entries [34]. The scanning electron microscope image (100 times magnification) of the sample was processed by Photoshop, and the binary image recognized by computer was obtained. The pre-processed image was shown in Appendix B Figure A1.

3.2.4. Calculation of Fractal Dimension

This experiment used the box-counting method and Matlab programming to calculate the kapok fibre composite material's fractal dimension on the processed binary image (see Appendix A for the program) [35]. Appendix B Figure A2 showed the calculation results of the fractal dimension of composite materials with different volume densities, kapok fibre mass fractions, and thicknesses.

The square of linear correlation coefficient of each simulated composite material image was above 0.99 in 13 and 0.98 in 3. According to the calculation results of the fractal dimension of composite material volume density, fibre mass fraction, and thickness, the composite material's pore size had fractal characteristics, and the fractal dimension of the sample hole was represented by the slope of the straight line.

3.3. Relationship between Fractal Dimension and Various Factors

3.3.1. Influence of Volume Density on Fractal Dimension

Table 1 showed the relevant parameters of samples at different densities. Based on the table data, the researcher drew the relationship curve between the fractal dimension of the model and the maximum sound-absorption coefficient under various volume densities, as shown in Figure 8. Figure 8 could see that when the sample's volume density increased, the corresponding fractal dimension and the maximum sound-absorption coefficient usually showed a decreasing trend. In exploring the effect of volume density on the sound-absorption performance, as the volume density increased, the sound-absorbing curve of the peaks moved to low frequency; when the sample was 0.172 g/cm^3, volume density samples corresponding to the hole shape and aperture size combination made throughout the sound-absorbing performance were better in the absorption spectrum, the maximum sound-absorption coefficient of wave in the test frequency range, and the average sound-absorption coefficient was higher, but did not affect the overall trend.

Table 1. Relevant parameters of samples at different volumetric densities.

Sample Number	Composites Volume Density g·cm^{-3}	Maximum Sound Absorption Coefficient	Average Sound Absorption Coefficient	Fractal Dimension	The Square of the Correlation Coefficient R^2
1#	0.139	0.940	0.326	1.888	0.998
2#	0.155	0.710	0.303	1.794	0.996
3#	0.172	0.800	0.365	1.874	0.998
4#	0.188	0.750	0.347	1.836	0.997
5#	0.205	0.680	0.323	1.855	0.998
6#	0.211	0.630	0.300	1.582	0.980

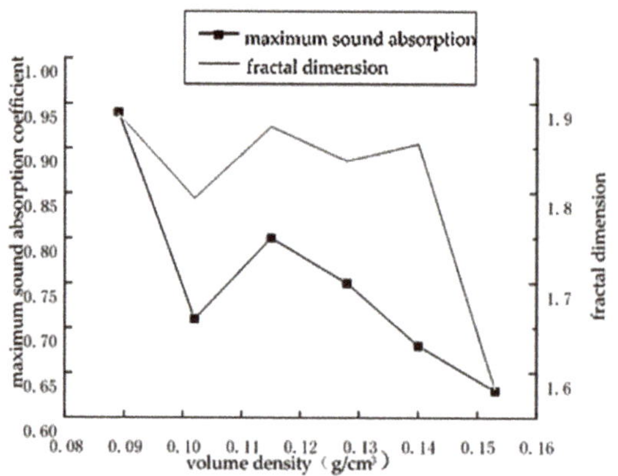

Figure 8. Relationship between fractal dimension and maximum sound-absorption coefficient of samples at different volumetric densities.

3.3.2. Influence of Mass Fraction of Kapok Fibre on Fractal Dimension

Table 2 showed the relevant parameters of different kapok fibre mass fraction samples. Based on the table's data, the relationship between the fractal dimension of models with different kapok fibre mass fractions and the maximum sound-absorption coefficient was drawn, as shown in Figure 9. As the kapok fibre's mass fraction increased, the sound-absorption coefficient and fractal dimension decreased gradually, indicating that the complexity of the internal structure of the composites decreased. In other words, the complexity of the kapok fibre composite material's internal system decreased as the fibre mass fraction increased.

Table 2. Relevant parameters of samples under different kapok fibre mass fractions.

Sample Number	Quality Fraction of Kapok Fibre %	Maximum Sound Absorption Coefficient	Average Sound Absorption Coefficient	Fractal Dimension	The Square of the Correlation Coefficient R^2
7#	30	0.940	0.308	1.774	0.995
8#	40	0.860	0.318	1.713	0.990
9#	50	0.770	0.322	1.745	0.994
10#	60	0.750	0.360	1.712	0.992

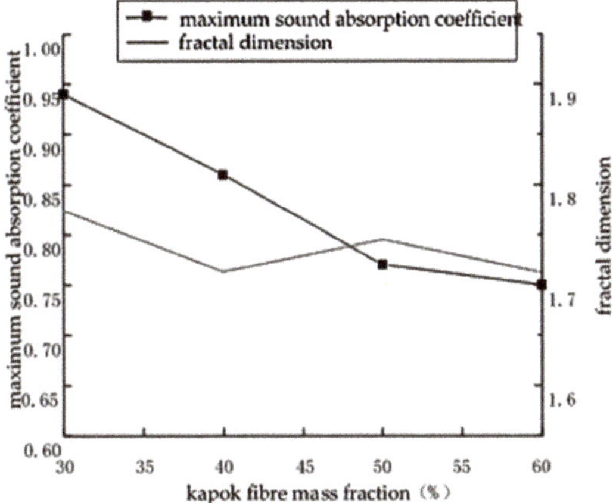

Figure 9. Relationship between sample fractal dimension and maximum sound-absorption coefficient under different kapok fibre mass fractions.

3.3.3. Influence of Thickness on Fractal Dimension

Table 3 showed the relevant parameters of specimens with different thicknesses. The relationship between the fractal dimension of models with different thicknesses and the resonant sound-absorption frequency was drawn based on the table's data, as shown in Figure 10.

Since the thickness significantly influenced the composite material's sound-absorption performance, as the thickness increased, the sound-absorbing curve expanded toward low frequencies. The relationship between the resonance sound-absorbing rate corresponding to the inflexion point was explained by the sound-absorption coefficient and the fractal dimension. The figure showed that the fractal dimension had a good correlation with the sound-absorbing resonance frequency. It showed that the turning point of the sound-absorbing performance of composite materials could be studied through the fractal dimension. That was, the resonance frequency point at which the sound-absorption performance of the material was greatly improved. When the composite's thickness increased, the fractal dimension of the response decreased. The resonance frequency corresponding to the turning point moved to the low frequency; this was because as the composite material's thickness increased, the length of the channel through which the sound waves enter the interior increased, and the hole depth increased, reducing its fractal dimension.

Table 3. Relevant parameters of specimens under different thicknesses.

Sample Number	Thickness cm	Resonant Sound-Absorption Frequency	Average Sound-Absorption Coefficient	Fractal Dimension	The Square of the Correlation Coefficient R^2
11#	0.5	6300	0.182	1.580	0.989
12#	1	4000	0.327	1.815	0.992
13#	1.5	2000	0.400	1.715	0.997
14#	2	1000	0.445	1.697	0.992
15#	2.5	800	0.465	1.697	0.992
16#	3	630	0.525	1.599	0.984

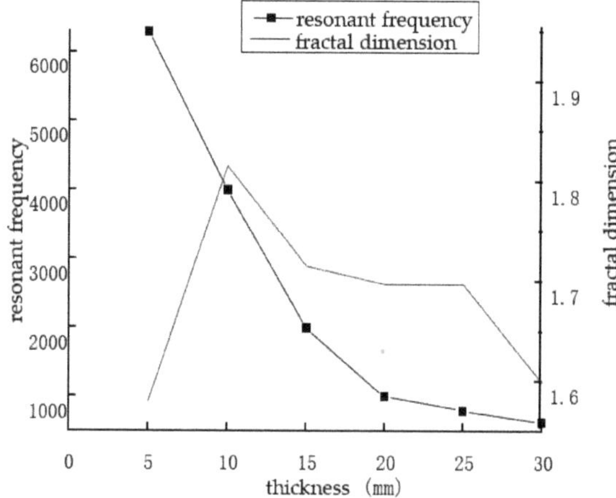

Figure 10. Relationship between fractal dimension of specimen and resonant frequency under different thicknesses.

3.3.4. Relationship between Fractal Dimension and Average Sound-Absorption Coefficient

Figure 11 showed the relationship between the fractal dimension of samples 1#–17# and the average sound-absorption coefficient.

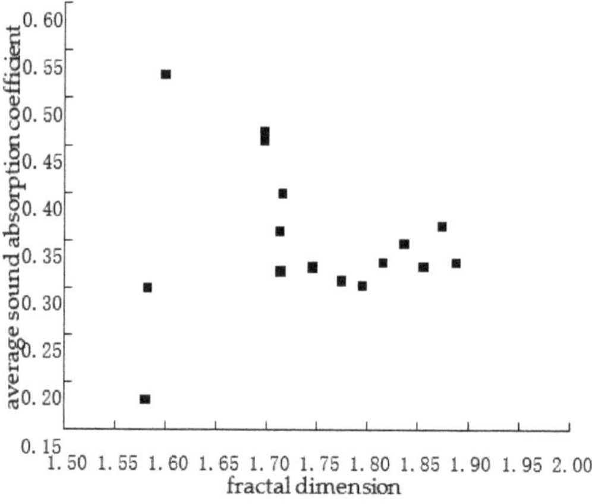

Figure 11. Scatter plot.

Due to the influence of deviation during image adoption, some images with considerable variation were selected for fitting, and Figure 12 showed the fitted curve. The curve obtained was as follows:

$$Y = 4.6 \times X^2 - 16.58X + 15.26 \tag{5}$$

In the formula, Y was the average sound-absorption coefficient and X was the fractal dimension.

As the thickness of the material increased, the resonance absorption frequency corresponding to the sound-absorption coefficient's inflexion point moved to lower frequencies. The fractal dimension of the corresponding response increased. As shown in Figure 13, the fractal dimension was matched with the resonance absorption frequency. The suitable curve was shown below:

$$Y = -2.37 \times 10^{-8}X^2 + 1.56 \times 10^{-4}X + 1.54 \tag{6}$$

In the formula: Y was the fractal dimension; X was the resonant frequency.

This was in accordance with Equation (5), and designed the sound-absorbing performance parameters of composite materials. If the average sound-absorption coefficient was 1, the material's fractal dimension was 2.187 or 1.418. According to the fact that the fractal dimension was about 1.4, the composites had a simple internal structure and a small sound-absorption coefficient. Therefore, the fractal dimension was chosen as 2.187. The larger fractal dimension indicated that the complexity of the internal system of the composites was enhanced.

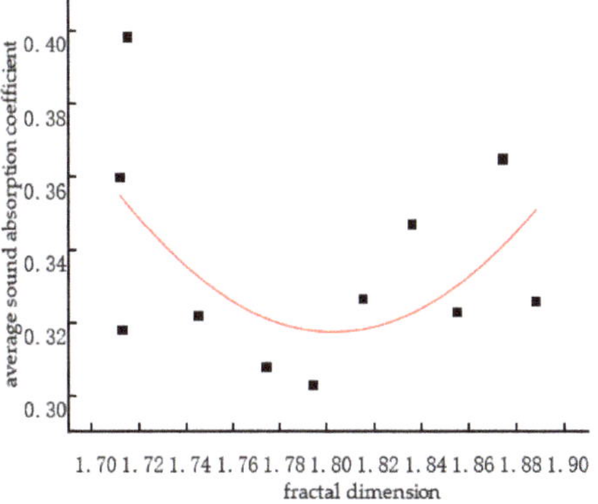

Figure 12. Fitting curve of average sound-absorption coefficient and fractal dimension.

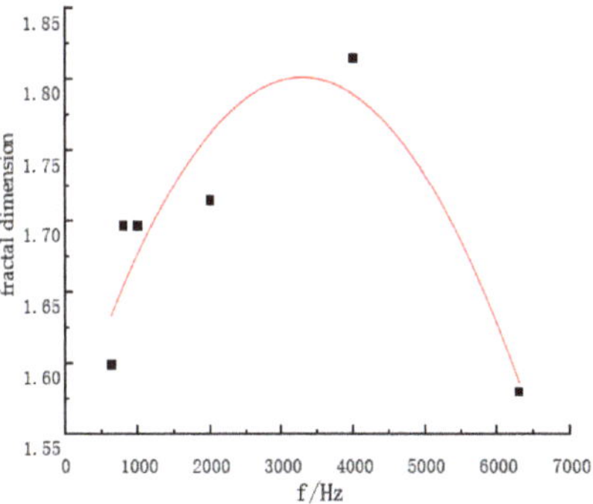

Figure 13. Fitting curve of fractal dimension and resonance sound-absorption frequency.

This was in accordance with Equation (6), and designed the sound-absorbing performance parameters of composite materials. If a 125 Hz resonant sound-absorption frequency was needed, the fractal dimension of the material was 1.555. The experiment proved that with the decrease in the fractal dimension, the thickness of composite material increased, and the resonance frequency corresponding to the turning point moved to low frequency.

4. Conclusions

The composite material was prepared from kapok fibre and polycaprolactone by the hot-pressing method. The following conclusions were drawn:

(1) Through single-factor experiments, composite material determined the optimal process conditions as follows. The composite material's volume density was 0.172 g/cm^3, the mass fraction of kapok fibre was 40%, and the composite material's thickness was 2.0 cm. Therefore, this experiment obtained a porous sound-absorbing material with a high sound-absorption coefficient and absorption bandwidth. The maximum sound-absorption coefficient was 0.830, and the average sound-absorption coefficient was 0.520.

(2) The box-counting method was used to calculate the kapok fibre/polycaprolactone sound-absorbing composite material's fractal dimension. The results showed that the kapok fibre/polycaprolactone composite material had strong fractal characteristics. By fitting the fractal dimension and average sound-absorption coefficient of the kapok fibre composites, the fitting curve was $Y = 4.6 \times X^2 - 16.58X + 15.26$. The right angle of fractal dimension and resonance sound-absorption frequency was $Y = -2.37 \times 10^{-8} \times X^2 + 1.56 \times 10^{-4}X + 1.54$, which provided the theoretical basis for the study parameter design of the sound-absorption performance of kapok fibre composite material, and had important guiding significance.

(3) Kapok fibre sound-absorption composite material had the advantages of low production cost, light weight, simple process, and recycling. Therefore, this sound-absorbing composite material could be widely used in the building environment. In this paper, only the sound-absorption performance of kapok fibre sound-absorption composite material was studied. In practical application, other properties such as mechanical properties, antibacterial properties, thermal insulation properties, and so on need to be considered.

Author Contributions: Data curation, D.Z. and Y.T.; Software, X.Z.; Writing—original draft, D.Z.; Writing—review & editing, L.L. All authors have read and agreed to the published version of the manuscript.

Funding: This research was funded by Science and Technology Innovation Fund Project of Dalian, Grant No. 2019J12SN71.

Institutional Review Board Statement: Not applicable.

Informed Consent Statement: Not applicable.

Conflicts of Interest: The authors declare no conflict of interest.

Appendix A

1. The calculation program of background elimination

```
I1 = imread ('Original image.tif');
I2 = imread ('Background images.tif');
I1 = rgb2gray (I1); % Image grayscale
I2 = rgb2gray (I2);
Img = imsubtract (I2, I1); Background differentiation
Img = 255/max (max (Img)).*Img;
Img = imadjust (Img, (0.3, 0.9)); % local image enhancement, x value according to the
actual situation
```

```
threshold = graythresh (Img); % the between-cluster variance method of OSTU threshold
figure;
subplot (131), imshow (I1);
subplot (132), imshow (I2);
subplot (133), imshow (Img);
2. Image enhancement calculation procedures
I = imread ('Original image.tif'); % read in grayscale
I = rgb2gray (I)
Subplot (2,1,2),imhist (I),title ('gray histogram'); % for gray histogram
3. Image binarisation calculation procedures
I = imread ('*.bmp');
BW = im2bw (I, 0.43); % Transformation threshold converted gray image into a binary image
figure;
imshow (BW);
title ('Binary Image');
B = size (BW, 1); % returns the number of rows of I
if mod (log2(B),1) > 0; % the mod function is a complementary function. Returns the remainder after dividing two numbers. The sign of the result is the same as the divisor
error('The size of image must be 2^n');
end
t = log2(B); % t is 2 to the power of t, which is the number of boxes
s = 2^(1:t); % s is the box size
Nr = zeros(1,t); % create a 1 line, zero matrix t column
for k = 1:t;
d = s(k);
h = B/d;
For m = 1:h %Cycle by line
For n = 1:h % Cycle by column
A = BW (d*(m − 1)+(1:d),d*(n − 1) + (1:d));
mx = max(A(:));
nr = mx;
Nr(k) = Nr(k) + nr;
end
end
end
r = d./s;
x = log(r);
y = log(Nr);
plot(x,y,'+');
lsline; p = polyfit(x,y,1);
Xlabel ('ln (1/epsilon)');
Ylabel ('N (epsilon)');
Dm = p;
hold on; z = y./x;
```

Appendix B

1. Binarisation images of composites with different volume density, mass fraction, and thickness

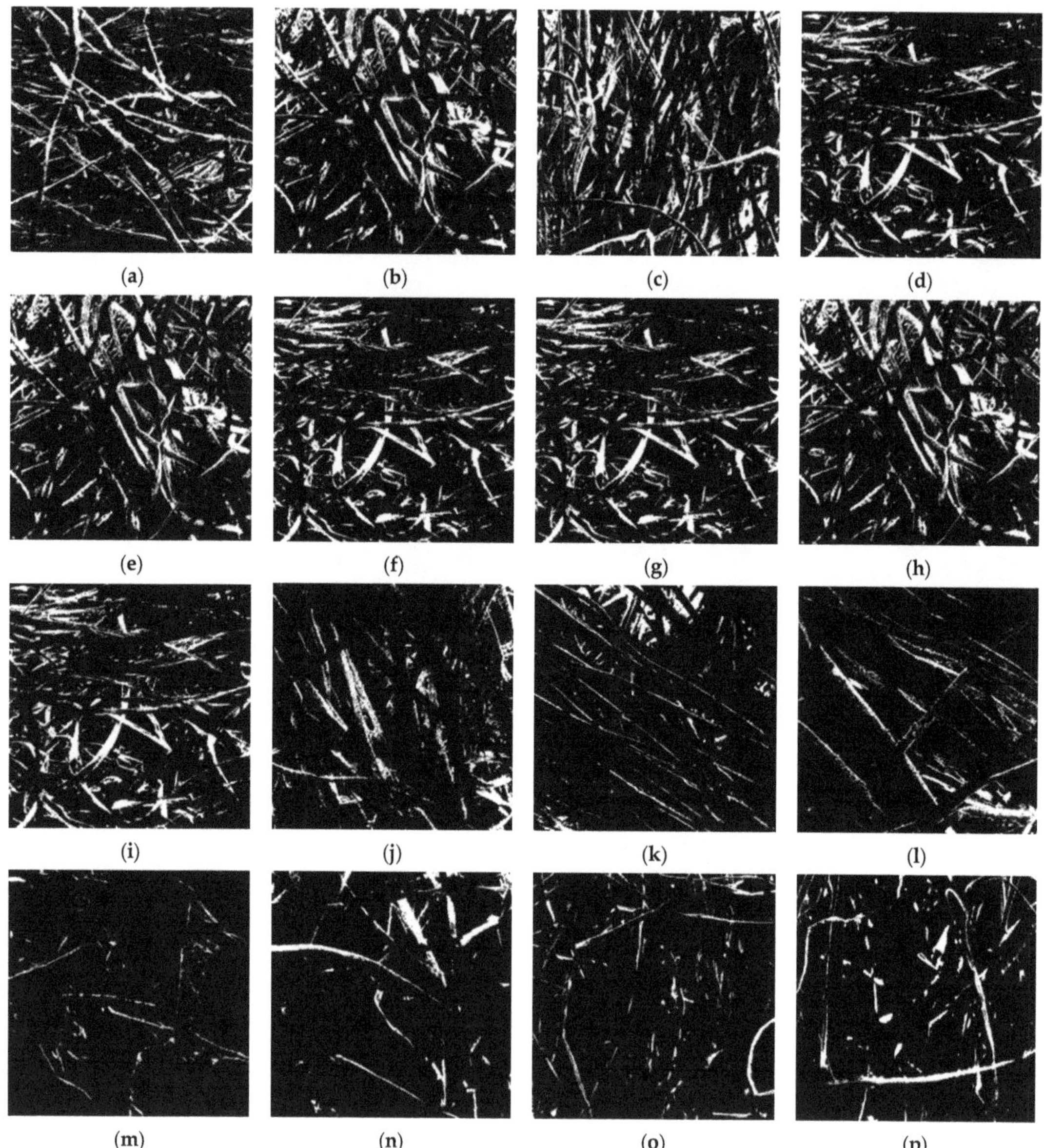

Figure A1. (**a**) Volume density 0.139 g/cm^2; (**b**) volume density 0.155 g/cm^2; (**c**) volume density 0.172 g/cm^2; (**d**) volume density 0.188 g/cm^2; (**e**) volume density 0.205 g/cm^2; (**f**) volume density 0.211 g/cm^2; (**g**) kapok fibre 30%; (**h**) kapok fibre 40%; (**I**) kapok fibre 50%; (**j**) kapok fibre 60%; (**k**) thickness of 5 mm; (**l**) thickness of 1.0 cm; (**m**) thickness of 1.5 cm; (**n**) thickness of 2.0 cm; (**o**) thickness of 2.5 cm; (**p**) thickness of 3.0 cm.

2. Calculation results of the fractal dimension of composite materials with different volume density, mass fraction, and thickness.

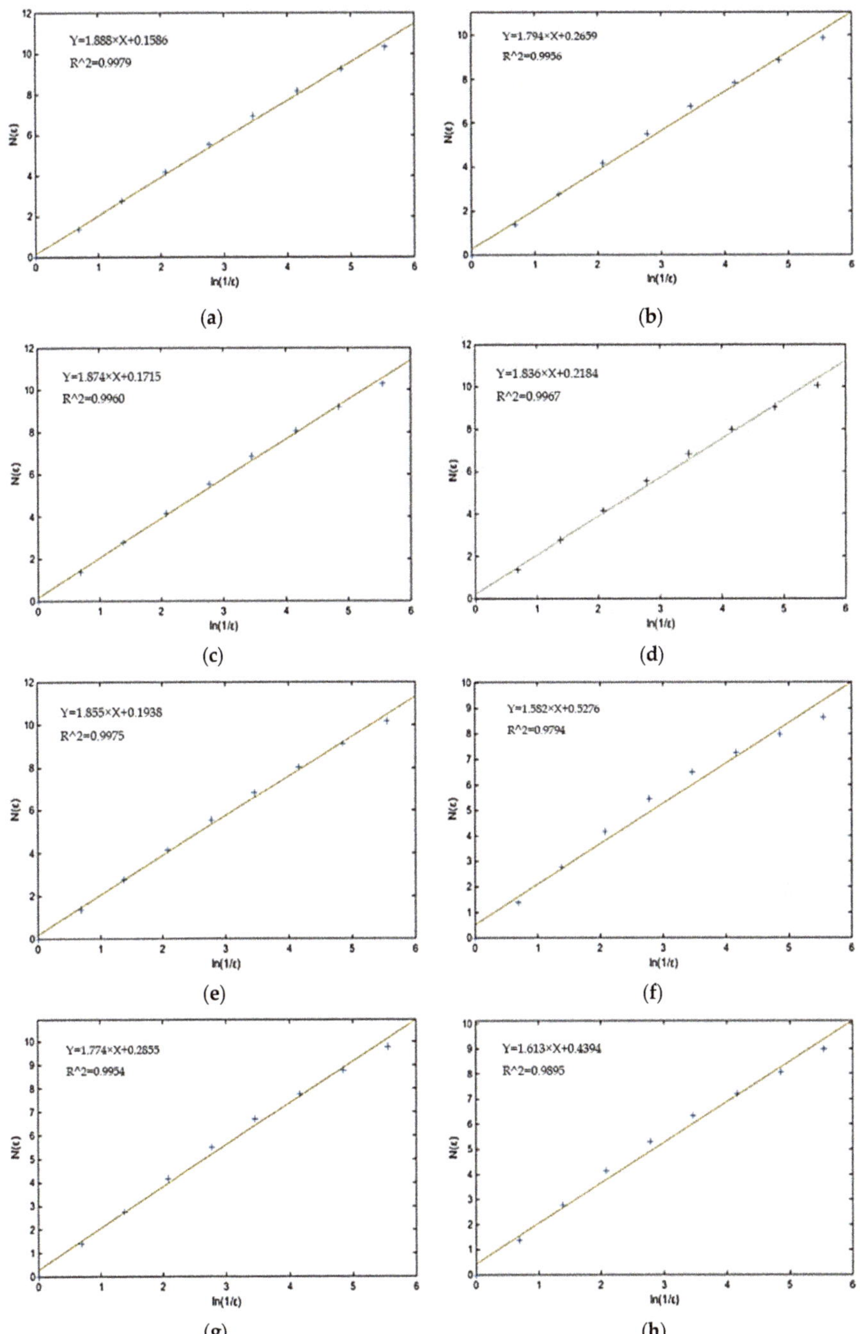

Figure A2. *Cont.*

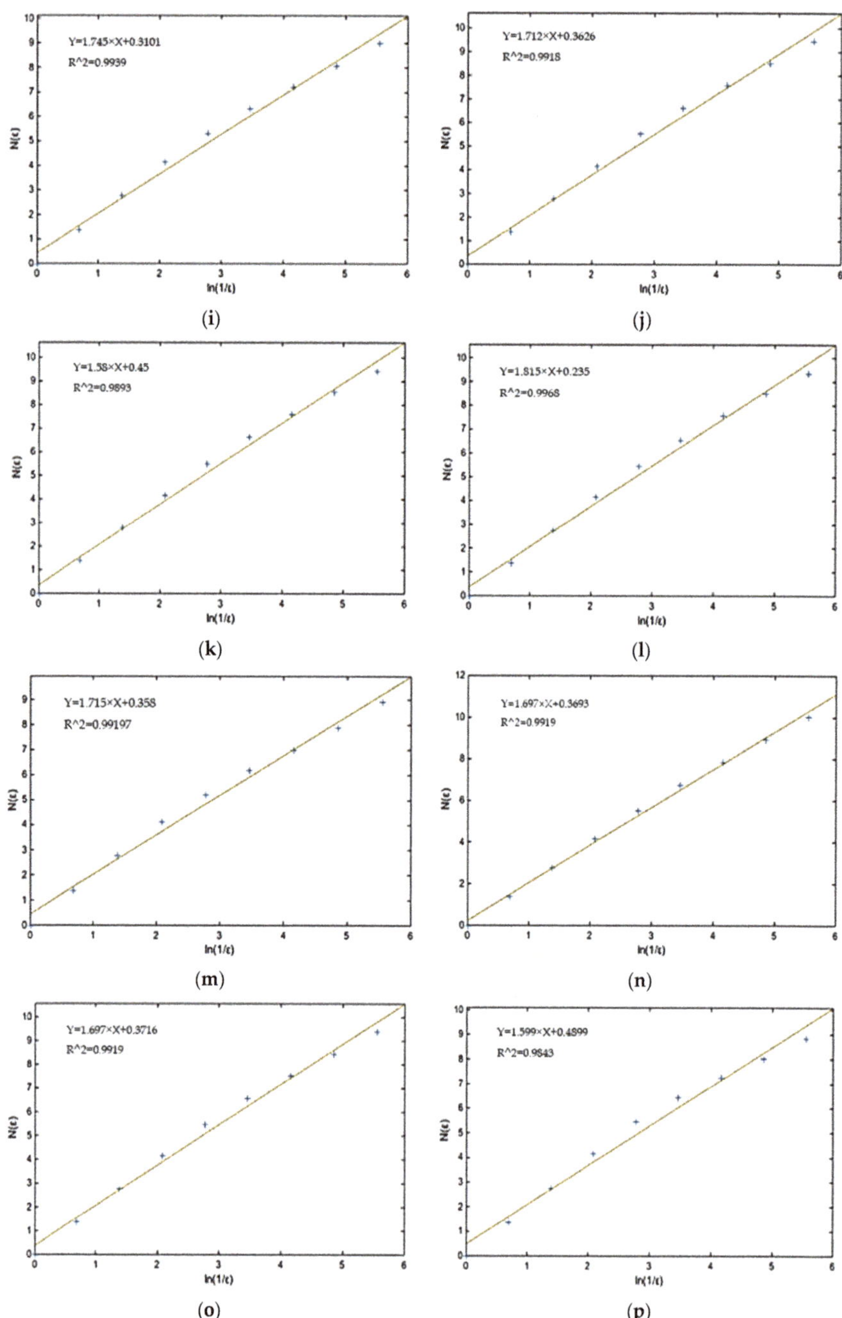

Figure A2. (**a**) Volume density 0.139 g/cm^2; (**b**) volume density 0.155 g/cm^2; (**c**) volume density 0.172 g/cm^2; (**d**) volume density 0.188 g/cm^2; (**e**) volume density 0.205 g/cm^2; (**f**) volume density 0.211 g/cm^2; (**g**) kapok fibre 30%; (**h**) kapok fibre 40%; (**I**) kapok fibre 50%; (**j**) kapok fibre 60%; (**k**) thickness of 5 mm; (**l**) thickness of 1.0 cm; (**m**) thickness of 1.5 cm; (**n**) thickness of 2.0 cm; (**o**) thickness of 2.5 cm; (**p**) thickness of 3.0 cm.

References

1. Yuan, Y. Analysis of Hazard and Control of Urban Noise Pollution. *Reg. Control* **2020**, *13*, 61–62.
2. He, Q.; Lin, L.; Lu, C.Y.; Lu, W.J.; Jing, X.F.; Chen, W. Research Development of Noise Reduction Materials. *Smart Grid* **2016**, *4*, 966–971.
3. Tudor, E.M.; Dettendorfer, A.; Kain, G.; Barbu, M.C.; Ren, R.; Kristak, L. Sound-Absorption Coefficient of Bark-Based Insulation Panels. *Polymers* **2020**, *29*, 1012. [CrossRef]
4. Tudor, E.M.; Kristak, L.; Barbu, M.C.; Gergel, T.; Nemec, M.; Kain, G.; Ren, R. Acoustic Properties of Larch Bark Panels. *Forests* **2021**, *12*, 887. [CrossRef]
5. Smardzewski, J.; Batko, W.; Kamisinski, T.; Flach, A. Experimental Study of Wood Acoustic Absorption Characteristics. *Holzforschung* **2013**, *68*, 467–476. [CrossRef]
6. Asdrubali, F.; Schiavoni, S.; Horoshenkov, K.V. A Review of Sustainable Materials for Acoustic Applications. *Build. Acoust.* **2012**, *19*, 283–312. [CrossRef]
7. Fouladi, M.H.; Ayub, M.; Nor, M. Analysis of Coir fibre Acoustical Characteristics. *Appl. Acoust.* **2011**, *72*, 35–42. [CrossRef]
8. Liang, L.S.; Guo, W.L.; Zhang, Y.; Li, L.P.; Du, J.J. Research Progress of New Sound Absorption Materials and Sound Absorption Model. *J. Funct. Mater.* **2020**, *51*, 19–25.
9. Arenas, J.P.; Asdrubali, F. Eco-Materials with Noise Reduction Properties. In *Handbook of Ecomaterials*; Martínez, L., Kharissova, O., Kharisov, B., Eds.; Springer: Cham, Switzerland, 2018; pp. 1–26.
10. Song, J.B.; Liu, X.X.; Yuan, G.P.; Fu, F.; Yang, W.B. Rheological and Mechanical Properties of HDPE /Tea stalk /CNTs Composites. *Plast. Ind.* **2015**, *4*, 79–82.
11. Chu, J.Y.; Liao, S.Q.; Li, H.G. Performance and Application Progress of Kapok Fibre. *Ind. Text.* **2018**, *36*, 6–10.
12. Xiang, H.F.; Wang, D.; Liu, H.C.; Zhao, N.; Xu, J. Investigation on Sound Absorption Properties of Kapok Fibres. *Chin. J. Polym. Sci.* **2013**, *31*, 521–529. [CrossRef]
13. Su, Y.F.; Ge, M.Q. Kapok Fibre and its aPPLICATION in Textile. *J. Fash. Technol.* **2018**, *3*, 9–13.
14. Zou, J.; Yao, H. Building a Modern Socialist Country in an All-Round Way: A NEW STRATEGIC GOAL of "Four Comprehensiveness". *J. Communist Party Sch. Shijiazhuang* **2021**, *23*, 10–14.
15. Xu, Y.J. Kapok and Wool Composite Thermal Insulation Building Materials. *Ind. Text.* **1999**, *4*, 23.
16. Liu, X.T.; Li, L.; Yan, X. Carbon Absorbing Properties of WOVEN Composites. *Mater. Sci. Eng. B* **2012**, *55*, 550–559.
17. Makki, A.I.; Oktariani, E. Acoustic Absorptive Properties of Kapok Fibre, Kapok Fibre Layered Tricot Fabric and Kapok Fibre Layered Double Weave Fabric. *J. Phys.* **2019**, *1381*, 1–4.
18. Azieyanti, N.A.; Faris, R. Acoustic Isolation of Jute and Kapok Fibre Reinforcement Polypropylene Composites. *IOP Conf. Ser. Mater. Sci. Eng.* **2020**, *932*, 012096. [CrossRef]
19. Liu, X.T.; Tang, X.N.; Deng, Z.M. Sound Absorption Properties for Multi-Layer of Composite Materials using Nonwoven Fabrics with Kapok. *J. Ind. Text.* **2020**, *152808372090492*. [CrossRef]
20. Yu, B.M. Analysis of Heat and Mass Transfer in Fractal Medium. *J. Eng. Thermophys.* **2003**, *3*, 481–483.
21. Chen, W.H.; Chen, T.N.; Wang, X.P.; Zhang, C. Study on Fractal Model of Flow Resistance of Fibre Porous Metal. *J. Xi'an Jiaotong Univ.* **2015**, *49*, 132–137.
22. Fu, X.W. Thermal Conductivity Fractal Model of Down Fibre Aggregate. Master's Thesis, Donghua University, Shanghai, China, 2012.
23. Lyu, L.H.; Liu, Y.J.; Bi, J.H.; Guo, J. Sound Absorption Properties of DFs/EVA Composites. *Polymers* **2019**, *11*, 811. [CrossRef]
24. Xu, Z.C. Research on Architectural Space Construction Based on Fractal Theory. Master's Thesis, Nanjing University of the Arts, Nanjing, China, 2015.
25. Liu, Y.J. Preparation and Sound Absorption Performance of the Poplar Seed Fibre/Polycaprolactone Composite Materials. Master's Thesis, Dalian Polytechnic University, Dalian, China, 2020.
26. Qian, J.M.; Li, X.X. Sound Absorption Mechanism of Polymer Matrix Composite Foams. *Noise Vib. Control* **2000**, *2*, 41–43.
27. Su, W.; Li, X.Y. Feasibility of Absorbent Materials used for Road Noise-Deafening Barrier. *China Fashion Inst. Technol.* **2009**, *2*, 8–12.
28. Wang, H.; Ma, Y.X.; Jin, X.Y. Study on Structure and Sound Absorption Performance of Compound Needled Nonwovens. *Nonwovens* **2009**, *17*, 31–34.
29. Liu, Y.F.; Jeng, D.S. Pore Structure of Grain-Size Fractal Granular Material. *Materials* **2019**, *12*, 2053. [CrossRef]
30. Li, Y. Study on Fractal Dimension Algorithm and its Evolution of Historical Buildings. *Shanxi Archit.* **2020**, *46*, 36–38.
31. Yang, S. Structural Characteristics of Fibre Aggregates and Their Acoustic Absorption Performance. Ph.D. Thesis, Donghua University, Shanghai, China, 2011.
32. Lyu, L.H.; Tian, Y.Y.; Lu, J.; Xiong, X.Q.; Guo, J. Flame-Retardant and Sound-Absorption Properties of Composites Based on Kapok Fibre. *Materials* **2020**, *13*, 2845. [CrossRef] [PubMed]
33. Liu, L.B. Study on PCL and MPEG-PCL Block Copolymer Microspheres. Master's Thesis, Shanghai Jiao Tong University, Shanghai, China, 2009.
34. Liu, Y.J.; Lyu, L.H.; Guo, J.; Wang, Y. Sound Absorption Performance of the Poplar Seed Fibre/PCL Composite Materials. *Materials* **2020**, *13*, 1465. [CrossRef]
35. Wang, J.C.; Zhang, X.F.; Zhou, J.H.; Mei, C.Z.; Wang, X.D. Carbonation Depth Model of Waste Fibre Recycled Concrete Based on Fractal Theory. *Build. Struct.* **2019**, *49*, 137–141.

Article

Enhanced Antibacterial Efficiency of Cellulosic Fibers: Microencapsulation and Green Grafting Strategies

Dorra Dridi [1,†], Aicha Bouaziz [2,3,†], Sondes Gargoubi [4,*], Abir Zouari [5], Fatma B'chir [6], Aghleb Bartegi [2], Hatem Majdoub [7] and Chedly Boudokhane [5]

[1] Unit of Analysis and Process Applied to the Environment (UR17ES32) Issat Mahdia, Department of Environmental Sciences & Nutrition, University of Monastir, Monastir 5000, Tunisia; dorra.dridi.jeddi@gmail.com
[2] Bio-Resources, Integrative Biology & Valorization (BIOLIVAL, LR14ES06), Higher Institute of Biotechnology of Monastir, University of Monastir, Monastir 5000, Tunisia; bouaziz.aicha@gmail.com (A.B.); a.bartegi@gmail.com (A.B.)
[3] Higher School of Health Sciences and Techniques of Sousse, University of Sousse, Sousse 4054, Tunisia
[4] Textile Engineering Laboratory—LGTex, Textile Department, University of Monastir, Monastir 5000, Tunisia
[5] Research Unity of Applied Chemistry and Environment, Faculty of Science of Monastir, University of Monastir, Monastir 5000, Tunisia; zouariabir2@gmail.com (A.Z.); chimi.tex@planet.tn (C.B.)
[6] Research Department, Institute National de Recherche et d'Analyse Physico-Chimique-Pôle Technologique de Sidi Thabet, Sidi Thabet 2020, Tunisia; bchirbnfatma@gmail.com
[7] Laboratory of Advanced Materials and Interfaces, Faculty of Sciences of Monastir, University of Monastir, Monastir 5000, Tunisia; hatemmajdoub.fsm@gmail.com
* Correspondence: gargoubisondes@yahoo.fr
† D. Dridi and A. Bouaziz contributed equally to this work as first authors.

Citation: Dridi, D.; Bouaziz, A.; Gargoubi, S.; Zouari, A.; B'chir, F.; Bartegi, A.; Majdoub, H.; Boudokhane, C. Enhanced Antibacterial Efficiency of Cellulosic Fibers: Microencapsulation and Green Grafting Strategies. *Coatings* **2021**, *11*, 980. https://doi.org/10.3390/coatings11080980

Received: 8 July 2021
Accepted: 12 August 2021
Published: 18 August 2021

Publisher's Note: MDPI stays neutral with regard to jurisdictional claims in published maps and institutional affiliations.

Abstract: We report an analysis of chemical components of essential oils from barks of *Ceylon cinnamon* and cloves of *Syzygium aromaticum* and an investigation of their antibacterial activity. The components of oils were determined by using Gas Chromatography/Mass Spectrometry (GC-MS) analysis, and the antimicrobial activity was assessed by the disk diffusion test. The synergic effect of essential oils mixture (cinnamon oil and clove oil) was evaluated. Antimicrobial properties were conferred to cellulosic fibers through microencapsulation using citric acid as a green binding agent. Essential oil mixture was encapsulated by coacervation using chitosan as a wall material and sodium hydroxide as a hardening agent. The diameter of the produced microcapsules varies between 12 and 48 µm. Attachment of the produced microcapsules onto cotton fabrics surface was confirmed by Attenuated Total Reflectance-Fourier Transformed Infrared (ATR-FTIR) spectroscopy, optical microscopy and Scanning Electron Microscopy (SEM) analysis. The results show that microcapsules were successfully attached on cotton fabric surfaces, imparting antibacterial activity without significantly affecting their properties. The finished cotton fabrics exhibited good mechanical properties and wettability.

Keywords: cotton; microcapsules; chitosan; essential oil; antibacterial activity

1. Introduction

One year ago, the coronavirus disease of 2019 (COVID-19) crisis forced us into confinement. The unexpected pandemic highlighted how infectious diseases can affect health and economies worldwide. As most countries slowly move toward recovery, considerable attention must be paid to the emergence of multi-resistant microorganisms. The fact that some infectious diseases can no longer be treated with antibiotics depicts an alarming issue in the health care sector [1]. This trend reinforces the need for improved measures to protect and save lives. The resistance to antimicrobial agents is due to the intensive and inappropriate use of antibiotics, self-drug addiction, hospital-acquired infections, unsafe disposal of hospital and biomedical wastes and the use of antibiotics in veterinary medicine [2].

41

The demand for antimicrobial agents with high disinfection properties is on the rise. Effective and safe microbe-resistant materials are highly demanded to fight against the emergence of microbial contamination [3,4].

Plant-based antimicrobial substances are emerging as promising tools in the fight against multi-resistant pathogens, as reported by numerous recent studies [5]. Plants contain valuable components, such as glycosides, tannins, steroids, terpenoids, saponins, alkaloids, polyphenols, flavonoids, quinones and coumarins [6]. These substances are used as a defensive response to microorganisms, insects and herbivores [7].

Among the plant-derived products, essential oils are one of the most diverse classes of specialized metabolites that play an important role in the defense against microbial attacks [8]. Essential oils present a complex mixture of numerous volatile molecules with low molecular weight, including terpenoides, terpenes and other aliphatic and aromatic substances [9]. These constituents exhibit a broad spectrum of antimicrobial activities [10]. In addition, essential oils are biodegradable and safe for human health [11].

Increasing demand for effective and low-toxicity antimicrobial agents depicts essential oils as promising alternatives in a variety of domains [12]. Given the emergence of techniques such as microencapsulation and nanotechnology applications [13,14], the use of essential oils in the textile sector is emerging. These substances are used to impart either a pleasant smell or antimicrobial activity to the textile materials [13].

Microencapsulation and nano-finishing provide good efficacy for treated textiles. In comparison to conventional techniques, the large surface area of micro/nano particles ensures better affinity for textile substrates and leads to a successful and durable functionalization [15].

Despite numerous studies showing the antimicrobial action of essential oils when applied to textiles, the synergistic effect of these substances for textile applications has not been reported [16].

In the current work, we report an environmentally friendly finishing method to functionalize cotton fabric surfaces with antibacterial activity based on the synergic effect of two essential oils attached via a microencapsulation technique with citric acid as a binding agent using the padding method.

2. Materials and Methods

2.1. Materials

Essential oils were obtained from the bark of *Ceylon cinnamon* and cloves of *Syzygium aromaticum*, which were purchased from spice shops. Cinnamon and clove are commonly used spices which have been well recognized for their antimicrobial activities [17,18].

Chitosan (low molecular weight), acetic acid and NaOH were purchased from Sigma Aldrich (Sigma-Aldrich Chemie GmbH, Steinheim, Germany). Citric acid and Tween 20 (surfactant) were donated by the S2C Company (S2C, Sousse, Tunisia). The desized, bleached and mercerized woven cotton fabric used during the experiments was bought from the local market with a surface area of 280 g·m^{-2}.

2.2. Methods

2.2.1. Essential Oil Extraction

Hydro-distillation was used to extract essential oils. The plant material (mashed cinnamon barks/cloves) was laid out with water, boiled and then concentrated by an evaporator with heat at low pressure. The essential oil was transferred through the steam toward a condenser where the oil was separated from the cooled steam using a separatory funnel and collected in a receiving container.

2.2.2. Microcapsule Preparation

Chitosan–essential oil microcapsules were prepared by coacervation technique as described in the literature with some modifications [19]. Chitosan solution was prepared by dissolving 1 g of chitosan in 100 mL of 1% (*v*/*v*) aqueous acetic acid solution. The resulting system was left under magnetic agitation for 3 h. Chitosan was used to form the wall membrane of microcapsules.

An amount of 20 mL of a mixture of cinnamon and clove essential oil (1:1 *v/v*) was added to the prepared chitosan solution. Then, 0.1 mL of surfactant (Tween 20) was added to the system.

To form an oil–water emulsion, the system was emulsified at a speed rate of 900 rpm using an Ultra-Turrax (IKA, Staufen, Germany). Coacervation was induced by adding the obtained emulsion to a solution (250 mL) containing sodium hydroxide (0.2 M) and ethanol (4/1). The stirring speed was reduced to 400 rpm. The microcapsules were washed with ultrapure water and collected by decantation.

2.2.3. Green Grafting of the Microcapsules

A green binding agent (citric acid) was used to attach the shell material based on chitosan to the cellulosic fibers through ester bonds. The padding drying method was applied as reported in previous studies, with a few rectifications.

The cotton samples were immersed in an aqueous solution (pH = 5) containing 100 g·L^{-1} of the microcapsule emulsion, 30 g·L^{-1} of citric acid and 2 g·L^{-1} of monosodium phosphate (a catalyst). The system was heated at 50 °C for 5 min. Finally, samples were washed with ultrapure water and padded to obtain a wet pickup percentage of 80%. Drying was carried out at 90 °C for 5 min and thermocondensation at 120 °C for 5 min.

2.3. FTIR Analysis

In order to evidence the crosslinking of microcapsules on cotton fabrics, Attenuated Total Reflectance-Fourier Transformed Infrared (ATR-FTIR) spectra were recorded using a spectrophotometer (ATR-FTIR, Perkin Elmer, Waltham, MA, USA). The spectra were collected in transmittance mode with a spectral resolution of 32 scans and 4 cm^{-1} over the range of 4000–400 cm^{-1}.

2.4. GC-MS Analysis

Extracted oils were analysed by Gas Chromatography/Mass Spectrometry (GC/MS) provided with HP-5MS fused−5% phenyl methyl siloxane capillary column (30 m, 0.25 mm, 0.25 μm) (Agilent Technologies, Santa Clara, CA, USA) and interfaced with a PerkinElmer Turbo mass spectrometer (Software version 4.1) (PerkinElmer, Inc., Waltham, MA, USA). The column operation conditions were adjusted in the following manner: the sample solution injected was 1 μL, the carrier gas (helium) was adjusted at 1.0 mL/min, the injector was set at 250, the ion source temperature was regulated at 200 °C, the split ratio was 100:1, the electron impact ionization was of 70 eV and the interface line temperature was set at 300 °C. The operating temperature started at 110 °C, then was gradually raised to 180 °C with a gradient of 4 °C/min, then 220 °C with a gradient of 2 °C/min, and reached 300 °C at a gradient of 20 °C/min.

2.5. Optical Microscopy

In order to confirm that essential oils were successfully microencapsulated, an optical microscope (Leica DM 500, Leica Microsystems, Heerbrugg, Switzerland) was employed.

The optical microscope is equipped with a digital camera controlled by analysis software and different objectives. Microscopy technique was also considered to control the size and size distribution of microcapsules.

2.5.1. Scanning Electron Microscopy

Morphological analysis was performed using a high-resolution scanning electron microscope (FEIQ250, Thermo-Fisher ESEM, Waltham, MA, USA). The cotton samples were coated with a conductive gold layer (a thin layer) by vapor deposition. Imaging parameters were regulated at low working voltages (5–10 kV) to save the original state of the textile material.

2.5.2. Mechanical Properties

Tensile strength tests were conducted using a dynamometer (Lloyd LR 5 K, Lloyd Instruments Ltd., Largo, FL, USA). Samples were cut to a gauge length of 75 mm, and tests were performed as stated by NF EN ISO 13934-2 standards using a load cell of 5 kN and an extension speed of 50 mm·min^{-1}.

2.5.3. Surface Wettability

The contact angle aspects were recorded using a Drop Shape Analyzer (DSA25, Kruss, Germany). A droplet of 2 µL of ultra-pure water was placed on the fabric surface immobilized on the stage. For each sample, the test was repeated three times in different places.

2.5.4. Washing Durability

To evaluate the washing durability, a domestic washing cycle was conducted as described in our previous publication [20]. A qualitative evaluation of the microcapsules crosslinking was realized based on optical microscope photographs.

2.6. Antibacterial Activity

The agar diffusion method was applied to evaluate essential oil activity. The cell density was adjusted to approximately 10^6 CFU/mL. An amount of 100 mL of the inoculi was spread over each petri dish containing the nutrient agar (Mueller–Hinton Agar). A paper filter disc (5 mm) impregnated with 20 µL/disc of the essential oils (cinnamon oil, cloves oil and oil mixture 50/50) was placed on the surface of the petri dishes, which were then incubated at 37 °C for 24 h. Each test was repeated twice.

The same method was applied to evaluate the activity of treated textiles against bacteria. All the tests were performed using the strains *Staphylococcus aureus* (ATCC 25923) and *Escherichia coli* (ATCC 8739). An autoclave was used, before experiments, for instruments and fabrics sterilization by heated steam at 120 °C for 20 min.

3. Results and Discussion

3.1. Essential Oil Extraction and Characterization

The hydro-distillation of the barks of Ceylon *cinnamon* and cloves of *Syzygium aromaticum* produced essential oils in yields of 1.8% and 10.5%, *v/w*, respectively, based on fresh weight.

The phytochemical composition of the two extracted oils was determined using a gas chromatograph coupled with a mass selective detector.

The composition of clove and cinnamon essential oils were found to be diverse. The major active constituents of clove essential oil were eugenol (83.55%) and phenol, 2-methoxy-4-(2-propenyl)-, acetate (14.92%), while the main active constituents of cinnamon essential oil were cinnamaldheyde (28.37%) and cinnamyl acetate (22.63%).

3.2. Morphology and Size of Microcapsules

Figure 1a shows an optical photograph of essential oil-loaded microcapsules. Figure 1b shows the optical microscopy of a microcapsule on the cellulosic fiber surface. The prepared microcapsules clearly show regular and spherical structures without fragments.

Figure 1b evidences the crosslinking of microcapsules on the fabric surface conserving their spherical shape. Figure 1c shows the cellulosic fiber, after a cycle of domestic washing, on which we can clearly identify an attached microcapsule.

Finally, Figure 1d shows the SEM images, at different magnifications, of a ruptured microcapsule after lightly rubbing the fabric. The crushed microcapsule presents a rough shell wall.

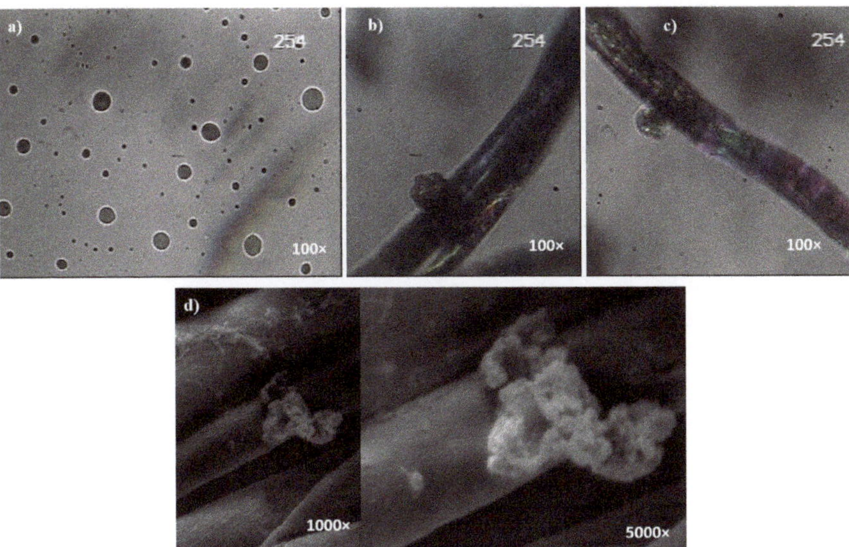

Figure 1. Optical photographs of (**a**) microcapsule solution, (**b**) microcapsules on the surface of the cellulosic fiber, (**c**) microcapsules on the surface of the cellulosic fiber after washing and (**d**) SEM images of ruptured microcapsule at different magnifications (the inset in **d**) is the ruptured microcapsule at 1000× magnification).

The microcapsule size distribution was evaluated by optical microscopy. The size of 80 microcapsules was determined. Figure 2 shows the size distribution of prepared microcapsules is between 12 and 48 μm, and the mean diameter is 22 μm. It was reported that the non-uniformity of size distribution is due to the stirring process. The water–oil emulsion flows rapidly around the agitator. The microcapsules with larger size are formed far away from the agitator, while small microcapsules are formed near the agitator [21]. In addition, the agglomeration of the oil phase enhances the non-uniformity of the microcapsules' size [22].

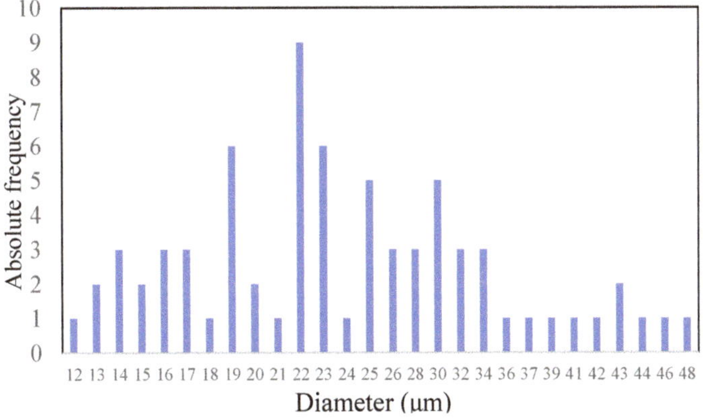

Figure 2. Microcapsule size distribution.

3.3. ATR-FTIR Results

In order to evaluate the success of the attachment of microcapsules on the surface of the cellulosic fibers through the citric acid, the control fabric (untreated cotton) and the treated fabric were characterized by ATR-FTIR (Figure 3).

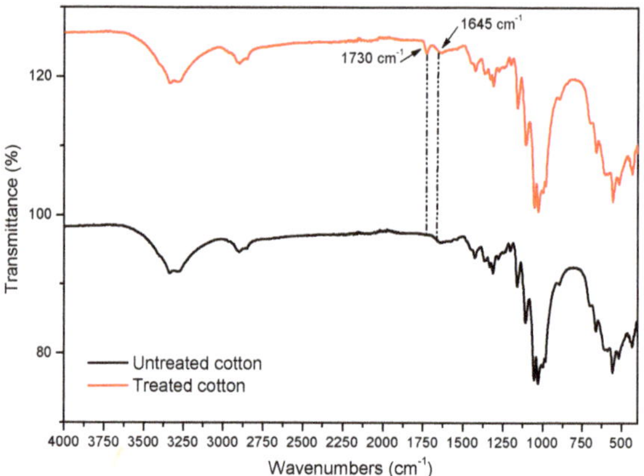

Figure 3. FTIR spectra of untreated cotton fabric and cotton fabric treated with microcapsules.

The FTIR spectra show prominent peaks of cellulose. The stretching vibration of the hydroxyl group gives a wide peak at 3300 cm^{-1}. The band at 2890 cm^{-1} is characteristic of CH stretching vibration. Typical bands observed in the region of 1650–898 cm^{-1} are assigned to cellulose. The characteristic peaks of –CH$_2$ and –CH, –OH and C–O bonds in cellulose gives the bands at 1428, 1368, 1317, 1061 and 895 cm^{-1} [20].

The appearance of a new peak at 1730 cm^{-1} was revealed on the spectrum of the treated cotton fabric. This peak is not present in the untreated sample spectrum and it corresponds to C=O ester stretching, which evidences the successful interaction between the chitosan (the wall of microcapsules) and cellulosic fibers through ester bond formation.

In addition, the small intensity of the peak at 1645 cm^{-1} is attributed to the NH group bending vibration, resulting from the chemical reaction between the NH$_2$ groups of chitosan in the microcapsule walls and the COOH groups of the acid.

The ATR-FTIR results are in agreement with optical microscopy and SEM photographs of treated samples where the attached microcapsules were clearly seen (Figure 1).

3.4. Mechanical Properties

Maximum breaking load and extension were the characteristic parameters considered to evaluate mechanical properties. The two parameters were extracted from the load versus extension curves. As shown in Table 1, the maximum breaking load of untreated fabric is lower than that of the treated one. This result is due to the fact that the grafting process was conducted under acidic conditions (pH = 5). Highly acidic mediums are known to reduce the mechanical properties of cellulose by affecting its crystalline structure. The reduction rates depend on the pH value. High acidic pH values result in an important reduction in mechanical properties.

The extension indicates the percentage of strain at maximum applied force. High extension values evidence a stretch behavior of the textile substrate. Microcapsule grafting slightly reduced the stretch rate of the fabric. However, this reduction is not considered sufficient to affect the textile properties.

Table 1. Mechanical properties.

Mechanical Property	Untreated Cotton	Treated Cotton
Maximum breaking load (N)	430 ± 26	374 ± 18
Extension (%)	23 ± 0.6	22 ± 0.5

$\pm$ Standard deviation.

3.5. Surface Wettability

Dynamic contact angle was used to evaluate the wetting properties of the treated fabrics. Due to the high hydrophilicity of textiles, it is difficult to evaluate their static contact angles. The time required by the fabric to totally absorb the water droplet was considered. The longer the required time, the lower the sample wettability. Figure 4 shows the wetting times. It is clear that the treated fabric presents a more hydrophilic surface compared to the control fabric. A short wetting time was recorded when the water droplet fully spread on the surface. This result can be explained by the fact that the chitosan (microcapsule wall)-treated fabric presents an important percent in oxygen and more C=O or C–O bonds, which are associated with the hydrophilic property.

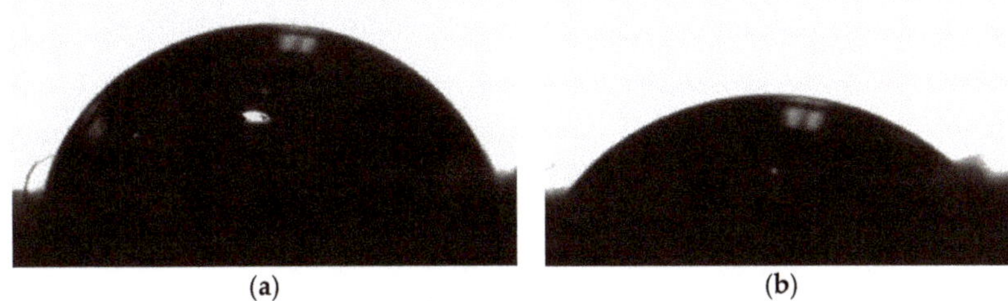

(a)   (b)

Figure 4. Water droplets on the surface of (**a**) untreated fabric, and (**b**) treated fabric after 3 s of contact.

3.6. Antibacterial Activity Evaluation

As shown in Figure 5, the extracted oils showed a remarkable antimicrobial activity. They were effective against both *E.coli* and *S.aureus* strains. A clear zone of inhibition was detected. Average distance of peripheral inhibition zone (clear zone around the disc) was measured (Table 2). An additive effect was observed when testing the activity of the oil mixture. The combined effect was greater than individual effects.

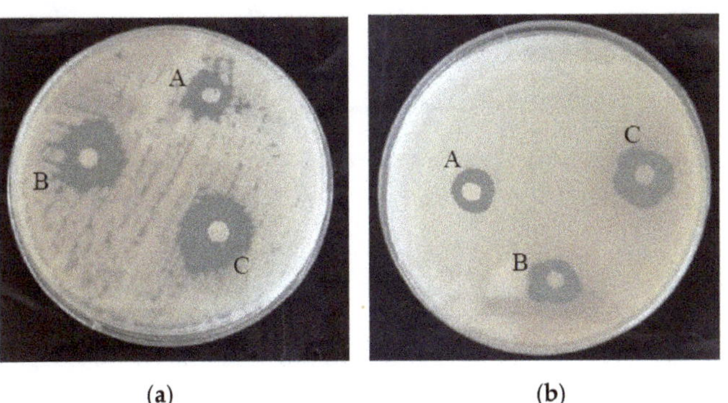

(a)   (b)

Figure 5. Antibacterial activity of essential oils in solution using agar diffusion method. A: cloves oil, B: cinnamon oil and C: oil mixture ((**a**): *S.aureus*/(**b**): *E.coli*).

Table 2. Peripheral inhibition zones values (mm).

Bacteria Strain	Cinnamon Oil	Clove Oil	Oil Mixture
E.coli	2.5 ± 53	1.75 ± 0.46	3.25 ± 0.46
S.aureus	3 ± 0.75	1.75 ± 0.46	3.5 ± 0.53

± Standard deviation.

Figure 6 shows that untreated samples did not present any activity since they did not show an inhibition zone around the sample.

Antibacterial activity was detected for individual oil and the mixture. A slight difference was seen between samples against the same bacteria strain and the same sample against the two bacteria strains.

The antibacterial effect exhibited by the microcapsules is predominantly due to the encapsulated essential oils during their release through the microcapsule wall and not from the chitosan itself. It is well known that chitosan is an antibacterial agent [23,24]. However, during the grafting of microcapsules onto cotton fibers, most of the positive amino groups of chitosan were complexed with negative carboxylic groups of the crosslinking agent.

Cinnamon and clove oils affect the envelope structure of Gram-positive and Gram-negative bacteria. The major antibacterial agents of these oils penetrate through the bacteria cell wall and damage the cytoplasmic membrane [25,26].

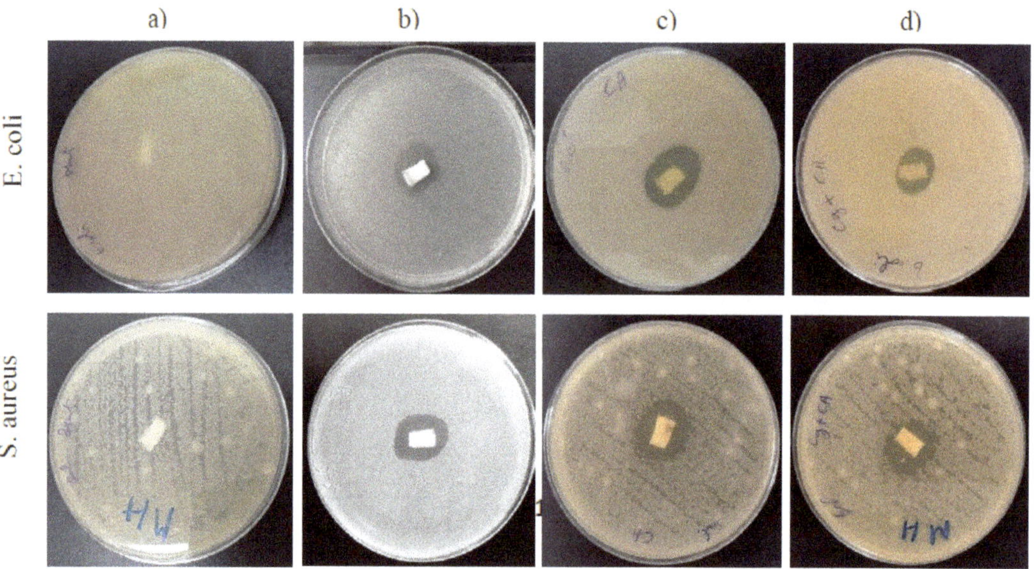

Figure 6. Antibacterial activity results of the (**a**) untreated cotton, (**b**) cotton grafted with clove oil microcapsules, (**c**) cotton grafted with cinnamon oil microcapsules and (**d**) cotton grafted with oil mixture microcapsules.

Comparing results of the inhibition obtained from the clear zones, indifference was observed when using the oil mixture. There was no interaction between one another. The combined effect was the same as when the two oils were individually applied.

Cinnamon oil exhibits good antibacterial activity. However, it is known to be an irritant oil. Combination of cinnamon oil with clove oil can avoid irritant and contact allergy reactions by using lower concentrations of the irritant oil while conserving its important activity against bacteria.

4. Conclusions

Essential oils are a family of natural compounds that are well known for their biological activities. They are applied to many treatments. The goal of this work was to apply a mixture of essential oils (cinnamon oil and clove oil) for the functionalization of cellulosic fibers. Essential oils were extracted by hydro-distillation. The extracted oils were characterized by GC/MS. This technique revealed the chemical composition of each oil. Chitosan was used to produce essential oil microcapsules, which were then grafted on the cotton fabric surface via citric acid. Some textile properties were evaluated, such as wettability and mechanical properties

Finally, antibacterial activity of different samples was assessed. It was concluded that essential oil combination is a promising tool to provide antimicrobial agents with a broad spectrum of activity. Moreover, the encapsulation avoids the rapid evaporation and degradation of their active components. The application of encapsulated oil mixture enhances the efficacy against multi-resistant pathogens.

Author Contributions: Conceptualization, D.D., A.B. (Aicha Bouaziz) and S.G.; methodology, D.D., A.B. (Aicha Bouaziz) and S.G.; experiments, A.B. (Aicha Bouaziz), D.D., S.G., F.B. and A.Z.; validation, H.M., C.B. and A.B. (Aghleb Bartegi); analysis, A.B. (Aicha Bouaziz), D.D., S.G., F.B., A.Z. and H.M.; investigation, D.D., A.B. (Aicha Bouaziz) and S.G.; resources, C.B. and H.M.; writing—original draft preparation, D.D., A.B. (Aicha Bouaziz), S.G. and H.M.; writing—review and editing, D.D., A.B. (Aicha Bouaziz), S.G. and H.M.; supervision, A.B. (Aghleb Bartegi), H.M. and C.B.; project administration, S.G. and C.B. All authors have read and agreed to the published version of the manuscript.

Funding: This research received no external funding.

Institutional Review Board Statement: Not applicable.

Informed Consent Statement: Not applicable.

Data Availability Statement: Data sharing is not applicable to this article.

Conflicts of Interest: The authors declare no conflict of interest.

References

1. Pacios, O.; Blasco, L.; Bleriot, I.; Fernandez-Garcia, L.; González Bardanca, M.; Ambroa, A.; López, M.; Bou, G.; Tomás, M. Strategies to combat multidrug-resistant and persistent infectious diseases. *Antibiotics* **2020**, *9*, 65. [CrossRef] [PubMed]
2. Srivastava, J.; Chandra, H.; Nautiyal, A.R.; Kalra, S.J. Antimicrobial resistance (AMR) and plant-derived antimicrobials (PDA ms) as an alternative drug line to control infections. *3 Biotech* **2014**, *4*, 451–460. [CrossRef] [PubMed]
3. Chen, C.K.; Lee, M.C.; Lin, Z.I.; Lee, C.A.; Tung, Y.C.; Lou, C.W.; Law, W.C.; Chen, N.T.; Lin, K.Y.A.; Lin, J.H. Intensifying the antimicrobial activity of poly [2-(tert-butylamino) ethyl methacrylate]/polylactide composites by tailoring their chemical and physical structures. *Mol. Pharm.* **2018**, *16*, 709–723. [CrossRef] [PubMed]
4. Hoque, J.; Prakash, R.G.; Paramanandham, K.; Shome, B.R.; Haldar, J. Biocompatible injectable hydrogel with potent wound healing and antibacterial properties. *Mol. Pharm.* **2017**, *14*, 1218–1230. [CrossRef]
5. Subramani, R.; Narayanasamy, M.; Feussner, K.D. Plant-derived antimicrobials to fight against multi-drug-resistant human pathogens. *3 Biotech* **2017**, *7*, 1–15. [CrossRef]
6. Ladan, Z.; Amupitan, J.; Oyewale, O.; Ayo, R.; Temple, E.; Ladan, E. Phytochemical screening of the leaf extracts of Hyptis spicigera plant. *Afr. J. Pure Appl. Chem.* **2014**, *8*, 83–88.
7. Santamaria, M.E.; Martínez, M.; Cambra, I.; Grbic, V.; Diaz, I. Understanding plant defence responses against herbivore attacks: An essential first step towards the development of sustainable resistance against pests. *Transgenic Res.* **2013**, *22*, 697–708. [CrossRef]
8. Trifan, A.; Luca, S.V.; Greige-Gerges, H.; Miron, A.; Gille, E.; Aprotosoaie, A.C. Recent advances in tackling microbial multidrug resistance with essential oils: Combinatorial and nano-based strategies. *Crit. Rev. Microbiol.* **2020**, *46*, 338–357. [CrossRef]
9. Akthar, M.S.; Degaga, B.; Azam, T. Antimicrobial activity of essential oils extracted from medicinal plants against the pathogenic microorganisms: A review. *J. Issues ISSN* **2014**, *2350*, 1588.
10. Xianfei, X.; Xiaoqiang, C.; Shunying, Z.; Guolin, Z. Chemical composition and antimicrobial activity of essential oils of Chaenomeles speciosa from China. *Food Chem.* **2007**, *100*, 1312–1315. [CrossRef]
11. Walentowska, J.; Foksowicz-Flaczyk, J. Thyme essential oil for antimicrobial protection of natural textiles. *Int. Biodeterior. Biodegrad.* **2013**, *84*, 407–411. [CrossRef]
12. Jugreet, B.S.; Suroowan, S.; Rengasamy, R.; Mahomoodally, M.F. Chemistry, bioactivities, mode of action and industrial applications of essential oils. *Trends Food Sci. Technol.* **2020**, *101*, 89–105. [CrossRef]

13. Ghayempour, S.; Montazer, M. Micro/nanoencapsulation of essential oils and fragrances: Focus on perfumed, antimicrobial, mosquito-repellent and medical textiles. *J. Microencapsul.* **2016**, *33*, 497–510. [CrossRef] [PubMed]
14. Lammari, N.; Louaer, O.; Meniai, A.H.; Elaissari, A. Encapsulation of essential oils via nanoprecipitation process: Overview, progress, challenges and prospects. *Pharmaceutics* **2020**, *12*, 431. [CrossRef]
15. Sayed, U.; Sharma, K.; Parte, S. Application of essential oils for finishing of textile substrates. *J. Text. Eng. Fash. Technol.* **2017**, *1*, 42–47. [CrossRef]
16. Tawiah, B.; Badoe, W.; Fu, S. Advances in the development of antimicrobial agents for textiles: The quest for natural products. Review. *Fibres Text. East. Eur.* **2016**, *3*, 136–149. [CrossRef]
17. Singh, R. Antimicrobial activity of selected medicinal plants cinnamon and clove. *Int. J. Res. Pharm. Pharm. Sci.* **2018**, *3*, 48–50.
18. Adesiji, Y.; Alli, O.; Shittu, A.; Oluremi, A. Antibacterial activity of cinnamon and clove oil on different food borne bacterial isolates. *Nig. J. Pure Appl. Sci.* **2015**, *28*, 2610–2616.
19. Sharkawy, A.; Fernandes, I.P.; Barreiro, M.F.; Rodrigues, A.E.; Shoeib, T. Aroma-loaded microcapsules with antibacterial activity for eco-friendly textile application: Synthesis, characterization, release, and green grafting. *Ind. Eng. Chem. Res.* **2017**, *56*, 5516–5526. [CrossRef]
20. Bouaziz, A.; Dridi, D.; Gargoubi, S.; Zouari, A.; Majdoub, H.; Boudokhane, C.; Bartegi, A. Study on the Grafting of Chitosan-Essential Oil Microcapsules onto Cellulosic Fibers to Obtain Bio Functional Material. *Coatings* **2021**, *11*, 637. [CrossRef]
21. Yuan, L.; Sun, T.; Hu, H.; Yuan, S.; Yang, Y.; Wang, R.; Lyu, C.; Yang, F.; Lyu, X. Preparation and characterization of microencapsulated ethylenediamine with epoxy resin for self-healing composites. *Sci. Rep.* **2019**, *9*, 1–10.
22. Javid, A.; Raza, Z.A.; Hussain, T.; Rehman, A. Chitosan microencapsulation of various essential oils to enhance the functional properties of cotton fabric. *J. Microencapsul.* **2014**, *31*, 461–468. [CrossRef]
23. Hoque, J.; Adhikary, U.; Yadav, V.; Samaddar, S.; Konai, M.M.; Prakash, R.G.; Paramanandham, K.; Shome, B.R.; Sanyal, K.; Haldar, J. Chitosan derivatives active against multidrug-resistant bacteria and pathogenic fungi: In vivo evaluation as topical antimicrobials. *Mol. Pharm.* **2016**, *13*, 3578–3589. [CrossRef] [PubMed]
24. Chen, C.-K.; Liao, M.-G.; Wu, Y.-L.; Fang, Z.-Y.; Chen, J.-A. Preparation of highly swelling/antibacterial cross-linked N-maleoyl-functional chitosan/polyethylene oxide nanofiber meshes for controlled antibiotic release. *Mol. Pharm.* **2020**, *17*, 3461–3476. [CrossRef] [PubMed]
25. Zhang, Y.; Liu, X.; Wang, Y.; Jiang, P.; Quek, S. Antibacterial activity and mechanism of cinnamon essential oil against *Escherichia coli* and *Staphylococcus aureus*. *Food Control* **2016**, *59*, 282–289. [CrossRef]
26. Weerakkody, N.S.; Caffin, N.; Turner, M.S.; Dykes, G.A. In vitro antimicrobial activity of less-utilized spice and herb extracts against selected food-borne bacteria. *Food Control* **2010**, *21*, 1408–1414. [CrossRef]

coatings

Article

Continuous Mechanical Extraction of Fibres from Linseed Flax Straw for Subsequent Geotextile Applications

Saif Ullah Khan [1,2,3], Laurent Labonne [1], Pierre Ouagne [3] and Philippe Evon [1,*]

[1] Laboratoire de Chimie Agro-industrielle, Université de Toulouse, INRAE, ENSIACET, 31030 Toulouse, France; saifullah.khan@toulouse-inp.fr (S.U.K.); laurent.labonne@toulouse-inp.fr (L.L.)
[2] Department of Textile Engineering, Balochistan University of Information Technology, Engineering and Management Sciences, Quetta 87300, Pakistan
[3] Laboratoire Génie de Production, Université de Toulouse, ENIT, 65016 Tarbes, France; pierre.ouagne@enit.fr
* Correspondence: philippe.evon@toulouse-inp.fr

Abstract: Linseed flax is a multipurpose crop. It is cultivated for its seeds and particularly for its oil. The main contributors for this crop are Canada, France and Belgium. In general, straws of linseed flax are buried in the fields or burnt. However, these solutions are not good practices for the environment and from an economical point of view. In this study, straws of linseed flax (six batches in total) with different dew retting durations and harvesting techniques were studied to possibly use them for producing innovative geotextiles. Two different fibre extraction processes were investigated. A first process (A) involved horizontal breaker rollers and then a breaking card. A second one (B) consisted in using vertical breaker rollers, and an "all fibre" extraction device (fibre opener) followed by sieving. The chemical composition of fibres in parietal constituents appeared to be globally equivalent to the one of textile flax with a pectic content decreasing as a function of the dew retting duration. This contributed to an increase in the cellulose content. The fibre content was situated in a range from 29% to 33%, which corresponds to a good yield for linseed flax fibre. The level of purity can reach values of up to 90% for method A (without extra-sieving) and 96% for method B (with extra-sieving), and the length of the fibres (larger for method A than for method B) and their tensile properties make them suitable for structural geotextile yarn manufacturing.

Keywords: linseed flax; straw; fibre mechanical extraction; shives; mean fibre length; mean fibre diameter; tensile properties; geotextiles

Citation: Khan, S.U.; Labonne, L.; Ouagne, P.; Evon, P. Continuous Mechanical Extraction of Fibres from Linseed Flax Straw for Subsequent Geotextile Applications. *Coatings* **2021**, *11*, 852. https://doi.org/10.3390/coatings11070852

Received: 28 June 2021
Accepted: 13 July 2021
Published: 15 July 2021

Publisher's Note: MDPI stays neutral with regard to jurisdictional claims in published maps and institutional affiliations.

1. Introduction

Between 2016 and 2018, an average of 97,700 ha/year of textile flax was cultivated in France with an average straw yield of 6 t/ha [1]. So, about 600 ktons of straw are globally available on this market. This economic activity represents about 12,000 direct jobs in in the growing and scutching sectors [2]. The long line fibres (about 25% of the straw mass) are used for fine and delicate textiles and structural composite materials, which are in high demand. In 2015, China was importing about 140 ktons of these fibres, representing approximately the total amount of the French production [3]. Even if the long textile flax are also produced in Belgium and the Netherlands, the demand is continuously increasing from China and Europe, but the land available for the production of textile flax cannot globally be extended in a great extent. The demand for tow fibres (scutching and hackling tows) is also very high for the manufacturing of coarser yarns for home textiles or technical applications such as in the composite industry for medium load bearing applications or for injected parts [4].

To answer the high flax fibre demand (both long line and tows), other sources of fibres may be promoted. Linseed, or oilseed flax may be one of them, as at the present time the valorization of this straw and its associated fibres only concerns low added value applications such as insulations or paper, as the traditional linseed flax straw is often short in length (40–60 cm) with a high ramification, leading to a low straw yield (0.4 t/ha) available for fibre production [5]. Rennebaum et al. [6] showed that much higher straw yields can be obtained if no chemical shortener is used. A straw yield of 5.1 t/ha was obtained in their study, and this indicates that the linseed flax straw in their study consisted of between 25% and 30% fibre (in weight), and so could be a large source of fibre. The tensile properties of linseed flax fibres were also investigated by Pillin et al. [7]. Their work showed that the tensile properties of elementary fibres extracted manually are equivalent to the ones of textile flax extracted using the same process. During the continuous mechanical extraction of linseed flax fibres, Ouagne et al. [8] showed that the process parameter should be chosen with care as this may degrade the mechanical potential of the fibres if the extraction process is too aggressive.

An "all fibre" extraction set-up needs to be used as the linseed flax stems are randomly aligned on the ground and collected as such in round bales. In those conditions, it is not possible to use the same device as the one used for textile flax (scutching and hackling). Other devices exist to extract fibres from linseed flax or hemp stems. These ones are often based on hammer mills, which is a very efficient process permitting to break the woody part of the stem and to separate globally the shives from the fibres [9]. The process is very efficient but also very aggressive, and it was demonstrated that the fibres extracted using this process show relatively poor mechanical properties [10]. It is therefore important to minimize the impact of the fibre extraction on the mechanical potential of the fibres, especially when a geotextile valorization is targeted [11,12]. With hammer mill fibre extraction, the fibres are damaged and numerous defects such as kink-bands that are at the origin of strength and modulus decreases may be observed [13,14]. Two other processes which are expected to minimise their impact on the fibre mechanical properties will be tested in this study.

During fibre extraction, most of the shives in the straw are separated [8]. As for shives from textile flax, they could be used as animal litters because of their high water absorbency [15]. Recent works have nevertheless shown that shives from linseed flax could be also used to replace wood particles in the manufacture of boards [15,16]. Extrusion-refining pre-treatment of shives increases their mean aspect ratio and, consequently, the mechanical strength of the boards, and proteins from linseed cake can even be added as naturel binder to further increase this strength [16]. On the basis of their usage properties, such materials could find various applications, e.g., intermediate containers, furniture, domestic flooring, shelving, general construction, etc. In addition, as they do not emit formaldehyde, they are much more environmentally and human-health friendly than the wood-based boards currently found in the market, e.g., medium-density fibreboard (MDF), chipboard, oriented strand board (OSB) or plywood [16].

The aim of the present work is to study and to compare two different fibre extraction methods used to obtain technical fibres from various batches of oleaginous flax straw having different characteristics (e.g., cultivation locations, cutting height, and dew retting duration), for subsequent geotextile applications. Fibre content, fibre purity, and the mechanical properties of mechanically extracted fibres are determined, and they are discussed in relation to field parameters such as dew retting performed to ease the fibre extraction and to determine if the geotextile targeted application may seriously be considered to increase the income of the farmers besides the ones of seeds (for human or industrial oils) and shives (for animal litters and boards) [15,16].

2. Materials and Methods

2.1. Plant Material

Straws of oleaginous flax (*Linum usitatissimum* L.) used in this study were supplied by Ovalie Innovation (Auch, France). They were from the Everest variety, and the plants were cultivated in the south-west part of France. All the straw samples tested were collected using a combine harvester, and their harvesting took place after the seed one. Six different batches were used, with different dew retting durations (i.e., from 0 to 6 weeks) and with different harvesting methods (i.e., high cut or low cut of the straw, and disconnection or not of the shredder of the combine harvester during the straw harvest). The details of those six batches are given in the Table 1. On the one hand, the cutting height of the straw directly influenced the quantity of straw harvested in the field. For a high cut (i.e., 30 cm above the ground), the length of the harvested stems was on average 35 cm. For a low cut (i.e., only 5 cm above the ground), it was around 60 cm in length. On the other hand, the objective of the disconnection of the shredder of the combine harvester during the straw harvesting was to better preserve the fibre integrity.

Table 1. Harvesting details of all the batches used in this study.

Batch Number	Cutting Type	Disconnection of the Shredder	Dew Retting Duration before Harvesting
1	High cut	Yes	6 weeks
2	High cut	Yes	No dew retting
3	Low cut	No	No dew retting
4	Low cut	Yes	No dew retting
5	High cut	No	2 weeks
6	Low cut	No	No dew retting

Those six batches were cultivated during the summer 2018 in two different locations. However, both were relatively close to each other (14 km). Firstly, batches number 1 to 4 were collected in the city of Pavie (43°62′ N, 0°57′ E). For this first location, after the seeds were harvested on July 11, the straw batches number 2 to 4 were harvested on July 12 and the straw batch number 1 was harvested on August 22, i.e., after six weeks of dew retting. Secondly, batches number 5 and 6 were collected in the city of Seissan (43°49′ N, 0°55′ E). For this second location, the seed harvesting took place on August 12, and straw batches number 5 and 6 were made on August 28 (dew retting during two weeks) and August 16 (no dew retting), respectively.

2.2. Manual Extraction of Fibres

For all the batches of oleaginous flax straw treated in this study, around fifty stems were chosen randomly, and those were used to separate manually fibres from shives. At the end of the manual extraction, the mass content of fibres inside each batch was determined. In parallel, the fibres extracted were used for determining their chemical composition.

2.3. Chemical Composition of Fibres

The moisture content of the fibrous materials was determined according to the ISO 665:2000 standard [17].

The chemical composition of fibres was determined from those which were extracted manually. Once extracted, the fibres were cut with a pair of scissors to a length of about 4–5 mm. Then, they were grinded finely using a Foss Tecator Cyclotec 1093 mill (Foss, Hillerød, Denmark) fitted with a 1 mm sieve. The obtained powder was the test sample used for the chemical characterizations.

Firstly, the three main parietal constituents, i.e., cellulose, hemicelluloses, and lignins, were evaluated using the ADF-NDF (ADF for Acid Detergent Fibre, and NDF for Neutral Detergent Fibre) method of Van Soest and Wine [18,19]. A Foss Tecator FT122 Fibertec hot extractor (Foss, Hillerød, Denmark) was used for the hot extractions, and a Foss Tecator FT121 Fibertec cold extractor (Foss, Hillerød, Denmark) was used for the cold ones.

Secondly, pectins were also quantified inside the manually extracted fibres. To begin, a non-normalized two-step extraction procedure was conducted on the test sample for analyses using each time a Foss Tecator FT122 Fibertec hot extractor:

(i)   An alcohol-insoluble solid fraction was first obtained after 3 min boiling in ethanol (95°), filtration, rinsing with ethanol (80°), and then drying.

(ii)   To continue, the total pectins were extracted, and their acid hydrolysis into galacturonic acid was realized; for these two simultaneous actions, 30 min boiling in 100 mL HCl 0.05 M was required, and the extraction was conducted twice.

Once the extract containing galacturonic acid recovered, it was then analysed using the Blumenkrantz and Asboe-Hansen colorimetric method for quantifying galacturonic acid [20]. Pectins are in fact polymers of galacturonic acid, i.e., polygalacturonic acids.

For each replication, the test sample mass was around 1 g for the determination of cellulose, hemicelluloses and lignins, and it was around 500 mg for measuring pectins. All determinations were conducted in duplicate. Results were expressed as mean values $\pm$ standard deviations. They were mentioned as a percentage of the dry matter mass of the fibres.

2.4. Mechanical Extraction of Fibres

Two different methodologies were used in this study to extract mechanically fibres from the oleaginous flax straw batches, i.e., the method A, and the method B. The two next paragraphs describe them in more details.

2.4.1. Method A

The method A consisted in the succession of (i) horizontal breaker rollers (Hemp-Act, Lacapelle-Marival, France), (ii) a thresher (Hemp-Act, Lacapelle-Marival, France), and then (iii) a breaking card (Hemp-Act, Lacapelle-Marival, France). In the horizontal rollers, the straw samples were broken, and shives were partly removed by gravity at the bottom of the device. The latter was in fact the assembly of five horizontal rollers in a staggered arrangement. At the output of these horizontal rollers, the fibrous material was then fed to the thresher, where additional shives were removed by intense threshing. Lastly, a breaking card from Russian origin inspired from the classical Mackie breaking cards was used for an additional removing of shives plus the alignment of fibres. At the outlet of the breaking card, a web was formed, and it was rolled up continuously.

2.4.2. Method B

The method B consisted in the succession of (i) vertical breaker rollers (Taproot Fibre Lab, Port Williams, NS, Canada), (ii) an "all fibre" extraction device (Laroche, Cours, France), and then (iii) a sieving machine (Ritec, Signes, France). To begin, the straw samples were processed in the vertical breaker rollers of a lab-scale device (processing machine) after stems (around 1 kg) were packed inside small bags having an appropriate volume. Here, the main purpose of this step was to break the stems, and to remove part of the shives inside the oleaginous flax straw before mechanically extracting the fibres. There were five vertical breaker rollers, and an efficient breaking of the stems was obtained thanks to their specific groove structure, to their staggered position, and also to an optimized spacing between them. After the breaker rollers partially broke the straw samples, the latter were kept in a climatic chamber at 25 °C and 90% relative humidity (RH) during seven to ten days to favour the adsorption of water by fibres until equilibration (i.e., up to a constant weight). Because this conditioning step allowed an increase in the moisture content of the fibrous material up to 14–19% in weight, it was helpful for the next fibre extraction step to better preserve the fibre length during their mechanical extraction. Indeed, with an increased moisture content, Ouagne et al. [8] evidenced that the rigidity of fibres is reduced, thus leading to less breakages when extracting them mechanically.

Then, the fibrous material was treated using a Laroche Cadette 1000 "all fibre" extraction device (Laroche, Cours, France). This industrial equipment was used to extract fibres in a mechanical way, and it is presented in Figure 1. With a 1 m width, this machine allowed the treatment of a 175 kg/h inlet flow rate of the fibrous material collected at the outlet of the vertical breaking rollers, corresponding to a 3.5 m/min speed for the feed belt.

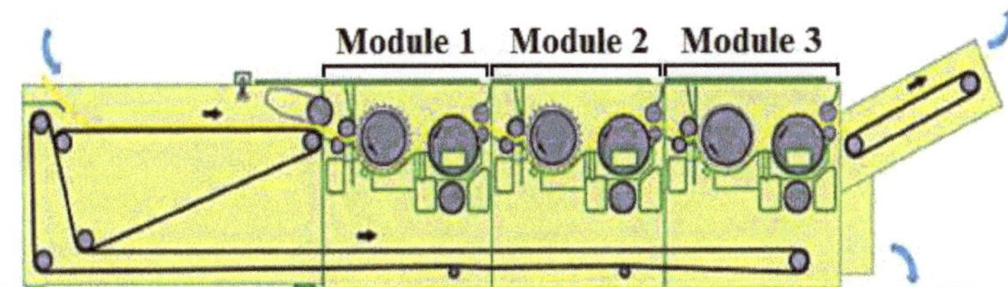

Figure 1. Laroche Cadette 1000 "all fibre" extraction device used in this study (from Laroche Cadette company website, https://laroche.fr, accessed on 21 June 2021).

This tearing machine has a double ability, i.e., the opening and the cleaning of natural fibres, as well as the formation of laps. It is composed of three successive modules. A pair of rollers, one smooth and the other grooved (made of rubber), is placed at the inlet of each module, thus ensuring a regular feeding of the raw material. In each module, a first cylinder equipped with nails (i.e., the extracting roller or fibre extraction roller) was the place where fibres were extracted. Its rotation speed can vary from 750 rpm to 1800 rpm. In addition, at the level of each of the three modules, shives can be (at least partially) eliminated by gravity, thanks to a trap door situated under the extracting roller. In this study, the opening of the trap door was set at its maximum to favour the most possible the removal of shives.

A perforated cylinder is also positioned at the end of each module. It is equipped with a ventilation system having a motor with a 2865 rpm maximum rotation speed. The perforated cylinder has three functions: (i) removing dust from the fibrous material by aspiration, (ii) forming a fibrous lap, and (iii) passing it on to the next module or the machine outlet.

The operating parameters used in the present study for the "all fibre" extraction device are presented in Table 2. They were the same as those used in [8], as the extraction of fibres from oleaginous flax straw was considered in that previous study as satisfactory for the two batches tested. Finally, the speed of the output belt was 1.8 m/min.

Table 2. Operating parameters used for the "all fibre" extraction device.

Operating Parameter	Module 1	Module 2	Module 3
Transmission speed of the lap (m/min)	2.2 (from module 1 to module 2)	1.5 (from module 2 to module 3)	
Rotation speed of the extracting roller (rpm)	725	725	725
Rotation speed of the motor for the aspiration at the level of the perforated cylinder (rpm)	1500	2000	2000

Lastly, after the mechanical extraction of fibres using the "all fibre" extraction device, the fibrous lap obtained was sieved using a Ritec 600 (Ritec, Signes, France) vibrating sieving machine fitted with a 12 mm sieve. Here, the objective was to eliminate as much as possible the dust and the residual shives still trapped inside the lap, to obtain oleaginous flax fibres with a much higher purity.

2.5. Purity in Fibres of the Different Fibrous Materials

As the laps collected after the mechanical extraction of fibres did not consist only in bast fibres but also in some remaining shives and dust, the content in those impurities inside the different fibrous materials (including the starting oleaginous flax straw batches) was evaluated. For each fibrous material, a 50 g test sample mass was collected, and shives and dust were removed thanks to a mechanical sieving operation conducted during 5 min.

Then, the residual shives that remained trapped were collected manually. The content of impurities inside the fibrous material, and thus the real bast fibre content could then be calculated, and they were expressed as a percentage in weight.

2.6. Morphological Analysis of the Mechanically Extracted Fibre Bundles

The morphological analysis of the vegetal fractions obtained at the outlet of the fibre extraction devices used in this study were carried out on all the treated batches. A specific mass (around 10 g) of the extracted fibres was randomly picked from different places inside the laps, and around two hundred fibre bundles were measured in length.

Each bundle was attached to one end and then linearly extended in order to know its actual length, which was measured between its two extremities using a double decimetre with a 0.25 mm accuracy. The average fibre length was then calculated, and also the corresponding standard deviation.

2.7. Tensile Testing on Elementary Fibres

Diameter, tensile strength and elastic modulus of elementary fibres were measured using a Diastron Ltd. Lex 820 (Diastron Ltd., Andover, UK) automated high-precision extensometer. For this purpose, elementary fibres were firstly separated, and they were then glued on a 12 mm gauge length. To begin, the samples were placed on a peripheral equipment for the diameter measurement. This portion of the Diastron testing machine measured the diameter of each elementary fibre along ten different places throughout the whole fibre length. The diameter was measured with an accuracy of 0.01 μm.

Then, the same fibre sample was positioned in the high-precision extensometer of the testing machine for measuring tensile strength and elastic modulus. Here, a stepping motor was used for traction up to the breakage of the individual fibre, and a 20 N capacity load cell was used for measuring the force applied to the sample analysed. The extensometer was used for failing at low strain values. The accuracy for the obtained displacement was 1 μm, and the tensile test was conducted at a 1 mm/min speed. A 20 ms periodicity was chosen for recording the measuring points, i.e., the force and the displacement. The calculation of the tensile strength and elastic modulus thus became possible without using any supplementary strain measurement device. For each batch, forty samples of individual fibres were tested, and the mean value and standard deviation were then calculated for both tensile strength and elastic modulus.

3. Results

3.1. Contents in Fibres and Shives inside the Straw Batches

Table 3 mentions the contents of fibres and shives inside the six straw batches tested in this study. A manual extraction of the fibres inside the straw samples was conducted to obtain these results. For batches number 1 to 5, the content of fibres inside straw varied from 29% to 33%, and these results were in perfect accordance with those of a previous study where the fibre content inside linseed flax straw was found to be around 30% as well [21]. In another study [6], the fibre content inside technical stems was a little lower (from 23% to 29%), depending on different factors such as the variety of oleaginous flax.

Table 3. Contents in fibres and shives in the different straw batches tested (determination through manual extraction) (% in weight).

Batch Number	Fibre Content	Shives Content
1	30.1	69.9
2	28.9	71.1
3	33.3	66.7
4	32.8	67.2
5	32.2	67.8
6	45.9	54.1

When comparing batches number 3 and 4 with batches number 1 and 2, all coming from the same location, a low cut used during the stem harvesting had a slightly positive effect in the fibre content of the starting straw, i.e., 33% instead of 30–31%. Because the fibre proportion along the stem is known to be a little more important at its bottom [22], choosing a low cut for the combine harvester thus favours the collection of a higher proportion of fibres. However, such a setting for the harvesting machine may cause machine breakage if stones are present in the field that are caught at the same time. The harvesting must therefore be done with great care in this case.

With a fibre content surprisingly much higher (i.e., 46%), the straw batch number 6 was the exception in this study. Here, the reason is presumably due to cultivation considerations. Indeed, for this specific batch, no plant-growth regulator was used on the plot during the oleaginous flax development, and stems partly suffered lodging before the seed harvesting. The subsequent straw harvesting using the combine harvester was thus much more difficult in that case, and it is reasonable to assume that some part of the shives were in fact directly lost on the field during the harvesting of the straw. Indeed, for this specific batch, it was observed before the manual extraction of fibres that stems were already partially broken, and shives were absent in some places along them.

In fact, with a 46% fibre content inside straw, batch number 6 was more in accordance with results on textile flax presented in another study [23], where the crude fibre content varied from 36% to 50%, depending on many factors such as the soil type, and the nature of the fertilizer applied. Nevertheless, one should be aware that this batch is not representative of what could be produced on average at an industrial scale.

3.2. Chemical Composition of Bast Fibres

With cellulose, hemicelluloses and lignins, pectins are one of the main constituents of the plant cell walls. They are also the predominant compound within the middle lamella. In fact, pectins have the function to hold the cells of plant tissue together. In linseed flax stems, pectins are more specifically responsible for "cementing" the bast fibres together. Thus, by reducing the pectin content, this definitely helps to separate the fibres from each other, and both their extraction and opening thus become facilitated. For fibrous plants such as textile flax, linseed flax and hemp, this is the retting operation that allows the removing of pectins before conducting the extraction of fibres, and the evolution of their content over time is a clue to estimate its efficiency.

Table 4 shows the evolution of the pectin content inside manually extracted bast fibres as a function of the dew retting duration. Dew retting is a natural retting process conducted directly on the ground before the straw harvesting. It involves in fact two different actions, both resulting in the progressive dissolution of the pectic binder over time, i.e., a chemical action (thanks to the morning dew) and a biological one (thanks to the microorganisms, fungi and bacteria in soil).

Table 4. Chemical composition of manually extracted fibres as a function of the dew retting duration (from 0 to 6 weeks) (% of dry matter).

Dew Retting Duration	Hemicellu-Loses	Lignins	Cellulose	Pectins
No dew retting (i.e., batch number 2)	12.9 ± 0.5	5.3 ± 0.4	62.1 ± 0.4	4.4 ± 0.2
2 weeks	6.6 ± 0.1	7.2 ± 0.4	65.5 ± 0.4	4.4 ± 0.1
4 weeks	8.5 ± 0.5	6.1 ± 0.3	66.9 ± 0.5	3.6 ± 0.2
6 weeks (i.e., batch number 1)	10.3 ± 0.0	5.5 ± 0.1	67.0 ± 0.1	3.7 ± 0.2

Results are presented as mean values ± standard deviations.

For this part of the work, batches number 2 and 1 were analysed, corresponding to no dew retting and to a six weeks dew retting duration, respectively. To complete the analysis, two stem samples were also collected in the same plot during the retting period, i.e., after two weeks and four weeks of dew retting, respectively. Table 4 also mentions the evolutions in the content for the three other parietal constituents (i.e., hemicelluloses, lignins, and cellulose).

From the results in this table, it is clear that the dew retting played an important role in the reduction of the pectin content, which was only 3.7% after six weeks of dew retting instead of 4.4% immediately after the seed harvesting (i.e., no dew retting). The pectin content remained unchanged after two weeks, and a four weeks dew retting duration was at least necessary to see a reduction in the pectin content inside the manually extracted bast fibres. In parallel, from zero to six weeks, their cellulose content increased progressively (i.e., from 62.1% to 67.0%), indicating that retting was effective but rather slow. The content of hemicelluloses was also lowered after retting, whatever its duration. In contrast, because lignins are biopolymers with a high hydrophobic character due to their polyphenol structure, their content inside bast fibres logically tended to increase during retting.

3.3. Purity of Extracted Fibres from Process A

The A methodology for the extraction of fibres from straw used a succession of three different operations, i.e., (i) horizontal breaker rollers, (ii) a thresher, and then (iii) a breaking card. No sampling was made from this methodology after the breaker rollers and the thresher, meaning that the purity of the extracted fibres was measured only inside the final fibrous material, collected at the outlet of the breaking card in the form of a rolled web. The purity of the obtained webs in fibres is mentioned in Table 5, and it varied from 66% to 90%.

Table 5. Comparison of the final purities in fibres (%) for the two processes tested (i.e., A and B).

Batch Number	1	2	3	4	5	6
Process A	89.9	66.9	79.2	66.2	84.7	87.8
Process B	89.3	91.9	88.5	81.9	93.8	96.1

The web originating from batch number 4 showed the minimum value for fibre purity (i.e., 66%). As a reminder, that straw batch was not dew retted, and the linseed flax straw was harvested using a low cut setting, with disconnection of the combine harvester shredder (Table 1). This resulted in longer stems with shives still well glued by the pectic cement to the bast fibres. Undoubtedly, this had reduced the chance to remove in a proper way shives from the extracted fibres. The web purity in fibres was in the same order of magnitude (i.e., 67%) for batch number 2. In the same way, this straw batch was not dew retted, and it was also harvested with the shredder of the combine harvester disconnected. On the contrary, the batch number 1 showed the maximal purity level (i.e., 90%) for the obtained web at the outlet of the breaking card. Because this batch was dew retted during six weeks, shives were less attached to the fibres. In addition, straw was harvested using a high cut in that case. Stems collected on the field were thus shorter, and shives were probably less retained by the fibre web during its formation in the inside of the breaking card. With that 90% purity, additional spinning and weaving operations should be perfectly possible to manufacture geotextiles. Inside the obtained webs, straw batches number 5 and 6 had quite comparable fibre purities to that of batch number 1, i.e., 85% and 88%, respectively. Additionally, this means that the corresponding extracted fibres would be also suitable for subsequent geotextile applications.

3.4. Purity of Extracted Fibres from Process B

As a reminder, the B methodology for the extraction of fibres from the straw involved three successive steps, i.e., (i) vertical breaker rollers for the stem breaking and the removing of part of shives, (ii) the "all fibre" extraction device for extracting fibres (plus the loss of more shives by gravity at the bottom of each of the three extracting rollers), and (iii) a sieving extra-step which consisted in eliminating as much as possible the remaining shives to increase more the purity in fibres of the final lap. Table 6 mentions the purity of the different fibrous materials in fibres after the vertical breaker rollers, after the "all fibre" extraction device, and also after the sieving extra-step. The impurities in the fibrous lap after sieving also appear in Table 6. For the purpose of comparison between the two tested processes, the final purity in fibres for process B is also mentioned in Table 5.

Table 6. Purity of the fibrous materials in fibres during process B (i.e., after the vertical breaker rollers, for the fibre lap at the outlet of the "all fibre" extraction device, and after sieving), and amount of impurities in the fibrous lap after sieving (% in weight).

Batch Number	After the Vertical Breaker Rollers	After the "All Fibre" Extraction Device	After Sieving	Impurities [1] in the Fibrous Lap after Sieving
1	45.1	68.2	89.3	10.7
2	46.2	70.7	91.9	8.1
3	47.1	62.6	88.5	11.5
4	44.9	64.7	81.9	18.1
5	51.6	68.4	93.8	6.2
6	66.0	68.9	96.1	3.9

[1] Shives and dust amount.

After the fibre extraction step using the "all fibre" machine, a 63–71% purity in fibres was achieved instead of 29–46% in the starting materials. The minimal purities (i.e., 63% and 65% in weight, respectively) were observed in the case of the batches number 3 and 4, which were the non-retted batches and had a low cut at harvesting. Because stems were in that cases longer (i.e., around 60 cm) and still contained their pectic cement, the extraction of fibres was more difficult, and this resulted in a reduced fibre purity level. The separation of shives from fibres was disadvantaged due to the presence of more pectins (Table 4), and the shive particles really separated from the fibres remained also easier trapped inside the fibrous material made of longer stems. In parallel, when comparing batches number 3 and 4, the disconnection of the shredder of the combine harvester at the moment of the straw harvesting (case of batch number 4) had in fact no real importance for the purity in fibres at the "all fibre" device outlet. In fact, a low cut at harvesting should be privileged to increase the crop yield in the field, even if it is necessary in that case to pay attention to the increased risks of machine breakage.

The highest purity in fibres at the outlet of the "all fibre" extraction device (i.e., 71%) was obtained for the batch number 2. Because the latter was harvested using a high cut at the straw harvesting, stems collected were shorter, and woody parts were thus more easily removed (when harvesting straws just as when extracting fibres) as the stem entanglement was less significant. For the three other batches (i.e., batches number 1, 5 and 6), the fibre purities were quite similar and median, with contents of fibres between 68% and 70%. In particular, batches retted before the straw harvesting (i.e., batches number 1 and 5) did not reveal especially higher purity values after the "all fibre" extraction device. This means that even if the shives were more easily detached from the bast fibres once the straw had been retted, the "all fibre" extraction device was not able to remove them in totality from the inside of the obtained lap. In fact, part of them remained trapped inside the fibrous lap, and this is the reason why a sieving extra-step was applied to it to eliminate the remaining shives as much as possible.

After sieving, the purity of all batches was markedly increased, and it ranged from 82% to 96%, which was considered as a good material purity for future geotextile applications. In fact, such values confirmed the real efficiency of the sieving extra-step for reaching a better fibre purity of the final lap. In addition, the obtained purities were much better than those obtained from the A methodology using the breaking card as fibre extraction mode (i.e., maximum value of 90%). In the case of process B, the minimal fibre purity (i.e., 82%) was observed for batch number 4, and it was in fact exactly the same situation when using the A methodology for fibre extraction (Table 5). Although this purity would be enough for further processing in geotextile applications, the most likely reasons for such low purity may be the cutting type (i.e., a low one) on the straws on the field and the disconnection of the combine harvester shredder during the straw harvesting. Those two harvesting settings both contributed to stems with higher length and thus in a more important entangling of fibres once extracted, and this probably unfavoured the removing of shives at sieving.

Batches number 5 and 6 showed the best values for purity (i.e., 94–96%). Here, it must be kept in mind that those two batches originated from a different location than the four others. Contrary to batches number 1 to 4, no plant-growth regulator was used during the linseed flax cultivation in that second location. It is thus reasonable to assume that both soil and climate conditions, and cultivation mode were the most likely reasons to explain such differences in fibre purity.

3.5. Morphological Analysis of the Extracted Fibre Bundles

Length of the extracted fibre bundles was measured after having randomly taking a handful of them from each batch. Then, around two hundred bundles were analysed for their lengths, and the results (i.e., mean values ± standard deviations) are given in Table 7. Here, the average length gives a good indication of the ease of subsequently obtaining a yarn, after passing over a drawing frame. For its part, Figure 2 represents the frequency distribution curve of the length of fibre bundles extracted from batch number 6 using the two extracting methodologies (i.e., A and B).

Table 7. Length of fibre bundles (cm) inside all the final laps produced.

Batch Number	Process A	Process B
1	11.7 ± 6.2	9.4 ± 4.6
2	10.9 ± 5.9	6.4 ± 3.2
3	10.7 ± 5.6	7.7 ± 4.0
4	10.6 ± 5.7	8.3 ± 3.9
5	11.1 ± 5.3	8.1 ± 4.1
6	13.4 ± 7.8	7.9 ± 3.5

Results are presented as mean values ± standard deviations.

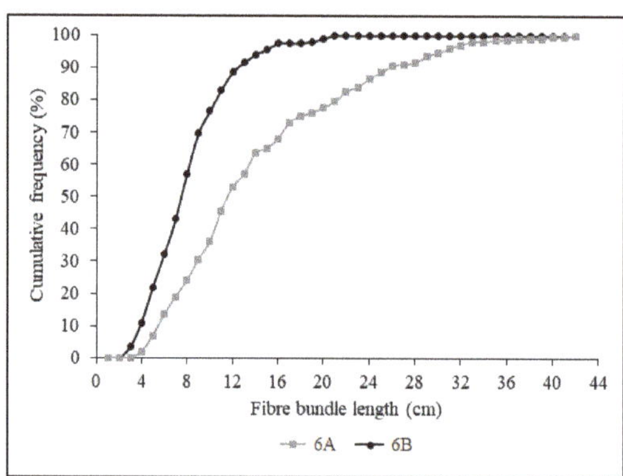

Figure 2. Frequency distribution curve of the length of fibre bundles extracted from batch number 6 using the A and B extracting methodologies.

As a first result, longer fibre bundles were obtained in the present study in comparison with other values from linseed flax fibres in the literature [8,24], regardless of the linseed flax straw batch and the extraction method used.

Additionally, Figure 2 perfectly illustrates the fact that process A is more accurate for better preservation of fibre length, and this was true for all six batches tested in this study, even though Figure 2 only shows the results obtained from batch number 6. Only 13% and up to 32% fibre bundles extracted from batch number 6 were observed less than 6 cm in length for processes A and B, respectively, which is interesting for further spinning stages.

Portions of 36% and 76%, respectively, of fibre bundles were less than 10 cm in length, and at least half of the fibre bundles (i.e., 50% and 62%, respectively) had a length comprised between 6 and 14 cm, which should be really interesting for subsequent spinning.

An increase in the standard deviations was nevertheless observed in the case of the fibre bundles extracted through process A (Table 7), thus indicating that a larger dispersion in length existed for the bundles extracted from the breaking card.

3.6. Mechanical Properties of Elementary Fibres inside the Extracted Bundles

The tensile properties of the individual fibres inside the extracted bundles are mentioned in Table 8. The individual fibres analysed were separated with care from the bundles using a method presented in [14]. Fibre bundles originating from the six straw batches used, and from both methodologies (i.e., A and B), were immersed in water at 30 °C for 72 h. Here, the objective was to facilitate the manual extraction of the elementary fibres from the bundles, and to extract them with the highest care.

Table 8. Tensile properties, minimal diameter, maximal diameter, and cross section area of individual fibres inside all the final laps produced.

Characteristic	1A	1B	2A	2B	3A	3B	4A	4B	5A	5B	6A	6B
Tensile strength (MPa)	456 ± 308	694 ± 374	832 ± 578	626 ± 419	778 ± 516	660 ± 383	751 ± 454	939 ± 525	701 ± 402	827 ± 490	874 ± 560	880 ± 1077
Young's modulus (GPa)	29.5 ± 14.1	39.4 ± 20.1	41.9 ± 24.0	35.3 ± 19.3	43.5 ± 22.4	36.7 ± 15.3	43.6 ± 23.8	46.2 ± 21.7	40.9 ± 21.7	50.6 ± 23.0	46.1 ± 24.8	63.7 ± 58.2
Minimal diameter (μm)	15.1 ± 5.7	16.2 ± 5.9	13.5 ± 5.7	15.4 ± 5.7	13.7 ± 5.5	16.1 ± 5.0	11.5 ± 3.8	12.1 ± 4.4	10.0 ± 3.1	10.5 ± 3.9	12.7 ± 5.8	13.0 ± 6.5
Maximal diameter (μm)	34.4 ± 16.2	33.4 ± 12.8	26.2 ± 10.9	31.8 ± 12.8	27.8 ± 11.1	31.2 ± 13.2	23.5 ± 9.1	29.2 ± 8.1	28.2 ± 10.4	25.2 ± 8.9	28.8 ± 12.0	30.4 ± 12.4
Cross section area (μm^2)	462 ± 338	466 ± 307	323 ± 253	431 ± 290	348 ± 271	442 ± 315	248 ± 201	306 ± 175	243 ± 159	241 ± 210	334 ± 302	364 ± 333

Results are presented as mean values ± standard deviations.

For all the treated samples, the section of the individual fibres had the form of an oval shape and not of a cylindrical one. Indeed, minimal and maximal diameters measured were always different. The first one ranged from 10.0 to 16.2 μm, whereas the second one ranged from 23.5 to 34.4 μm. The cross section area of the manually extracted individual fibres was thus between 241 and 466 μm^2, and this would correspond to diameters ranging from 17.5 to 24.4 μm if considering the cross section of all fibres as an ideal cylinder.

The mechanical results obtained for the elementary fibres inside the extracted bundles showed that a reduced cross-sectional area led to better tensile properties (Table 8). In addition, because the obtained diameter values were in the same order of magnitude as those mentioned in other studies for individual fibres from linseed [8] and textile [23] flax, it is reasonable to assume that, in the present study, fibres manually extracted from bundles using the hot water procedure were indeed elementary fibres.

Values for the average tensile strength ranged from 456 to 939 MPa. The minimal value (i.e., 456 MPa) was for the 1A sample, corresponding to the batch number 1 treated with the A methodology. In parallel, the maximal one (i.e., 939 MPa) was obtained from sample 4B (i.e., straw batch number 4 using the B methodology). Similarly, Young's modules ranged from 29.5 to 63.7 GPa (samples 1A and 6B, respectively).

4. Discussion

4.1. Contents in Fibres and Shives inside the Straw Batches, and Chemical Composition of Bast Fibres

The contents of fibres inside straw batches in this study (from 29% to 33% for batches number 1 to 5) are quite coherent in comparison with other data in the literature also dealing with oleaginous flax [21]. The batch number 6 was the only straw batch for which the fibre content was surprisingly much higher (i.e., 46%), and the reason was presumably due to cultivation considerations as mentioned above. Using oleaginous flax straw for its richness in bast fibres (most often around 30% in weight with the exception of the 46% fibre content in the case of batch number 6, which is however not representative with what could be obtained at industrial scale) was thus quite possible, the straw possibly becoming in that way a supplementary added-value product of the linseed flax cultivation in addition to the seeds.

From the results in Table 4 presenting the chemical composition of bast fibres, it is obvious that dew retting contributed to a reduction in the pectin content. However, the duration for dew retting was probably not long enough in that case. The cultivation place in France (Gers department) is also known to be a quite sunny and hot region in summer, and with low rainfall. Surely, this can explain, at least in part, why the reduction in the pectin content was not so important. For future works, because dew retting is really of key importance, an increase in its duration will thus need to be tested in order to identify the optimal retting duration. Choosing agricultural plots at the bottom of the valleys, where the morning dew is stronger and more persistent over time, would also be better.

Jute and flax fibres [25] are subjected to the dew retting mostly. Pectin content is removed in dew retting due to bacterial activity. Dew retting depends on soil fungi colonization on stem/bast to degrade pectin and hemicelluloses by releasing polygalacturunase and xylanase.

In a review, ref. [26] reported the changes in flax fibre during dew retting process. Cell wall composition was directly related to the microbial colonization. Partial damage and fibre bundle decohesion was observed due to fungal hyphase and parenchyma on the epidermis and around the fibre during dew retting. Furthermore, a decrease on the primary cell wall of polysaccharides was noticed on the fourteenth day of retting due to higher enzymatic activities. This spreading of microbial colonization went towards the inner core of the stem after 6 weeks retting.

4.2. Comparison of the Purity of Extracted Fibres from Processes A and B

When comparing the two extracting methodologies, i.e., A and B, the batch number 4 resulted in both cases in the minimal fibre purity. This means that long stems for which no dew retting was applied before their harvesting were an unfavourable situation for reaching high fibre purity (Table 5). In contrast, batches number 5 and 6—for which no plant-growth regulator was used during the linseed flax cultivation—had high purity levels (up to 96% for batch number 6 treated using process B) once fibres were extracted, for both methodologies used.

In addition, except for batch number 1, laps produced using the B methodology (i.e., the "all fibre" extraction device plus the sieving extra-step) had all a more important fibre purity than the webs collected at the outlet of the breaking card for process A. This highlights with no doubt the real interest of the sieving extra-step conducted at the end of the B methodology. For future work, it could thus be interesting to apply the same sieving treatment to webs produced using the breaking card, on the condition of unwrapping them before sieving. In conclusion, producing geotextiles should be accessible with fibre purities of at least 85%, and such purity was in fact reached in eight out of twelve cases, including five out of six cases for laps originating from process B.

4.3. Morphology of the Extracted Fibre Bundles

The average fibre length for all batches (11–13 cm for process A, and 8–9 cm for process B) (Table 7) was increased as compared to previous values in the literature also obtained from linseed flax fibres, i.e., only 2 cm [24] and around 5 cm [8]. In the case of process B, this increase in fibre length in comparison with previous results [8] was probably partly due to the rewetting of the raw batches inside the humid climatic chamber before extracting fibres. With higher moisture levels (14–19%) for straw after conditioning instead of 9–10.5% before conditioning, the fibres thus became more flexible (i.e., less brittle), and their breakage during the severe extraction action in the "all fibre" extraction machine tended to be reduced. The same results were observed in Ouagne et al. [8] but to a lesser extent. The overall fibre length for process B varied from 6.4 cm to 9.4 cm, which was considered as good enough for subsequent spinning process before obtaining the geotextiles. When looking more precisely to the length values for batches number 1 to 4, it has to also be mentioned here that a six weeks dew retting duration (case of batch number 1) tended to favour a better preservation of the bundle length, even if the differences were not statistically significant. On the contrary, for batches number 5 and 6 coming from the same plot, the lengths of bundles were the same, meaning that a two weeks dew retting was probably too short for generating longer fibre bundles once extracted.

Fibre bundles produced through the breaking card (i.e., process A) were much longer, and their average length varied from 10.6 cm to 13.4 cm. However, the associated standard deviations were even higher. In fact, the horizontal breaker rollers and/or the garniture of the breaking card (i.e., geometry, height and diameter of nails, and distance between them) used in process A were much more suitable to better preserve the fibre bundle length, which is surely a much less aggressive mechanical extraction technique in comparison with the "all fibre" machine used for process B. In addition, the interest of the dew retting step before the straw collection was less evident in the case of process A.

In fact, the highest average length of extracted fibre bundles coming from the A methodology was obtained for batch number 6, which was a non-retted batch. However, because a low cut setting was used at harvesting for that batch, the longer stems collected on the field favoured higher bundle length once extracted.

As a conclusion, with the longer fibres obtained from the breaking card (especially those from batch number 6), the subsequent spinning step will be easier from the A-based rolled web. Indeed, less twist should be required in that case, and there should have also more friction between the fibres, thus favouring better maintenance of fibre bundles in the form of slivers or yarns at weaving.

4.4. Mechanical Properties of Elementary Fibres inside the Extracted Bundles

Table 8 mentions the tensile properties of elementary fibres inside the produced laps. Here, the influences of (i) the batch type, and (ii) the mechanical extraction methodology on these properties were not easy to discuss. Nonetheless, when calculating a mean value of the average tensile strengths, it was 732 and 771 MPa for processes A and B, respectively. In parallel, this mean value was 41 and 45 GPa, respectively, for the average Young's modules. It is thus reasonable to consider that both mechanical extraction methodologies (i.e., the breaking card for process A, and the "all fibre" extraction device for process B) resulted in quite equivalent tensile properties for the individual fibres inside the extracted bundles. Such overall average values for tensile strengths and Young's modules will undoubtedly authorize the use of the extracted bundles in geotextile applications. Indeed, for comparison, a previous study dealing with the mechanical extraction of fibres from oleaginous flax straw revealed much lower tensile strengths for the individual fibres (i.e., only 323–377 MPa mean values) for the same application in geotextiles [8].

As a reminder, batch number 1 was dew retted for six weeks before the straw harvesting. However, the tensile properties of the individual fibres obtained from that batch were the minimal ones for process A (i.e., horizontal breaker rollers and breaking card route), and they were median for process B (i.e., vertical breaker rollers plus "all fibre" extraction device plus sieving). Whereas a long-term dew retting was expected to favour the tensile properties of the extracted fibres, the results of the present study from batch number 1 did not confirm surprisingly this assumption. On the contrary, the batch number 5 which was dew retted during only two weeks resulted in much higher tensile properties for both extraction processes tested. Nonetheless, the plots were not the same (i.e., differences in the soil composition and, to a lesser extent, in climatic conditions) for batches number 1 and 5. In addition, no plant-growth regulator was used in the case of batch number 5. It is thus difficult to conclude at this point on the differences observed in the tensile properties between these two batches.

However, in the case of textile flax, it should be borne in mind that dew retting can last up to three months depending on the climatic conditions and industrial requirements [27]. In addition, it is mainly grown in the Northern part of France, a region known to be much wetter and more prone to morning dew than the Gers department where the linseed flax straws in this study were produced. It is therefore reasonable to assume that the dew retting durations tested here (six weeks max) were simply not sufficient. For future work, longer durations should thus be tested. Due to the low rainfall observed in summer in south-western France, periods of at least three months should be considered. The plots should also be chosen carefully, and those at the bottom of valleys should be preferred because of their higher humidity.

In conclusion for this study, the elementary linseed flax fibres inside the extracted bundles thus revealed an average tensile strength of 750 MPa and an average Young's modulus of 43 GPa. This was in perfect accordance with recent results on the tensile properties of hemp fibres (660 and 38 GPa for tensile strength and Young's modulus, respectively) [14] and nettle ones (712–812 MPa and 36–53 GPa) [28], both intended for the production of load-bearing composite materials. As geotextiles require lower mechanical properties, the use of the linseed flax fibres resulting from this study should thus be possible for such an application.

The fibre extraction processes used in this study were selected to work with bast fibre type plants such as hemp, flax or nettle. As mentioned in the text, it is important that a minimum level of retting is performed so that the natural cements binding the fibres together and the ones binding the fibres to the rest of the plant are degraded. This favours the fibre division and the separation between the fibres and the woody core of the plant and the bark. As these machines are mainly dedicated to the fibre extraction of European bast fibres, one can imagine that they could also operate with other types of bast fibre plants such as kenaf or jute for example. Of course, modification of some of the settings would be necessary to accommodate the changes in plant diameter, or woody core breaking strength.

Both the machines were designed to operate in a dry state. If dried, resources such as banana trunks or sisal leaves could probably be used with the breaking rollers and the breaking card. However, some modifications in the corrugated rollers and of course in the process settings would probably be necessary. In any case, dedicated machines already exist for these kinds of resources [29,30], and the machines presented in this study should be dedicated to the use of retted bast fibre plants.

5. Conclusions

Linseed flax is a dual-purpose crop. Even if it is firstly cultivated for its seeds, its straw can be also useful, thus possibly contributing to an additional source of income for farmers due to its high bast fibre content. In this study, the straw batches tested had a significant amount of fibre, the latter varying from 29% to 46% of the straw dry mass. Depending on the dew retting duration of oleaginous flax straw on the field before its harvesting, the cellulose content inside bast fibres was between 62% when no dew retting was applied and 67% when a dew retting duration of six weeks was chosen. Simultaneously, their pectin content decreased from 4.4% to 3.7%. Dew retting appeared as an important parameter for the subsequent mechanical extraction of fibres and their opening. It favoured the extraction of fibres, contributing at the same time to a better preservation of the fibre length. Two different methodologies were also tested in this study for the fibre mechanical extraction. The first one consisted in the use of a breaking card, and the second one in the use of an "all fibre" extraction device. As the second methodology was completed by a sieving extra-step for better purifying the obtained fibre lap, it resulted in the production of purer fibres as compared to the first extraction method. However, the purity of the fibres extracted thanks to the breaking card was considered as good enough for geotextile applications, and this purity may be further increased by additional work through sieving and/or their treatment using new breaker rollers. For all the straw batches tested, the breaking card resulted in much longer fibre bundles, and this could be attributed to a too strong severity of the "all fibre" extraction device when used. Looking at their length and their tensile properties, the extracted fibre bundles obtained from both methodologies would be of sufficient quality for further spinning into yarns. Thus, they appeared as promising candidates for the manufacture of innovative and renewable geotextiles.

For future work, much longer dew retting durations (at least three months or even more) should be applied as this is known to have a beneficial effect on the ease of the mechanical extraction of bast fibres. A low cutting height of the straw should also be favoured as it increases straw yields, especially if no shortener has been added during plant cultivation. The shredder of the combine harvester should also be disconnected as it allows a better preservation of the mechanical strength of the fibres at the moment of their harvesting. Lastly, with regard to the fibre extraction stage, a breaking card can produce longer fibre bundles, which can be an advantage for subsequent spinning operations. However, a sieving extra-step should be carried out in order to eliminate the residual shives present in the extracted fibres as much as possible.

Author Contributions: Conceptualization, P.E. and P.O.; methodology, P.E., P.O. and L.L.; validation, P.E. and P.O.; formal analysis, P.E. and P.O.; investigation, S.U.K. and L.L.; resources, P.E. and P.O.; writing—original draft preparation, S.U.K.; writing—review and editing, P.E. and P.O.; supervision, P.E. and P.O.; project administration, P.E. and P.O. All authors have read and agreed to the published version of the manuscript.

Funding: This research was funded by Région Occitanie, France, grant number MP0013559.

Institutional Review Board Statement: Not applicable.

Informed Consent Statement: Not applicable.

Data Availability Statement: Data is contained within the article.

Acknowledgments: The authors would like to express their sincere gratitude to Ovalie Innovation (Auch, France) for supplying the batches of oleaginous flax straw used for the purpose of this study.

Conflicts of Interest: The authors declare no conflict of interest.

References

1. Fibres Recherche et Développement. Available online: https://www.f-r-d.fr/etudes/mémento-2020 (accessed on 21 June 2021).
2. Linen & Hemp Community. Available online: http://blog.europeanflax.com/infographie-barometre-2015-du-lin-europeen-fibre-verte-et-innovante (accessed on 21 June 2021).
3. Heller, K.; Sheng, Q.C.; Guan, F.; Alexopoulou, E.; Hua, L.S.; Wu, G.W.; Jankauskienė, Z.; Fu, W.Y. A comparative study between Europe and China in crop management of two types of flax: Linseed and fibre flax. *Ind. Crops Prod.* **2015**, *68*, 24–31. [CrossRef]
4. Bourmaud, A.; Le Duigou, A.; Baley, C. Mechanical performance of flax-based biocomposites. In *Woodhead Publishing Series in Composites Science and Engineering, Biocomposites: Design and Mechanical Performance*; Misra, M., Pandey, J.K., Mohanty, A.K., Eds.; Woodhead Publishing: Sawston, UK, 2015; Volume 14, pp. 365–399.
5. Zuk, M.; Richter, D.; Matuła, J.; Szopa, J. Linseed, the multipurpose plant. *Ind. Crops Prod.* **2015**, *75*, 165–177. [CrossRef]
6. Rennebaum, H.; Grimm, E.; Warnstorff, K.; Diepenbrock, W. Fibre quality of linseed (*Linum usitatissimum* L.) and the assessment of genotypes for use of fibres as a by-product. *Ind. Crops Prod.* **2002**, *16*, 201–215. [CrossRef]
7. Pillin, I.; Kervoelen, A.; Bourmaud, A.; Goimard, J.; Montrelay, N.; Baley, C. Could oleaginous flax fibers be used as rein-forcement for polymers? *Ind. Crops Prod.* **2011**, *34*, 1556–1563. [CrossRef]
8. Ouagne, P.; Barthod-Malat, B.; Evon, P.; Labonne, L.; Placet, V. Fibre extraction from oleaginous flax for technical textile applications: Influence of pre-processing parameters on fibre extraction yield, size distribution and mechanical properties. *Procedia Eng.* **2017**, *200*, 213–220. [CrossRef]
9. Xu, J.; Chen, Y.; Laguë, C.; Landry, H.; Peng, Q. Analysis of energy requirement for hemp fibre decortication using a hammer mill. *Can. Biosyst. Eng.* **2012**, *54*, 2.1–2.8. [CrossRef]
10. Placet, V.; Trivaudey, F.; Cisse, O.; Gucheret-Retel, V.; Boubakar, M.L. Diameter dependence of the apparent tensile modulus of hemp fibres: A morphological, structural or ultrastructural effect? *Compos. Part A Appl. Sci. Manuf.* **2012**, *43*, 275–287. [CrossRef]
11. Ghosh, M.; Choudhury, P.K.; Sanyal, T. Suitability of natural fibres in geotextile application. In Proceedings of the Indian Geotechnical Conference, Guntur, India, 17–19 December 2009; pp. 449–501.
12. Renouard, S.; Hano, C.; Ouagne, P.; Doussot, J.; Blondeau, J.-P.; Lainé, E. Cellulose coating and chelation of antibacterial compounds for the protection of flax yarns against natural soil degradation. *Polym. Degrad. Stab.* **2017**, *138*, 12–17. [CrossRef]
13. Baley, C. Influence of kink bands on the tensile strength of flax fibers. *J. Mater. Sci.* **2004**, *39*, 331–334. [CrossRef]

14. Grégoire, M.; Barthod-Malat, B.; Labonne, L.; Evon, P.; De Luycker, E.; Ouagne, P. Investigation of the potential of hemp fibre straws harvested using a combine machine for the production of technical load-bearing textiles. *Ind. Crops Prod.* **2020**, *145*, 111988. [CrossRef]

15. Evon, P.; Barthod-Malat, B.; Grégoire, M.; Vaca-Medina, G.; Labonne, L.; Ballas, S.; Véronèse, T.; Ouagne, P. Production of fiberboards from shives collected after continuous fiber mechanical extraction from oleaginous flax. *J. Nat. Fibers* **2018**, *16*, 453–469. [CrossRef]

16. Evon, P.; Labonne, L.; Khan, S.U.; Ouagne, P.; Pontalier, P.-Y.; Rouilly, A. Twin-screw extrusion process to produce renewable fiberboards. *J. Vis. Exp.* **2021**, *167*, e62072. [CrossRef]

17. ISO. *ISO 665:2000, Oilseeds—Determination of Moisture and Volatile Matter Content*; International Organization for Standardi-zation: Geneva, Switzerland, 2000.

18. Van Soest, P.J.; Wine, R.H. Use of detergents in the analysis of fibrous feeds. IV. Determination of plant cell-wall constituents. *J. Assoc. Off. Anal. Chem.* **1967**, *50*, 50–55. [CrossRef]

19. Van Soest, P.J.; Wine, R.H. Determination of lignin and cellulose in acid-detergent fiber with permanganate. *J. Assoc. Off. Anal. Chem.* **1968**, *51*, 780–785. [CrossRef]

20. Blumenkrantz, N.; Asboe-Hansen, G. New method for quantitative determination of uronic acids. *Anal. Biochem.* **1973**, *54*, 484–489. [CrossRef]

21. Diederichsen, A.; Ulrich, A. Variability in stem fibre content and its association with other characteristics in 1177 flax (*Linum usitatissimum* L.) genebank accessions. *Ind. Crops Prod.* **2009**, *30*, 33–39. [CrossRef]

22. Sankari, H.S. Linseed (*Linum usitatissimum* L.) Cultivars and breeding lines as stem biomass producers. *J. Agron. Crop Sci.* **2000**, *184*, 225–231. [CrossRef]

23. Alvarez, J.M. Influence of soil type and natural Zn chelates on flax response, tensile properties and soil Zn availability. *Plant Soil* **2009**, *328*, 217–233. [CrossRef]

24. Kymäläinen, H.-R.; Hautala, M.; Kuisma, R.; Pasila, A. Capillarity of flax/linseed (*Linum usitatissimum* L.) and fibre hemp (*Cannabis sativa* L.) straw fractions. *Ind. Crops Prod.* **2001**, *14*, 41–50. [CrossRef]

25. Paridah, M.T.; Basher, A.B.; SaifulAzry, S.; Ahmed, Z. Retting process of some bast plant fibres and its effect on fibre quality: A review. *BioResources* **2011**, *6*, 5260–5281. [CrossRef]

26. Lee, C.; Khalina, A.; Lee, S.; Liu, M. A comprehensive review on bast fiber retting process for optimal performance in fibers reinforced polymer composites. *Preprints* **2019**, *2019110310*, 1–40.

27. USRTL, Industrie Française du Lin. Available online: http://www.usrtl-ifl.fr/spip.php?article34 (accessed on 21 June 2021).

28. Jeannin, T.; Yung, L.; Evon, P.; Labonne, L.; Ouagne, P.; Lecourt, M.; Cazaux, D.; Chalot, M.; Placet, V. Native stinging nettle (*Urtica dioica* L.) growing spontaneously under short rotation coppice for phytomanagement of trace element contaminated soils: Fibre yield, processability and quality. *Ind. Crops Prod.* **2020**, *145*, 111997. [CrossRef]

29. Snyder, B.J.; Bussard, J.; Dolak, J.; Weiser, T. A portable sisal decorticator for kenyan farmers. *Int. J. Serv. Learn. Eng. Humanit. Eng. Soc. Entrep.* **2006**, *1*. [CrossRef]

30. Tenerife, P.; De la Cruz, A.; Arce, A.; Pabularcon, M.; Ortega, K.; Rafallo, R. Design and development of banana fiber de-corticator with wringer. *Int. J. Recent Technol. Eng.* **2019**, *8*, 82–84.

 *coatings*

Article

Fibre Individualisation and Mechanical Properties of a Flax-PLA Non-Woven Composite Following Physical Pre-Treatments

Maxime Gautreau [1], Antoine Kervoelen [2], Guillaume Barteau [2], François Delattre [3], Thibaut Colinart [2], Floran Pierre [4], Maxime Hauguel [4], Nicolas Le Moigne [5], Fabienne Guillon [1], Alain Bourmaud [2] and Johnny Beaugrand [1,*]

1 Biopolymères Interactions Assemblages (BIA), INRA, Rue de la Géraudière, F-44316 Nantes, France; maxime.gautreau@inrae.fr (M.G.); fabienne.guillon@inrae.fr (F.G.)
2 Research Institute Dupuy De Lôme (IRDL), University Bretagne Sud, UMR CNRS 6027, F-56100 Lorient, France; antoine.kervoelen@univ-ubs.fr (A.K.); guillaume.barteau@univ-ubs.fr (G.B.); thibaut.colinart@univ-ubs.fr (T.C.); alain.bourmaud@univ-ubs.fr (A.B.)
3 Unité de Chimie Environnementale et Interactions sur le Vivant, Littoral Côte d'Opale University, UR 4492, UCEIV, SFR Condorcet FR CNRS 3417, 145 Avenue Maurice Schumann, 59140 Dunkerque, France; francois.delattre@univ-littoral.fr
4 Eco-Technilin SAS, ZA Caux Multipôles, 76190 Valliquerville, France; floran@eco-technilin.com (F.P.); maxime@eco-technilin.com (M.H.)
5 Polymers Composites and Hybrids (PCH), IMT Mines Ales, 30100 Alès, France; nicolas.le-moigne@mines-ales.fr
* Correspondence: johnny.beaugrand@inrae.fr

Citation: Gautreau, M.; Kervoelen, A.; Barteau, G.; Delattre, F.; Colinart, T.; Pierre, F.; Hauguel, M.; Le Moigne, N.; Guillon, F.; Bourmaud, A.; et al. Fibre Individualisation and Mechanical Properties of a Flax-PLA Non-Woven Composite Following Physical Pre-Treatments. *Coatings* **2021**, *11*, 846. https://doi.org/10.3390/coatings11070846

Received: 23 June 2021
Accepted: 11 July 2021
Published: 14 July 2021

Publisher's Note: MDPI stays neutral with regard to jurisdictional claims in published maps and institutional affiliations.

Abstract: Pre-treatments for plant fibres are very popular for increasing the fineness of bundles, promoting individualisation of fibres, modifying the fibre-matrix interface or reducing water uptake. Most pre-treatments are based on the use of chemicals and raise concerns about possible harmful effects on the environment. In this study, we used physical pre-treatments without the addition of chemical products. Flax tows were subjected to ultrasound and gamma irradiation to increase the number of elementary fibres. For gamma pre-treatments, a 20% increase in the number of elementary fibres was quantified. The biochemical composition of pre-treated flax tows exhibited a partial elimination of sugars related to pectin and hemicelluloses depending on the pre-treatment. The hygroscopic behaviour showed a comparable decreasing trend for water sorption-desorption hysteresis for both types of pre-treatment. Then, non-woven composites were produced from the pre-treated tows using poly-(lactid) (PLA) as a bio-based matrix. A moderate difference between the composite mechanical properties was generally demonstrated, with a significant increase in the stress at break observed for the case of ultrasound pre-treatment. Finally, an environmental analysis was carried out and discussed to quantitatively compare the different environmental impacts of the pre-treatments for composite applications; the environmental benefit of using gamma irradiation compared to ultrasound pre-treatment was demonstrated.

Keywords: flax tows; ultrasound; gamma treatment; DVS; environmental analysis; mechanical properties; composite materials

1. Introduction

The intensification of global warming coupled with a gradual depletion of petro-sourced resources is pushing our societies and industries to use an increasing number of biomass resources in an ecologically responsible approach. The composite materials sector is not an exception to this trend; for several decades, plant fibres, especially flax, have offered a solution to replace predominant synthetic reinforcements, such as glass fibres. Indeed, flax fibres have some advantages: a lower density [1], specific mechanical properties equivalent to glass fibres [2,3] and a natural and renewable origin. Generally, the use of flax fibres makes it possible to reduce the carbon footprint of performance composite materials [4], thus enables one to obtain a less impactful life cycle.

Flax processing provides a range of products of different qualities. At the scutching step, good-quality scutched fibres are produced but also low-cost and qualitative scutching tows, which comprise a mixture of elementary fibres, fibre bundles and shives [5]. In industry, they are mainly used for the production of non-woven preforms. These products, manufactured at low cost [6] and in high volume, are particularly popular in the automotive sector to design and produce non-structural interior composite parts [7]. Currently, such preforms are made using a mix of plant fibres and thermoplastic fibres, such as poly-(propylene) (PP), which has good mechanical properties as well as a low melting temperature, which is an important parameter to prevent plant fibre damage during the processing stage [8,9]. To improve the end of life of these materials [10], it is interesting to study the coupling of flax fibres with a biobased and biodegradable polymer matrix such as poly-(lactid) (PLA) or poly-(β-hydroxybutyrate) (PHB).

Cellulose is the main component of flax fibres, which varies in content between 50% and 80% depending on the variety considered [11]. Other non-cellulosic components are hemicelluloses (between 11% and 20%), pectins (2% and 8%) and lignins (approximately 1%–3%). The structure of an elementary flax fibre is characterised by a primary wall, a secondary wall divided into several sub-layers, generally 3, denoted S1, S2 and S3, and a cavity in its centre called the lumen [12]. The thickest S2 sub-layer (5–10 μm) largely influences the mechanical properties of the fibre. The cohesion of elementary fibres within fibre bundles is ensured by the pectin-rich middle lamella [13]. However, hemicelluloses and lignins can also be found in this latter part. Non-wovens are made using tows, which contain both elementary fibres and bundles but also some woody sticks to a certain extent; these materials are named shives and are considered contaminants by materials engineers. The shives composition and mechanical properties strongly differ from those of technical fibres [14,15], and contain a reduced fraction of cellulose and more lignin.

For industrial materials, mechanical properties are among the predominant characteristics determining future functionalities and uses. Non-woven composites have interesting mechanical properties; according to Ashby material charts, they are superior to injection-moulded composites thanks to longer fibres but inferior to unidirectional composites (UDs) [16]. In general, in plant fibre-reinforced composites, individualisation of fibre bundles into elementary fibres is one of the key parameters to obtain good mechanical properties. Indeed, the middle lamella, which ensures fibre cohesion within fibre bundles, is an area of weakness due to its poor mechanical properties [17]. In addition, an aspect ratio (length/diameter) for the fibres of greater than 10 is considered the minimum value for efficient stress transfer during mechanical stress loading [18]. As an example, a positive impact of individualisation was observed on the mechanical properties of flax/epoxy UD composites [19]. These different elements show that by damaging or solubilising the middle lamella, an increase in elementary fibre content, and, at the same time, a decrease in fibre bundle content are promoted. At the composite scale, this modification of the architecture of the fibres allows for a larger specific fibre/matrix interfacial area as well as an optimised aspect ratio, and; therefore, better mechanical properties.

To induce this individualisation for fibre bundles, several types of pre-treatments exist, potentially used as alternatives or post-assisting to retting [20]. First, chemical pre-treatments, such as the use of chelating agents such as ethylenediaminetetraacetic acid (EDTA), are frequently used for the defibrization of plant fibres. Chelation removes calcium ions from the pectins of the middle lamella [21]. Alkaline treatments of the NaOH or KOH type are commonly used to eliminate non-cellulosic components of the fibre, such as hemicelluloses and pectins [22]. However, using chemical agents used to modify plant fibres penalise their eco-friendly aspect and can possibly reduce fibre and composite performance [23]. In this regard, physical and chemical-free pre-treatments are needed to increase the individualisation for fibre bundles. Ultrasonic treatment of hemp fibres has been shown to reduce the diameter of fibre bundles by removing pectic and hemicellulosic structures [24]. This elimination of cell-wall polymers is possibly due to the phenomenon of cavitation; the asymmetric implosion of bubbles near the material results in turbulence, which can damage certain binding constituents, such as those in the middle lamella in particular. Ultrasound treatment allows further changes to the fibre. In fact, ultrasonically treated fibres show a significant reduction in water sorption [25] as well as a delignification process [26]. Other physical pre-treatments that do not involve immersion in a solvent such as gamma irradiation can be used to promote individualisation for fibre bundles. The use of gamma irradiation for lignocellulosic fibres shows interesting effects. Scanning Electron Microscope (SEM) observations showed that the primary wall and the middle lamella of henequen fibres are strongly impacted by gamma irradiation [27]. In addition, non-cellulosic polymers such as low DP hemicelluloses, impurities and waxes undergo degradation at low irradiation doses (<10 kGy) [28]. Therefore, it appears to be possible to preferentially degrade the pectins in the middle lamella while preserving the cellulose structure and thermal and mechanical properties of plant fibres, providing that low irradiation doses are used (<30 kGy) [28]. Gamma irradiation can decrease water uptake for fibres treated at low doses thanks to the removal of hydrophilic components; jute fibres irradiated and then immersed in water for 30 days indicate a 14% drop in water absorption [29]. Gamma irradiation can also promote fibre/matrix interfacial adhesion by modifying the surface topography of plant fibres and enhancing their roughness [27], thus allowing better mechanical interlocking and a potential increase in interphase properties with a polymer matrix.

The aim of this work is to apply free-of-chemicals pre-treatments to maximise the environmental benefit of flax technical fibres. Two physical pre-treatments: ultrasound and gamma irradiation were applied to flax tows with the aim of increasing the specific surface area and aspect ratio of the flax reinforcements. The efficiency of these physical pre-treatments was evaluated by monitoring the individualisation into elementary fibres and the reduction in the size of the fibre bundles by means of a dynamic morphological analysis. In addition, the chemical composition and water uptake of (un)treated tows was evaluated. Then, treated flax tows were used to produce non-woven flax-PLA composites, and the impact of pre-treatments on their microstructure and mechanical properties was studied. Finally, an environmental analysis of the various treatments was undertaken to assess whether or not the environmental impact of the pre-treatments is justified by the potential gain in the mechanical properties of the related composites.

2. Materials and Methods

2.1. Materials

2.1.1. Flax Tows

Flax tows (*Linum usitatissimum* L.) from the Bolchoï variety were supplied by Depestele (Bourguebus, France). Flax plants were cultivated in 2017 (Normandy, France) and dew-retted in the field. These plants had a length of approximately 80 mm and a shive rate less than 5%. This sample was named Native, no-treated flax tows.

2.1.2. Flax-PLA Non-Woven Preforms

The non-woven preforms were composed of 50:50 wt% of flax tows and PLA fibres and manufactured in October 2020 by the EcoTechnilin Company (Yvetot, France). PLA fibres had a length of 60 mm and a diameter of approximately 30 µm. Three non-woven preforms were produced with the different batches of flax: untreated, gamma irradiation and ultrasound pre-treated. All non-woven preforms had a basis weight of approximately 100 g/m^2.

2.1.3. Composite Manufacturing

The flax-PLA non-woven composites were produced by hot pressing using a Darragon 35T hydraulic press. The dimensions of the composite plates were 160 × 160 × 1.5 mm^3.

The time-temperature-pressure cycle used is described as follows: Heat up to 190 °C, 3 min of preheating on contact, followed by the application of 50 bar pressure for 3 min at 190 °C, and, finally, a 15 min cooling under pressure to reach 10 °C.

2.2. Physical Pre-Treatments

2.2.1. Ultrasound

Ultrasound pre-treatment were carried out using a semi-pilot tank (SinapTec, Lezennes, France) at the Université du Littoral (Dunkerque, France). The pre-treatment parameters are an output power of 2500 W and a frequency of 20 kHz. The semi-pilot tank can contain a volume of 50 L of water. Ultrasonic pre-treatment requires immersion in a solvent for the propagation of ultrasonic waves. The pre-treatment was carried out by immersion in water during 30 min. This duration of the pre-treatment was established by preliminary tests, which showed the best individualisation for this duration (data not shown). For ultrasound pre-treatment, 2.5 kg of tows was used. The ultrasound pre-treated tows were named US 30 min. A chiller with an air-cooled refrigerating unit and circulation pump (Unichiller 025 OLE, Huber, Offenburg, Germany) was used to maintain the water temperature at approximately 25 °C during the pre-treatment. After the pre-treatment, the flax tows were dried overnight in an oven (BD 56, Binder GmbH, Tuttlingen, Germany) at 60 °C.

2.2.2. Gamma Irradiation

Gamma irradiation was carried out industrially by the company Ionisos (Dagneux, France). To preserve the fibre structure and physical properties [28], a low dose of 5–7 kGy was applied to approximately 2.5 kg of flax tows; this sample was named Gamma. Gray is the unit derived from the international system of units for absorbed dose. A Gray represents the energy of ionising radiation from one joule to a homogeneous medium with a mass of one kilogram.

2.3. Characterisation

2.3.1. Particle Size Analysis

The morphology of the particles was studied with an automed dynamic morphological analyser (SympaTec GmbH, Clausthal, Germany) called QICPIC and further use in the text as its model name by measuring the particle diameter (DIFI). DIFI was defined as the quotient of the total projection area and the sum of all the lengths of the branches of the fibre backbone:

$$\text{Diameter} = \frac{\text{total projection area}}{\text{sum of the lengths of all the tendrils}} \tag{1}$$

Flax tows were studied in a water solution using a dispersion unit, LIXELL (SympaTec GmbH, Clausthal, Germany), to prevent particle aggregation. Each sample was weighed to approximately 50 mg and then pre-dispersed in 5 mL of ethanol and then 45 mL of distilled water before undergoing further dispersion using an ultrasound probe (Vibra-Cell^{TM} 75022, Bioblock Scientific, Hampton, NH, USA). The solution was finally dispersed in 950 mL of distilled water under magnetic stirring. All samples were analysed using a M7 lens, which allows precise measurement of particles with dimensions ranging from 4.2 to 8665 μm (ISO). The number of particles analysed varied between hundreds of thousands to a million depending on the samples. For each sample, measurements were carried out in triplicate to ensure the reproducibility of the data. PAQXOS software (SympaTec GmbH, Clausthal, Germany) was used to calculate the particle diameter (DIFI).

2.3.2. SEM Observations

SEM images of fibres and composites were recorded by using a Jeol JSM IT-500 HR (JEOL Ltd., Tokyo, Japan). All samples were spin coated with gold for 180 s using an Edwards Scancoat Six device before observations. Secondary emission electrons were used, with an accelerating voltage of 3.0 kV.

For the composite cross-section observations, a section of composite was inserted into a mixture of 34 g of catalyser and 100 g of epoxy resin. The sample was placed overnight in an oven at 50 °C to promote crosslinking. Then, the surface of the sample, which was subsequently characterised using a SEM was polished using an automatic polisher (TegraForce-5, Struers, Copenhagen, Denmark). Several polishes were carried out with different polishing times and grains: 60 s/500, 90 s/800, 120 s/1000, 150/1200, 180 s/2000 and 210 s/4000.

2.3.3. Monosaccharide Composition

The quantification and identification of the neutral monosaccharides constituting flax tows was carried out by gas chromatography after acid hydrolysis and conversion of the monomers into alditol acetates [30]. A DB 225 capillary column (J&W Scientific, Folsorn, CA, USA; temperature 205 °C, carrier gas H_2) was used to perform the chromatography. For calibration, a range of glucose solutions and inositol as an internal standard were used. Approximately 12 g of fibres in total were tested, and the measurements were performed in duplicate.

2.3.4. Water Sorption/Desorption

The sorption and desorption isotherms for water were established with a dynamic vapour sorption device (IGAsorpt-HT, Hiden Isochema, Warrington, UK). The flax tows were cut to approximately 3–4 mm, and 25 mg was placed in a microbalance located in the hermetic reactor. Prior to adsorption, the samples were dried at 105 °C for 1 h. Inside the reactor, the temperature and relative humidity (RH) were controlled. The sorption/desorption sequence was programmed as follows: Increase from 0% to 90% RH, and then decrease to 0% RH at 20% RH steps. For each RH step, the sample mass was continuously measured until equilibrium was reached (i.e., when the mass variation became less than 0.1 µg/min over a 1 h window).

2.3.5. Mechanical Properties of the Composite Plates

The specimens were cut from the manufactured composite plates with a milling machine. For each plate, five tensile (20 × 8 mm^2), four bending (80 × 10 mm^2) and two SEM (20 × 20 mm^2) test specimens were cut. The tensile properties of the flax-PLA non-woven composites were determined in accordance with the ISO 527 standard. The uniaxial tensile tests are performed in the static state during loading. An MTS Criterion Model 42 machine (MTS, Eden Prairie, MN, USA) with an MTS extensometer (nominal length 25 mm) was used with a 5 kN load cell (MTS, Eden Prairie, MN, USA) and a constant displacement rate of 1 mm/min. Load cell indicated the maximum force that the (traction) cell can provide. Five specimens were tested for each composite.

The bending properties of flax-PLA non-woven composites were determined according to the ISO 178 standard using an MTS Synergie 1000 R/T apparatus (MTS, Eden Prairie, MN, USA). The three-point bending tests were carried out at a constant displacement rate of 1 mm/min with a 250 N load cell. Four specimens were tested for each composite.

2.4. Environmental Analysis

The purpose of this study was to determine the environmental impacts of two physical treatments applied to flax tows: ultrasound (see Section 2.2.1) and gamma irradiation (see Section 2.2.2). Here, a simplified environmental analysis was conducted and not a full LCA (ISO 14044). The model was realised with the software Simapro. The Ecoinvent $V_{3.5}$ database was used. The system boundaries are based on the cradle-to-gate approach and are represented in Figures 1 and 2. The locations of the production sites were included. The flax tows data were obtained from the company Depestele, located in Normandy, France as part of the program FARBioTY. The model works by mass allocation, which means that if a process has several products, the impact of each of the products is allocated in proportion to its mass. The studied impacts are abiotic depletion, global warming, ozone layer depletion, human toxicity, acidification and eutrophication based on the CML 2 baseline 2000 calculation method. The energy consumption, fossil and nuclear, was calculated with the Cumulative Energy Demand V1.11 method. Evaluations of indicators are available in the literature [31]. The results are presented as diagrams with a relative scale. Due to the uncertainties, for the same environmental indicator, if the difference between two studied pre-treatments is less than 5%, they are considered equal. Interventionary studies involving animals or humans, and other studies that require ethical approval, must list the authority that provided approval and the corresponding ethical approval code.

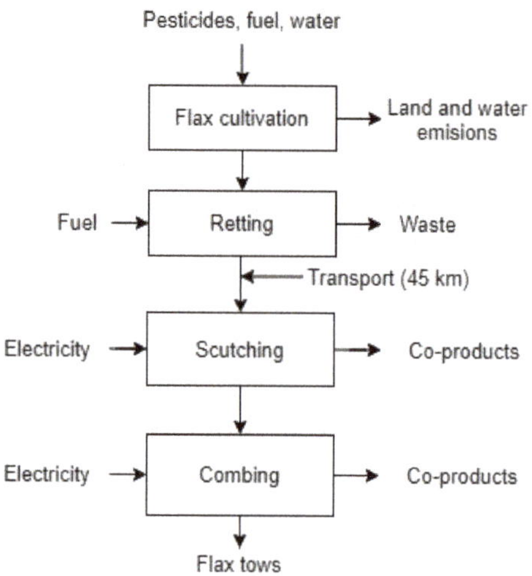

Figure 1. Flow chart of flax tows production.

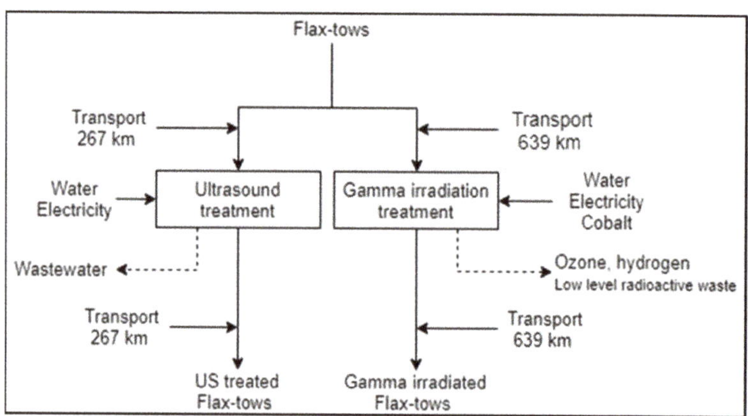

Figure 2. Flow chart for ultrasound and gamma irradiation pre-treatments.

3. Results

3.1. Individualisation of Pre-Treated Flax Tows

The distribution of the fibre element diameters for the Native, Gamma and US 30 min samples is illustrated in Figure 3. Figure 3b clearly shows the presence of larger fibre bundles for the US 30 min treated sample with a last decile of 159.9 ± 5.8 µm compared to 140.6 ± 15.9 µm for the Native sample. The lowest median diameter of 54.4.9 ± 3.0 µm was obtained for the gamma-treated sample, compared to a value of 66.8 ± 11.2 μm for the Native sample. To visualise the evolution of the smallest diameters for which very few differences are visible on box plots, another representation for the evolution of flax tow diameters is given in Figure 4.

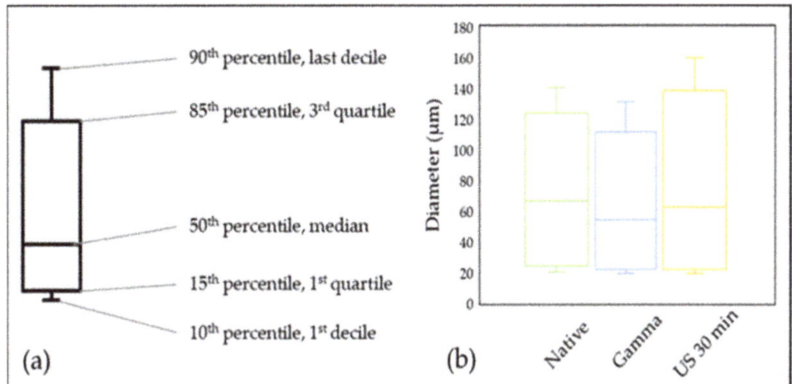

Figure 3. (**a**) Signification of a box-plot representation; (**b**) diameter distribution of native and pre-treated flax tows.

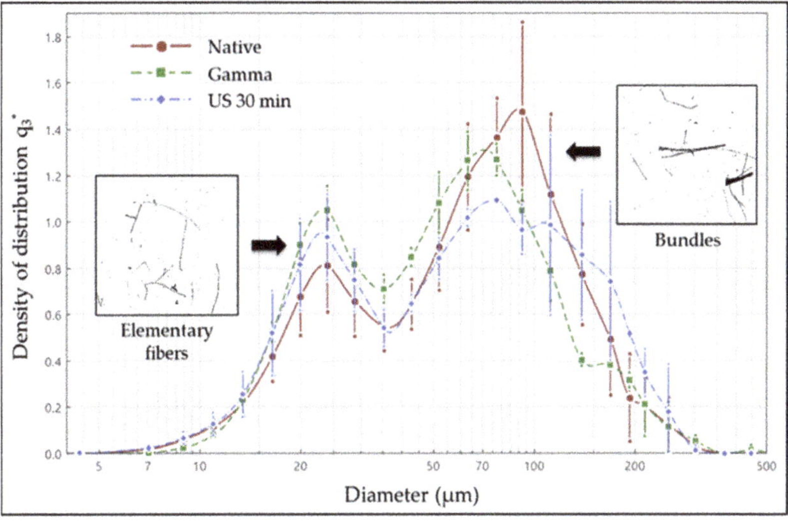

Figure 4. Diameter density of native and pre-treated flax tows and binary images of elementary fibres and fibre bundles in the two insets.

Figure 4 shows the distribution of the fibre element diameters for native and pre-treated tows treated with gamma irradiation and ultrasound, respectively. First, all the plots show a bimodal distribution of diameters with a first peak centred at approximately 25 μm and a second peak located towards diameters of 70–100 μm, corresponding to elementary fibre and fibre bundles, respectively. Representative picture illustrations obtained by the morphometric analyser (QICPIC) are encrusted in Figure 4. The red curve related to native tows (Native) has the highest peak attributed to fibre bundles with a density close to 1.5 for diameters of 90 μm. For the gamma pre-treatment (Gamma), the peak related to the fibre bundles shows a distribution density of 1.3 for a diameter of 70 µm. The density was 1.1 for a diameter of 75 µm for tows pre-treated by ultrasound (US 30 min). Therefore, a reduction in the diameters of the fibre bundles for the Gamma and US 30 min samples was observed. However, Figure 4 indicates a greater distribution density for the US 30 min sample than the Native sample for fibre bundles exceeding 150 µm in diameter, suggesting an agglomeration of fibres during or after ultrasound pre-treatment. Then, the peak at approximately 25 µm shows an opposite tendency to that for a greater diameter at 70–100 µm. In fact, this peak related to the elementary fibres indicates higher distribution densities for the pre-treated tows, with values of 1.05, 0.95 and 0.8 for the Gamma, US 30 min and Native samples, respectively; hence, highlighting the positive impact of both treatments on fibre individualisation. The gamma pre-treatment clearly results in a decrease in the median diameter (Figure 3b) due to the increase in the number of elementary fibres and the decrease in the number of fibre bundles. For the US 30 min sample, the median diameter is similar to that of the Native sample because the higher number of elementary fibres is counterbalanced by the appearance of fibre bundles with a diameter of more than 150 µm. Thus, gamma irradiation pre-treatment allows for a significant increase in the number of elementary fibres of approximately 20%. This increased number of elementary fibres is slightly less pronounced for ultrasound pre-treatment at +15%.

Figure 5 shows typical SEM images of the three flax tow samples. All images show fibre bundles, but the fibre individualisation differs depending on the sample considered. Indeed, the middle lamella for the Gamma and US 30 min samples is much less visible on the surface of the fibres (Figure 5b,c) than for the Native sample (Figure 5a). Pre-treatments appear to be effective in partially removing the middle lamella and promoting fibre individualisation; these qualitative observations support the diameter measurements obtained with the automated morphometric analyser. However, fibre individualisation was not fully achieved, even after the pre-treatments. Indeed, the middle lamella appears to be still present in the core of the fibre bundles. The distribution density for the fibre bundles (i.e., diameter of 70–100 µm) remains quite high. We can, nevertheless, suppose that, even if present, the middle lamella are weakened by the pre-treatments and might be more easily broken during composite manufacturing, especially if shearing is applied upon processing.

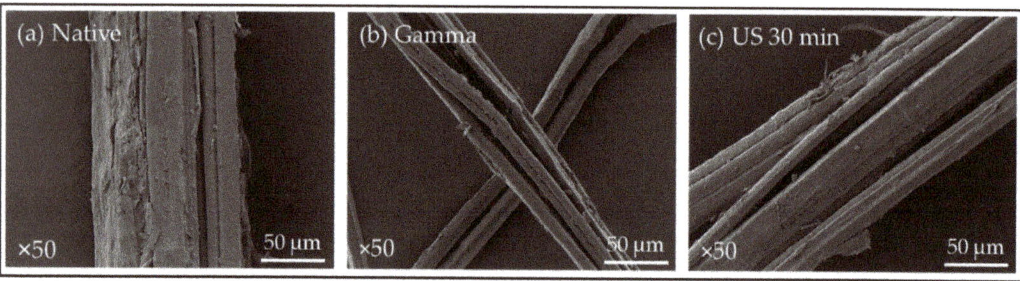

Figure 5. SEM observations (×50) of tows (**a**) Native; (**b**) Gamma; (**c**) US 30 min.

3.2. Monosaccharide Composition of Flax Tow Samples

The impact of the two pre-treatments on the monosaccharide composition was investigated. Figure 6 shows the percentages of each monosaccharide for the three samples, except glucose for the sake of readability (see Table 1 for glucose).

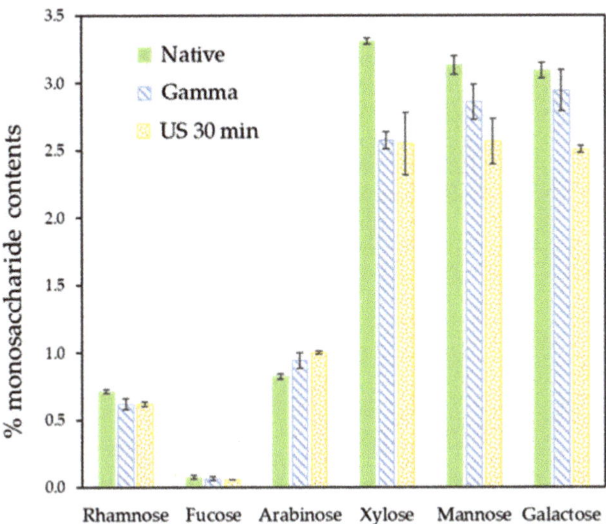

Figure 6. Monosaccharide contents for native and pre-treated flax tows.

Table 1. Glucose contents of Native, Gamma and ultrasound pre-treated flax tows.

Samples	Glucose (%)
Native	68.1 ± 0.9
Gamma	65.5 ± 1.2
US 30 min	64.4 ± 2.1

For rhamnose (Rha) and fucose (Fuc), few differences appeared between pre-treated and native tows. The Rha value ranges between 0.6% and 0.7% for the Gamma and Native samples, respectively. The value of Fuc is very low at approximately 0.065% for the three samples. Both Rha and Fuc monosaccharides are generally associated with pectic structures. The other four sugars quantified—arabinose (Ara), xylose (Xyl), mannose (Man) and galactose (Gal)—show more differences between the three samples. For Xyl, Man and Gal, the same trend is observed with a decrease in the percentage of monosaccharides for both pre-treatments. The most significant decrease occurs for Xyl, which shows a decrease from 3.3% to 2.5% for the Native and US 30 min samples, respectively (i.e., a 25% decrease). For Man, the values obtained for the Native, Gamma and US 30 min samples are 3.1%, 2.8% and 2.6%, respectively. Regarding Gal, the US 30 min sample has the smallest content, followed by the Gamma and Native samples at 2.5%, 2.8% and 3.1%, respectively. Of these three sugars associated with hemicellulosic structures, ultrasound pre-treatment shows the best impact with the lowest monosaccharide values. Finally, the opposite trend is observed for Ara; the highest Ara content is obtained for the US 30 min sample with a value of 1% versus 0.8% for the Native sample. This slight increase in Ara content actually corresponds to compensation for the loss of other monosaccharides.

The percentage of glucose in the three samples is shown in Table 1. The Native sample has the highest glucose level at 68.1%. The two pre-treated samples, Gamma and US 30 min, contain relatively similar glucose levels of 65.5% and 64.4%, respectively. The two pre-treatments do not appear to have too much impact on the glucose, which is generally associated with cellulose. However, this loss of glucose can also be associated with a decrease in hemicelluloses. In both cases, the structural integrity of the flax tow appears to have been preserved. Therefore, the pre-treatments do not damage the structural polysaccharides too much.

3.3. Hygroscopic Behaviour of Pre-Tretaed Flax Tows

Figure 7 shows the sorption-desorption behaviour of water vapour for the different samples. Differences are visible in Figure 7a,b comparing the pre-treated tow with the native tow. The Native sample has a greater sorption capacity than the Gamma and US 30 min samples. Indeed, for a relative humidity (RH) equal to 90%, the mass difference for the Native, Gamma and US 30 min samples is 17.8%, 15.3% and 15.7%, respectively. In addition, the sorption-desorption behaviour of the gamma and US 30 min pre-treated samples appears to be very similar, as shown in Figure 7c.

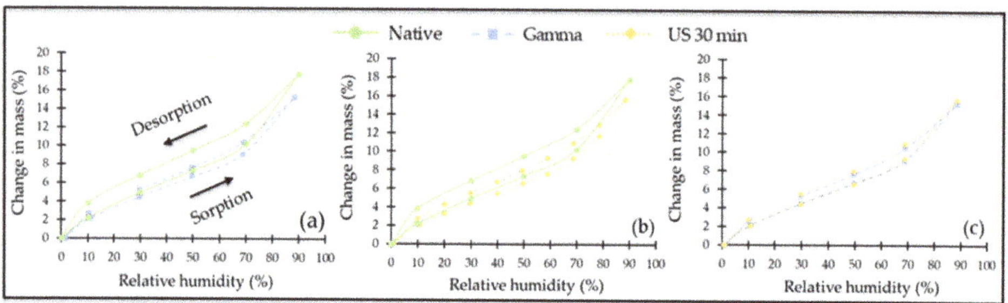

Figure 7. Comparison of sorption-desorption curves for (**a**) Native and Gamma samples; (**b**) Native and US 30 min samples; (**c**) Gamma and US 30 min samples.

Natural fibres belong to the type IV material classification of the International Union of Pure and Applied Chemistry (IUPAC). It is known that the sorption and desorption curves are not superimposable for the same sample [32]. The gap between these two curves (i.e., the hysteresis area) is calculated by subtracting the desorption curve area from the sorption curve area. The hysteresis area data are shown in Table 2. The hysteresis area for the native sample is 149 versus roughly half this value (73) for the Gamma sample and a much lower value (96) for the US 30 min sample. A decreased hysteresis area indicates significant structural changes. The gamma irradiation pre-treatment appears to have the most noticeable effect on decreasing the water uptake and hysteresis between the sorption-desorption curves. The elimination of pectic and hemicellulosic compounds due to the two pre-treatments is responsible for the hygroscopic behaviour of flax fibre [21], which explains the two phenomena mentioned above.

Table 2. Hysteresis area for the three different samples: Native, Gamma and US 30 min.

Samples	Hysteresis Area
Native	149
Gamma	73
US 30 min	96

3.4. SEM Observations for a Flax-PLA Preform

Figure 8 shows the structure of non-woven preforms. Flax tows are mainly composed of flax fibre bundles but also of low amounts of residual shives originating from the woody part of the stem (Figure 8a). The flax tows used have a contaminant level (shives) less than 5%. Figure 8b illustrates the flax and PLA fibres that constitute the flax-PLA preforms used in the manufacture of non-woven flax-PLA composites. During composite manufacturing, flax and PLA fibres tend to align in the machine direction (Figure 8c). This is why the mechanical behaviour of non-woven composites was tested using both the machine and cross directions.

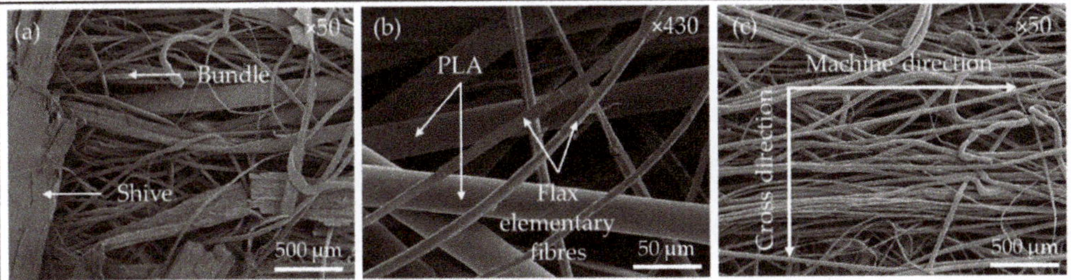

Figure 8. SEM observations for (**a**) Flax tows; (**b**,**c**) Flax-PLA preform.

3.5. Porosity Content for Flax-PLA Interface of Non-Woven Compositesgf

The cross-sections of the non-woven composites were investigated using SEM, as shown in Figure 9. By eye, contrasting porosities are observed among the different non-woven composites. Figure 9a,b illustrate the presence of porosities inside the Native and Gamma composites, respectively. In contrast, the US 30 min sample does not present porosity (Figure 9c).

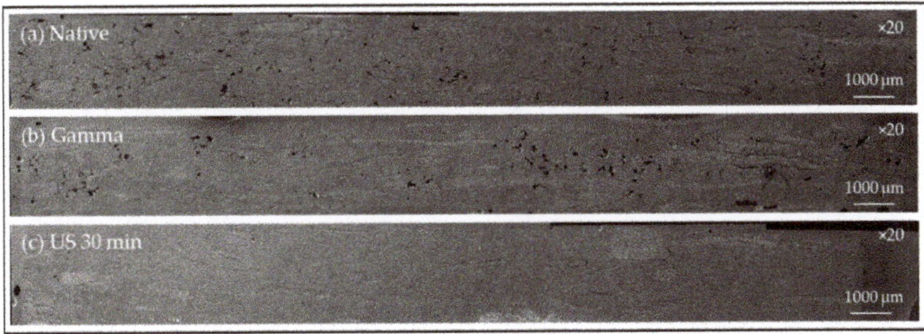

Figure 9. SEM observations of non-woven composites (**a**) Native; (**b**) Gamma; (**c**) US 30 min.

To quantify the porosity content, an image analysis of the cross-section was performed, and the results obtained are summarised in Table 3.

Table 3. Porosity content of the non-woven composites according to the card machine orientation.

Samples	Main Flax Direction	Porosity Content (%)
Native	Machine direction	2.8 ± 0.9
	Cross direction	5.1 ± 1.8
Gamma	Machine direction	2.9 ± 1.7
	Cross direction	4.7 ± 2.5
US 30 min	Machine direction	0.4 ± 0.1
	Cross direction	0.5 ± 0.3

Image analysis confirms that the US 30 min composite has very little porosity with values of 0.4% and 0.5% in the machine and cross directions, respectively. On the other hand, the two other non-woven composites, Native and Gamma, show porosities of 2.8% and 2.9% in the machine direction, respectively. These porosities are even greater in the cross direction, with values reaching 5.1% and 4.7% for the Native and Gamma samples, respectively.

3.6. Mechanical Properties of Non-Woven Composites

The results from machine and cross direction tensile tests on non-woven composites are summarised in Figure 10. For the machine direction, Young's modulus for the Native sample is the highest at 15.9 GPa (Figure 10a) compared to 13.1 and 13.7 GPa for the Gamma and US 30 min samples, respectively. On the other hand, the US 30 min sample shows the highest strength at 116 MPa, corresponding to an increase in stress of more than 10% at break compared to the Native sample (Figure 10b). For the Gamma sample, the stress at break is lower than that of the Native sample, with a decrease of approximately 13% from 106.3 to 94.5 MPa. Potential degradation of fibres due to exposure to gamma irradiation may result in poor mechanical properties. Regarding the strain at break (Figure 10c), there were no differences observed between the three samples tested, with a relatively large standard deviation of approximately 1%. For the cross direction, the trends are almost the same as those observed in the machine direction. The US 30 min sample has a Young's modulus of 7.3 GPa, which is equivalent to that of the Native sample (Figure 10a). For the stress at break, an increase of 12% is observed for the US 30 min sample compared to the Native sample with values of 54.1 and 48.7 MPa, respectively (Figure 10b). The stress at break for the Gamma sample is still below that for the Native sample. Finally, the strain at break does not show any change, with an average value of approximately 1.1% (Figure 10c).

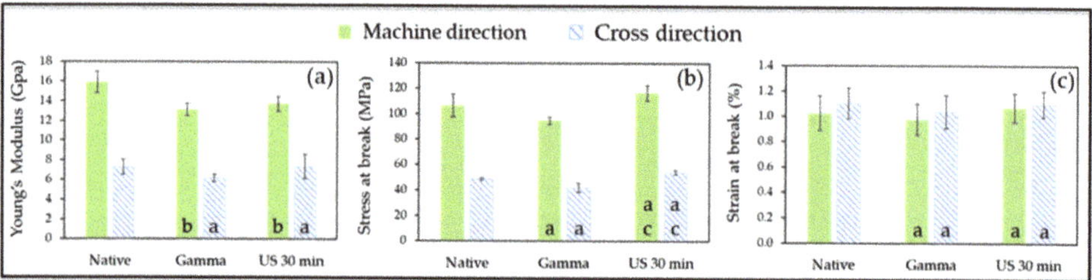

Figure 10. Tensile test results for (**a**) Young's modulus; (**b**) stress at break; (**c**) strain at break; Gamma column: If value is different compared to Native (*p* value < 0.05), letter b; otherwise letter a; US 30 min column: If value is different compared to Native, letter b otherwise letter a and if value is different compared to Gamma, letter c, otherwise left blank.

The results from the three-point bending tests for the non-woven composites are shown in Figure 11. The flexural modulus of the pre-treated tows, Gamma and US 30 min samples is lower than that for the Native composites with values of 7.1, 6.6 and 7.6 GPa, respectively (Figure 11a). Figure 11b indicates that the stress at break is very similar for the different composites, at approximately 140 MPa. On the other hand, the strain at break shows a value of 2.6% for the Gamma composite, which is the highest value obtained (Figure 11c). The strain at break is similar for the other two composites, Native and US 30 min, with values of 2.5% and 2.5%, respectively. The results in the cross direction are somewhat different. Indeed, the flexural modulus of the US 30 min composite is slightly higher than that of the Native composite at 5.2 and 5.0 GPa, respectively (Figure 11a). The Gamma composite shows a much lower flexural modulus of 3.0 GPa. The same trend is observed for the stress at break with values of 96.0, 90.1 and 53.8 MPa obtained for the US 30 min, Native and Gamma composites (Figure 11b). As for the tensile tests, the Gamma composites show the lowest mechanical properties, which can be attributed to degradation of the fibres by the pre-treatment. Finally, the strain at break shows relatively similar results in Figure 11c.

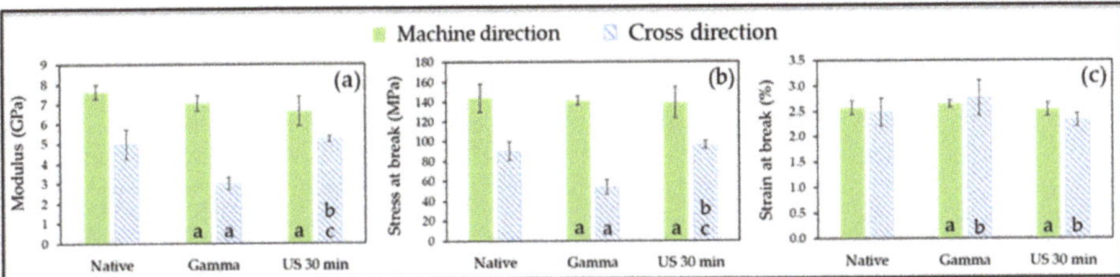

Figure 11. Bending tests results for (**a**) modulus; (**b**) stress at break; (**c**) strain at break ; Gamma column: If value is different compared to Native (*p* value < 0.05) letter b; otherwise letter a; US 30 min column: If value is different compared to Native, letter b otherwise letter a. In addition, if value is different compared to Gamma, letter c, otherwise left blank.

SEM observations of the fracture surface were performed for the different composites (Figure 12). Figure 12a shows the residual area of fibre bundle debonding generated by poor interfacial adhesion between the fibres and the matrix for the native non-woven composite. Another type of void is observable in Figure 12b again for the native composite. This is attributed to porosity because the surface of the matrix is much smoother than that found previously without any trace of fibre bundles. Figure 12c,d show the fibre–matrix interface for the US 30 min and Gamma composites, respectively. The US 30 min composite contains many elementary fibres that appear to be very well anchored within the matrix with narrow interstices between the two components. For the Gamma composite, a fibre bundle is perfectly visible in the centre of the image (Figure 12d). The fibre–matrix interface again looks good with no predominance of debonding. In general, voids produced by the debonding of a fibre bundle or an elementary fibre are mainly present for the Native composite, suggesting a possible positive effect of the treatments on polymer-fibre interfacial adhesion.

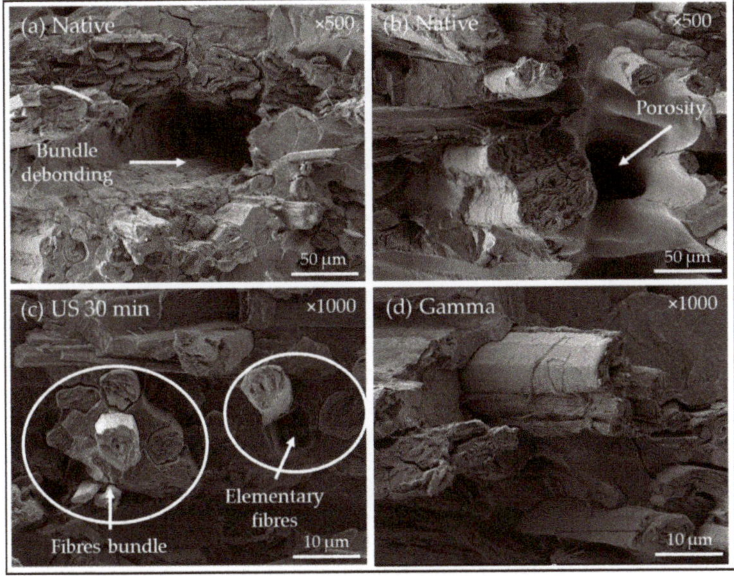

Figure 12. SEM observations of the interface of non-woven composites: (**a**) Debonding of fibre bundles; (**b**) porosity; (**c,d**) fibre bundles and elementary fibres in the US 30 min and Gamma composites, respectively.

3.7. Environmental Analysis of Pre-Treatments

Simplified environmental analysis was carried out to evaluate the environmental impacts of the two pre-treatments on flax tows. Figure 13 presents the results from the environmental analysis of flax tows pre-treated by ultrasound (US 30 min) and gamma irradiation (Gamma). Table 4 gathers the associated environmental impact values for native flax tows and energy consumption (fossil and nuclear) for the three studied tow batches.

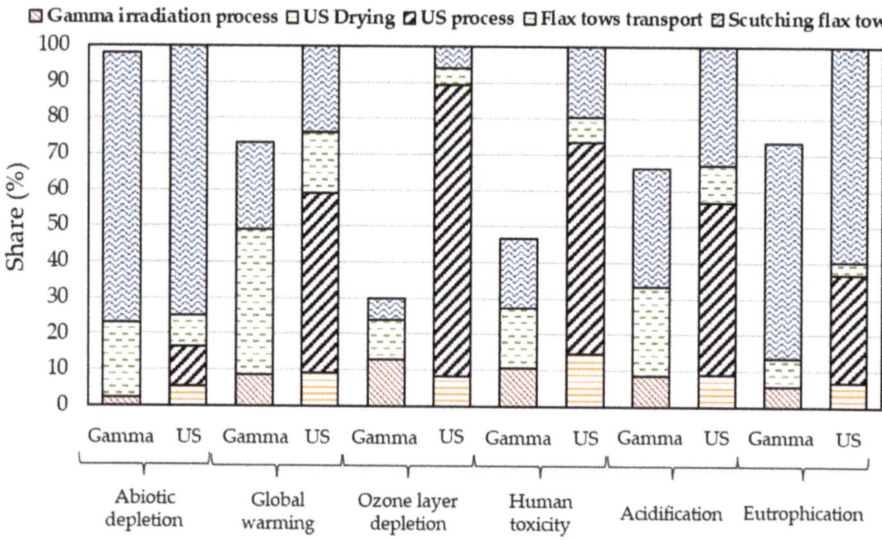

Figure 13. Normalised comparison of two fibre pre-treatments (Gamma and US 30 min) for six impact indicators decomposed according to the sources—scutching flax tows are included.

Table 4. Environmental impacts for the production of scutching native flax tows, flax tows pre-treated using the Gamma and US 30 min processes.

Impacts Category	Units	Untreated	Gamma	US 30 min
Abiotic depletion	kg Sb eq./kg	1.88×10^{-5}	2.46×10^{-5}	2.51×10^{-5}
Global warming (GWP100a)	kg CO_2 eq./kg	1.24×10^{-1}	3.76×10^{-1}	5.14×10^{-1}
Ozone layer depletion (ODP)	kg CFC 11 eq./kg	2.13×10^{-8}	1.05×10^{-7}	3.50×10^{-7}
Human toxicity	kg 1.4 DB eq./kg	1.08×10^{-1}	2.57×10^{-1}	5.51×10^{-1}
Acidification	kg SO_2 eq./kg	6.54×10^{-4}	1.32×10^{-3}	1.99×10^{-3}
Eutrophication	kg PO_4 eq./kg	8.08×10^{-4}	9.92×10^{-4}	1.35×10^{-3}
Energy consumption, fossil	MJ/kg	1.43	5.31	6.95
Energy consumption, nuclear	MJ/kg	0.78	6.24	38.2

Analysis of the results highlights the environmental interest of gamma irradiation compared to ultrasound. From Figure 13, for gamma irradiation compared to US 30 min, we observe the following: Abiotic depletion is equivalent, global warming is reduced by 27%, ozone layer is depleted by 70%, human toxicity by 53%, acidification by 34% and eutrophication by 27%. This difference is mainly due to the US process: Driven by its higher power consumption (Table 4), the US process gives considerably higher values for all the indicators than Gamma process. Furthermore, for the case of US 30 min, the flax tows must be dried. Despite this, it should be noted that transport is more impactful for Gamma than US 30 min composites because of the distance travelled from the tow production site to the treatment site (2 × 639 km versus 2 × 267 km). These observations are visible in the energy consumption of each treatment. Fossil energy consumption is similar for the two pre-treatments (5.31 MJ/kg for Gamma versus 6.95 MJ/kg for US 30 min). The US 30 min process consumes much more energy (mainly in the form of electricity usage) than the gamma process. Because the French electricity mix is composed of 70% nuclear power, nuclear energy consumption is significantly more important for US 30 min than gamma irradiation (38.22 against 6.24 MJ/kg). In addition, as shown in Table 4, for all the indicators, untreated flax tows are less impacted than flax tows treated by US 30 min or Gamma. In addition, US 30 min- and gamma-treated flax tows use four times more fossil energy and nine to 45 times more nuclear energy than untreated flax tows. The energy consumption for the untreated flax tows is of the same order of magnitude as reported in the literature [33].

To consider the mechanical aspect of the untreated and pre-treated flax tows, the previous results were weighted by the experimental maximum stress at break values for the PLA composite reinforced by the respective flax tows. To that end, a weight value (given in Table 5) was assigned to each case. The results are shown in Figure 14.

Table 5. Maximum stress values of PLA composite reinforced by respective flax-tows and corresponding weight values of scutching native and pre-treated flax tows by Gamma and US 30 min.

Flax-Tow Pre-Treatment	Maximum Stress (MPa)	Standard Deviation (MPa)	Weight Value
Native	106.3	8.8	1.09
Gamma	94.6	3.1	1.23
US 30 min	116.4	6.0	1.00

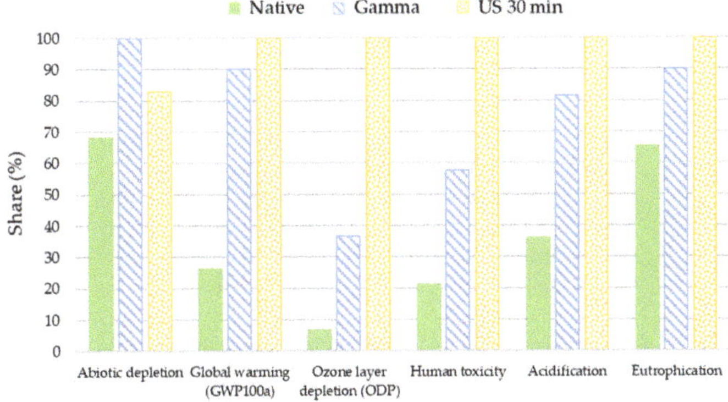

Figure 14. Normalised comparison of scutching native flax tows, flax tows pre-treated by Gamma and US 30 min for six impact indicators—weighted by the experimental maximum stress values obtained for the PLA composite reinforced by respective flax-tows.

The weight value was lower for the US 30 min composite compared with the untreated gamma composite. These differences result in a smaller gap between the US 30 min and Gamma processing than that shown in Figure 13. Despite this, Gamma irradiation remains less impactful than ultrasound pre-treatment, except for abiotic depletion. Moreover, even with a weight value based on maximal stress, the untreated flax tows show less environmental impact than that for the two studied pre-treatments.

4. Discussion

The impact of the two pre-treatments, Gamma irradiation and ultrasound applied to flax tows, was investigated at the tow scale and then at the composite material scale.

The increased individualisation of raw fibres, and; therefore, the increased number of elementary fibres after the two pre-treatments corroborates the biochemical results with an elimination of the hemicellulosic and pectic compounds belonging to the middle lamella and/or the primary wall [34]. Indeed, it has been shown that gamma irradiation strongly degrades the middle lamella as well as the outer layer of fibres corresponding to the primary wall, even for low gamma irradiation doses, <10 kGy [28]. Even if only partial, the elimination of compounds ensuring the cohesion of fibre bundles explains the decrease in the number of fibre bundles observed in the morphological analysis. Thus, the differences in sorption-desorption capacity in DVS illustrated in Figure 7 can be explained by the removal of hydrophilic components, leading to individualisation of the pre-treated tows. Moisture absorption is greater for a fibre bundle than for an elementary fibre. Indeed, the pectins present in the middle lamella consist of carboxyl groups known to be highly polar and capable of forming hydrogen bonds with polar solvents such as water [35]. In our study, this can be explained by the native fibres, which have more fibre bundles. At the fibre scale, hemicelluloses contribute to the hydrophilic nature of fibres. Loss of hemicelluloses following pre-treatment affects the hygroscopic behaviour of fibres by reducing their ability to interact with water molecules. Indeed, hemicellulosic compounds are composed of hydrophilic hydroxyl groups. Amorphous hemicelluloses are largely responsible for the hydrophilic nature of flax fibres and their capacity to sorb water within the cell walls [36]. The presence of hysteresis on the sorption-desorption curves indicates the presence of mesopores in the structure of the fibre [37]. Indeed, flax fibres have a porous structure with many exchange surfaces. In addition, during desorption, water can be trapped inside the mesopores of the fibres. The decrease in the area of the hysteresis for the two pre-treatments could then suggest a phenomenon based on pore closure generated by collapse resulting from the pre-treatments. Coupled with the removal of hydrophilic compounds, the decrease in the hysteresis area also indicates a decrease in the number of water sorption and retention sites [32]. Hysteresis is particularly associated with relaxation during sorption. Its reduction; therefore, indicates a significant change in structure, probably linked to the elimination of certain hygroscopic compounds.

Regarding the mechanical properties of a non-woven flax-PLA composite, the maximum strength was measured to be 90.4 ± 7.8 MPa [38]. Apart from the 10% increase in the tensile strength for the US 30 min composite, no significant gain for the Young's modulus and the elongation at break was observed for the US 30 and Gamma composites. In other words, the increase in the number of elementary fibres and in the specific interfacial area between the matrix and the flax fibres had no significant positive impact or was not sufficient to improve the final mechanical properties of the non-woven composites. These results may appear to be surprising at first. Indeed, fibre individualisation is considered an important parameter in the final mechanical properties of a composite material. It has been shown that greater individualisation generates better strength at break for a UD flax-epoxy composite for fibre volume fractions of 22%, 42% and 51% [19]. The two pre-treatments increase the number of fibre–matrix interfaces within the composite material, and, potentially, the quality of the fibre–matrix interface and load transfer between the matrix and the reinforcements during mechanical stress. Questions about the impact of the pre-treatments on the fibre–matrix interfacial shear strength are not addressed here. This interface can be characterised through different techniques and at different scales: Micro or macro scale corresponding here to the fibre scale or composite scale, respectively. The fibre–matrix interface could be tested at the microscale via microdroplet tests to obtain the interfacial shear strength (IFSS) [39]. In addition, tests have been performed at the interface between flax fibres and PLA with an IFSS value that has been measured for flax fibres and PLA [40]. IFSS was determined to be equal to 15.6 ± 2.7 MPa. It would then be interesting to compare this value with that for Native and pre-treated flax tows, and the poor mechanical properties of the Gamma composite may be potentially due to a poor-quality fibre–matrix interface. However, this may also be due to the degradation of fibres during pre-treatment which induces poor mechanical properties. Indeed, the changes occurring at the surface of the irradiated tows can induce changes in the adhesion between the fibre and polymer matrix. In addition, contact angle tests can allow better characterisation of the surface of flax fibres via determination of the surface energies. It is possible to perform these tests with several solvents, such as water (to assess the hydrophilic or hydrophobic character), diiodomethane or even hexane [41].

However, it is also important to note that the mechanical properties are always better after US pre-treatment in the cross direction for both tensile and flexural tests. Because flax fibres are preferably oriented in the machine direction, they are oriented perpendicular to the axis of mechanical stress. In other words, in the cross direction, the fibre–fibre interfaces are more loaded; that is, the polymers belong to the middle lamella and the primary wall. Better mechanical properties in the cross direction result in stronger fibre-fibre interfaces equivalent to more elementary fibres, and; therefore, better individualisation. Thus, in our case, the positive impact of US is more pronounced in the transverse direction due to the strong impact of the quality of the interface on the transverse properties. When longitudinal properties are considered, fibre properties are more impactful due to the high orientation level of our materials [42]. This difference between transverse and longitudinal behaviour is clearly confirmed by the composite mechanical performances.

Obviously, the mechanical properties are much better in the longitudinal direction of the fibres than in the transverse direction, confirming literature data and the impact of fibre properties according to the considered direction. For example, for a UD flax-PLA composite, Young's modulus in the longitudinal and transverse directions is estimated to be 20.1 ± 2.8 and 4.2 ± 0.4 GPa, respectively [40]. Here, the best value for Young's modulus in the longitudinal direction for Native composites is 15.9 ± 1.0 GPa, which, interestingly, is not far from the UD value. On the other hand, in the cross direction, Young's modulus for the Native and US 30 min composites is equal to 7.3 ± 0.8 and 7.4 ± 1.3 GPa, respectively; these low values confirm the minor contribution of flax fibres in this direction and the high impact of the interface.

Generally, interfacial properties and fibre orientation highly influence the composite mechanical properties, but the porosity content and intrinsic properties of fibres must also be considered to better understand the moderate impact of fibre individualisation on the final composite performance. It would then be interesting to assess the fibre properties after the treatment stage. Indeed, a possible counterbalance effect is possible between the increase in individualisation and a potential degradation of fibre properties. Tensile tests for single fibres would make it possible to observe whether or not pre-treatments results in degradation of mechanical properties. It may also be interesting to focus on the microstructure of the fibres to observe any reorganisation of the wall polymers and the creation of cavities. The most suitable technique for this type of scale analysis is nuclear magnetic resonance, possibly coupled with mechanical Atomic Force Microscope Peakforce investigations, to assess the impact of treatments on cell wall properties and ultrastructure. Finally, one must keep in mind the impact of porosity on composite mechanical properties; a very low porosity content is achieved for US-treated composites, which also has an impact on performance and can partially explain the improved mechanical properties for this composite. This can raise debate and questions about the exact impact of treatments. For the two other materials, the porosity content is high and suggests possible packing issues and limits in fibre content, especially when low-pressure processing tools, such as thermocompression, are used.

The environmental analysis applied to the two pre-treatments shows a significant impact of these pre-treatments through different environmental indicators. In addition, the gain provided by the pre-treatments for the mechanical properties of non-woven composites remains low compared to the impact on the environment. In other words, our pre-treatments do not currently appear to be a credible and environmentally friendly alternative to increase the mechanical properties of composite materials. However, gamma pre-treatment shows lower environmental indicators. Thus, pre-treatments that do not involve immersion, and; therefore, do not involve a drying step, are favoured. Nevertheless, the range of pre-treatments available without immersion and without the use of chemicals is not that wide, strongly restricting choice. In addition, this type of pre-treatment can instead be used to improve the fibre–matrix interface by considerably modifying the surface of natural fibres. For example, there exists plasma pre-treatment, UV pre-treatment or steam explosion, among others [22,43].

5. Conclusions

The impact of ultrasound and gamma irradiation pre-treatment was investigated at the scale of tows and non-woven composite materials. Morphological analysis showed better individualisation for pre-treated tows, with a 20% increase in elementary fibres for gamma irradiation and a more than 15% increase for ultrasound pre-treatment. Both pre-treatments partially removed the middle lamella responsible for the cohesion of the fibre bundles, as targeted. Consequently, hemicellulosic and pectic components known to belong to the middle lamella were also removed, which also impacted the sorption-desorption behaviour of flax tow. Both pre-treatments showed a decrease in water uptake behaviour. At the non-woven composite scale for the tensile test, a 10% increase in stress at break for the US 30 min composite in the machine direction was observed. However, no improvement in Young's modulus or elongation at break was observed for Gamma and US 30 min composites compared to the Native composite. Fine quantification of the porosity in the non-woven composites showed that the ultrasound pre-treatment displays a lower porosity content, arguably leading to the observation of a slight mechanical improvement for non-woven flax-PLA composites. Finally, environmental analysis showed that if any additional pre-treatment weighs down the environmental balance compared to native tows, gamma irradiation is much more suitable than US processing.

Author Contributions: Conceptualization, M.G., F.G., A.B. and J.B.; Investigation, M.G., G.B. and T.C.; Supervision, F.G., A.B. and J.B.; Validation, F.G., A.B. and J.B.; Writing—original draft, M.G.; Writing—review & editing, M.G., A.K., G.B., F.D., T.C., F.P., M.H., N.L.M., F.G., A.B. and J.B. All authors have read and agreed to the published version of the manuscript.

Funding: This research was funded by by Interreg V.A Cross-Channel Programme through, FLOWER: Flax composites, LOW weight, End of life and Recycling (http://flower-project.eu), Grant No. 23) and Pays de la Loire region.

Institutional Review Board Statement: Not Applicable.

Informed Consent Statement: Not applicable.

Data Availability Statement: Not Applicable.

Acknowledgments: We thank Sylviane Daniel (INRAe Nantes) for the monosaccharide results and analysis. We also thank Antohny Magueresse (Univ Bretagne Sud) for the different SEM observations and Bastien Watbled (Littoral Côte d'Opale University) for its availability during ultrasound pre-treatment.

Conflicts of Interest: The authors declare no conflict of interest.

References

1. Le Gall, M.; Davies, P.; Martin, N.; Baley, C. Recommended flax fibres density values for composite property predictions. *Ind. Crop. Prod.* **2018**, *114*, 52–58. [CrossRef]
2. Lefeuvre, A.; Bourmaud, A.; Morvan, C.; Baley, C. Tensile properties of elementary fibres of flax and glass: Analysis of reproducibility and scattering. *Mater. Lett.* **2014**, *130*, 289–291. [CrossRef]
3. Beaugrand, J.; Berzin, F. Lignocellulosic fiber reinforced composites: Influence of compounding conditions on defibrization and mechani-cal properties. *J. Appl. Polym. Sci.* **2013**, *128*, 1227–1238. [CrossRef]
4. Joshi, S.V.; Drzal, L.T.; Mohanty, A.K.; Arora, S. Are natural fiber composites environmentally superior to glass fiber reinforced composites? *Compos. Part A Appl. Sci. Manuf.* **2004**, *35*, 371–376. [CrossRef]
5. Nuez, L.; Gautreau, M.; Mayer-Laigle, C.; d'Arras, P.; Guillon, F.; Bourmaud, A.; Baley, C.; Beaugrand, J. Determinant morphological features of flax plant products and their contribution in injection moulded composite reinforcement. *Compos. Part C* **2020**, *3*, 100054.
6. Van de Velde, K.; Kiekens, P. Thermoplastic polymers: Overview of several properties and their consequences in flax fibre reinforced composites. *Polym. Test.* **2001**, *20*, 885–893. [CrossRef]
7. Bledzki, A.K.; Faruk, O.; Sperber, V.E. Cars from bio-fibres. *Macromol. Mater. Eng.* **2006**, *291*, 449–457. [CrossRef]

8. Bourmaud, A.; Shah, D.U.; Beaugrand, J.; Dhakal, H.N. Property changes in plant fibres during the processing of bio-based composites. *Ind. Crop. Prod.* **2020**, *154*, 112705. [CrossRef]
9. Ramakrishnan, K.; Le Moigne, N.; De Almeida, O.; Regazzi, A.; Corn, S. Optimized manufacturing of thermoplastic biocomposites by fast induction-heated compression moulding: Influence of processing parameters on microstructure development and mechanical behaviour. *Compos. Part A Appl. Sci. Manuf.* **2019**, *124*, 105493. [CrossRef]
10. Kushwaha, G.S.; Sharma, N.K. Green initiatives: A step towards sustainable development and firm's performance in the automobile industry. *J. Clean. Prod.* **2016**, *121*, 116–129. [CrossRef]
11. Bourmaud, A.; Beaugrand, J.; Shah, D.U.; Placet, V.; Baley, C. Towards the design of high-performance plant fibre composites. *Prog. Mater. Sci.* **2018**, *97*, 347–408. [CrossRef]
12. Wang, C.; Wang, N.; Liu, S.; Zhang, H.; Zhi, Z. Investigation of microfibril angle of flax fibers using X-ray diffraction and SEM. *J. Nat. Fibers* **2018**. [CrossRef]
13. Zamil, M.S.; Geitmann, A. The middle lamella-more than a glue. *Phys. Biol.* **2017**, *14*, 11. [CrossRef] [PubMed]
14. Nuez, L.; Beaugrand, J.; Shah, D.U.; Mayer-Laigle, C.; Bourmaud, A.; D'arras, P.; Baley, C. The potential of flax shives as reinforcements for injection moulded polypropylene composites. *Ind. Crops Prod.* **2020**, *148*, 13. [CrossRef]
15. Buranov, A.U.; Mazza, G. Extraction and characterization of hemicelluloses from flax shives by different methods. *Carbohydr. Polym.* **2010**, *79*, 17–25. [CrossRef]
16. Shah, D. Natural fibre composites: Comprehensive ashby-type materials selection charts. *Mater. Des.* **2014**, *62*, 21–31. [CrossRef]
17. Melelli, A.; Arnould, O.; Beaugrand, J.; Bourmaud, A. The middle lamella of plant fibers used as composite reinforcement: Investigation by atomic force microscopy. *Moleculars* **2020**, *25*, 632. [CrossRef]
18. Bourmaud, A.; Ausias, G.; Lebrun, G.; Tachon, M.-L.; Baley, C. Observation of the structure of a composite polypropylene/flax and damage mechanisms under stress. *Ind. Crop. Prod.* **2013**, *43*, 225–236. [CrossRef]
19. Coroller, G.; Lefeuvre, A.; Le Duigou, A.; Bourmaud, A.; Ausias, G.; Gaudry, T.; Baley, C. Effect of flax fibres individualisation on tensile failure of flax/epoxy unidirectional composite. *Compos. Part A Appl. Sci. Manuf.* **2013**, *51*, 62–70. [CrossRef]
20. Le Moigne, N.; Sonnier, R.; El Hage, R.; Rouif, S. Surfaces and Interfaces in Natural Fibre Reinforced Composites: Fundamentals, Modifications and Characteriza-tion. In *Surfaces and Interfaces in Natural Fibre Reinforced Composites: Fundamentals, Modifications and Characterization*; Springer: Berlin, Germany, 2018; p. 1.
21. Liu, M.; Meyer, A.S.; Fernando, D.; Silva, D.A.S.; Daniel, G.; Thygesen, A. Effect of pectin and hemicellulose removal from hemp fibres on the mechanical properties of unidirectional hemp/epoxy composites. *Compos. Part A Appl. Sci. Manuf.* **2016**, *90*, 724–735. [CrossRef]
22. Kalia, S.; Kaith, B.; Kaur, I. Pretreatments of natural fibers and their application as reinforcing material in polymer composites-A review. *Polym. Eng. Sci.* **2009**, *49*, 1253–1272. [CrossRef]
23. Fernandez, J.A.; Le Moigne, N.; Caro-Bretelle, A.S.; El Hage, R.; Le Duc, A.; Lozachmeur, M.; Bono, P.; Bergeret, A. Role of flax cell wall components on the microstructure and transverse mechanical behaviour of flax fabrics rein-forced epoxy biocomposites. *Ind. Crops Prod.* **2016**, *85*, 93–108. [CrossRef]
24. Borsa, J.; László, K.; Boguslavsky, L.; Takacs, E.; Rácz, I.; Tóth, T.; Szabó, D. Effect of mild alkali/ultrasound treatment on flax and hemp fibres: The different responses of the two substrates. *Cellulose* **2016**, *23*, 2117–2128. [CrossRef]
25. Ghosh, R.; Ramakrishna, A.; Reena, G. Effect of air bubbling and ultrasonic processing on water absorption property of banana fibre-vinylester composites. *J. Compos. Mater.* **2013**, *48*, 1691–1697. [CrossRef]
26. Subhedar, P.B.; Gogate, P.R. Alkaline and ultrasound assisted alkaline pretreatment for intensification of delignification process from sustaina-ble raw-material. *Ultrason. Sonochem.* **2014**, *21*, 216–225. [CrossRef]
27. Choi, H.Y.; Han, S.O.; Lee, J.-S. The Effects of Morphological Properties of Henequen Fiber Irradiated by EB on the Mechanical and Thermal Properties of Henequen Fiber/PP Composites. *Compos. Interfaces* **2009**, *16*, 751–768. [CrossRef]
28. Le Moigne, N.; Sonnier, R.; El Hage, R.; Rouif, S. Radiation-induced modifications in natural fibres and their biocomposites: Opportunities for controlled physi-co-chemical modification pathways? *Ind. Crop. Prod.* **2017**, *109*, 199–213. [CrossRef]
29. Khan, R.A.; Khan, M.A.; Khan, A.H.; Hossain, M.A. Effect of gamma radiation on the performance of jute fabrics-reinforced polypropylene composites. *Radiat. Phys. Chem.* **2009**, *78*, 986–993.
30. Quéméner, B.; Bertrand, D.; Marty, I.; Causse, M.; Lahaye, M. Fast data preprocessing for chromatographic fingerprints of tomato cell wall polysaccharides using chemometric methods. *J. Chromatogr. A* **2007**, *1141*, 41–49. [CrossRef]
31. Hauschild, M.G.M.; Guinee, J.; Heijungs, R.; Huijbregts, M.; Jolliet, O.; Margni, M.; De Schryver, A.; Pennington, D.; Pant, R.; Sala, S.; et al. *Recommendations for Life Cycle Impact Assessment in the European Context—Based on Existing Environmental Impact Assessment Models and Factors (International Reference Life Cycle Data System—ILCD Handbook)*; EUR 24571 EN; Luxembourg Publications Office of the European Union: Luxembourg, 2011.
32. Salmén, L.; Larsson, P.A. On the origin of sorption hysteresis in cellulosic materials. *Carbohydr. Polym.* **2018**, *182*, 15–20. [CrossRef]
33. Le Duigou, A.; Davies, P.; Baley, C. Environmental impact analysis of the production of flax fibres to be used as composite material reinforcement. *J. Biobased Mater. Bioenergy* **2011**, *5*, 153–165. [CrossRef]
34. Lefeuvre, A.; Baley, C.; Morvan, C. Analysis of Flax Fiber Cell-Wall Non-Cellulosic Polysaccharides Under Different Weather Conditions (Marylin Variety). *J. Nat. Fibers* **2018**, *15*, 539–544. [CrossRef]
35. Celino, A.; Fréour, S.; Jacquemin, F.; Casari, P. The hygroscopic behavior of plant fibers: A review. *Front. Chem.* **2014**, *2*, 12.

36. Hill, C.A.S.; Norton, A.; Newman, G. The water vapor sorption behavior of natural fibers. *J. Appl. Polym. Sci.* **2009**, *112*, 1524–1537. [CrossRef]
37. Padovani, J.; Legland, D.; Pernes, M.; Gallos, A.; Thomachot-Schneider, C.; Shah, D.; Bourmaud, A.; Beaugrand, J. Beating of hemp bast fibres: An examination of a hydro-mechanical treatment on chemical, structural, and nanomechanical property evolutions. *Cellulose* **2019**, *26*, 5665–5683. [CrossRef]
38. Pantaloni, D.; Shah, D.; Baley, C.; Bourmaud, A. Monitoring of mechanical performances of flax non-woven biocomposites during a home compost degradation. *Polym. Degrad. Stab.* **2020**, *177*, 109166. [CrossRef]
39. Miller, B.; Muri, P.; Rebenfeld, L. A microbond method for determination of the shear strength of a fiber/resin interface. *Compos. Sci. Technol.* **1987**, *28*, 17–32. [CrossRef]
40. Pantaloni, D.; Rudolph, A.L.; Shah, D.U.; Baley, C.; Bourmaud, A. Interfacial and mechanical characterisation of biodegradable polymer-flax fibre composites. *Compos. Sci. Technol.* **2020**, *201*, 108529. [CrossRef]
41. Pucci, M.F.; Liotier, P.J.; Seveno, D.; Fuentes, C.; Van Vuure, A.; Drapier, S. Wetting and swelling property modifications of elementary flax fibres and their effects on the Liquid Composite Molding process. *Compos. Part A Appl. Sci. Manuf.* **2017**, *97*, 31–40. [CrossRef]
42. Gager, V.; Legland, D.; Bourmaud, A.; Le Duigou, A.; Pierre, F.; Behlouli, K.; Baley, C. Oriented granulometry to quantify fibre orientation distributions in synthetic and plant fibre composite preforms. *Ind. Crop. Prod.* **2020**, *152*, 7. [CrossRef]
43. Liu, M.; Thygesen, A.; Summerscales, J.; Meyer, A.S. Targeted pre-treatment of hemp bast fibres for optimal performance in biocomposite materials: A review. *Ind. Crop. Prod.* **2017**, *108*, 660–683. [CrossRef]

Article

Preparation and Application of Efficient Biobased Carbon Adsorbents Prepared from Spruce Bark Residues for Efficient Removal of Reactive Dyes and Colors from Synthetic Effluents

Glaydson Simões dos Reis [1,*], Sylvia H. Larsson [1], Mikael Thyrel [1], Tung Ngoc Pham [2], Eder Claudio Lima [3], Helinando Pequeno de Oliveira [4] and Guilherme L. Dotto [5]

[1]   Biomass Technology Centre, Department of Forest Biomaterials and Technology, Swedish University of Agricultural Sciences, SE-901 83 Umeå, Sweden; sylvia.larsson@slu.se (S.H.L.); mikael.thyrel@slu.se (M.T.)
[2]   Faculty of Chemistry, The University of Danang-University of Science and Technology, 54 Nguyen Luong Bang, Lien Chieu, Da Nang 550000, Vietnam; pntung@dut.udn.vn
[3]   Institute of Chemistry, Federal University of Rio Grande do Sul (UFRGS), Av. Bento Gonçalves 9500, Porto Alegre 91501-970, RS, Brazil; profederlima@gmail.com
[4]   Institute of Materials Science, Federal University of Sao Francisco Valley, Juazeiro 48920-310, BA, Brazil; helinando.oliveira@univasf.edu.br
[5]   Chemical Engineering Department, Federal University of Santa Maria (UFSM), Santa Maria 97105-900, RS, Brazil; guilherme_dotto@yahoo.com.br
*   Correspondence: glaydsonambiental@gmail.com or glaydson.simoes.dos.reis@slu.se

Citation: dos Reis, G.S.; Larsson, S.H.; Thyrel, M.; Pham, T.N.; Claudio Lima, E.; de Oliveira, H.P.; Dotto, G.L. Preparation and Application of Efficient Biobased Carbon Adsorbents Prepared from Spruce Bark Residues for Efficient Removal of Reactive Dyes and Colors from Synthetic Effluents. *Coatings* **2021**, *11*, 772. https://doi.org/10.3390/coatings11070772

Academic Editors: Philippe Evon and Artur P. Terzyk

Received: 20 May 2021
Accepted: 25 June 2021
Published: 27 June 2021

Publisher's Note: MDPI stays neutral with regard to jurisdictional claims in published maps and institutional affiliations.

Abstract: Biobased carbon materials (BBC) obtained from Norway spruce (Picea abies Karst.) bark was produced by single-step chemical activation with $ZnCl_2$ or KOH, and pyrolysis at 800 °C for one hour. The chemical activation reagent had a significant impact on the properties of the BBCs. KOH-biobased carbon material (KOH-BBC) had a higher specific surface area (S_{BET}), equal to 1067 m^2 g^{-1}, larger pore volume (0.558 cm^3 g^{-1}), more mesopores, and a more hydrophilic surface than $ZnCl_2$-BBC. However, the carbon yield for KOH-BBC was 63% lower than for $ZnCl_2$-BBC. Batch adsorption experiments were performed to evaluate the ability of the two BBCs to remove two dyes, reactive orange 16 (RO-16) and reactive blue 4 (RB-4), and treat synthetic effluents. The general order model was most suitable for modeling the adsorption kinetics of both dyes and BBCs. The equilibrium parameters at 22 °C were calculated using the Liu model. Upon adsorption of RO-16, Q_{max} was 90.1 mg g^{-1} for $ZnCl_2$-BBC and 354.8 mg g^{-1} for KOH-BBC. With RB-4, Qmax was 332.9 mg g^{-1} for $ZnCl_2$-BBC and 582.5 mg g^{-1} for KOH-BBC. Based on characterization and experimental data, it was suggested that electrostatic interactions and hydrogen bonds between BBCs and RO-16 and RB-4 dyes played the most crucial role in the adsorption process. The biobased carbon materials showed high efficiency for removing RO-16 and RB-4, comparable to the best examples from the literature. Additionally, both the KOH- and $ZnCl_2$-BBC showed a high ability to purify two synthetic effluents, but the KOH-BBC was superior.

Keywords: biobased carbon materials; meso- and microporous carbons; dye adsorption; chemical adsorption; electrostatic interactions

1. Introduction

Biomass is a renewable and widespread resource that, if utilized sustainably, can help to reduce the emission of carbon dioxide that directly affects global warming [1]. Agricultural and forestry residues and by-products from biobased industries can be used as feedstock for energy production and material applications to replace fossil fuel sources [2,3].

Norway spruce (*Picea abies* (L.) Karst.) is one of the most common and economically valuable trees for the European forest industry as it is widely distributed from central to boreal and eastern Europe [4]. Together with pine and birch, spruce is the most common tree species in Sweden; these three combined comprise more than 90% of standing volume [4].

The Swedish annual forest harvest amounts to approximately 90 Mm^3 standing volume [5], and they are economically very important for sawmill and paper and pulp industries. However, in the production of sawn timber, pulp, and paper, only the stem wood is used—the remaining components can be considered industrial by-products. Around 10–15% of the feedstock volume delivered to the forest industries consists of bark currently mainly utilized as fuel and other low-value applications [5,6]. Consequently, research to employ bark as a precursor for value-added and eco-friendly material products is motivated.

Using biomass to produce biobased carbon materials (BBC) such as biochar (BC), activated carbon (AC), carbon composite materials (CCM) is an application with great potential. It reduces fossil carbon use and can provide new types of functionalities [6–14]. BBC is the oldest, most common, and efficient material for removing pollutants from aqueous media [7–13]. Besides its chemical stability, surface functionalities, high porosity, and specific surface area are essential characteristics for efficient application in the adsorption process [8–16]. However, high-purity activated carbons are expensive; therefore, the use of other biobased carbon materials can be explored as adsorbents for the removal of pollutants and micropollutants [8,10–16].

Adsorption is seen as one of the most suitable treatment methods for tackling pollutants from contaminated water and wastewaters due to its simple operating conditions, high efficiency, and low-cost employment. To design a very efficient adsorption process, the BBC must be prepared to achieve suitable properties.

The BBC properties are highly dependent on pyrolysis conditions and activation methods [14,17]. For instance, chemical activation can create carbon materials with ultra-high BET surface area (S_{BET}) and porosities because of extensive micro and mesoporosity development. Each pore structure has a specific role in the adsorption process, e.g., the micropore structure contributes significantly to the S_{BET} values and the adsorption of small-sized contaminants (e.g., metallic species and small organic molecules) [13,17]. Mesopores are essential as vectors to the surface areas within the carbon material particle, and their respective quantities are primarily dependent on the pyrolysis conditions and activation method. The mesopore structure is vital for larger-molecule adsorption, which is the case for dyes and colored effluents.

It is estimated that over 10,000 different dyes and pigments are used in the food, leather, cosmetics, and textile industries, e.g., only the textile industry consumes up to 200,000 tons of dyes yearly, thereby generating large amounts of colored effluents [18]. These colored effluents are, if not adequately treated, discharged into the environment, where they are potentially harmful to the aquatic systems and ecosystem integrity. Besides, many dyes are reported as mutagenic and carcinogenic [19]. Therefore, these effluents must be treated before their discharge into the environment, and the adsorption process using biomass-activated carbon is one the most suitable treatment process [12–16].

The purpose of this study was to investigate the potential of spruce bark residues as a precursor to producing efficient carbon-based materials by pyrolysis, using KOH (KOH-BBC) and ZnCl$_2$ (ZnCl$_2$-BBC) as chemical activators. The effect of the chemical reagents on the BBC characteristics such as morphology, specific surface area and porosity, surface chemistry, BBC composition, hydrophobicity index, carbon yield, and adsorption of two dyes and different synthetic effluents were evaluated.

2. Materials and Methods

2.1. Preparation of BBCs

Norway spruce (*Picea abies* Karst.) bark was delivered from a pulp and paper mill in northeast Sweden and prepared at the Biomass Technology Centre (BTC), Swedish University of Agricultural Sciences, Umeå, Sweden. The wet bark was dried in a custom-made plane drier at 40 °C, shredded with a screen size of 15 mm (Lindner Micromat 2000, Lindner-Recyclingtech Gmbh), hammer-milled with a screen size of 4 mm (Bühler DFZK 1, Bühlergroup), representatively sampled according to ISO 18135:2017, and cutting-milled with a screen size of 200 μm using a Fritsch Pulverisette 14 mill (FRITSCH GmbH, Germany).

The pyrolysis was done in a single pyrolysis-step preparation according to a previously reported procedure [20–24]. First, 15.0 g of the spruce bark was mixed in a weight ratio of 1:1 with each chemical activation agent (KOH and $ZnCl_2$). During the mixing, about 40.0 mL of water was added to form homogeneous pastes. These two pastes were dried in an oven at 105 °C for 24 h. The dried pastes were placed in a metallic crucible and treated thermally in a conventional high-temperature oven under a nitrogen flow of 600 mL min^{-1}. They were heated from 20 to 800 °C at a rate of 10 °C min^{-1} and held at 800 °C for 60 min. The oven was turned off to cool down the pyrolyzed samples while the nitrogen flow was kept, and when the temperature dropped to 200 °C, the nitrogen flow was shut off. The pyrolyzed materials were milled with a screen size of 200 μm and completely leached out by conventional leaching using 0.1 M HCl, under reflux, for 2 h for KOH-BBC and using 1.0 M HCl, under reflux, for 2 h for $ZnCl_2$-BBC [13,15–17]. The BBC preparation procedure is summarized in Figure 1.

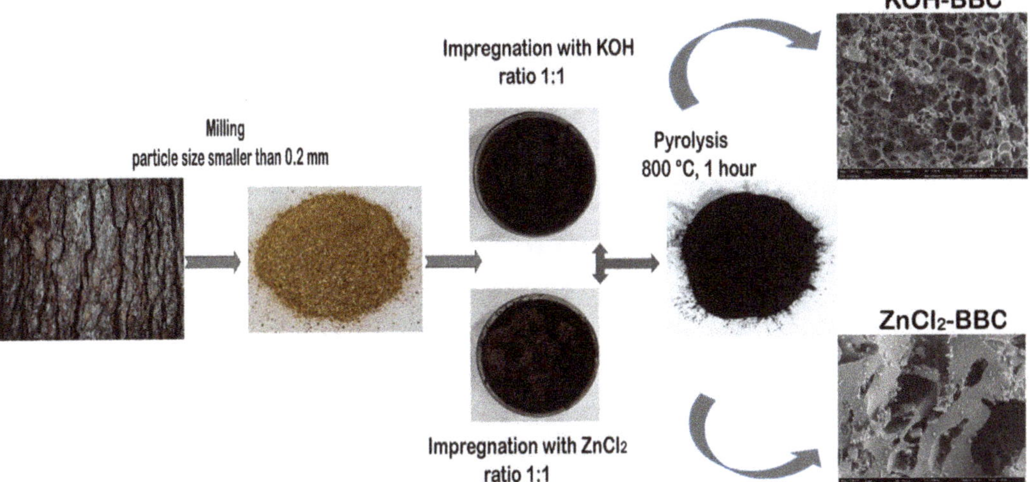

Figure 1. The BBC preparation procedure.

2.2. BBC Characterization

2.2.1. Textural Properties

The adsorbent's textural properties, especially for biobased carbon materials, are crucial for evaluating their applications and potential application efficiencies. Common BBCs are almost always heterogeneous, having an unknown range of pore sizes and a range of pore shapes, blocked and network pores [12,14,15].

The samples' surface morphology was observed with scanning electron microscopy (SEM) (55-VP, Supra, Zeiss), using an acceleration voltage of 20 kV and magnification ranging from 100 to 20,000.

N_2 adsorption/desorption isotherm analysis (Tristar 3000 apparatus, Micrometrics Instrument Corp., Norcross, GA, USA) was performed to quantify the porosity (by DFT method) and surface area (BET method). Before the analysis, the sample was degassed at 180 °C for 3 h in an N_2 atmosphere. The specific surface area was calculated in the relative pressure interval of 0.05–0.3 using the Brunauer–Emmett–Teller (BET) method [16,22]. Mesopore size and distribution were calculated by the Barrett–Joyner–Halenda (BJH) method from desorption curves while the micropore area values were calculated by the t-plot method [16,20,22]. The percentage of the mesopore and micropore areas were calculated based on the S_{BET} values [22].

2.2.2. Elemental Analysis, Yield (%), Raman Spectroscopy, and Zeta Potential

The elemental analysis was carried out to evaluate the volatiles and fixed carbon contents and quantify the elemental composition of the BBCs, respectively. The analysis was made using a CHN Perkin Elmer M CHNS/O Analyzer, model 2400.

The yield (%) was calculated from the dry matter quota of the biomass precursor after and before activation.

Raman spectroscopy is widely applied to characterize BBCs. It is applied to obtain structural information on the bulk of carbon materials. Raman spectra were recorded on a Renishaw inVia Raman spectrometer (Renishaw, Kingswood, UK) at 633 nm HeNe laser.

Zeta-potential was performed to obtain the charge (whether positive or negative) of the BBCs. It was determined at pH 7 using a potential analyzer (Zetasizer Nano ZS90, Malvern Panalytical, Malvern, UK).

2.2.3. Water Vapor Sorption and Hydrophobicity/Hydrophilicity

The BBCs' water vapor sorption isotherms and the hydrophobicity/hydrophilicity index (HI) are used to determine the properties of the BBC/water interface and the water molecules' ability to attach to the BBC surface, which both may influence dye adsorption.

The BBCs' H_2O vapor adsorption isotherms were determined by dynamic vapor sorption (DVS Advantage, Surface Measurement Systems) at 25 °C, where RH was varied from 0 to 95% and back in 5% steps. The hydrophobicity/hydrophilicity index (HI) was performed according to a method previously reported in the literature [23]: 0.3 g of each BBC was placed into 5 mL beakers and inserted into plugged 1.5 L E-flasks with saturated atmosphere solvent vapor (water or n-heptane) using 80 mL of each solvent. The beakers were placed in the center of the E-flasks to avoid contact with the flask walls. After 24 h, the beakers were removed and weighed. The weight gained was used to calculate the maximum vapor adsorption.

2.3. Dye Adsorption Analysis

2.3.1. Batch Adsorption Studies

Aliquots of 20.00 mL of 30.00–1000.0 mg L^{-1} of RB-4 and RO-16 were added to 50.0 mL Falcon flat tubes containing 30 mg (dosage of 1.5 g L^{-1}) of each BBC [20,24,25]. The Falcon tubes containing RB-4 or RO-16 and BBCs were agitated in a shaker model TE-240 between 0.1–12 h to obtain the kinetics data. Afterward, to separate the dyes and BBCs, the flasks were centrifuged. After adsorption, the residual solutions of RB-4 and RO-16 were quantified using a UV-Visible spectrophotometer (Shimadzu 1800) at a maximum wavelength of 595 and 494 nm, respectively.

The amount of RO-16 adsorbed by the BBCs and the percentage of removal were calculated using Equations (1) and (2), respectively [20,24,25]:

$$q = \frac{\left(C_0 - C_f\right)}{m} \cdot V \tag{1}$$

$$\% \, Removal = 100 \cdot \frac{\left(C_0 - C_f\right)}{C_0} \tag{2}$$

where q is the amount of selected dye uptake by the BBCs (mg g^{-1}); C_0 is the initial dye concentration in contact with BBCs (mg L^{-1}), C_f is the final concentration (mg L^{-1}) after adsorption, V is the volume of dye solutions (L) in contact with the BBCs, and m is the BBC mass (g).

2.3.2. Adsorption Kinetics and Equilibrium Analysis

Adsorption kinetics provides information on the adsorption rate, the adsorbent's performance, and the mass transfer mechanisms [20,23–25]. Knowing the adsorption kinetics is crucial for designing efficient adsorption systems.

The RO-16 adsorption kinetics of the KOH-BBC and $ZnCl_2$-BBC samples were evaluated at two initial concentrations: 500 and 700 mg L^{-1}. The suitability of different models for predicting the adsorption kinetics was assessed by analyzing R^2_{adj} and SD values. Pseudo-first-order (PFO) model, pseudo-second-order (PSO) model, and general order models were used to evaluate the kinetic adsorption process [23–25].

The mathematical representations of pseudo-first-order, pseudo-second-order, and general order are shown in Equations (3)–(5), respectively.

$$q_t = q_e \cdot [1 - \exp(-k_1 \cdot t)] \tag{3}$$

$$q_t = \frac{k_2 \cdot q_e^2 \cdot t}{1 + q_e \cdot k_2 \cdot t} \tag{4}$$

$$q_t = q_e - \frac{q_e}{\left[k_N \cdot (q_e)^{n-1} \cdot t \cdot (n-1) + 1 \right]^{1/(n-1)}} \tag{5}$$

Equilibrium isotherms are used to determine the adsorption affinity and dye removal mechanisms of the adsorption systems [8,23–25]. Each adsorption system (individual adsorbent material and adsorbate) has a unique isotherm, and the quantity of adsorbed adsorbate on an adsorbent depends on both the BBC's and the solution's properties. Therefore, equilibrium studies are mandatory to evaluate and establish adsorbent efficiency.

The equilibrium process was analyzed by Langmuir, Freundlich, and Liu's models. The fit quality was assessed through statistical indicators such as R^2, R^2_{adj}, and SD. See further details about these indicators in references [13,14,20,23].

Langmuir, Freundlich, and Liu's models are shown in Equations (6)–(8), respectively.

$$q_e = \frac{Q_{max} \cdot K_L \cdot C_e}{1 + K_L \cdot C_e} \tag{6}$$

$$q_e = K_F \cdot C_e^{1/nF} \tag{7}$$

$$q_e = \frac{Q_{max} \cdot (K_g \cdot C_e)^{nL}}{1 + (K_g \cdot C_e)^{nL}} \tag{8}$$

Detailed information about all these equations can be found in the literature [10,23,24].

2.3.3. Preparation of the Dyeing Synthetic Effluents

De-ionized water was used for the preparation of all solutions used in the dye adsorption experiments. RB-4 ($C_{23}H_{14}N_6Cl_2O_8S_2$) and RO-16 ($C_{20}H_{17}N_3O_{10}S_3Na_2$) were obtained from Sigma Aldrich, Sweden. The stock solution was prepared by dissolving the dye in distilled water to 2.00 g L^{-1}. Working solutions were obtained by diluting the dye stock solution to the required concentrations without adjusting the pH.

Synthetic effluents with different compositions (see Table 1) were prepared to test the BBCs' applicability for treating real effluents.

2.3.4. Analytical Control of the Measurements and Statistical Evaluation of Nonlinear Models

The adsorption equations were fitted using the nonlinear approach obtained by the Simplex method and the successive interactions of the Levenberg–Marquardt algorithm [10,15–17,21]. This fitting was acquired by the nonlinear fitting facilities of the Microcal Origin 2020 software, and they were used to fit the kinetic and equilibrium data. The determination coefficient (R^2), adjusted determination coefficient (R^2_{adj}), and the standard deviation of the residues (SD) were employed to analyze the suitability of the models [10,15–17,21–25].

Table 1. Effluent compositions and concentrations.

Compounds	Concentration (mg L^{-1})		λ_{max} (nm)
Effluent	A	B	
RO-16	50	50	494
RB- 4	50	50	595
Methylene Blue	50	50	668
Bismarck Brown	50	-	468
Crystal Violet	50	-	590
Methyl Red	-	50	507
Methyl Orange	-	50	522
Phenol Red	-	50	550
Sodium Dodecyl	25	25	-
Sodium sulfate	25	25	-
Ammonium chloride	20	25	-
Sodium acetate	20	25	-
pH	5.1	4.9	-

Residual standard deviation measures the difference between the theoretical and experimental amounts of dyes removed from solutions. The R^2, R^2_{adj}, and SD are given in Equations (9)–(11), respectively [21–25].

$$R^2 = \left(\frac{\sum_i^n \left(q_{i,exp} - \overline{q}_{i,exp} \right)^2 - \sum_i^n \left(q_{i,\,exp} - q_{i,\,model} \right)^2}{\sum_i^n \left(q_{i,exp} - \overline{q}_{i,exp} \right)^2} \right) \tag{9}$$

$$R^2_{adj} = 1 - \left(1 - R^2 \right) \cdot \left(\frac{n-1}{n-p-1} \right) \tag{10}$$

$$SD = \sqrt{ \left(\frac{1}{n-p} \right) \cdot \sum_i^n \left(q_{i,\,exp} - q_{i,\,model} \right)^2 } \tag{11}$$

where $q_{i,model}$ represents the individual theoretical q value predicted by the model. $q_{i,exp}$ represents each experimental q value. $\overline{q}_{exp}$ is the average of the experimental q values. n and p represent the number of experiments and model parameters, respectively.

3. Results and Discussion

3.1. BBC Characteristics

3.1.1. Textural Properties and Porosity

The SEM images show remarkable differences between both microstructures (see Figure 2). The ZnCl$_2$-BBC has a dense structure, with more elongated cavities and holes of different sizes and shapes (Figure 2A,C) that should have been formed during the leaching step with 6.0 mol L^{-1} HCl. Additionally, it is observed that ZnCl$_2$-BBC presents a rough surface.

KOH-BBC (Figure 2B,D) presents ordered macropore structure and holes with lower diameter in its surface, which should be attributed to the lower concentration of HCl (1.0 mol L^{-1}) used in the leaching step. Both BBC presents irregular particle size and rough surface.

The ZnCl$_2$-BBC was prepared by catalyzed dehydration and elimination of carbonyl and hydroxyl groups during the heat treatment [26,27], and that the ZnCl$_2$ (due to its low melting point at 290 °C and the boiling point at 732 °C) is fused into the biomass matrix, thereby creating a denser structure and a microporous network [27,28]. On the other hand, KOH activation provokes the breakage of C–O–C and C–C bonds, creating pores and well-developed porosity [26]. Additonally, an uneven distribution of KOH in the bark matrix can promote hyperactivation during the pyrolysis process, resulting in pore wall demolition and widening of the micropores into mesopores [26]. These structural transformations are beneficial for the BBC's physical adsorption of RB-4 and RO-16 and effluents treatment.

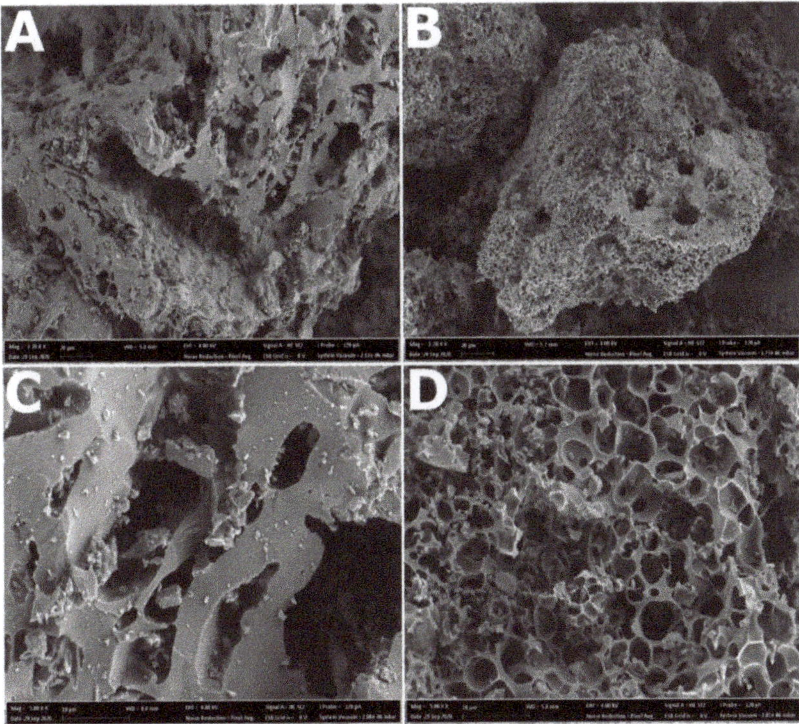

Figure 2. SEM images of BBCs: (**A**) ZnCl$_2$-BBC at 1.2 K of magnification, (**B**) KOH-BBC at 1.2 K of magnification, (**C**) ZnCl$_2$-BBC at 5 K of magnification, (**D**) KOH-BBC at 5 K of magnification.

The N$_2$ isotherms for ZnCl$_2$-BBC and KOH-BBC (Figure 3) can be ascribed to a type I isotherm. A type I isotherm (also mentioned as Langmuir isotherm) is typical for microporous materials (with pore diameter <2 nm) [16]. Higher amounts of N$_2$ are adsorbed at low relative pressures for microporous materials, and when it is close to 1, the curve may reach a limiting value or rise if larger pores are present [16].

Although both BBCs exhibited a Type I isotherm, the adsorbed N$_2$ volumes differed significantly (Figure 3). The KOH-treated BBCs had an almost 30% higher S$_{BET}$ (1067 m^2 g^{-1}) than the ZnCl$_2$-BBC (754 m^2 g^{-1}) (Table 2). The external surface area and the micropore and mesopore volumes agree with these results. Hence, it can be concluded that KOH-activation produced a BBC with better textural properties and better adsorption performance than ZnCl$_2$-activation. The percentage of mesopores in KOH-BBC (49.29%) is higher when compared with the ZnCl$_2$-BBC sample (43.51%), while the share of micropores is higher in ZnCl$_2$-BBC (56.49%) when compared to KOH-BBC (50.71%) (See Table 2). However, both micro and mesoporous materials are highly efficient to adsorb organic molecules with small sizes and, therefore, suitable for adsorption of RB-4 (size of 1.59 nm) and RO-16 (size of 1.68 nm).

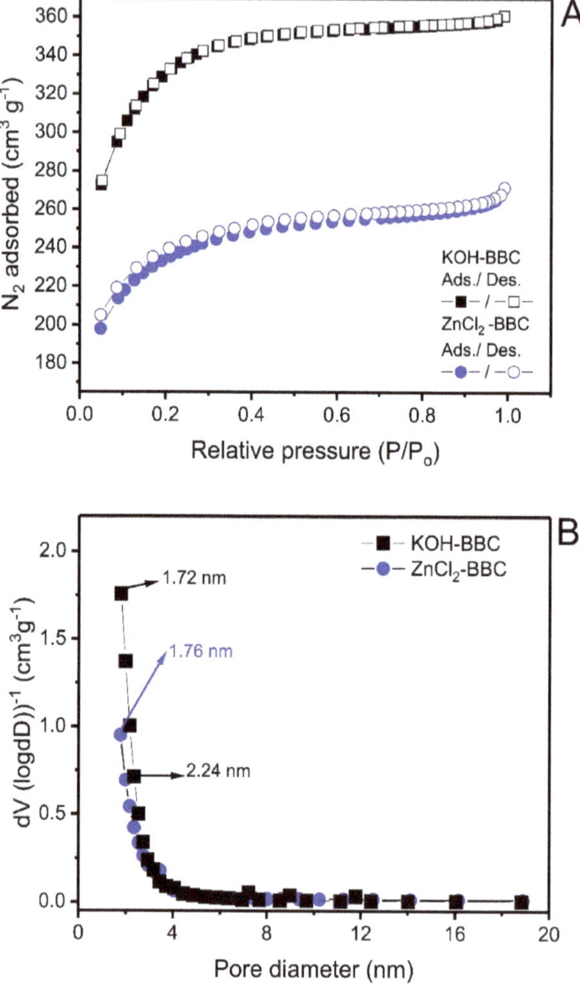

Figure 3. N_2 isotherm curves for $ZnCl_2$-BBC and KOH-BBC (**A**) and pore distribution curves (**B**).

Table 2. Textural properties of the activated carbons.

Samples	$ZnCl_2$-BBC	KOH-BBC
Parameters		
S_{BET} (m^2 g^{-1})	754	1067
External surface area (m^2 g^{-1})	328	526
% of mesopore area (%)	43.51	49.29
t-plot Micropore area (m^2 g^{-1})	425.8	541.2
% of micropore area (%)	56.49	50.71
Total pore volume (cm^3 g^{-1})	0.4205	0.5585
t-plot micropore volume (cm^3 g^{-1})	0.2172	0.2776
% of micropore volume (%)	51.65	49.70
Volume of mesopores (cm^3 g^{-1})	0.2033	0.2809
Average pore size (nm)	2.231	2.093

The pore size distributions derived from the BJH plots of both BBC samples are displayed in Figure 3B. The chemical activation seemed to affect the pore structure of the BBC samples. BBC-KOH showed a much higher distribution of the pores in the range of large micropores or small mesopores, 1.72–2.24 nm (see the line with squares). According to the BJH plots, both samples possess large quantities of micropores and homogeneous and small mesopores. The creation of large micropores and small mesopores enhanced the BBC-KOH sample, which is in good agreement with the porosity data.

Literature data reveals significant variance in S_{BET} values depending on the type of biomass and preparation conditions (Table 3). For instance, Sipola et al. [8] prepared activated carbon from scots pine (*Pinus sylvestrus*) and spruce (*Picea* spp.) barks for wastewater purification and found specific surface areas ranging from 200 to 600 $m^2\,g^{-1}$. In another work [9], the spruce bark porous materials were produced and employed in methylene blue dye adsorption. The materials presented S_{BET} ranging from 351 to 1275 $m^2\,g^{-1}$ and were successfully employed in the dye removal from aqueous solutions. In addition, a specifically high S_{BET} (2330 $m^2\,g^{-1}$) was achieved with rice plant residue as a biomass precursor. However, in that case, a highly complex preparation procedure was required: First, pre-carbonization at 500 °C for one hour followed by NaOH washing; Secondly, BBC was mixed with KOH at a ratio of 1:4 (biomass: KOH) and pyrolyzed at 800 °C for 30 min and then, followed by HCl washing to remove the inorganic compounds. Consequently, due to the cumbersome procedure, the high S_{BET} comes with a high cost. The S_{BET} values of the $ZnCl_2{}^-$ and KOH-activated Norway spruce bark BBCs are comparable with BBCs from several other biomass precursors, but in this case, the manufacturing method is simple, and the feedstock material highly available and cheap.

Table 3. Comparison of KOH-BBC and ZnCl2-BBC preparation methods and S_{BET} for a variety of biomass precursors.

Adsorbent	Activation Reagent	Preparation Conditions	S_{BET} ($m^2\,g^{-1}$)	Ref.
Scots pine bark	Steam + N_2	Firstly, the biomass was carbonized using slow pyrolysis at 475 °C for 3 h. Afterward, heated at 800 °C for 3.5 h under steam activation [steam (30 and 40%) + N_2 (66 and 300 L/h)].	539–603	[8]
Norway spruce bark	Steam + N_2	Firstly, the biomass was carbonized using slow pyrolysis at 475 °C for 3 h. Afterward, heated at 800 °C for 3.5 h under steam activation [steam (30 and 40%) + N_2 (66 and 300 L/h)].	187–369	[8]
Norway spruce bark	Steam + N_2	The biomass was heated at 600 °C for 2 h under steam activation (steam + N_2).	351	[9]
Norway spruce bark	$ZnCl_2$	A mixture of $ZnCl_2$ and biomass powder at ratio 2.0:1.0 ($ZnCl_2$:biomass) and pyrolyzed at 600 °C for 2 h. Afterward, it was washed with HCl to remove the inorganic compounds.	1495	[9]
Tea leave residue	KOH	A mixture of KOH and tea powder (2:1) and pyrolyzed at 900 °C for 60 min. Afterward, it was washed with HCl to remove the potassium compounds and further pyrolyzed at 1200 °C.	912	[28]
Palm shell	KOH+ $ZnCl_2$	Pre-carbonization of biomass at 400 °C for 2 h. Afterward, a mixture of biomass and both KOH (75%) and $ZnCl_2$ (25%) at the final ratio of biomass: chemical activator 1:4. The mixture was then pyrolyzed at 850 °C for 1 h and washed with HCl.	1295	[29]
Garlic peel	KOH	First, it was hydro-carbonized and then chemically activated by KOH (ratio 2:1, KOH: biomass) and pyrolyzed at 600 °C at 4 °C/min under N_2 flow for 2 h.	947	[30]

Table 3. *Cont.*

Adsorbent	Activation Reagent	Preparation Conditions	S_{BET} (m^2 g^{-1})	Ref.
Rice plants	KOH	The biomass was Pre-carbonized at 500 °C for 1 h, followed by NaOH washing. Afterward, the pyrolyzed BBC was mixed with KOH at ratio 1:4. The mixture was then pyrolyzed at 800 °C for 30 min and then washed with HCl.	2330	[31]
Brazil nutshells	ZnCl$_2$	A mixture of ZnCl$_2$ and biomass powder at ratio 1.5:1.0 (ZnCl$_2$:biomass) and pyrolyzed at 600 °C for 30 min. Afterward, it was washed with 6.0 M HCl to remove the inorganic compounds.	1457	[32]
Sewage sludge	ZnCl$_2$	A mixture of ZnCl$_2$ and biomass powder at ratio 0.5:1.0 (ZnCl$_2$:biomass) and pyrolyzed at 500 °C for 15 min. Afterward, it was washed with HCl to remove the inorganic compounds.	679	[33]
Coconut shell	ZnCl$_2$	Blending coconut shell powder and ZnCl$_2$ at ratio 1:3 in 50 mL of 3 M FeCl$_3$ solution. Afterward, heated at 900 °C for 1 h under an inert atmosphere. Afterward, it was washed with HCl to remove the inorganic compounds.	1874	[34]
Norway spruce bark	ZnCl$_2$	ZnCl$_2$ and biomass powder mixture at ratio 1.0:1.0 (ZnCl$_2$:biomass) and pyrolyzed at 800 °C for 60 min. Afterward, it was washed with 6.0 M HCl to remove the inorganic compounds.	754	This work
Norway spruce bark	KOH	A mixture of KOH and biomass powder at ratio 1.0:1.0 (KOH: biomass) and pyrolyzed at 800 °C for 60 min. Afterward, it was washed with 1.0 M HCl to remove the inorganic compounds.	1067	This work

3.1.2. Elemental Analysis, Carbon Yield, Raman Spectroscopy, Zeta-Potential, and FTIR

The carbon content of the spruce bark ZnCl$_2$- and KOH-activated BBCs was 94.8% and 91.6%, respectively (see Table 4). These values are very high compared to literature; Correa et al. [35] produced several BBCs from different biomasses, and the carbon content varied from 76.9 to 87.8%, while Duan et al. [36] obtained 82.66% of carbon content in BBC made from coconut shells. High carbon content can reflect good adsorption efficiency because hydrophobic interactions of the aromatics of BBC can interact with organic molecules. In addition, high carbon content means less ash content, and ashes in the BBC reduce S_{BET} and functional groups, which hinder the adsorption process. Concerning the oxygen content, KOH-BBC presented higher content when compared to ZnCl$_2$-BBC; this can positively influence the carbons' hydrophilicity index and the water/dye adsorption behavior [35].

$$HI = \frac{\frac{amount\ of\ water\ vapor\ (mg)}{mass\ of\ BBC\ (g)}}{\frac{amount\ of\ h-heptane\ vapor\ (mg)}{mass}\ of\ BBC\ (g)} \tag{12}$$

The *BBC* yield from pyrolysis and KOH activation was approximately one-third of the ZnCl$_2$ treatment (Table 4). This result indicates a strong reaction between bark and KOH during the pyrolysis process. Via breakage of C–O–C and C–C bonds, KOH can play a catalytic role in the material's volatilization, leading to a low carbon yield [37]. Impregnation with ZnCl$_2$ results in degradation of the cellulosic material that, combined with the dehydration during carbonization, leads to charring and aromatization of the carbon skeleton. These pyrolytic conditions inhibit the formation of tar and reduce mass loss [38]. As a result, BBC production by ZnCl$_2$ activation generally provides higher yields than when using other chemical reagents [38].

Table 4. Properties and elementary analysis of activated carbons.

Samples	ZnCl$_2$-BBC	KOH-BBC
Parameters		
HI (H$_2$O/n-heptane)	1.19	1.29
Zeta potential (mV)	−19.4	−20.5
pH	5.1	6.0
Carbon content (%)	94.8	91.6
Nitrogen content (%)	0.51	0.29
Hydrogen (%)	1.2	1.6
Oxygen (%)	2.5	5.3
Ash (%)	0.99	1.21
BBC yield (%)	38.1	14.2

The Raman spectra of the ZnCl$_2$- and KOH-BBCs are shown in Figure 4. The D and G bands indicate the degree of defective structure and the activated carbons' graphitization, respectively [28,30]. These bands' position, area, and intensity can also show differences in the structural characteristics [34,39]. Both samples' D- and G-bands are located at around 1340 and 1593 cm^{-1}, corresponding to the defect/disorder-induced structures in the BBCs' graphite layers and the vibration of sp2-bonded carbon atoms in a two-dimensional hexagonal lattice, respectively [28,34]. The relative strength intensity (I$_D$/I$_G$) represents the degree of defect in the BBCs—higher values indicate more defects [30]. The obtained I$_D$/I$_G$ values were 1.1 and 0.99 for ZnCl$_2$-BBC and KOH-BBC, respectively, i.e., the graphitization level in the KOH-BBC was slightly higher than in the ZnCl$_2$-BBC [28,30].

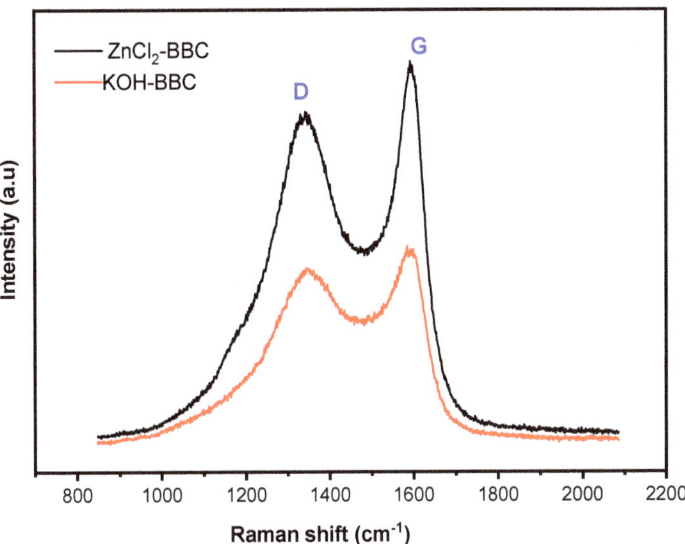

Figure 4. Raman spectra of BBC samples.

The Zeta-potential of both BBCs were negative, with a slightly higher value for the KOH-BBC (see Table 4). The negative charging comes from COO–, –COH–, and –OH– functionalities that can positively affect the adsorption process [40].

FTIR was employed to identify the presence of the functional groups on BBCs samples. The FTIR spectra of the BBC samples are presented in Figure 5. It is possible to identify that the different chemical treatments affected the chemical functionalities on the BBCs. In KOH-BBC, the presence of peaks in between 4000–3600 cm^{-1} represents the O–H stretching vibration in carboxyl and phenol groups [10–12,15]. The sample treated with

KOH also exhibited a sharper and broader transmittance band at 3410–3535 cm^{-1} when compared with the ZnCl$_2$-treated sample, which is assigned to the O–H stretching mode of hydroxyl groups and adsorbed water [11,12,15]. The peaks at 2948 cm^{-1} (asymmetric) and 2875 cm^{-1} (symmetric) are related to the CH– stretching and appeared only in the sample treated with ZnCl$_2$. A new peak at 2373 cm^{-1} is observed only in KOH-BBC, which is assigned to hydrogen-bonded OH. The peaks at around 1542–1574 cm^{-1} are assigned to the asymmetric stretching of O=C of carboxylates. The band at 1138–1160 cm^{-1} are related to CO– of alcohols, and at around 963–1009 cm^{-1} to the OCC—a stretch of an ester is identified [10–12,15]. These functional groups on BBCs surfaces are often related to a good adsorption efficiency process [10–12,15].

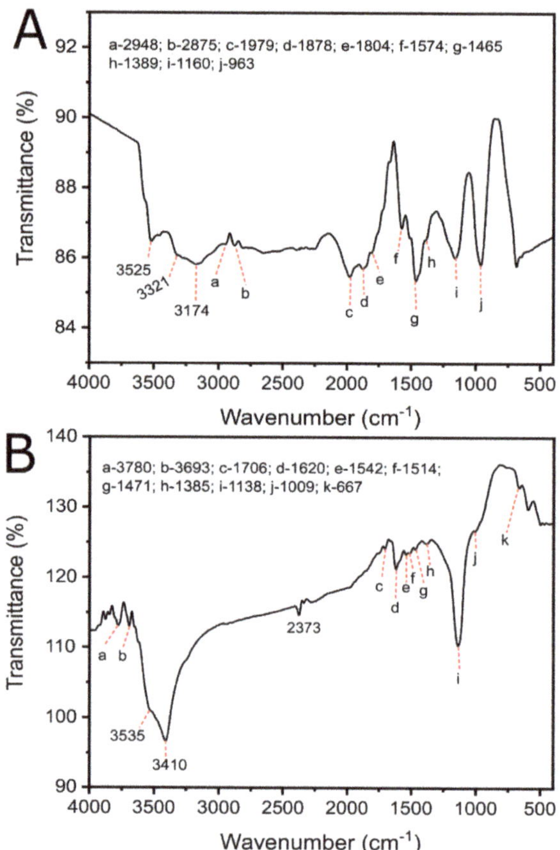

Figure 5. FTIR spectra of BBC samples (**A**) ZnCl$_2$-BBC and (**B**) KOH-BBC.

3.1.3. Water Vapor Adsorption Isotherms, Hydrophilicity Index (HI)

Water vapor adsorption isotherms for both BBCs are shown in Figure 6. According to the IUPAC classification [41], both isotherms are very close to type V, characterized by low levels of water uptake at low relative pressures and the presence of a hysteresis loop over the majority of the pressure range. Adsorption of water vapor was higher for KOH-BBC than for ZnCl$_2$-BBC (see Table 4), indicating a more hydrophilic surface for KOH-BBC than ZnCl$_2$-BBC.

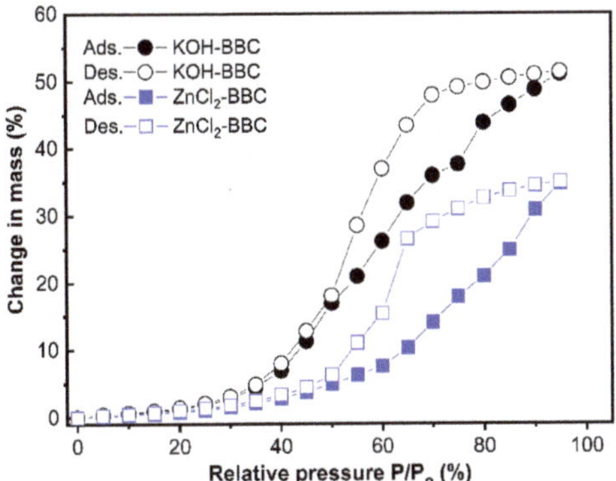

Figure 6. Water sorption isotherms for KOH-BBC and ZnCl$_2$-BBC samples at 25 °C.

The N$_2$ and H$_2$O isotherms differ both in type and shape. Although there is a non-existing correlation between these two techniques, it is worth pointing out that N$_2$ adsorption generated type I isotherms, while H$_2$O adsorption yields isotherms of type V. Different on isotherm curves may be because the process is complex and does not depend only on the porosity. The adsorption of water vapor on biomass materials is known to be dependent on surface chemistry. BBC materials have plenty of surface functional groups, which initiate predominant water adsorption through the hydrogen bonding between a water molecule and surface functional groups.

3.2. Dye Adsorption Analysis

Adsorption Kinetics

The kinetic curves and their parameters are shown in Figure 7 and Table 5, respectively.

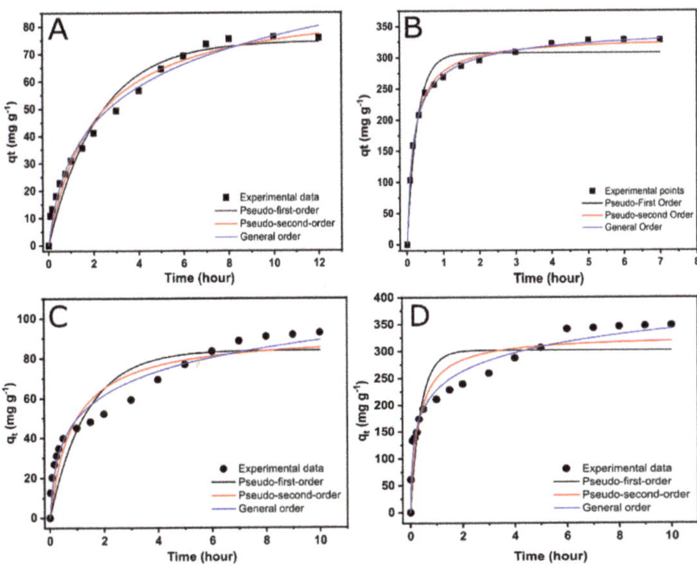

Figure 7. Kinetics of adsorption curves for uptake of RO-16 onto ZnCl$_2$-BBC (**A**) and uptake of RO-16 onto KOH-BBC (**B**), uptake of RB-4 onto ZnCl$_2$-BBC (**C**), uptake of RB-4 onto KOH-BBC (**D**). Initial pH of 5.5 and 4.0 for RO-16 and RB-4, respectively, the adsorbent dosage of 1.5 g L^{-1}. The temperature was 22 °C.

Table 5. Kinetic parameters of RO-16 and RB-4 adsorption onto the BBC samples.

Model	RO-16 Initial Concentration (1000 mg L^{-1})		RB-4 Initial Concentration (1000 mg L^{-1})	
	ZnCl$_2$-BBC	KOH-BBC	ZnCl$_2$-BBC	KOH-BBC
Pseudo-first order				
q_1 (mg g^{-1})	74.71	307.5	84.01	301.7
k_1 (min^{-1})	0.4529	3.502	0.7381	2.489
R^2	0.9639	0.9606	0.8502	0.8182
R^2_{adj}	0.9614	0.9573	0.8408	0.8068
SD (mg g^{-1})	5.107	20.22	11.79	46.03
Pseudo-second order				
q_2 (mg g^{-1})	89.96	332.7	92.62	329.1
k_2 (g mg^{-1} min^{-1})	0.00566	0.01539	0.01277	0.009700
R^2	0.9783	0.9963	0.9102	0.9019
R^2_{adj}	0.9768	0.9960	0.9046	0.8958
SD (mg g^{-1})	3.957	6.217	9.126	33.81
General order				
q_n (mg g^{-1})	78.98	356.3	136.6	355.2
k_n (min^{-1} (g mg^{-1})$^{n-1}$)	1.114×10^{-6}	4.140×10^{-4}	4.964×10^{-7}	2.595×10^{-5}
n (-)	22.69	2.6270	33.08	40.08
R^2	0.9852	0.9985	0.9629	0.9799
R^2_{adj}	0.9838	0.9983	0.9580	0.9831
$t_{0.5}$ (hour)	1.46	0.24	1.57	0.43
$T_{0.95}$ (hour)	6.00	2.98	6.95	3.21
SD (mg g^{-1})	3.311	4.049	8.861	10.22

The general order model had the highest R^2_{adj} and lowest SD values for both dyes on both BBCs (Table 5) and was, therefore, considered as the most suitable model type. The general order kinetic equation gives different values for n (order of adsorption rate) when both dyes—RB-4 and RO-16—concentrations change. Hence, it is hard to make an accurate comparison of the model's kinetic parameters. Therefore, $t_{0.5}$ and $t_{0.95}$ were utilized to compare the RO-16 and RB-4 adsorption kinetics on the ZnCl$_2$-BBC and KOH-BBC carbons. The $t_{0.5}$ and $t_{0.95}$ represent the time (h) when 50% and 95% of saturation (q_e) is achieved, respectively [23–25]. For RO-16 on the ZnCl$_2$-BBC and KOH-BBC samples, $t_{0.5}$ was 1.46 and 0.24 h, respectively, while $t_{0.95}$ was 6.00 and 2.98 h. For RB-4 on the ZnCl$_2$-BBC and KOH-BBC samples, $t_{0.5}$ was 1.57 and 0.43 h, respectively, while $t_{0.95}$ was 6.95 and 3.21 h, respectively (Table 5).

Due to the BBCs' textural properties and chemical surface features, the KOH-BBC had faster kinetics compared to ZnCl$_2$-BBC (Table 5), when the values of $t_{0.5}$ and $t_{0.95}$ are considered. KOH-BBC exhibited a much higher S_{BET} and higher amount of micro and mesopores (see Table 2), and this could also be the reason for the better efficiency in the adsorption process. The RB-4 and RO-16 have molecular sizes of 1.59 and 1.68 nm (see Figure 3B), respectively, and are, therefore, readily adsorbed in micro- (<2 nm) and mesopores (2–50 nm). KOH-BBC also has a more hydrophilic surface (Table 2 and Figure 6), which increases the bulk solution's dispersion and the contact between the dyes and available adsorption sites on the KOH-BBC surface.

The adsorption work was further continued by establishing the contact times such as 6.5 and 3.5 h for ZnCl$_2$-BBC and KOH-BBC for RO-16, respectively; and 7.5 and 3.6 h for ZnCl$_2$-BBC and KOH-BBC for RO-16, respectively. The established contact times were slightly higher than the $t_{0.95}$ to ensure that the adsorption process had enough time to reach the equilibrium between the dyes and the BBCs.

3.3. Equilibrium of Adsorption

The equilibrium curves and their parameters are shown in Figure 8 and Table 6, respectively.

materials. BBC can be regenerated and reused many times without losing adsorption performance [7,29]. However, after being fully saturated, its final disposal or other utilization must be considered once they no longer can be regenerated for water treatment application [7]. The main employed methods to manage used BBC are landfill disposal and incineration [23,24]. However, in some cases, used adsorbents are used as soil fertilizer [23,24], depending of the type of the adsorbate loaded on the BBc surface. These methods are influenced by some factors such as, cost of the adsorbent, type and toxicity of the pollutant, costs involved with the methods including the cost of the combustion and incineration plant, and fees for disposal. Although landfills have typically been used for the disposal of sorbents, as well as soil fertilizers, these methods might have subsequent pollution risk when toxic compounds leach from adsorbents into the soil.

5. Conclusions

The spruce bark BBCs were produced using $ZnCl_2$ and KOH as the activation agents. The BBC characteristics were strongly dependent on the type of activating agent. KOH-BBC had a higher S_{BET} (1067.2 m^2 g^{-1}) and a larger pore volume (0.5584 cm^3 g^{-1}) than $ZnCl_2$-BBC. However, the KOH-BBC had a more developed aromatic structure. KOH treatment generated a BBC with a more well-developed porosity and a higher number of mesopores than $ZnCl_2$-BBC. Additionally, KOH-BBC had a less hydrophobic surface and a higher H and O content than $ZnCl_2$-BBC. However, the carbon yield for KOH-activation was 63% lower than for $ZnCl_2$-activation. For both dyes' adsorption on both BBCs, the general-order model and the Liu model exhibited the best fitness for adsorption kinetics and equilibrium, respectively. The equilibrium Q_{max} at 22 °C was for RO-16 on KOH-BBC and $ZnCl_2$-BBC 354.8 and 90.1 mg g^{-1}, respectively, and for RB-4 582.5 and 332.9 mg g^{-1}. Based on characterization and experimental data, it was suggested that electrostatic interactions and hydrogen bonds between BBCs and RO-16 and RB-4 dyes played the most important role in the adsorption process. In an analysis of removing two synthetic effluents, both BBCs had good outcomes in the percentage; the BBC made with KOH had much better performances. We have shown that efficient and low-priced BBCs can be produced from Norway spruce bark through simple activation procedures. These results call for further studies on underlying mechanisms and how to optimize the treatment procedures for different applications.

Author Contributions: Conceptualization, G.S.d.R.; investigation G.S.d.R.; formal analysis, G.S.d.R. and T.N.P.; data curation, G.S.d.R.; writing—original draft preparation, G.S.d.R.; funding acquisition, S.H.L.; writing—review and editing, S.H.L., M.T., H.P.d.O., E.C.L. and G.L.D. All authors have read and agreed to the published version of the manuscript.

Funding: This research was funded by the Treesearch Postdoctoral program, Bio4Energy—a Strategic Research Environment appointed by the Swedish government, and the Swedish University of Agricultural Sciences.

Institutional Review Board Statement: Not applicable.

Informed Consent Statement: Not applicable.

Data Availability Statement: Not applicable.

Acknowledgments: Lima thanks CNPq, CAPES, and FAPERGS for supporting his researches. The authors are grateful to ChemAxon for giving us an academic research license for the Marvin Sketch software, Version 21.3.0 (http://www.chemaxon.com), 2021 used for molecule physical-chemical properties.

Conflicts of Interest: The authors declare no conflict of interest.

References

1. Kaza, S.; Yao, L.; Bhada-Tata, P.; Woerden, V.F. *What a Waste 2.0: A Global Snapshot of Solid Waste Management to 2050*; International Bank for Reconstruction and Development; The World Bank: Washington, DC, USA, 2018.
2. Kang, C.; Huang, Y.; Yang, H.; Yan, X.F.; Chen, Z.P. A Review of Carbon Dots Produced from Biomass Wastes. *Nanomaterials* **2020**, *10*, 2316. [CrossRef]
3. Hira, S.A.; Yusuf, M.; Annas, D.; Hui, H.S.; Park, K.H. Biomass-Derived Activated Carbon as a Catalyst for the Effective Degradation of Rhodamine B dye. *Processes* **2020**, *8*, 926. [CrossRef]
4. Aldea, J.; Ruiz-Peinado, R.; del Río, M.; Pretzsch, H.; Heym, M.; Brazaitis, G.; Jansons, A.; Metslaid, M.; Barbeito, I.; Bielak, K.; et al. Species stratification and weather conditions drive tree growth in Scots pine and Norway spruce mixed stands along with Europe. *Ecol. Manag.* **2021**, *481*, 118697. [CrossRef]
5. Anerud, E.; Routa, J.; Bergström, D.; Eliasson, L. Fuel quality of stored spruce bark—Influence of semi-permeable covering material. *Fuel* **2020**, *279*, 118467. [CrossRef]
6. Corsi, I.; Winther-Nielsenb, M.; Sethic, R.; Punta, C.; Della Torre, D.; Libralato, G.; Lofrano, G.; Sabatini, L.; Aiello, M.; Fiordi, L.; et al. Ecofriendly nanotechnologies and nanomaterials for environmental applications: Key issue and consensus recommendations for sustainable and eco-safe nano remediation. *Ecotoxicol. Environ. Saf.* **2018**, *154*, 237–244. [CrossRef]
7. Marsh, H.; Reinoso, F.R. *Activated Carbon*; Elsevier: Amsterdam, The Netherlands, 2006.
8. Siipola, V.; Pflugmacher, S.; Romar, H.; Wendling, L.; Koukkari, P. Low-Cost Biochar Adsorbents for Water Purification Including Microplastics Removal. *Appl. Sci.* **2020**, *10*, 788. [CrossRef]
9. Varila, T.; Brännström, H.; Kilpeläinen, P.; Hellström, J.; Romar, H.; Nurmi, J.; Lassi, U. From Norway spruce bark to carbon foams: Characterization and applications. *BioResources* **2020**, *15*, 3651–3666.
10. Thue, P.S.; Umpierres, C.S.; Lima, E.C.; Lima, D.R.; Machado, F.M.; Reis, G.S.d.; Silva, R.S.; Pavan, F.A.; Tran, H.N. Single-step pyrolysis for producing magnetic activated carbon from tucumã (*Astrocaryum aculeatum*) seed and nickel(II) chloride and zinc(II) chloride. Application for removal of nicotinamide and propranolol. *J. Hazard. Mater.* **2020**, *398*, 122903. [CrossRef]
11. He, X.; Zhu, J.; Wang, H.; Zhou, M.; Zhang, S. Surface Functionalization of Activated Carbon with Phosphonium Ionic Liquid for CO_2 Adsorption. *Coatings* **2019**, *9*, 590. [CrossRef]
12. Fröhlich, A.C.; dos Reis, G.S.; Pavan, F.A.; Lima, E.C.; Foletto, E.L.; Dotto, G.L. Improvement of activated carbon characteristics by sonication and its application for pharmaceutical contaminant adsorption. *Environ. Sci. Pollut. Res.* **2018**, *25*, 24713–24725. [CrossRef]
13. Reis, G.S.d.; Adebayo, M.A.; Lima, E.C.; Sampaio, C.H.; Prola, L.D.T. Activated carbon from sewage sludge for preconcentration of copper. *Anal. Lett.* **2016**, *49*, 541–555. [CrossRef]
14. Reis, G.S.D.; Wilhelm, M.; Silva, T.C.A.; Rezwan, K.; Sampaio, C.H.; Lima, E.C.; Souza, S.M.A.G.U. The use of the design of experiments for the evaluation of the production of surface-rich activated carbon from sewage sludge via microwave and conventional pyrolysis. *Appl. Therm. Eng.* **2016**, *93*, 590. [CrossRef]
15. Kasperiski, F.M.; Lima, E.C.; Umpierres, C.S.; Reis, G.S.d.; Thue, P.S.; Lima, D.R.; Dias, S.L.P.; Saucier, C.; da Costa, J.B. Production of porous activated carbons from Caesalpinia ferrea seed pod wastes: Highly efficient removal of captopril from aqueous solutions. *J. Clean. Prod.* **2018**, *197*, 919–929. [CrossRef]
16. Cunha, M.R.; Lima, E.C.; Cimirro, N.F.G.M.; Thue, P.S.; Dias, S.L.P.; Gelesky, M.A.; Dotto, G.L.; Reis, G.S.d.; Pavan, F.A. Conversion of Eragrostis plana Nees leaves to activated carbon by microwave-assisted pyrolysis for the removal of organic emerging contaminants from aqueous solutions. *Environ. Sci. Pollut. Res.* **2018**, *25*, 23315–23327. [CrossRef]
17. Leite, A.B.; Saucier, C.; Lima, E.C.; dos Reis, G.S.; Umpierres, C.S.; Mello, B.L.; Shirmardi, M.; Dias, S.L.P.; Sampaio, C.H. Activated carbons from avocado seed: Optimisation and application for removal several emerging organic compounds. *Environ. Sci. Pollut. Res.* **2018**, *25*, 7647–7661. [CrossRef] [PubMed]
18. Chequer, F.M.D.; de Oliveira, G.A.R.; Anastacio Ferraz, E.R.; Carvalho, J.; Zanoni, M.V.B.; Oliveira, D.P. Textile dyes: Dyeing process and environmental impact. In *Eco-Friendly Textile Dyeing and Finishing*; Gunay, M., Ed.; InTech: London, UK, 2013. [CrossRef]
19. Carneiro, P.A.; Umbuzeiro, G.A.; Oliveira, D.P.; Zanoni, M.V.B. Assessment of water contamination caused by a mutagenic textile effluent/dyehouse effluent bearing disperse dyes. *J. Hazard. Mater.* **2010**, *174*, 694–699. [CrossRef] [PubMed]
20. Lima, V.V.C.; Dalla Nora, F.B.; Peres, E.C.; Reis, G.S.; Lima, É.C.; Oliveira, M.L.S.; Dotto, G.L. Synthesis and characterization of biopolymers functionalized with APTES (3–aminopropyltriethoxysilane) for the adsorption of sunset yellow dye. *J. Environ. Chem. Eng.* **2019**, *7*, 103410. [CrossRef]
21. Dos Reis, G.S.; Adebayo, M.A.; Sampaio, C.H.; Lima, E.C.; Thue, P.S.; de Brum, I.A.S.; Dias, S.l.P.; Pavan, F.A. Removal of Phenolic Compounds from Aqueous Solutions Using Sludge-Based Activated Carbons Prepared by Conventional Heating and Microwave-Assisted Pyrolysis. *Water Air Soil Pollut* **2017**, *228*, 33. [CrossRef]
22. Umpierres, C.S.; Thue, P.S.; Lima, E.; dos Reis, G.S.; de Brum, I.A.S.; de Alencar, W.A.; Dias, S.L.P.; Dotto, G.L. Microwave-activated carbons from tucumã (*Astrocaryum aculeatum*) seed for efficient removal of 2-nitrophenol from aqueous solutions. *Environ. Technol.* **2018**, *39*, 1173–1187. [CrossRef]
23. Dos Reis, G.S.; Lima, E.C.; Sampaio, C.H.; Rodembusch, F.S.; Petter, C.O.; Cazacliu, B.G.; Dotto, G.L.; Hidalgo, G.E.N. Novel kaolin/polysiloxane based organic-inorganic hybrid materials: Sol−gel synthesis, characterization, and photocatalytic properties. *J. Solid State Chem.* **2018**, *260*, 106–116. [CrossRef]

24. Dos Reis, G.S.; Thue, P.S.; Cazacliu, B.G.; Lima, E.C.; Sampaio, C.H.; Quattrone, M.; Ovsyannikova, E.; Kruse, A.; Dotto, G.L. Effect of concrete carbonation on phosphate removal through adsorption process and its potential application as fertilizer. *J. Clean Prod.* **2020**, *256*, 120416. [CrossRef]

25. Dos Reis, G.S.; Cazacliu, B.G.; Correa, C.R.; Ovsyannikova, E.; Kruse, A.; Sampaio, C.H.; Lima, E.C.; Dotto, G.L. Adsorption and recovery of phosphate from aqueous solution by the construction and demolition wastes sludge and its potential use as phosphate-based fertilizer. *J. Environ. Chem. Eng.* **2020**, *8*, 103605. [CrossRef]

26. Ma, Y. Comparison of Activated Carbons Prepared from Wheat Straw via $ZnCl_2$ and KOH Activation. *Waste Biomass Valor.* **2017**, *8*, 549–559. [CrossRef]

27. Ma, J.; Yang, H.; Li, L.; Xie, X.; Liu, B.; Zhang, L.; Zhang, L. Synthesis of aligned ZnO submicron rod arrays by heating zinc foil covered with $ZnCl_2$ solution. *Acta. Chimi. Sin.* **2009**, *67*, 1515–1522.

28. Song, X.; Ma, X.; Li, Y.; Ding, L.; Jiang, R. Tea waste-derived microporous active carbon with enhanced double-layer supercapacitor behaviors. *Appl. Surf. Sci.* **2019**, *487*, 189–197. [CrossRef]

29. Efeovbokhan, V.E.; Alagbe, E.E.; Odika, B.; Babalola, R.; Oladimeji, T.E.; Abatan, O.G.; Yusuf, E.O. Preparation and characterization of activated carbon from plantain peel and coconut shell using biological activators. *J. Phys. Conf. Ser.* **2019**, *1378*, 032035. [CrossRef]

30. Huang, G.G.; Liu, Y.F.; Wu, X.-X.; Cai, J.-J. Activated carbons prepared by the KOH activation of a hydrochar from garlic peel and their CO_2 adsorption performance. *New Carbon Mater.* **2019**, *34*, 247–257. [CrossRef]

31. Cuong, D.V.; Liu, N.-L.; Nguyen, V.A.; Hou, C.-H. Meso/micropore-controlled hierarchical porous carbon derived from activated biochar as a high-performance adsorbent for copper removal. *Sci. Total Environ.* **2019**, *692*, 844–853. [CrossRef]

32. Lima, D.R.; Bandegharaei, A.H.; Thue, P.S.; Lima, E.C.; de Albuquerque, Y.R.T.; dos Reis, G.S.; Umpierres, C.S.; Dias, S.L.P.; Tran, H.N. Efficient acetaminophen removal from water and hospital effluents treatment by activated carbons derived from Brazil nutshells. *Colloids Surface A* **2019**, *583*, 123966. [CrossRef]

33. Dos Reis, G.S.; Mahbub, M.K.B.; Wilhelm, M.; Lima, E.C.; Sampaio, C.H.; Saucier, C.; Dias, S.L.P. Activated carbon from sewage sludge for removal of sodium diclofenac and nimesulide from aqueous solutions. *Korean J. Chem. Eng.* **2016**, *33*, 3149–3161. [CrossRef]

34. Sun, L.; Tian, C.; Li, M.; Meng, X.; Wang, L.; Wang, R.; Yin, J.; Fu, H. From coconut shell to porous graphene-like nanosheets for high-power supercapacitors. *J. Mater. Chem. A* **2013**, *1*, 6462–6470. [CrossRef]

35. Correa, C.R.; Otto, T.; Kruse, A. Influence of the biomass components on the pore formation of activated carbon. *Biomass Bioenergy* **2017**, *97*, 53–64. [CrossRef]

36. Duan, X.-H.; Srinivasakannan, C.; Yang, K.-B.; Peng, J.-H.; Zhang, L.-B. Effects of Heating Method and Activating Agent on the Porous Structure of Activated Carbons from Coconut Shells. *Waste Biomass Valor.* **2012**, *3*, 131–139. [CrossRef]

37. Ravichandran, P.; Sugumaran, P.; Seshadri, S.; Basta, A.H. Optimizing the route for the production of activated carbon from *Casuarina equisetifolia* fruit waste. *R. Soc. Open Sci.* **2018**, *5*, 171578. [CrossRef] [PubMed]

38. Caturla, F.; Molina-Sabio, M.; Rodriguez-Reinoso, F. Preparation of activated carbon by chemical activation with $ZnCl_2$. *Carbon* **1991**, *29*, 999–1007. [CrossRef]

39. Ucar, S.; Erdem, M.; Tay, T.; Karago, S. Preparation and characterization of activated carbon produced from pomegranate seeds by $ZnCl_2$ activation. *Appl. Surf. Sci.* **2009**, *255*, 8890–8896. [CrossRef]

40. Xu, D.; Cao, J.; Li, L.; Howard, A.; Yu, K. Effect of pyrolysis temperature on characteristics of biochars derived from different feedstocks: A case study on ammonium adsorption capacity. *Waste Manag.* **2019**, *87*, 652–660. [CrossRef]

41. Thommes, M.; Kaneko, K.; Neimark, A.V.; Olivier, J.P.; Rodriguez-Reinoso, F.; Rouquerol, J. Physisorption of gases, with special reference to the evaluation of surface area and pore size distribution (IUPBBC technical report). *Pure Appl. Chem.* **2015**, *87*, 1051–1069. [CrossRef]

42. Machado, F.M.; Bergmann, C.P.; Lima, E.C.; Royer, B.; de Souza, F.E.; Jauris, I.M.; Calvete, T.; Fagan, S.B. Adsorption of Reactive Blue 4 dye from water solutions by carbon nanotubes: Experiment and theory. *Phys. Chem. Chem. Phys.* **2012**, *14*, 11139–11153. [CrossRef]

43. Vakili, M.; Rafatullah, M.; Ibrahim, M.H.; Adullah, A.Z.; Salamatinia, B.; Gholami, Z. Chitosan hydrogel beads impregnated with hexadecyl amine for improved reactive blue-4 adsorption. *Carbo. Polym.* **2016**, *137*, 139–146. [CrossRef]

44. Dison, S.P.F.; Tanabe, E.H.; Bertuol, D.A.; dos Reis, G.S.; Lima, E.C.; Dotto, G.L. Alternative treatments to improve the potential of rice husk as adsorbent for methylene blue. *Water Sci. Technol.* **2017**, *75*, 296–305.

45. Teixeira, R.A.; Lima, E.C.; Benetti, A.D.; Thue, P.S.; Cunha, M.R.; Cimirro, N.F.G.M.; Sher, F.; Dehghani, M.H.; dos Reis, G.S.; Dotto, G.L. Preparation of hybrids of wood sawdust with 3-aminopropyl-triethoxysilane. Application as an adsorbent to remove Reactive Blue 4 dye from wastewater effluents. *J. Taiwan Inst. Chem. Eng.* **2021**. [CrossRef]

46. Marrakchi, F.; Khandaya, W.A.; Asif, M.; Hameed, B.H. Cross-linked chitosan/sepiolite composite for the adsorption of methylene blue and reactive orange 16. *Inter. J. Biol. Macro.* **2016**, *93*, 1231–1239. [CrossRef]

47. Marrakchi, F.; Ahmed, M.J.; Khanday, W.A.; Asif, M.; Hameed, B.H. Mesoporous carbonaceous material from fish scales as a low-cost adsorbent for reactive orange 16 adsorption. *J. Taiwan Inst. Chem. Eng.* **2017**, *71*, 47–54. [CrossRef]

48. Calvete, T.; Lima, E.C.; Cardoso, N.F.; Vaghetti, J.C.P.; Dias, S.L.P.; Pavan, F.A. Application of carbon adsorbents prepared from Brazilian-pine fruit shell for the removal of reactive orange 16 from aqueous solution: Kinetic, equilibrium, and thermodynamic studies. *J. Environ. Manag.* **2010**, *91*, 1695–1706. [CrossRef]

49. Akbar, A.A.M.; Karthikeyan, R.K.; Sentamil, S.M.; Mithilesh, K.R.; Madhangi, P.; Maheswari, N.; Janani, S.G.; Padmanaban, V.C.; Singh, R.S. Removal of Reactive Orange 16 by adsorption onto activated carbon prepared from rice husk ash: Statistical modelling and adsorption kinetics. *Sep. Sci. Technol.* **2020**, *55*, 26–34.

50. Shah, J.A.; Butt, T.A.; Mirza, C.R.; Shaikh, A.J.; Khan, M.S.; Arshad, M.; Riaz, N.; Haroon, H.; Gardazi, S.M.H.; Yaqoo, K.; et al. Phosphoric Acid Activated Carbon from Melia azedarach Waste Sawdust for Adsorptive Removal of Reactive Orange 16: Equilibrium Modelling and Thermodynamic Analysis. *Molecules* **2020**, *25*, 2118. [CrossRef] [PubMed]

51. Malakootian, M.; Heidari, M.R. Reactive orange 16 dye adsorption from aqueous solutions by psyllium seed powder as a low-cost biosorbent: Kinetic and equilibrium studies. *Appl. Water Sci.* **2018**, *8*, 212. [CrossRef]

52. Auta, M.; Hameed, B.H. Optimized and functionalized paper sludge activated with potassium fluoride for single and binary adsorption of reactive dyes. *J. Ind. Eng. Chem.* **2014**, *20*, 830–840. [CrossRef]

53. Ramachandran, P.; Vairamuthu, R.; Ponnusamy, S. Adsorption isotherms, kinetics, thermodynamics and desorption studies of reactive orange 16 on activated carbon derived from *Ananas comosus* (L.) carbon. *J. Eng. Appl. Sci.* **2011**, *6*, 15–22.

54. Won, S.W.; Choi, S.B.; Yun, Y.S. Performance and mechanism in the binding of reactive orange 16 to various types of sludge. *Biochem. Eng. J.* **2006**, *28*, 208–214. [CrossRef]

55. Ahmad, M.A.; Rahman, N.K. Equilibrium, kinetics and thermodynamic of Remazol Brilliant Orange 3R dye adsorption on coffee husk-based activated carbon. *Chem. Eng. J.* **2011**, *170*, 154–161. [CrossRef]

56. Deshuai, S.; Zhongyi, Z.; Mengling, W.; Yude, W. Adsorption of reactive dyes on activated carbon developed from Enteromorpha prolifera. American. *J. Anal. Chem.* **2013**, *4*, 17–26.

57. Aguiar, J.E.; Bezerra, B.T.C.; Braga, B.M.; Lima, P.D.S.; Nogueira, R.E.F.Q.; de Lucena, S.M.P. Adsorption of anionic and cationic dyes from aqueous solution on non-calcined Mg-Al layered double hydroxide: Experimental and theoretical study. *Sep. Sci. Technol.* **2013**, *48*, 2307–2316. [CrossRef]

58. Periolatto, M.; Ferrero, F. Cotton filter fabrics functionalization by chitosan UV-grafting for removal of dyes. *Chem. Eng. Trans.* **2013**, *32*, 1–6.

59. Plum, L.M.; Rink, L.; Haase, H. The Essential Toxin: Impact of Zinc on Human Health. *Int. J. Environ. Res. Public Health* **2010**, *7*, 1342–1365. [CrossRef]

60. Wang, C.; Wu, D.; Wang, H.; Gao, Z.; Xu, F.; Jiang, K. A green and scalable route to yield porous carbon sheets from biomass for supercapacitors with high capacity. *J. Mater. Chem. A* **2018**, *6*, 1244. [CrossRef]

Article

Elaboration by Wrapping Process and Multiscale Characterisation of Thermoplastic Bio-Composite Based on Hemp/PA11 Constituents

Chaimae Laqraa *, Manuela Ferreira, Ahmad Rashed Labanieh and Damien Soulat

Centrale Lille, Laboratoire de Génie et Matériaux, Gemtex, Ensait, F-59000 Roubaix, France;
manuela.ferreira@ensait.fr (M.F.); ahmad.labanieh@ensait.fr (A.R.L.); damien.soulat@ensait.fr (D.S.)
* Correspondence: chaimae.laqraa@ensait.fr; Tel.: +33-7-67-49-99-22

Abstract: The present work investigates the potential of developing bio-composites based on thermoplastic polymers reinforced with natural fibres by using hybrid yarns. The hybrid yarns were produced by the wrapping technique, in which a multifilament of polyamide 11 (PA11) was wrapped around an untreated low-twisted hemp roving to produce a yarn with sufficient tenacity and stiffness for the next step of weaving. The tensile behaviour of the wrapped yarns was identified both in the dry- and thermo-state. Then, two different fabrics were woven and tested to study the influence of yarn densities and weave diagrams on the tensile and flexural properties. At this fabric scale, properties of fabrics made from hybrid yarns were compared with those of fabrics from a previous study made from 100% hemp roving. Composites made from these fabrics, with stacking of two cross-plies, were produced by thermocompression and characterised regarding mechanical strength.

Keywords: hybrid yarns; hemp; PA11; woven fabric; bio-based composite; mechanical characterisation

Citation: Laqraa, C.; Ferreira, M.; Rashed Labanieh, A.; Soulat, D. Elaboration by Wrapping Process and Multiscale Characterisation of Thermoplastic Bio-Composite Based on Hemp/PA11 Constituents. *Coatings* **2021**, *11*, 770. https://doi.org/10.3390/coatings11070770

Academic Editor: Fengwei (David) Xie

Received: 1 June 2021
Accepted: 24 June 2021
Published: 26 June 2021

Publisher's Note: MDPI stays neutral with regard to jurisdictional claims in published maps and institutional affiliations.

1. Introduction

Among plant fibres, flax (*Linum Usitatissimum* L.) and hemp (*Cannabis Sativa*) fibres are now the two most-produced bast fibres in Europe [1]. Due to their properties, such as important environmental advantages, good specific mechanical properties, and often a viable cost [1], these fibres have emerged as an alternative to synthetic fibres, and the use of plant fibre composites (PFCs) has become a market reality [2–4]. Despite numerous similarities (cell wall, thicknesses, and numbers of layers/sub-layers, biochemical composition, cellulose microfibril angle, MFA), these two bast fibres generally exhibit differences in their tensile properties and their global tensile behaviour. Flax fibres generally have slightly better tensile properties than hemp fibres, especially in terms of tensile strength and stiffness; on the other hand, they reach a lower tensile strain at failure than hemp [5]. If a lot of studies are conducted at the scale of fibres to explain these differences [1,5–7], few papers are dedicated to the development of hemp-continuous fabrics used as reinforcement of composite samples [4,8,9]. This deficit for hemp could be attributed to technological barriers, such as fibre separation and the alignment of fibres throughout the transformation process and consequently the unavailability of these products on an industrial scale [10]. Hemp fibres are naturally discontinuous; therefore, hemp reinforcements have so far been based on twisted yarns of staple fibres by means of long-staple spinning techniques, mainly ring spinning. However, the high twist level in the yarns leads to fibre misalignment in composite materials and thus reduced stiffness. Furthermore, the high twist level compacts the yarn section and reduces the inter-fibre gaps, making it very difficult for the resin to penetrate inside the yarn structure [11]. Therefore, the use of very low twisted yarns is advised for composite application [12,13]. However, a low twist implies poor inter-fibre cohesion, and the yarn loses its otherwise good weaveability properties which are due to good tenacity and low hairiness.

Along with this fibrous reinforcement, the challenge is to identify new combinations of raw materials for the production of green composites whose performance is good enough to propose their use in suitable applications. Among the various bio-based thermoplastic resins, polyamide 11 (PA11) is a semi-crystalline bio-polyamide produced using 11-aminoundecanoic acid derived from castor oil and has gained a special industrial interest due to a good combination of mechanical properties and chemical resistance [14]. In particular, PA11 exhibits good toughness, compared to other bio-based thermoplastic resins, such as, polylactic acid (PLA), which is often proposed as a matrix for bio-composites [15,16]. The natural fibre/PA11 combination has been used to study the performance of bio-composites. Haddou et al. [17] associated long bamboo fibres with PA11 films to analyse the tensile behaviour. Gourrier et al. [18] studied the tensile, impact, and thermal properties of unidirectional flax tape with PA11 films. In these studies, composites were made by film stacking, but other processes offer a route for efficient manufacturing of thermoplastic composites due to the reduced flow distance of resin in reinforcement to optimise impregnation. Awais et al. [19] compared tensile, flexural and impact behaviour of commingled fabrics (woven and knitted) based on jute/flax/hemp fibres with PP yarns. To improve the impregnation of thermoplastic resin into fibre yarns, Kobayashi et al. [20] used the micro-braiding method to mix hemp roving and PLA multifilament in the yarns. These micro-braided yarns were placed in a pre-heated moulding die for consolidation by compression moulding to produce composite specimens. Zhai et al. [21] compared yarn morphologies, structures, mechanical tensile properties, and braidabilities of commingled flax/PP yarns obtained by micro-braiding or wrapping methods. In the wrapping process [22,23], a thermoplastic multifilament is wrapped around hemp roving, resulting in increased inter-fibre friction and improved yarn cohesion. This manufacturing process was successfully used by Corbin et al. [8] to produce a commingled yarn based on hemp roving and PA12 multifilament and associated woven fabrics and composite samples. In all of these studies, although the mechanical and thermal properties of thermoplastic polymers are described, as in Di Lorenzo et al. [24], few papers [25] deal with the identification of the thermomechanical properties of commingled yarns, which can be essential to improve parameters of impregnation during the thermocompression process. This paper deals with hemp roving and PA11 multifilament and describes the wrapping process, and how the hybrid yarns are used to weave fabrics. Composite samples reinforced by these fabrics were manufactured by thermocompression. The tensile properties of these commingled yarns were studied according to temperature and strain rates, and the mechanical properties of the woven fabrics and the composite samples were identified.

2. Materials and Methods

2.1. Materials

An extruded PA11 multifilament yarn produced at GEMTEX Laboratory was used as the matrix material and as the wrapping material for the production of the hybrid yarn. The PA11 yarn was made from Rilsan® PA11 pellets supplied by Arkema, Colombes, France. This thermoplastic matrix has a density of 1.03 g/cm³, a melt temperature of 190–195 °C, and a glass transition temperature of 55–60 °C. Hemp roving with a twist of 37 turns per meter (TPM) was used as the core yarn of the hybrid yarns. This untreated roving was supplied by the Italian company Linificio e Canapificio Nazionale, Villa d'Almè, Italy. The main properties of these yarns are shown in Table 1.

Table 1. The main properties of the raw materials.

Yarns	Density (g/cm³)	Linear Density (Tex)	Twist Level (tpm)	Tenacity (cN/Tex)	Deformation at Break (%)
Hemp roving	1.50	312 ± 19	37 ± 2	10.09 ± 2.48	3.67 ± 0.42
PA11	1.03	111 ± 3	–	27.02 ± 0.95	35.02 ± 5.43

Their linear density was measured according to the NF G07-316 standard [26], the twist level according to the NF G07-079 standard [27], and tenacity at break according to the NF EN ISO 2062 standard [28].

2.2. Methods

2.2.1. Manufacturing of Hemp/PA11 Hybrid Yarns

The hemp/PA11 hybrid yarns were produced by the wrapping process, on a hollow spindle machine, Gualchieri e Gualchieri shown in Figure 1. In the wrapping process, the thermoplastic multifilament yarns (PA11) are wrapped around low twisted and untreated hemp roving. These wrapping yarns will melt during the thermocompression process, thus forming the matrix part of the composite material. The wrapping process is mainly used to increase the inter-fibre cohesion of the core yarn to sustain the tension loads applied during the transformation into reinforcement. Furthermore, the outer wrapped yarn protects the core yarn from rubbing aggression when passing machinery surfaces during these processes. These manufacturing loads represent the basic required weaveability conditions.

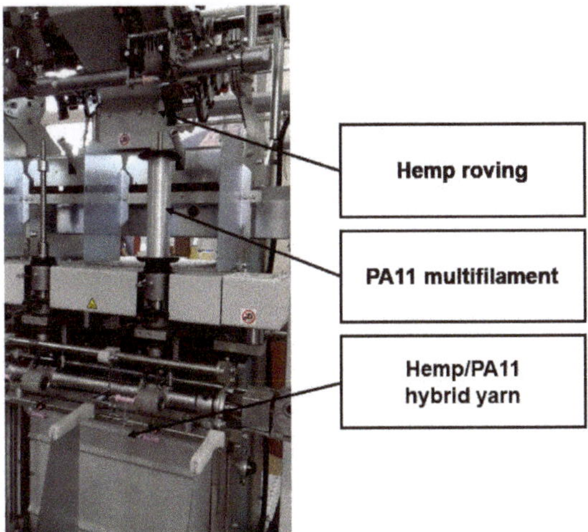

Figure 1. The wrapping process.

During production, the hemp roving passes through a roving condenser and drafting rollers and then is guided inside the hollow spindle together with the PA11. The PA11 is wrapped around the core roving by rotational movement of the hollow spindle to obtain the wrapped hemp/PA11 yarn. During the wrapping process, the ratio between hollow spindle rotational speed and yarn delivery speed determines the PA11 wrapping turns per meter and the mass proportion of two yarns for a given linear density. As output, to obtain the desired hybrid yarns with a mass ratio of hemp between 43% and 60% and a tenacity allowing its weaveability, the wrapping turn number and spindle speed are fixed at 500 tpm and 5000 turns/min, respectively. The linear densities of the hybrid yarns were measured according to the NF G07-316 standard [26], and the twist level according to the NF G07-079 standard [27].

2.2.2. Weaving Process with Hemp/PA11 Yarns

The hybrid yarns were woven on a Leclerc Weavebird manual loom in the GEMTEX laboratory (Roubaix, France). After the different preparation steps of the weaving process (warping and drawing-in), two woven fabrics with different weaving diagrams were produced, twill 6 weft effect and satin 6 weft effect. The two fabrics were woven with the same hybrid yarns in both directions and have the same warp density at 6 yarns/cm but different weft densities, namely, 11 yarns/cm for twill 6 and 6 yarns/cm for satin 6.

The textile properties of these fabrics in terms of yarn densities, areal density, thickness, and air permeability were identified and will be presented later. Areal density was measured according to the NF EN 12127 standard [29], thickness according to NF EN ISO 5084 standard [30], and air permeability according to the NF EN ISO 9237 [31]. The crimp level of warp and weft yarns after weaving was measured for both fabrics according to the NF ISO 7211-3 standard [32].

2.2.3. Composite Manufacturing

Composite plates based on hemp/PA11 satin 6 fabric and hemp/PA11 twill 6 fabric were manufactured by thermocompression moulding on an Agila Press 100 kN hot-press (Menen, Belgium). The fabrics were cut into 300×300 mm^2 squares and conditioned at a temperature of 23 °C and relative humidity of 50% for at least 24 h prior to the composite manufacturing. For the two types of fabrics, two cross-plies (0/90°) were stacked, as shown in Figure 2a. Then, they were placed between two Teflon-coated plates. The melting temperature of PA11 is 190 °C and the degradation temperature of hemp roving is 276 °C [8]; therefore, to produce the composite plates, the process temperature was fixed at 200 °C to preserve the hemp fibre properties and avoid its degradation. The composites were then prepared by pressing the layers at a temperature of 200 °C and according to the cycle presented in Figure 2b. These temperature and pressure cycles are specific to the combination of constituents involved (hemp/PA11) in the commingled yarn and to the stacking chosen in this study (two cross-plies of woven fabrics). Specimens for the mechanical testing were cut according to the appropriate standard, given later, and before the testing, they were conditioned for at least 24 h at 23 °C and 50% relative humidity.

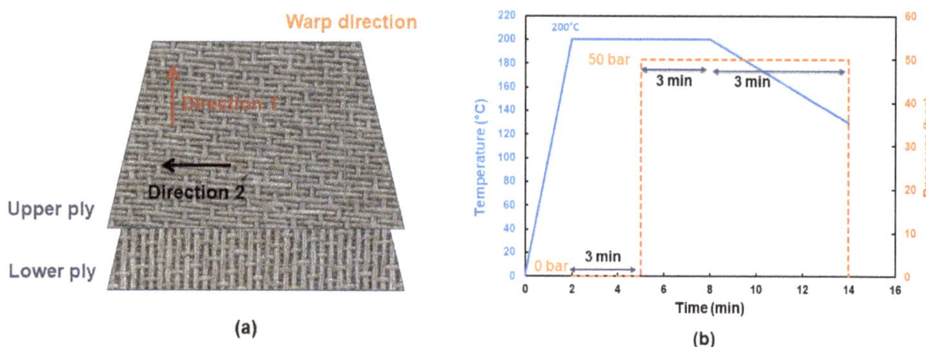

Figure 2. (**a**) The stacking of fabrics, adapt from: [8]; (**b**) the thermocompression moulding cycle.

2.3. Characterisation Steps

2.3.1. Properties of the Hybrid Yarns

First, dry-state tensile tests were performed on the hybrid yarns at different test speeds in order to investigate the influence of this parameter on the behaviour of the hybrid yarns. The tests were performed on an MTS Criterion 45 universal tensile apparatus (Eden Prairie, MN, USA) according to the NF EN ISO 2062 standard [28] with a gauge length of 200 mm, without pretension, and with three different test speeds: 20, 100, and 200 mm/min. For each crosshead speed, the test was repeated 20 times. In addition, tensile tests were also conducted on the hemp roving, but only at 200 mm/min test speed, to compare its behaviour with the hybrid yarns.

Then, thermo-state tensile tests were performed on these yarns in order to study the effect of temperature on the yarn properties and to understand the behaviour of these hybrid yarns at the chosen temperature. Tensile tests were then conducted on the same machine as the dry-state tests, using an isothermal oven. Due to the size of the oven, the gauge length of the yarn was 100 mm, the crosshead speed was 100 mm/min, and the temperature was set at 50, 70, 100, 120, and 150 °C. The single yarn was first inserted in the oven and clamped between the two jaws, and then the desired temperatures were set. Once the temperature was reached and stabilised, the test started.

2.3.2. Properties of the Produced Fabrics

Tensile tests were performed on the woven fabrics according to the NF EN ISO 13934-1 standard [33] and were carried out on an MTS Criterion 45 machine at ambient temperature. The two main directions of the woven fabrics were tested and for each direction, five samples were used, with a length of 300 mm and a width of 50 mm. For the tensile test, the gauge length was 200 mm, the test speed was 20 mm/min, and the preload wad 5 N. To present the properties of the two woven fabrics, we studied the maximum load/yarn and strain at maximum load for each fabric in order to eliminate the effects of the density of yarns. The bending rigidity of the fabrics was also identified by using a cantilever apparatus according to the ISO 4604 (05) standard [34], with the same samples used for the tensile tests (Figure 3). As described in the literature [35,36], the fabric sample is progressively advanced until the end, under its own weight, is in contact with the inclined plane at 41.5°. Then, the overhang length of the fabric is measured and the bending stiffness is computed according to Equation (1) as follows:

$$G = 9.81 \times m_s \times \left(\frac{l_m}{2}\right)^3 \tag{1}$$

where G is the bending stiffness coefficient (N·mm), m_s is the areal density of the fabric (g/m^2), and l_m is the overhanging length (m).

Figure 3. Cantilever test bench.

2.3.3. Properties of the Composite Plates

The tensile and flexural behaviours of hemp/PA11 composites were tested in an MTS Criterion 45 universal testing machine. The tensile properties of the composite specimens were performed according to the ASTM D3039-00 standard [37], while the flexural properties were tested according to the NF EN ISO 14125 standard [38]. The tensile tests were carried out over a gauge length of 150 mm and a test speed of 1 mm/min. The reported modulus was calculated (between 0% and 1% strain) for all composite samples. Thus, the strength and strain at break were calculated from the recorded force–displacement curves. For flexural behaviour, tests were performed at a test speed of 1 mm/min, with specimens of 25 mm width and 64 mm span length. For both tensile and flexural tests, five samples were tested in the two main directions of the composite plates which are direction 1 (associated with the warp direction of the upper ply) and direction 2 (associated with the weft direction of the upper ply).

3. Results and Discussion

3.1. Yarn Properties

3.1.1. Textile Properties of the Hybrid Yarns

The measured textile properties of the hybrid hemp/PA11 yarns and hemp roving are listed in Table 2. The hybrid yarn has a higher linear density than hemp roving, as a result of the addition of the thermoplastic multifilament PA11 (111 ± 3 Tex). Moreover, adding the wrapped PA11 multifilament leads to a decrease in the hybrid yarn hairiness. This decrease is due to the removal of most impurities during the wrapping process. Figure 4 presents the visual aspect of the core untreated hemp roving and the wrapped yarn. There are no more defects on the surface of the hybrid yarn in comparison to the hemp roving. After wrapping, the yarn structure becomes more compact and uniform, and its section is more circular. In addition, the wrapping process preserves the structure of the core roving and its properties by creating a mechanical bond between the two materials, unlike conventional methods used to improve the interfacial bond with polymer matrices, which use chemical, physical or biological treatments instead [39]. This hybrid yarn has, in weight, 50% of hemp fibre and 50% of PA11.

Table 2. The textile properties of the hybrid yarns and hemp roving.

Yarns	Linear Density (Tex)	Twist Level (tpm)	Hairiness (H ± sh)	Hemp Fibre Mass Fraction (%)
Hybrid yarns	486 ± 9	400	10.05 ± 2.86	50
Hemp roving	313 ± 19	37 ± 2	18.96 ± 3.31	100

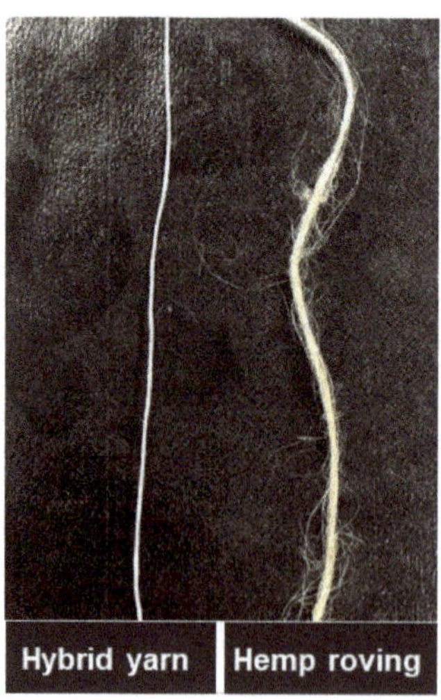

Figure 4. The visual aspect of a hybrid yarn and a hemp roving.

3.1.2. Dry-State Tensile Behaviour

- Comparison of Hemp Roving and Hemp/PA11 Hybrid Yarn Behaviour

Figure 5 shows the tenacity–strain curve of the hybrid yarn and the hemp roving at ambient temperature and at a speed of 200 mm/min. The behaviour of the hybrid yarn has the first peak of tenacity at a strain of around 3%, corresponding to the roving breakdown. At this first peak, tenacity reaches 8 cN/Tex. This peak is followed by a high deformation phase before the full breakdown of the hybrid yarn at a strain of 35% (Figure 6a), the same as that of the PA11 multifilament (Table 1). This part of the curve can be attributed to the elongation of the PA11 multifilament, with compacting of the hemp roving around which it is wrapped, as the load increases. By increasing the elongation, occasional slippage can occur between the PA11 filaments and the broken hemp roving, and the filaments can be stretched unevenly, leading to their rupture that appears on the tenacity–strain curve as a non-smooth part. In this part of the curve, some small breaks are detected by the load cell and involve a high standard deviation between samples on the measured load.

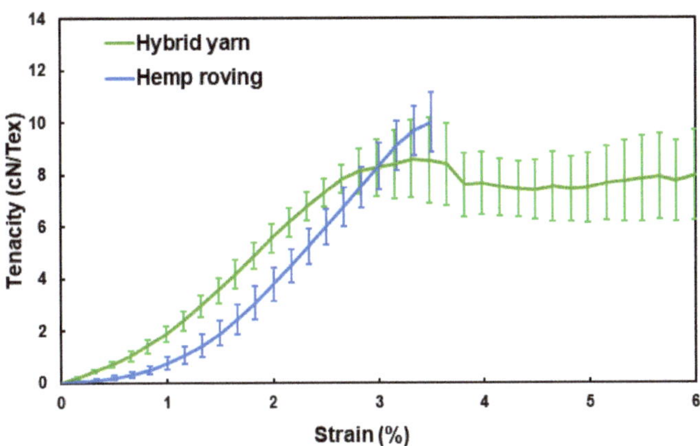

Figure 5. The tensile behaviour of the hybrid yarn and hemp roving alone.

From these results, it can be concluded that the wrapped multifilament PA11 does not induce a significant increase in the roving tenacity and rigidity. During the first phase (low tenacity), the roving curve corresponds to the fibre redressing in the axial load direction. This results from the roving tension applied during the wrapping process. Thus, even if the hybrid yarn did not reach the required tenacity for weaving, which is 15 cN/Tex, it can be woven because adding PA11 during the wrapping process increases the inter-fibre cohesion and protects them from damage during the weaving process. Weaveability of natural fibre reinforcement can be also improved by other methods, as described in [8,40,41].

- Tensile Behaviour of the Hybrid Yarns at Different Test Speeds

Figure 6a shows the tensile behaviour of the produced hybrid yarns for different test speeds (20, 100, and 200 mm/min) at ambient temperature. For these different speed tests, the shape of the tenacity–strain curve remains the same. High rigidity is observed in the first phase up to a peak, followed by a non-smooth part characterised by a high elongation and fluctuation of the tenacity around the tenacity of the first peak. However, a dependence of the tenacity at the first peak on the test speed can be seen in Figure 6b. At the 200 mm/min speed, the tenacity is 70% higher than at 20 mm/min. In addition, the behaviour of yarns at 100 and 200 mm/min is almost the same: in the range of 0 to 5% strain with a difference of 10% for the tenacity. These results match the results obtained in previous studies conducted on commingled glass/PP and on flax/PA12 yarns at different test speeds and in which the tenacity increases with increasing the test speed [25,35].

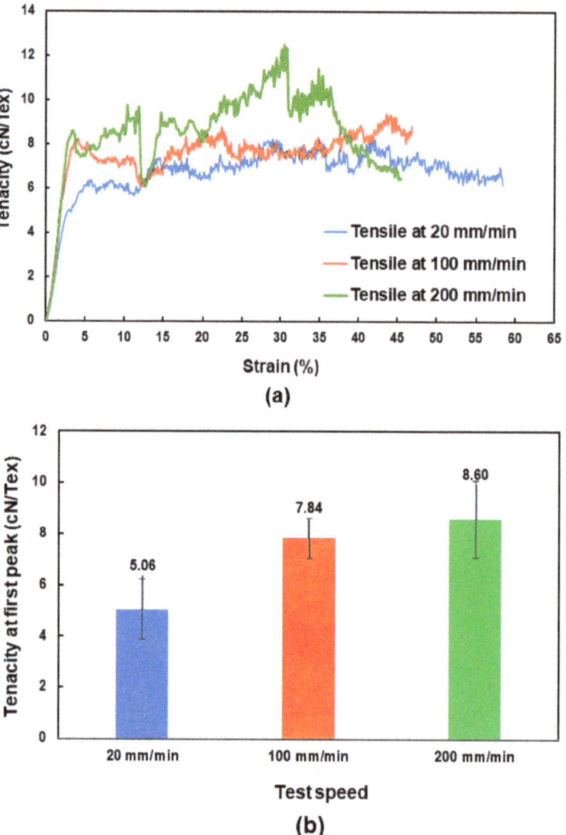

Figure 6. (**a**) Tenacity–strain curve of hybrid yarns at different test speeds; (**b**) tenacity at first peak at different test speeds.

3.1.3. Thermo-State Tensile Behaviour

The tensile behaviour of the hemp/PA11 yarns at different temperatures is shown in Figure 7a. The tensile tests for the different temperatures were performed at the same test speed of 100 mm/min. Depending on the temperature, the results obtained can be divided into three groups, namely, around the glass transition of PA11 (T = 50 °C), above the glass transition (T = 70, 100, and 120 °C), and below the melting temperature (T = 150 °C). The first phase of the different curves at the studied temperatures differs from the one at room temperature, as does the second phase. In contrast, the curves obtained at T = 100 and 120 °C show almost similar trends in the first and second phases. Hence, the tensile behaviour of these yarns depends strongly on the temperature setting. When the temperature is increased to 50 °C (upper glass transition), the tenacity of the wrapped yarn decreases as a result of the modification in the multifilament around the hemp roving, which becomes softer by passing the transition temperature and the interaction between hemp fibres and PA11 filaments changes. As the temperature increases (but is still far from the melting temperature), the PA11 filaments keep slipping over the core hemp roving. This is because adhesion between the filaments and fibres increases with increasing temperature. This increased adhesion strengthens the yarn and increases its tenacity, Figure 7b. Then, at T = 150 °C, the maximum tenacity of the first phase decreases significantly because the temperature nears the melting temperature, which strongly affects the structure of the hybrid yarn: the PA11 begins to stick locally to hemp fibres without creating a continuous medium to distribute the efforts between the hemp fibres. This was observed on the structure of the yarn once extracted from the climatic chamber after the tests. In the sections where the yarn had become solid in comparison to the original wrapped yarn, it accordingly displayed decreased extensibility.

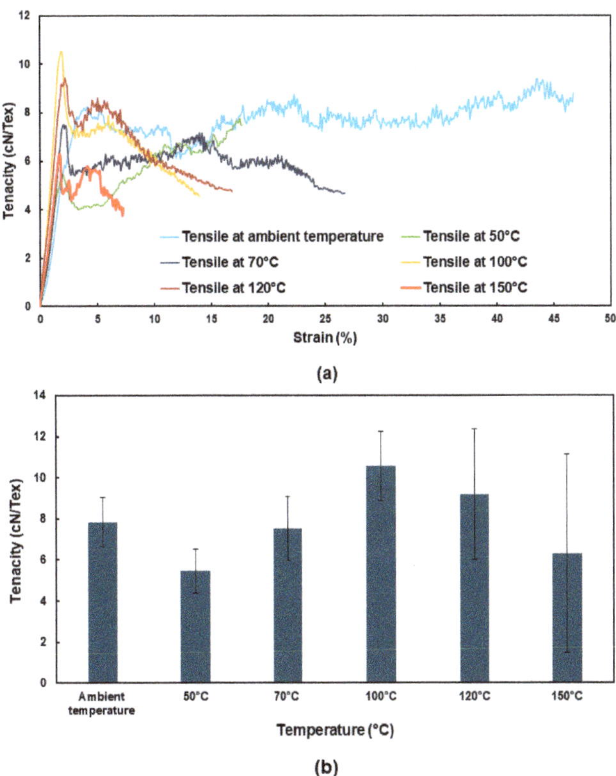

Figure 7. (**a**) Tenacity–strain curve of the hybrid yarns at different temperatures; (**b**) tenacity at first peak at different temperatures.

According to previous results obtained on flax/PA12 hybrid yarns [25], when the temperature increases and nears the melting value, the strain at break decreases, compared to the strain obtained at ambient temperature, whereas the deformation of hybrid yarns increases with increasing temperature above the melting temperature because below this temperature, the PA12 is fluid enough to increase the slippage between broken fibres. The same trend is observed in the presented results for tests below the melting temperature (Figure 7a). This can be attributed to the helical path of the wrapped PA11 filaments and the change in the interaction properties with the hemp fibres. At the ambient temperature, the spiral filaments stretch until they become aligned with the longitudinal axis of the yarn (unfolding). However, as the temperature increases, the PA11 filaments stick more locally to the hemp fibres, and that leads to concentrating the deformation at weak places on the wrapped yarn instead of having complete unfolding. Figure 7b shows the evolution of the tenacity at the first peak at different temperatures. As explained before, this tenacity increases initially with increasing temperature up to 100 °C and then decreases, but at higher temperatures (120 and 150 °C), the standard deviation is greater than at lower temperatures (50° and 70 °C), which is mainly due to the irregular behaviour of these yarns at high temperature.

3.2. Fabric Properties

3.2.1. Textile Properties

At this scale, the properties of the fabrics developed in this study will be compared to the properties of fabrics made in a previous study from 100% hemp roving. Table 3 shows the textile properties of these produced woven fabrics. For fabrics based on hybrid yarns, the twill 6 fabric has the higher weft density, its areal density and thickness are greater than satin 6. The areal density of this woven fabric depends mainly on the weft density, which, in turn, depends on how the fabric is packed during weaving and on the weave diagram. In addition, these fabrics are made of hybrid yarns that are heavier than conventional 100% natural fibre yarns. However, the air permeability of satin 6 is five times greater than that of twill 6, which highlights the higher inter-yarns gap, and it is correlated with a lower areal density, in comparison with twill 6, since it contains higher weft density. On the other hand, the fabrics in this study are heavier and thicker than the satin 6 and twill 6 of the previous study [42] manufactured with only hemp roving. This difference could be explained by the difference in the linear density of yarns which is higher for hybrid yarns (486 Tex) than for 100% hemp roving (259 Tex). In addition, even if the areal density of commingled fabrics is higher than that of 100% hemp fabrics, the air permeability is still higher. This could be due to the structure of the yarn used. In the case of hybrid yarns, the structure is more compact and the surface of the yarn contains less fibrils, compared to the structure of the roving, which is hairier, flat, and contains more fibrils.

Table 3. The textiles properties of the woven fabrics of this study and Corbin et al. [42] study.

Fabric Pattern	Twill 6 from Hybrid Yarns	Satin 6 from Hybrid Yarns	Satin 6 with 100% Hemp Roving [42]	Twill 6 with 100% Hemp Roving [42]
Warp density (yarns/cm)	6	6.4	6	6
Weft density (yarns/cm)	10	6	9.5	9.5
Areal density (g/m^2)	827 ± 15	583 ± 16	426 ± 8	402 ± 3
Thickness (mm)	2.19 ± 0.05	2.13 ± 0.04	1.55 ± 0.05	1.59 ± 0.06
Air permeability (L/m^2/s)	605 ± 91	3485 ± 239	401 ± 46	670 ± 115
Warp crimp (%)	0.67 ± 0.16	2.38 ± 1.11	2.24 ± 0.20	2.56 ± 0.34
Weft crimp (%)	2.65 ± 0.47	1.20 ± 0.30	1.90 ± 0.30	1.58 ± 0.22

3.2.2. Mechanical Properties

- Tensile Behaviour

Figure 8a shows the tensile properties of the two commingled fabrics and the two fabrics made from 100% hemp roving both in warp and weft direction. The tensile load is given in cN/yarn/Tex to remove the effects of the density and linear density of yarns and concentrate on the effect of the weave pattern and tenacity at the break of the roving. For the commingled satin 6 and twill 6 fabrics, a small difference in maximum tensile load is noted between the two directions. In terms of breaking strain, the weft direction of twill 6 is 55% higher than the warp direction, whereas the strain of satin 6 is similar for the two directions even if the crimp level is higher in the warp direction (Figure 8b). The high strain at break of twill 6 weft direction results from the higher crimp level of weft yarns as the warp yarns are under higher tension than weft yarns during the weaving process. Thus, the twill 6 fabric exhibits better tensile properties, both in warp and weft direction, than satin 6, and this difference can be explained by the difference of weave diagram between the structure and the arrangement of yarns inside the structure. In comparison to the previous study [42], it can be seen that twill 6 and satin 6 structures made from 100% hemp roving exhibit better maximum loads than twill 6 and satin 6 made from hybrid yarns, in the two directions for satin 6 structure and only in weft direction for twill 6 structure. That is mainly attributed to the low tenacity of these hybrid yarns (8 cN/Tex), compared to the tenacity of 100% hemp roving (24 cN/Tex) [42]. Furthermore, the higher hairiness level of hemp roving conduces higher inter-roving friction, leading to higher maximum loads. Strain at maximum load of the woven fabrics is balanced between the two directions for these structures except for commingled twill 6 structure, which has a high strain at break in the weft direction in comparison with the other structure.

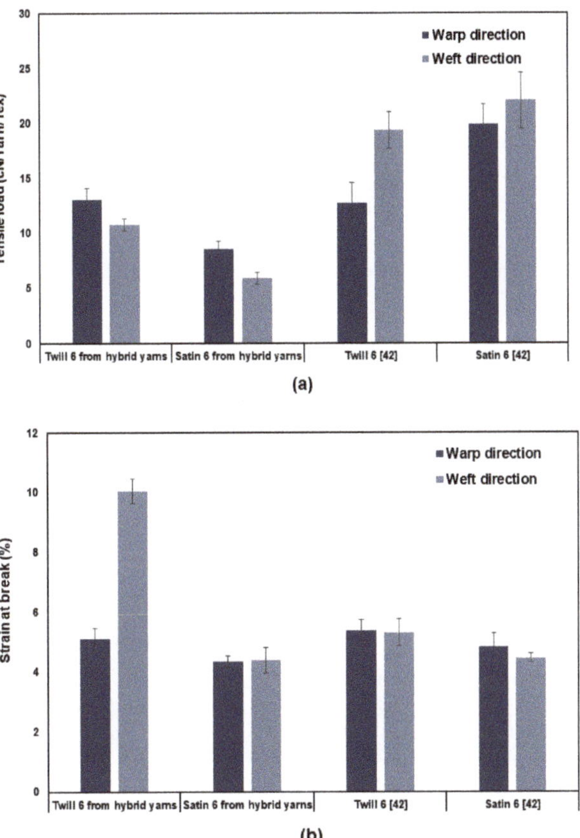

Figure 8. (**a**) The tensile load (in cN/yarn/Tex) of the woven fabrics; (**b**) the tensile strain at break of the woven fabrics.

- Flexural Behaviour

Figure 9 shows the flexural rigidity of the two fabrics. The two fabrics have the same warp density and differ only by their weft density and structure. Twill 6 exhibited better rigidity than satin 6 in the two directions, and this difference is mainly due to the high areal density of the fabric (42% higher than that of satin 6) and its yarn density in the weft direction. The flexural rigidity depends strongly on those two parameters. However, a high weft density and linear density result in a heavy fabric. The flexural rigidity also depends on the crimp level of yarns inside the structure. In the case of twill 6 fabric, the shrinkage of weft yarns is greater than warp yarns, and that led to higher rigidity in the weft direction of the fabric than in the warp direction. By contrast, for satin 6 fabric, the warp yarns exhibit higher crimp, which results in a higher rigidity in this direction than in the weft direction.

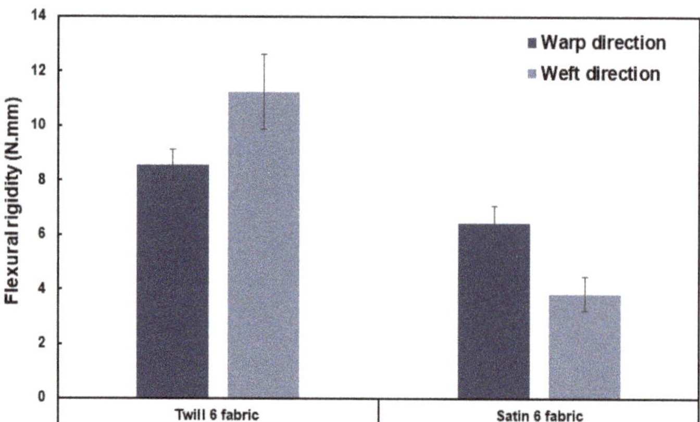

Figure 9. The flexural rigidity of the two fabrics.

3.3. Composite Properties

3.3.1. Composition of the Composite Plates

The obtained physical compositions of the two composite plates are summarised in Table 4. The two types of composites were produced by stacking two cross-plies. As a result, the thickness of twill 6 hemp/PA11 composite (C2) is 28% higher than composite made from satin 6 (C1). The weight of the two types of composites is the same and differs from that in the original (50% of hemp and 50% of PA11), which is explained by the loss of PA11 during the compression process. By contrast, the volume of fibre in the C2 is 22% greater than C1.

Table 4. The physical properties of the two composites.

Composite Plates	Thickness (mm)	Density (g/cm³)	Fibre Mass Fraction (%)	Fibre Volume Fraction (%)
Hemp/PA11 Satin 6 composite plate (C1)	1.01 ± 0.01	1.01 ± 0.01	52 ± 1	35.1 ± 0.1
Hemp/PA11 Twill 6 composite plate (C2)	1.39 ± 0.01	1.24 ± 0.03	52 ± 0.1	42.8 ± 1.2

3.3.2. Tensile Properties of the Composite Plates

The tensile strength, strain at break, and modulus of the two types of composites are shown in Figure 10. For each structure, the properties are almost the same for both directions of the composite plates, while at the same time, the tensile properties differ slightly per composite. The tensile stress and modulus of C1 exceed those of C2, and the opposite is the case for strain. Even if the fibre content of C1 composites is lower than that of C2, their tensile properties are higher. At the fabric scale, the twill 6 fabric exhibited better properties, both in tensile and flexural rigidity, whereas this is no larger than the case at the composite scale.

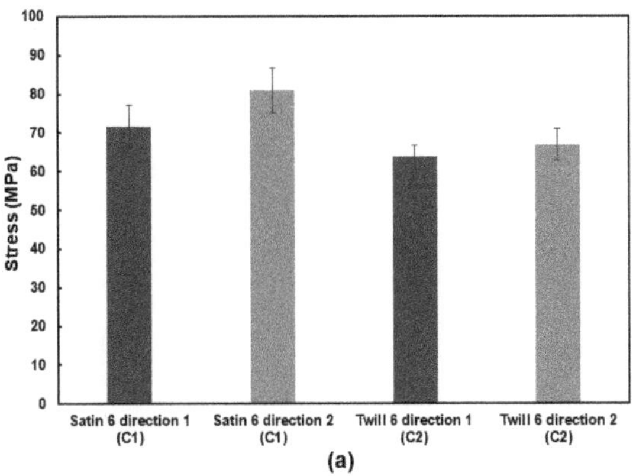

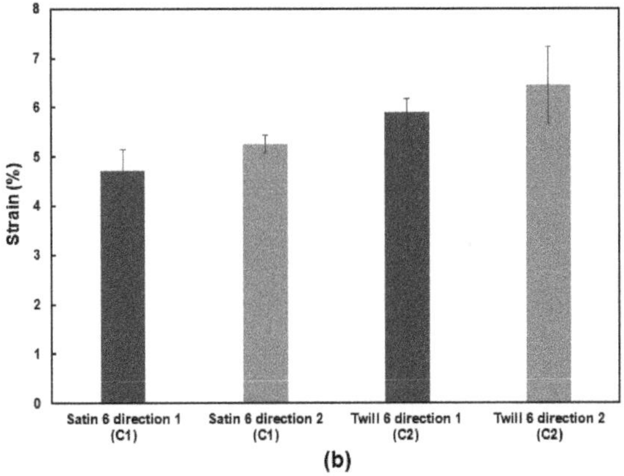

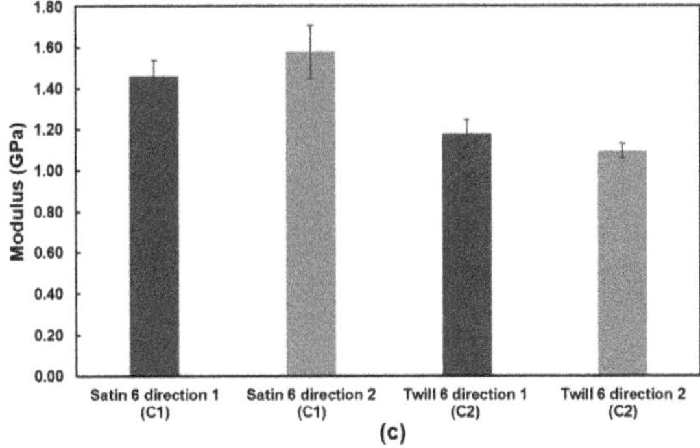

Figure 10. (**a**) Tensile strength of composite materials; (**b**) strain at the maximum strength; (**c**) modulus.

3.3.3. Flexural Behaviour of the Composite Plates

The results of the flexural testing are shown in Figure 11. For the satin 6 composites (C1), direction 2 presents better properties than direction 1, while for twill 6 (C2) the opposite is the case. The flexural strength is not balanced between the two directions of the composite plates even if the stacking is balanced. This difference is explained by the arrangement and the orientation of the yarns inside the structure of the composite. In the case of satin 6 composite plates (C1), the float yarns of direction 2 are located outside the specimen, and that provides additional rigidity to this direction. The same phenomenon happens to direction 1 of the twill 6 composite plates (C2). This behaviour has been identified in previous work within the same project [8] and confirmed in this study. The arrangement of yarns inside the composite plates depends strongly on the nature of the fabric structure used and on the way of stacking the different layers.

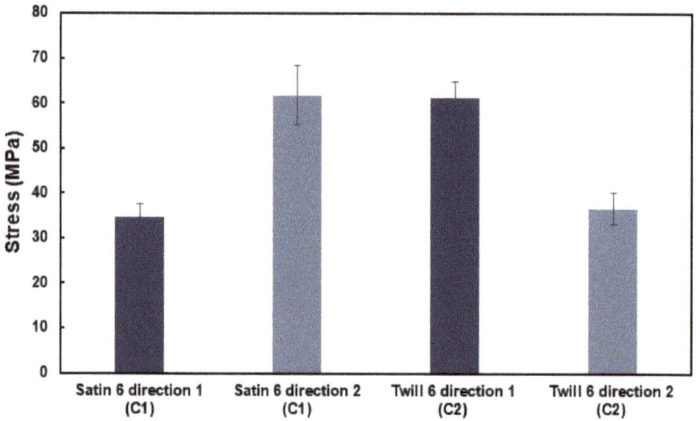

Figure 11. The flexural strength of the two composite materials.

4. Conclusions

This study investigated the multiscale analysis of bio-based composite materials made of hemp/PA11 commingled yarns. Understanding the different manufacturing stages and the influence of process parameters on yarn, fabric, and composite properties allows materials to be produced which better fit the final application requirements.

In this work, the hybrid yarns were produced by the wrapping process on a hollow spindle machine by wrapping a thermoplastic PA11 multifilament around an untreated hemp roving in order to produce yarns with sufficient tenacity (allowing them to be woven) and with a fibre content of no less than 40%. This process improves the hemp roving structure, which becomes more compact and less hairy.

At this yarn scale, tensile tests were initially conducted at different test speeds including the speed involved during fabric testing in order to investigate its influence on the mechanical properties, then at different temperatures including a temperature in the range of the glass transition temperature of the multifilament. Results from this test show the dependence of yarn properties on test speed and temperature, which is mainly due to the nature of the multifilament used.

Then, these yarns were used in weaving to produce two different fabrics with approximately the same warp density and different weft densities and weave diagrams. The textile and mechanical properties of these fabrics were determined, and the results show the dependence of these properties on the production parameters: preform properties in terms of maximum load and strain are either balanced or unbalanced between the warp and weft directions.

At the composite scale, tensile strength and stiffness for each structure are almost balanced between the two main directions of the composite plates. In addition, composite made from a satin 6 fabric shows an improvement of its mechanical properties even though at the fabric scale, the twill 6 fabric presented better properties.

The aim of future work will be the testing of other parameters at each scale in order to produce a light structure adapted to the end-use application. Moreover, the thermomechanical behaviour of hybrid yarns and commingled fabrics will be tested across a range of temperatures, including the melting temperature of PA11, in order to better understand the behaviour of these materials at high temperatures. Thus, the hydrophilic behaviour of these hybrid yarns will be characterised and compared with that of hemp roving [43].

Author Contributions: Investigation: C.L.; writing—original draft preparation: C.L.; validation; supervision: M.F., A.R.L. and D.S.; writing—review and editing: M.F., A.R.L. and D.S.; All authors have read and agreed to the published version of the manuscript.

Funding: This project was funded by "Bio-Based Industries Joint Undertaking" under the "European Union's Horizon 2020" Research and Innovation Program with Grant Agreement No. 744349—SSUCHY Project and Hauts de France Region.

Institutional Review Board Statement: Not applicable.

Informed Consent Statement: Not applicable.

Data Availability Statement: Data are contained within the article.

Acknowledgments: The authors gratefully acknowledge the Italian company Linificio and Canapificio Nazionale (LCN) for providing the hemp roving used in this study.

Conflicts of Interest: The authors declare no conflict of interest.

References

1. Bourmaud, A.; Beaugrand, J.; Shah, D.U.; Placet, V.; Baley, C. Towards the design of high-performance plant fibre composites. *Prog. Mater. Sci.* **2018**, *97*, 347–408. [CrossRef]
2. CELC. *Flax and Hemp Fiber Composites, a Market. Reality: The Biobased Solutions for the Industry*; JEC Group: Paris, France, 2018; ISBN 978-2-490-26300-4.
3. Mohanty, A.K.; Vivekanandhan, S.; Pin, J.-M.; Misra, M. Composites from renewable and sustainable resources: Challenges and innovations. *Science* **2018**, *362*, 536–542. [CrossRef] [PubMed]
4. Sala, B.; Gabrion, X.; Trivaudey, F.; Guicheret-Retel, V.; Placet, V. Influence of the stress level and hygrothermal conditions on the creep/recovery behaviour of high-grade flax and hemp fibre reinforced greenpoxy matrix composites. *Compos. Part A Appl. Sci. Manuf.* **2021**, *141*, 106204. [CrossRef]
5. Del Masto, A.; Trivaudey, F.; Guicheret-Retel, V.; Placet, V.; Boubakar, L. Investigation of the possible origins of the differences in mechanical properties of hemp and flax fibres: A numerical study based on sensitivity analysis. *Compos. Part A. Appl. Sci. Manuf.* **2019**, *124*, 105488. [CrossRef]
6. Baley, C.; Gomina, M.; Breard, J.; Bourmaud, A.; Davies, P. Variability of mechanical properties of flax fibres for composite reinforcement. A review. *Ind. Crop. Prod.* **2020**, *145*, 111984. [CrossRef]
7. Réquilé, S.; Mazian, B.; Grégoire, M.; Musio, S.; Gautreau, M.; Nuez, L.; Day, A.; Thiébeau, P.; Philippe, F.; Chabbert, B.; et al. Exploring the dew retting feasibility of hemp in very contrasting european environments: Influence on the tensile mechanical properties of fibres and composites. *Ind. Crop. Prod.* **2021**, *164*, 113337. [CrossRef]
8. Corbin, A.-C.; Ferreira, M.; Labanieh, A.R.; Soulat, D. Natural fiber composite manufacture using wrapped hemp roving with PA12. *Mater. Today Proc.* **2020**, *31*, S329–S334. [CrossRef]
9. Corbin, A.-C.; Sala, B.; Soulat, D.; Ferreira, M.; Labanieh, A.-R.; Placet, V. Development of quasi-unidirectional fabrics with hemp fiber: A competitive reinforcement for composite materials. *J. Compos. Mater.* **2021**, *55*, 551–564. [CrossRef]
10. Placet, V.; François, C.; Day, A.; Beaugrand, J.; Ouagne, P. Industrial hemp transformation for composite applications: Influence of processing parameters on the fibre properties. In *Advances in Natural Fibre Composites: Raw Materials, Processing and Analysis*; Springer: Berlin/Heidelberg, Germany, 2018; pp. 13–25. ISBN 978-3-319-64640-4.
11. Shah, D.U.; Schubel, P.J.; Clifford, M.J. Modelling the effect of yarn twist on the tensile strength of unidirectional plant fibre yarn composites. *J. Compos. Mater.* **2013**, *47*, 425–436. [CrossRef]
12. Goutianos, S.; Peijs, T.; Nystrom, B.; Skrifvars, M. Development of flax fibre based textile reinforcements for composite applications. *Appl. Compos. Mater.* **2006**, *13*, 199–215. [CrossRef]
13. Omrani, F.; Wang, P.; Soulat, D.; Ferreira, M. Mechanical properties of flax-fibre-reinforced preforms and composites: Influence of the type of yarns on multi-scale characterisations. *Compos. Part A Appl. Sci. Manuf.* **2017**, *93*, 72–81. [CrossRef]
14. Russo, P.; Simeoli, G.; Vitiello, L.; Filippone, G. Bio-polyamide 11 hybrid composites reinforced with basalt/flax interwoven fibers: A tough green composite for semi-structural applications. *Fibers* **2019**, *7*, 41. [CrossRef]
15. Lebaupin, Y.; Chauvin, M.; Hoang, T.-Q.T.; Touchard, F.; Beigbeder, A. Influence of constituents and process parameters on mechanical properties of flax fibre-reinforced polyamide 11 composite. *J. Thermoplast. Compos. Mater.* **2017**, *30*, 1503–1521. [CrossRef]
16. Oliver-Ortega, H.; Méndez, J.A.; Reixach, R.; Espinach, F.X.; Ardanuy, M.; Mutjé, P. Towards more sustainable material formulations: A comparative assessment of PA11-SGW flexural performance versus oil-based composites. *Polymers* **2018**, *10*, 440. [CrossRef]

17. Haddou, G.; Dandurand, J.; Dantras, E.; Maiduc, H.; Thai, H.; Giang, N.V.; Trung, T.H.; Ponteins, P.; Lacabanne, C. Mechanical properties of continuous bamboo fiber-reinforced biobased polyamide 11 composites. *J. Appl. Polym. Sci.* **2019**, *136*, 47623. [CrossRef]
18. Gourier, C.; Bourmaud, A.; Le Duigou, A.; Baley, C. Influence of PA11 and PP Thermoplastic polymers on recycling stability of unidirectional flax fibre reinforced biocomposites. *Polym. Degrad. Stab.* **2017**, *136*, 1–9. [CrossRef]
19. Awais, H.; Nawab, Y.; Amjad, A.; Anjang, A.; Md Akil, H.; Zainol Abidin, M.S. Effect of comingling techniques on mechanical properties of natural fibre reinforced cross-ply thermoplastic composites. *Compos. Part B Eng.* **2019**, *177*, 107279. [CrossRef]
20. Kobayashi, S.; Takada, K. Processing of unidirectional hemp fiber reinforced composites with micro-braiding technique. *Compos. Part A Appl. Sci. Manuf.* **2013**, *46*, 173–179. [CrossRef]
21. Zhai, W.; Wang, P.; Legrand, X.; Soulat, D.; Ferreira, M. Effects of micro-braiding and co-wrapping techniques on characteristics of flax/polypropylene-based hybrid yarn: A comparative study. *Polymers* **2020**, *12*, 2559. [CrossRef]
22. Alagirusamy, R.; Fangueiro, R.; Ogale, V.; Padaki, N. Hybrid yarns and textile preforming for thermoplastic composites. *Text. Prog.* **2006**, *38*, 1–71. [CrossRef]
23. Asghar, A.; Imad, A.; Nawab, Y.; Hussain, M.; Saouab, A. Effect of yarn singeing and commingling on the mechanical properties of jute/polypropylene composites. *Polym. Compos.* **2020**, *42*. [CrossRef]
24. Di Lorenzo, M.L.; Longo, A.; Androsch, R. Polyamide 11/poly(butylene succinate) bio-based polymer blends. *Materials* **2019**, *12*, 2833. [CrossRef] [PubMed]
25. Xiao, S.; Wang, P.; Soulat, D.; Gao, H. Thermo-mechanical characterisations of flax fibre and thermoplastic resin composites during manufacturing. *Polymers* **2018**, *10*, 1139. [CrossRef]
26. NF G07-316. *Textiles—Tests of Yarns—Determination of Linear Density*; AFNOR: La Plaine Saint-Denis, France, 1988.
27. NF G07-079. *Textiles—Testing Threads—Determining the Twisting of Threads by Untwisting/Retwisting with a Double Re-Test*; AFNOR: La Plaine Saint-Denis, France, 2011.
28. NF EN ISO 2062. *Textiles—Yarns from Packages—Determination of Single-End Breaking Force and Elongation at Break Using Constant Rate of Extension (CRE) Tester*; AFNOR: La Plaine Saint-Denis, France, 2010.
29. NF EN 12127. *Textiles—Fabrics—Determination of Mass per Unit Area Using Small Samples*; AFNOR: La Plaine Saint-Denis, France, 1998.
30. NF EN ISO 5084. *Textiles—Determination of Thickness of Textiles and Textile Products*; AFNOR: La Plaine Saint-Denis, France, 1996.
31. NF EN ISO 9237. *Textiles—Determination of Permeability of Fabrics to Air*; AFNOR: La Plaine Saint-Denis, France, 1995.
32. NF ISO 7211-3. *Textiles—Woven Fabrics—Construction—Methods of Analysis—Part 3: Determination of Crimp of Yarn in Fabric*; AFNOR: La Plaine Saint-Denis, France, 2017.
33. NF EN ISO 13934-1. *Tensile Properties of Fabrics—Part 1: Determination of Maximum Force and Elongation at Maximum Force Using the Strip Method*; AFNOR: La Plaine Saint-Denis, France, 2013.
34. ISO 4604. *Reinforcement Fabrics—Determination of Conventional Flexural Stiffness—Fixed-Angle Flexometer Method*; AFNOR: La Plaine Saint-Denis, France, 2011.
35. De Bilbao, E.; Soulat, D.; Hivet, G.; Gasser, A. Experimental study of bending behaviour of composite reinforcements. *Int. J. Mater. Form.* **2009**, *2*, 205–208. [CrossRef]
36. Bilisik, K. Bending behavior of multilayered and multidirectional stitched aramid woven fabric structures. *Text. Res. J.* **2011**, *81*, 1748–1761. [CrossRef]
37. ASTM D3039/D3039M-17. *Standard Test. Method for Tensile Properties of Polymer Matrix Composite Materials*; ASTM International: West Conshohocken, PA, USA, 2017.
38. NF EN ISO 14125. *Fibre-Reinforced Plastic Composites—Determination of Flexural Properties*; AFNOR: La Plaine Saint-Denis, France, 1998.
39. Tanasa, F.; Zănoagă, M.; Teacă, C.; Nechifor, M.; Shahzad, A. Modified hemp fibers intended for fiber-reinforced polymer composites used in structural applications—A review. I. Methods of modification. *Polym. Compos.* **2019**, *41*. [CrossRef]
40. Baley, C.; Gomina, M.; Breard, J.; Bourmaud, A.; Drapier, S.; Ferreira, M.; Le Duigou, A.; Liotier, P.J.; Ouagne, P.; Soulat, D.; et al. Specific features of flax fibres used to manufacture composite materials. *Int J. Mater.* **2019**, *12*, 1023–1052. [CrossRef]
41. Corbin, A.-C.; Soulat, D.; Ferreira, M.; Labanieh, A.; Gabrion, X.; Placet, V. Improvement of the weavability of natural-fiber reinforcement for composite materials manufacture. *RCMA* **2019**, *29*, 201–208. [CrossRef]
42. Corbin, A.-C.; Soulat, D.; Ferreira, M.; Labanieh, A.-R. Influence of process parameters on properties of hemp woven reinforcements for composite applications: Mechanical properties, bias-extension tests and fabric forming. *J. Nat. Fibers* **2020**, 1–13. [CrossRef]
43. Bismarck, A.; Aranberri, I.; Springer, J.; Lampke, T.; Wielage, B.; Stamboulis, A.; Shenderovich, I.; Limbach, H.-H. Surface characterization of flax, hemp and cellulose fibers; Surface properties and the water uptake behavior. *Polym. Compos.* **2002**, *23*, 872–894. [CrossRef]

Article

Development of Electrospun Films from Wastewater Treatment Plant Sludge

Gregor Lavrič [1], Aleksandra Miletić [2], Branka Pilić [2], Daša Medvešček [1], Saša Nastran [3] and Urška Vrabič-Brodnjak [4],*

[1] Pulp and Paper Institute, Bogišićeva ulica 8, 1000 Ljubljana, Slovenia; gregor.lavric@icp-lj.si (G.L.); dasa.medvescek@icp-lj.si (D.M.)
[2] Faculty of Technology, University of Novi Sad, Bulevar cara Lazara 1, 21000 Novi Sad, Serbia; alexm@uns.ac.rs (A.M.); brapi@uns.ac.rs (B.P.)
[3] JP Voka Snaga d.o.o., Vodovodna cesta 90, 1000 Ljubljana, Slovenia; sasa.nastran@vokasnaga.si
[4] Department of Textiles, Graphic Arts and Design, Faculty of Natural Sciences and Engineering, University of Ljubljana, Snežniška 5, 1000 Ljubljana, Slovenia
* Correspondence: urska.vrabic@ntf.uni-lj.si

Abstract: Electrospinning is a versatile method for producing continuous polymer nanofibers, including from wastewater treatment plant sludge (WTPS). In this context, purified WTPS was successfully used to produce electrospun fibers. The main objective of our research was to produce new, local, circular, renewable and environmentally friendly packaging material. The aim of the research was to purify and treat WTPS to make it suitable for the electrospinning process, thus producing a new material and chemically characterizing it in the first step. One of the major advantages of our process was that the electrospinning process could be carried out with water and ethylenediaminetetraacetic acid. The optimal viscosity was determined to be 20,000 mPas in order to produce sufficient nanofibers. Analyses such as Fourier-Transform Infrared Spectroscopy (FTIR) and ^{1}H-NMR (proton nuclear magnetic resonance) were used to determine the substances of unpurified and purified WTPS. The tensile properties, contact angle, surface properties and differential scanning calorimetry of the final material were determined and used. The ^{1}H-NMR analysis confirmed the presence of a small quantity of polyhydroxyalkanoates in the samples. Based on the properties, the final material was brittle and less stretchable compared to electrospun packaging films available in the market.

Keywords: wastewater treatment; electrospinning; differential scanning calorimetry; tensile properties; proton nuclear magnetic resonance spectroscopy; packaging

Citation: Lavrič, G.; Miletić, A.; Pilić, B.; Medvešček, D.; Nastran, S.; Vrabič-Brodnjak, U. Development of Electrospun Films from Wastewater Treatment Plant Sludge. *Coatings* **2021**, *11*, 733. https://doi.org/10.3390/coatings11060733

Academic Editor: Sandra Dirè

Received: 21 May 2021
Accepted: 17 June 2021
Published: 18 June 2021

Publisher's Note: MDPI stays neutral with regard to jurisdictional claims in published maps and institutional affiliations.

1. Introduction

The extensive consumption of natural resources in recent years is reflected in the increasing consumption of packaging [1], which consequently causes the increase of the amount of plastic waste. According to the World Economic Forum, there will be more plastic than fish in the oceans by 2050 if human habits persist [2]. From a different perspective, the current global health crisis has highlighted the importance of using local and easily accessible raw materials that do not have to travel long distances to reach the consumer. These facts indicate that all future packaging materials should be renewable, environmentally friendly and, if possible, made from alternative raw materials (e.g., industrial waste) [3,4]. In general, waste is described as a product or by-product substance derived from industrial or agricultural processes or other activities with end-use purposes [5]. Waste can be a direct result of processing technology or the product of secondary treatment of waste streams, for example, wastewater, which produces several types of wastewater treatment plant sludge.

Depending on the treatment stage, primary, secondary, tertiary and chemical sludges are produced. Primary sludge is formed during primary treatment (screening, grate removal, flotation, precipitation and sedimentation) when heavy solids, grease and oils are

125

separated from raw wastewater [6,7]. Usually, the primary sludge contains 2–9% solids. The remaining 90% (sometimes 99.5%) is water. Secondary sludge (waste activated sludge) is formed during biological treatment when the biodegradable organic content of wastewater is degraded by microorganisms. The total concentration of solids ranges from 0.8% to 3.3%, depending on the type of biological treatment process, with the remainder being water [7]. The organic fraction of waste activated sludge contains 50–55% carbon, 25–30% oxygen, 10–15% nitrogen, 6–10% hydrogen, 1–3% phosphorus and 0.5–1.5% sulfur [8]. Tertiary sludge is produced in advanced wastewater treatment stages when nutrients (nitrogen and phosphorus) need to be removed [6]. Usually, nutrient removal is carried out simultaneously with organic matter removal. Chemical sludge is produced by chemical processes carried out at the municipal wastewater treatment plant, such as chemically assisted primary sedimentation. In this process, an appropriate coagulant is added to the primary clarifier to reduce the organic load for further biological treatment. The qualitative and quantitative properties of the sludge depend on the reagent used and the dosage. Typical reagents are hydrated lime, ferric chloride, aluminum sulfate and chitosan. Chemical sludge may contain non-negligible quantities of metals, due in part to the inorganic coagulants used. Chemical sludge can also be produced by coagulation–flocculation of the supernatant thickener and by backwashing after sedimentation/chemical–physical treatment [9].

Due to legislation restricting the use of landfills and land as sludge disposal methods, many researchers have attempted to reuse and recycle sludge in the most sustainable possible way [9–25]. Taking into account that the organic components of sludge present a rich source vein in terms of energy and nutrients waiting to be tapped, various studies show that, in the context of the circular economy, an important advantage of energy and fuels derived from waste is that they can replace other energy resources and limit the associated CO_2 emissions. From a scientific point of view, the challenge of sewage sludge is one of the most studied in the last 30 years. The concepts of material and energy recovery, which are milestones of today's circular economy, have already been addressed by several working groups [19–22]. Most of the studies on the material utilization of wastewater sludge are based on the extraction of polyhydroxyalkanoates [9].

Over the years, it has been shown that wastes generated mainly from processes in the agricultural, food, textile and paper industries have a high end-use potential [10]. A good example of high-end potential from the agri-food sector is waste from coffee production, as described by Figueroa et al. and Malara et al. [11,12]. As a waste solution, an anthocyanin-based milk freshness indicator or sensor that could be used as an indicator of actual milk quality was shown by Weston et al. [13]. Researchers have also presented available technologies and materials for exploited cooking oil recycling, which has a significant impact on household waste solutions [14–16]. In recent years, substantial progress has been made in waste processing in the textile and paper industries [17–22].

In the past, studies on the utilization of wastewater treatment plant sludge (WTPS) for various purposes have been published [9,23–25]. Most of them are based on the extraction of polyhydroxyalkanoates (PHAs) from WTPS. PHAs are a class of bio-based and biodegradable polymers produced by bacterial fermentation of complex organic substrates [25]. They belong to the class of polyesters and can be thermoplasts or elastomers, and depending on the structure, they have similar properties to the conventional plastics, which makes PHAs suitable candidates for their substitution [25]. Despite the extensive literature review, no report of the direct use of WTPS for material preparation by electrospinning has been published so far. The materials produced by electrospinning technology are intended for a variety of applications in many fields such as medicine, pharmaceuticals, biotechnology, sustainable engineering materials and even packaging [26–28]. The advantage of the above technology lies in the production of nanofibers and their small diameter, large specific surface area and high porosity. Meanwhile, the rapid development of nanotechnology has enabled new applications for electrospun materials. In general, electrospinning technology is a new strategy for environmentally friendly nanomaterials with special properties and a promising solution for wastewater sludge as well. Accordingly, at least a partial presence of various biopolymers in a local WTPS was expected. With the main objective of producing a new, local, circular, renewable and environmentally friendly packaging material, the sludge was chemically characterized at the beginning of the research. The main advantage of our process is that the electrospinning solution was prepared using water and ethylenediaminetetraacetic acid (EDTA). However, the aim of the research was to clean and treat the wastewater treatment plant sludge in such a way that it is suitable for the electrospinning process, thus creating a new material.

2. Materials and Methods

2.1. Materials

Preparation of wastewater treatment plant sludge solution (WTPS): The WTPS was obtained from the main wastewater treatment plant of Ljubljana. The purification procedure was carried out according to a slightly modified method described by Fang and Jia in 1996 [29]. The purification was based on dissolution of WTPS in a 2% EDTA solution in water and centrifugation. Fifty grams of the absolute dry mass of WTPS was firstly cut down into pieces with a size of ~1 cm × 1 cm. The pieces were then dissolved in a 300 mL EDTA solution and stirred in a 500 mL beaker (500 rpm, 24 h). Centrifugation (16,800× g, 30 min) and separation of the usable supernatant were the final steps of the process. A series of solutions with different viscosities (20,000, 5000, and 2000 mPa) were prepared by water evaporation from the initial solution of WTPS by heating in a water bath (85 °C, ~240 min) while stirring at 250 rpm in order to optimize the process of electrospinning, since it was impossible to obtain continuous nanofibers from the initial solution.

2.2. Methods

2.2.1. Electrospinning Process

The electrospinning process was carried out using a Fluidnatek LE −10 electrospinning machine (Bioinicia, Valencia, Spain). Main process parameters: flow rate (µL/h), needle-to-collector distance (cm) and voltage (kV) were optimized to obtain a satisfactory nanofiber morphology, without visible defects, beads or drops within the fiber structure. The optimal voltage was 19 kV, flow rate was 200 µL/h and needle-to-collector distance was 15 cm.

2.2.2. Material Characterization

Viscosity Determination

The rheology (viscosity) of the solution was determined using a Viscotech 3000 viscometer (Viscotech Hispania, SL, El Vendrell-España, Spain) according to the Brookfield method. A 20 mL sample was analyzed at room temperature (22 ± 0.5 °C) in an appropriate container providing convenient spindle-to-wall distance.

Fourier-Transform Infrared Spectroscopy (FTIR)

Fourier-transform infrared spectra analysis of WTPS was performed using a Spectrum One ATR-FTIR spectrometer (Perkin Elmer, Waltham, MA, USA). Scans were performed between the infrared regions at 4000–400 cm^{-1}, with an average of 64 scans.

Proton Nuclear Magnetic Resonance Spectroscopy (^{1}H NMR)

Proton nuclear magnetic resonance spectroscopy of WTPS was performed using an Agilent Technologies DD2 spectrometer (Agilent Technologies, Santa Clara, CA, USA). The WTPS samples were dissolved in D_2O with trifluoroacetic acid (TFA) and deuterated chloroform ($CDCl_3$) for the purpose of analysis.

Differential Scanning Calorimetry (DSC)

Differential scanning calorimetry of purified WTPS was conducted with a differential scanning calorimeter (TA Instruments Q20, New Castle, DE, USA) within the range of

−60 to 150 °C. A small amount of sample (2–5 mg) was placed into a DSC pan and analyzed with a heat rate of 10 °C/min.

Tensile Properties

Tensile properties were determined using Zwick Roell Z010 (Zwick Roell, Ulm, Germany) tensile testing machine equipped with a 20 N measuring cell (Class 0.5, ISO 7500-1). The testing speed was set to 1 mm/min. Film strips of 1.5 cm in width and 10 cm in length were used. During the sample stretching, several load and elongation data per second were recorded until the breaking of a sample occurred.

Contact Angle

The wettability and water absorption of material were determined by measuring the contact angle of a 4 µL water drop using a Fibrodat 1100 (Fibro System AB, Molenbaan, The Netherlands) dynamic contact angle tester. The change in the contact angle in the first 30 s was measured and evaluated.

Surface Characterization with a Scanning Electron Microscope (SEM)

The surfaces/structures of electrospun materials were examined with electron scanning microscopy to acquire information about fiber arrangement and the material network in general. The micrographs were taken with a scanning electron microscope JSM-6060 LV (Jeol Ltd., Akishima, Japan). The instrument was operated at 10 kV at 1500× magnification.

3. Results and Discussion

In the first part of this study, WTPS obtained from a local wastewater treatment plant was chemically characterized then purified and prepared for the electrospinning process. This process produced a new thin material that could be used for packaging applications. In the final part of the study, the newly obtained material was characterized in terms of tensile strength and wettability.

3.1. FTIR and ^{1}H NMR Analysis of Wastewater Treatment Plant Sludge

In order to determine the presence of sodium polyacrylate in purified WTPS, FTIR spectra of purified WTPS were analyzed and compared to the sodium polyacrylate spectra (Figure 1).

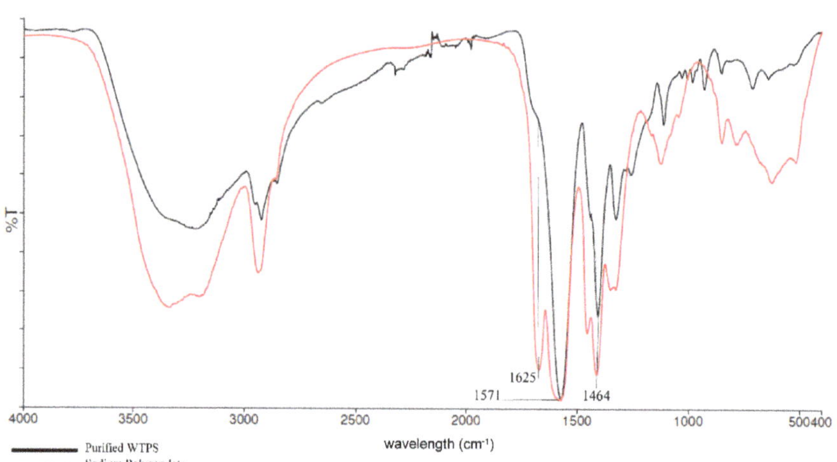

Figure 1. Fourier-transform infrared (FTIR) transmission spectra (%T) at the wavelength (cm^{-1}) of purified WTPS and sodium polyacrylate.

As shown, three peaks were observed in the pure sodium polyacrylate at 1464, 1571 and 1625 cm^{-1}. The peaks at 1464 and 1625 cm^{-1} are related to the bending of –CH2– and the C=O stretching mode in the carboxylic acid group of sodium polyacrylate [30,31]. From the spectra, it is evident that purified WTPS and sodium polyacrylate share common peaks at 1464, 1571, and 1625 cm^{-1} corresponding to the bands of sodium polyacrylate. As previously reported in the literature, the strong peak at 1571 cm^{-1} remains in the FITR spectra of WTPS, which is also masked by the carboxylate ions of WTPS and sodium polyacrylate [31]. At the same time, in the FTIR spectra of sodium polyacrylate and purified WTPS, the peaks in the range of 1625 and 1800 cm^{-1} are assigned to the carbonyl groups [30–32]. In the purified WTPS sample, the addition of sodium polyacrylate was detected as expected.

To further analyze the unpurified and purified WTPS, a comparison of the spectra was performed (Figure 2). As can be seen from the spectra, in unpurified WTPS, the peak at 1554 cm^{-1} is shifted to 1571 in purified WTPS. These two peaks correspond to the carboxylate ions [31]. In purified WTPS, more distinct peaks were observed at 2911, 1362, 1181, and 915 cm^{-1} wavelengths.

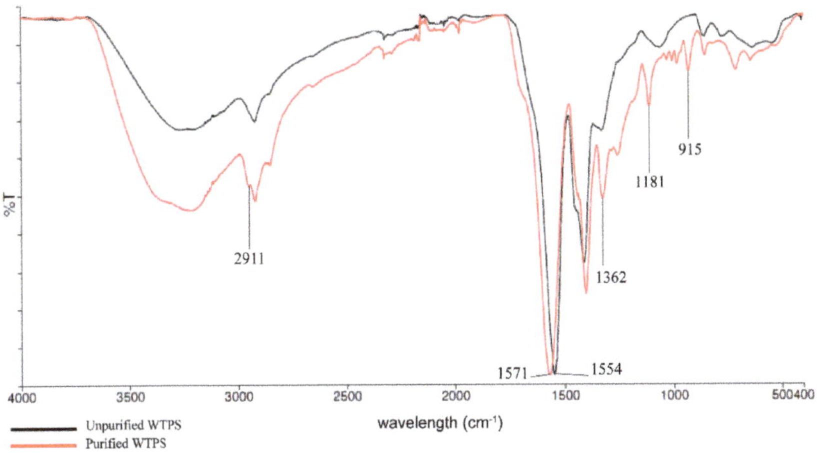

Figure 2. Fourier-transform infrared (FTIR) transmission spectra (%T) at the wavelength (cm^{-1}) of unpurified and purified WTPS.

Figure 3 illustrates the ^{1}H-NMR of purified WTPS and sodium polyacrylate. Sodium polyacrylate was used for the purposes of WTPS analysis. Therefore, the presence of this component was also detected in the purified WTPS sample. The characteristic solid peaks on the purified WTPS appeared between 1.6 and 2.6 ppm, which are also typical for sodium polyacrylate. A signal at 1.6 ppm belongs to the hydrogen of methylene. The spectra also show the resonance signal CH$_2$O–COOH bond at 2.580 ppm. As described in the literature on wastewater analysis, peaks between 3.3 and 4.6 ppm could correspond to glycine (3.5 ppm), glycerol (3.6 ppm), serine (3.7 ppm), 2-aminopropanol (3.9 ppm) and 3-hidroxibutyric acid (4.2 ppm) [33]. For the purified WTPS, the strong peaks were detected at 3.7 ppm corresponding to serine and 4.2 ppm corresponding to 3-hidroxibutyric acid.

Figure 4 presents ^{1}H-NMR spectra of unpurified WTPS and purified WTPS. The strong peak at 1.53 ppm corresponds to water in chloroform. The peaks of unpurified WTPS are stronger, indicating the addition of carboxylic acid in the sample compared to purified WTPS. At the same time, the minor addition of PHA was observed in both samples, as shown in Figure 4. PHA polymers contain hydrogen and carbon; therefore, typical peaks such as CH$_2$ at 1.35 ppm and CH$_3$ at 0.85 ppm were detected [34]. Typical peaks for PHA also include CH at 5.2 ppm and CH$_2$ at 2.55 ppm, which were not detected in our samples [35]. The peak CH_2 at 1.6 ppm was present but was also masked by the water and chloroform.

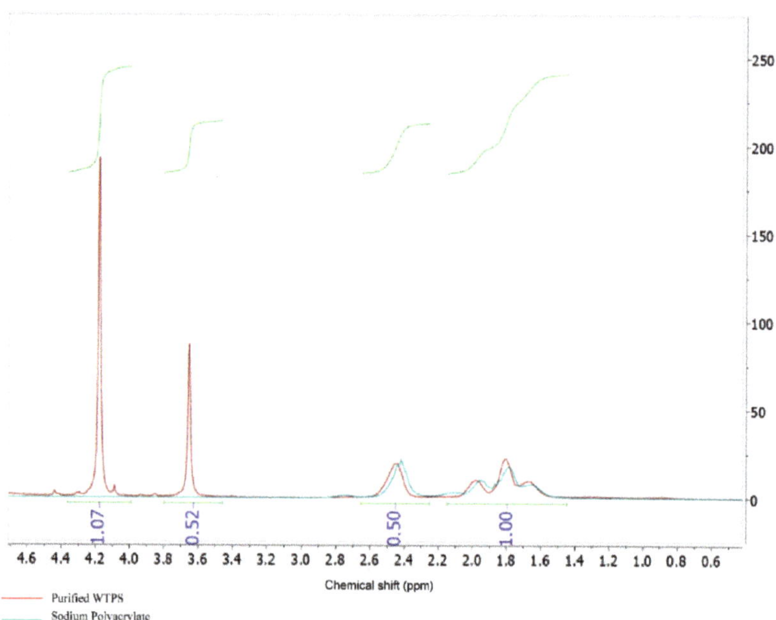

Figure 3. ^{1}H-NMR spectra of purified WTPS and sodium polyacrylate.

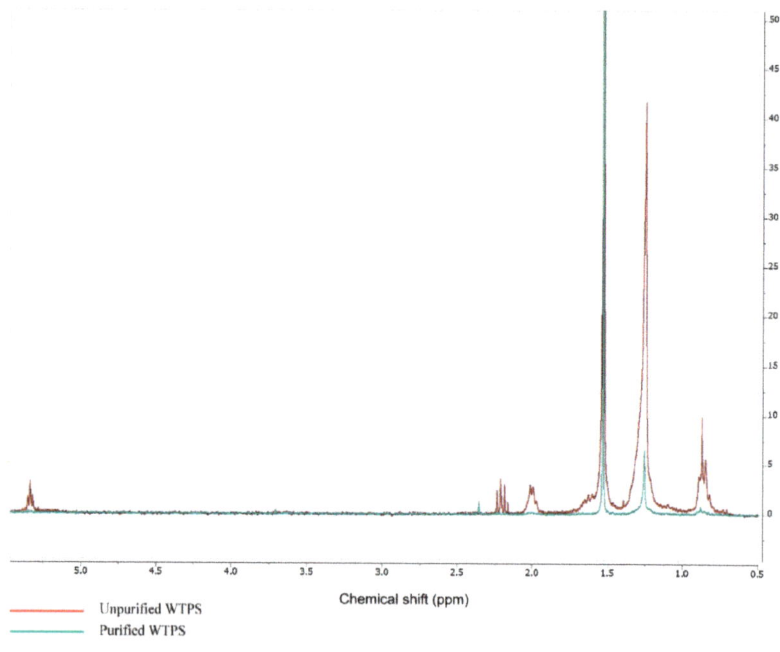

Figure 4. ^{1}H-NMR spectra of unpurified and purified WTPS.

In general, the quantitative estimate of PHA could be determined by the intensity ratio of the signals, and as before, the unpurified sample had higher peaks compared to the unpurified WTPS, especially in the detection of PHA groups. This could be due to the purification process and the reduction of the amount of PHA in the purified sample.

3.2. Determination of Optimal Viscosity and Electrospinning Parameters

The optimal viscosity for the WTPS suspension was experimentally determined to be 20,000 mPas (±500 mPas). This viscosity provided satisfactory nanofiber morphology without visible defects, beads or droplets within the fiber structure (Figure 5 with marked arrows). Purified WTPS with lower viscosities did not lead to uniform structure of the electrospun material. During the electrospinning process of such suspensions, undesirable electrospraying effects occurred in the form of tiny droplets. Higher viscosities (above 20,000 mPas), on the other hand, clogged the electrospinning device system and thus disrupted the production process.

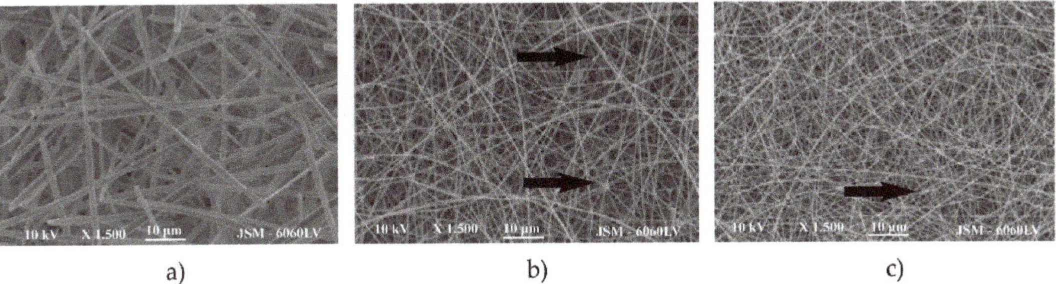

a)   b)   c)

Figure 5. SEM micrographs of final material (**a**) optimal-electrospun from suspension with a viscosity of 20,000 mPas; (**b**) electrospun from suspension with a viscosity of 5000 mPas; (**c**) electrospun from suspension with a viscosity of 2000 mPas at 1500× magnification.

As described previously, the optimal parameters of the electrospinning process were a voltage of 19 kV, a flow rate of 200 µL/h and a distance of 15 cm between needle and collector. Only the sample prepared from a suspension with optimal viscosity was further analyzed. Based on the results of the preliminary tests, it shows better performance compared to the other two (sample b and c).

3.3. Differential Scanning Calorimetry Characterization of the Final Material

The DSC heating curve is shown in Figure 6. The analysis was done in one heating cycle only, since the nanofiber morphology of the material is disrupted by the heating and further cooling, reheating serves no purpose. A small peak on the derivative curve at 62.60 °C indicates the glass transition temperature of the new material. PHAs usually have a bit lower glass transition temperature, but since this material contains a considerable number of impurities, the increase might be a consequence. Low-intensity peaks around 130 and 137 °C might also indicate the presence of impurities in the final material. The melting point of the material was detected at around 150 °C, which is in accordance with the research published by Lorini et al. [36].

DSC

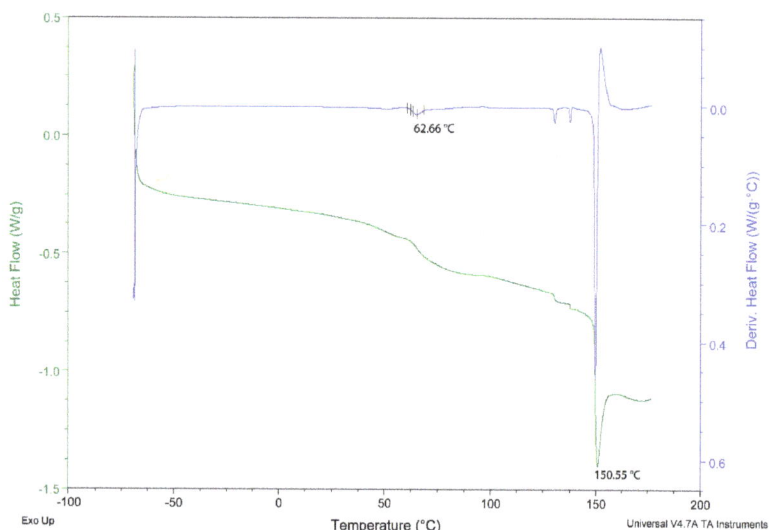

Figure 6. DSC of purified WTPS.

3.4. Tensile Properties and Contact Angle Analysis of Final Material

The results of the tensile analysis and contact angle analysis of the final material are summarized in Table 1 and presented in Figure 7. As can be seen from the results, the sample exhibited lower stress and strain than the results obtained in the literature [24–26]. The results confirmed a brittle sample, as only 0.422 N/mm^2 was applied to break it. The same trend was observed for strain, which was only 2.07%. This was more than 50% lower compared to the literature results. The tensile tests confirmed that purified WTPS was suitable for performing electrospun material, but it was very brittle. Therefore, one of the solutions could be a combination with other waste biopolymers or recycled polymers to obtain flexible material in further research.

Table 1. Stress–strain with standard deviation and contact angle measurements for final material.

Analysis Parameters	Sample	Results from the Literature: Bio-Based Electrospun Materials [37–39]
Stress (N/mm^2)	0.422 ± 0.053	1.5–1.85
Strain (%)	2.07 ± 0.129	4.71–13.8
Contact angle	0°	0.4°–22°

As shown in Table 1, the newly produced material reached a meagre value of the contact angle determined with water at a time of 1 s (0°). According to the solubility of the base material (WTPS) in the presence of EDTA in water, such a material property was expected. The presence of many accessible OH groups on the surface, in combination with other properties (size and interweaving of electrospun fibers), resulted in a material with high wettability. This property could prevent it from being used for packaging applications. However, with an appropriate blending method and/or surface treatment or coating, this could be drastically improved or adjusted. Since we wanted to present the properties of material derived only from WTPS in this study, these options were not used. Other chemicals that could improve this and possibly other properties of the material were also not used due to the preference for developing the new material in the most sustainable and environmentally friendly way possible. In the further development of the material, methods such as plasma and UV treatment and the possible use of other substances will also be used in accordance with these principles.

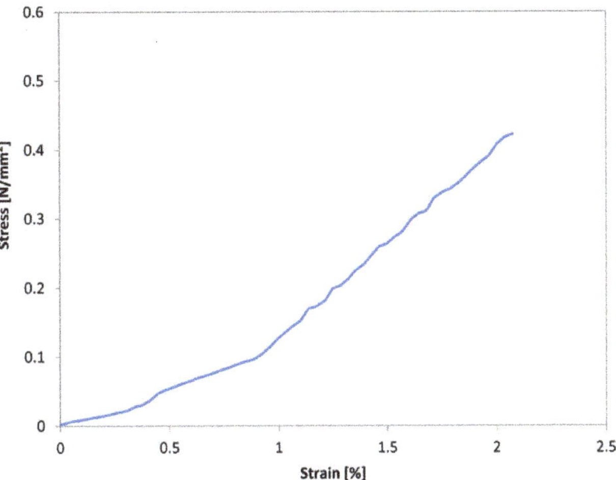

Figure 7. Stress–strain curve of the final material.

4. Conclusions

Biopolymers are currently in the spotlight in terms of their production and use, which is continuously increasing. This work evidences that it is possible to produce electrospun fibers from purified WTPS by using water and EDTA. Firstly, the viscosity was determined at 20,000 mPas, which led to satisfying nanofiber morphology, without visible defects, beads or drops within the fiber structure. Due to the presence of impurities in the WTPS, the glass transition temperature of the material was found to be at a higher temperature than usual for PHA, along with two low intensity peaks below the melting temperature. The melting point at 150 °C is within the range of temperatures characteristic for PHA polymers. The low amount of PHA was also confirmed in both samples. Due to the process of purification, the amount of PHA was lower in purified WTPS. However, all the analyzed properties showed that the final material was brittle and less stretchable compared to the packaging films on the market. Further research should be conducted on the possible addition of other biopolymers produced from alternatively sourced or recycled materials to purified WTPS to obtain a material with desired mechanical properties.

Author Contributions: Conceptualization, G.L. and U.V.-B.; methodology, G.L. and U.V.-B.; validation, G.L., A.M., B.P., S.N. and U.V.-B.; formal analysis, G.L., A.M. and U.V.-B.; investigation, G.L., A.M., D.M. and S.N.; writing—original draft preparation, G.L., A.M., B.P., S.N., D.M. and U.V.-B.; writing—review and editing, G.L., A.M., D.M. and U.V.-B. All authors have read and agreed to the published version of the manuscript.

Funding: The first author C.L. MSc extends his gratitude to COST action FP1405—Active and Intelligent Fiber-Based Packaging for providing the grant to support the Short Term Scientific Mission (STSM) research activities at the University of Novi Sad, Faculty of Technology, Novi Sad, Serbia.

Institutional Review Board Statement: Not applicable.

Informed Consent Statement: Not applicable.

Data Availability Statement: Not applicable.

Conflicts of Interest: The authors declare no conflict of interest.

References

1. Smithers Pira. Value Estimations & Global Packaging Trends in 2018 and Beyond | Smithers Pira. Available online: https://www.smitherspira.com/resources/2018/january/value-estimations-for-packaging-in-2018-and-beyond (accessed on 8 August 2020).
2. World Economic Forum. *The New Plastics Economy Rethinking the Future of Plastics*; World Economic Forum: Geneva, Switzerland, 2016; 14p.
3. Matthews, C.; Moran, F.; Jaiswal, A.K. A review on European Union's strategy for plastics in a circular economy and its impact on food safety. *J. Clean. Prod.* **2021**, *283*, 125263. [CrossRef]
4. Plastics Strategy. Available online: https://ec.europa.eu/environment/strategy/plastics-strategy_en (accessed on 24 April 2021).
5. Schettini, E.; Santagata, G.; Malinconico, M.; Immirzi, B.; Scarascia Mugnozza, G.; Vox, G. Recycled wastes of tomato and hemp fibres for biodegradable pots: Physico-chemical characterization and field performance. *Resour. Conserv. Recycl.* **2013**, *70*, 9–19. [CrossRef]
6. Manara, P.; Zabaniotou, A. Towards sewage sludge based biofuels via thermochemical conversion—A review. *Renew. Sustain. Energy Rev.* **2012**, *16*, 2566–2582. [CrossRef]
7. Suárez-Iglesias, O.; Collado, S.; Oulego, P.; Díaz, M. Graphene-family nanomaterials in wastewater treatment plants. *Chem. Eng. J.* **2017**, *313*, 121–135. [CrossRef]
8. Tyagi, V.; Lo, S. Sludge: A waste or renewable source for energy and resources recovery? *Renew. Sustain. Energy Rev.* **2013**, *25*, 708–728. [CrossRef]
9. Gherghel, A.; Teodosiu, C.; De Gisi, S. A review on wastewater sludge valorisation and its challenges in the context of circular economy. *J. Clean. Prod.* **2019**, *228*, 244–263. [CrossRef]
10. Väisänen, T.; Haapala, A.; Lappalainen, R.; Tomppo, L. Utilization of agricultural and forest industry waste and residues in natural fiber-polymer composites: A review. *Waste Manag.* **2016**, *54*, 62–73. [CrossRef]
11. Figueroa, G.A.; Homann, T.; Rawel, H.M. Coffee production wastes: Potentials and perspectives. *Austin Food Sci* **2016**, *1*, 1–5.
12. Malara, A.; Paone, E.; Frontera, P.; Bonaccorsi, L.; Panzera, G.; Mauriello, F. Sustainable exploitation of coffee silverskin in water remediation. *Sustainability* **2018**, *10*, 3547. [CrossRef]
13. Weston, M.; Phan, M.A.T.; Arcot, J.; Chandrawati, R. Anthocyanin-based sensors derived from food waste as an active use-by date indicator for milk. *Food Chem.* **2020**, *326*, 127017. [CrossRef] [PubMed]
14. Mannu, A.; Garroni, S.; Ibanez Porras, J.; Mele, A. Available technologies and materials for waste cooking oil recycling. *Processes* **2020**, *8*, 366. [CrossRef]
15. Yaakob, Z.; Mohammad, M.; Alherbawi, M.; Alam, Z.; Sopian, K. Overview of the production of biodiesel from waste cooking oil. *Renew. Sustain. Energy Rev.* **2013**, *18*, 184–193. [CrossRef]
16. Singhabhandhu, A.; Tezuka, T. The waste-to-energy framework for integrated multi-waste utilization: Waste cooking oil, waste lubricating oil, and waste plastics. *Energy* **2010**, *35*, 2544–2551. [CrossRef]
17. Shirvanimoghaddam, K.; Motamed, B.; Ramakrishna, S.; Naebe, M. Death by waste: Fashion and textile circular economy case. *Sci. Total Environ.* **2020**, *718*, 137317. [CrossRef] [PubMed]
18. Dissanayake, D.G.K.; Weerasinghe, D.U.; Thebuwanage, L.M.; Bandara, U.A.A.N. An environmentally friendly sound insulation material from post-industrial textile waste and natural rubber. *J. Build. Eng.* **2021**, *33*, 101606. [CrossRef]
19. Girard, E.B.; Kaliwoda, M.; Schmahl, W.W.; Wörheide, G.; Orsi, W.D. Biodegradation of textile waste by marine bacterial communities enhanced by light. *Environ. Microbiol. Rep.* **2020**, *12*, 406–418. [CrossRef]
20. Kumar, V.; Pathak, P.; Bhardwaj, N.K. Waste paper: An underutilized but promising source for nanocellulose mining. *Waste Manag.* **2020**, *102*, 281–303. [CrossRef]
21. Zakarya, I.A.; Fazhil, N.S.A.; Izhar, T.N.T.; Zaaba, S.K.; Jamaluddin, M.N.F. Municipal Solid Waste Characterization and Quantification as A Measure Towards Effective Solid Waste Management in UniMAP. *IOP Conf. Ser. Earth Environ. Sci.* **2020**, *616*, 12047. [CrossRef]
22. Zhang, Q.; Khan, M.U.; Lin, X.; Yi, W.; Lei, H. Green-composites produced from waste residue in pulp and paper industry: A sustainable way to manage industrial wastes. *J. Clean. Prod.* **2020**, *262*, 121251. [CrossRef]
23. Albuquerque, P.; Malafaia, C. Perspectives on the production, structural characteristics and potential applications of bioplastics derived from polyhydroxyalkanoates. *Int. J. Biol. Macromol.* **2018**, *107*, 615–625. [CrossRef]
24. Kumar, M.; Ghosh, P.; Khosla, K.; Thakur, I. Recovery of polyhydroxyalkanoates from municipal secondary wastewater sludge. *Bioresour. Technol.* **2018**, *255*, 111–115. [CrossRef]
25. Mannina, G.; Presti, D.; Montiel-Jarillo, G.; Suárez-Ojeda, M. Bioplastic recovery from wastewater: A new protocol for polyhydroxyalkanoates (PHA) extraction from mixed microbial cultures. *Bioresour. Technol.* **2018**, *282*, 361–369. [CrossRef]
26. Liu, H.; Gough, C.R.; Deng, Q.; Gu, Z.; Wang, F.; Hu, X. Recent advances in electrospun sustainable composites for biomedical, environmental, energy, and packaging applications. *Int. J. Biol. Macromol.* **2020**, *21*, 4019. [CrossRef]
27. Dodero, A.; Alloisio, M.; Vicini, S.; Castellano, M. Preparation of composite alginate-based electrospun membranes loaded with ZnO nanoparticles. *Carbohydr. Polym.* **2020**, *227*, 115371. [CrossRef] [PubMed]
28. de Souza, E.J.D.; Kringel, D.H.; Dias, A.R.G.; da Rosa Zavareze, E. Polysaccharides as wall material for the encapsulation of essential oils by electrospun technique. *Carbohydr. Polym.* **2021**, *265*, 118068. [CrossRef] [PubMed]
29. Fang, H.H.P.; Jia, X.S. Extraction of extracellular polymer from anaerobic sludges. *Biotechnol. Tech.* **1996**, *10*, 803–808. [CrossRef]

30. Hsiao, M.H.; Lin, K.H.; Liu, D.M. Improved pH-responsive amphiphilic carboxymethyl-hexanoyl chitosan-poly (acrylic acid) macromolecules for biomedical applications. *Soft Matter* **2013**, *9*, 2458–2466. [CrossRef]

31. Sun, Y.; Zhang, F.; Wu, D.; Zhu, H. Roles of polyacrylate dispersant in the synthesis of well-dispersed $BaSO_4$ nanoparticles by simple precipitation. *Particuology* **2014**, *14*, 33–37. [CrossRef]

32. Liu, S.; Gu, L.; Zhao, H.; Chen, J.; Yu, H. Corrosion resistance of graphene-reinforced waterborne epoxy coatings. *J. Mater. Sci. Technol.* **2016**, *32*, 425–431. [CrossRef]

33. Alves Filho, E.G.; Alexandre e Silva, L.M.; Ferreira, A.G. Advancements in waste water characterization through NMR spectroscopy. *Magn. Reson. Chem.* **2015**, *53*, 648–657. [CrossRef]

34. Tan, G.Y.A.; Chen, C.L.; Li, L.; Ge, L.; Wang, L.; Razaad, I.M.N.; Li, Y.; Mo, Y.; Wang, J.Y. Start a research on biopolymer polyhydroxyalkanoate (PHA): A review. *Polymers* **2014**, *6*, 706–754. [CrossRef]

35. Nkrumah-Agyeefi, S.; Scholz, C. Chemical modification of functionalized polyhydroxyalkanoates via "Click" chemistry: A proof of concept. *Int. J. Biol. Macromol.* **2017**, *95*, 796–808. [CrossRef] [PubMed]

36. Lorini, L.; Martinelli, A.; Pavan, P.; Majone, M.; Valentino, F. Downstream processing and characterization of polyhydroxyalkanoates (PHAs) produced by mixed microbial culture (MMC) and organic urban waste as substrate. *Biomass Convers. Biorefinery* **2021**, *11*, 693–703. [CrossRef]

37. de Oliveira Santos, R.P.; Ramos, L.A.; Frollini, E. Bio-based electrospun mats composed of aligned and nonaligned fibers from cellulose nanocrystals, castor oil, and recycled PET. *Int. J. Biol. Macromol.* **2020**, *163*, 878–887. [CrossRef] [PubMed]

38. Rossi, F.; Ramos, L.A.; Frollini, E. Renewable resources and a recycled polymer as raw materials: Mats from electrospinning of lignocellulosic biomass and PET solutions. *Polymers* **2018**, *10*, 538.

39. Santos, R.P.; Rodrigues, B.V.; Ramires, E.C.; Ruvolo-Filho, A.C.; Frollini, E. Bio-based materials from the electrospinning of lignocellulosic sisal fibers and recycled PET. *Ind. Crops Prod.* **2015**, *72*, 69–76. [CrossRef]

Article

Preparation of Nanochitin/Polystyrene Composite Particles by Pickering Emulsion Polymerization Using Scaled-Down Chitin Nanofibers

Ryuta Watanabe, Kakeru Izaki, Kazuya Yamamoto and Jun-ichi Kadokawa *

Graduate School of Science and Engineering, Kagoshima University, 1-21-40 Korimoto, Kagoshima 890-0065, Japan; k0984020@kadai.jp (R.W.); k3320769@kadai.jp (K.I.); yamamoto@eng.kagoshima-u.ac.jp (K.Y.)
* Correspondence: kadokawa@eng.kagoshima-u.ac.jp; Tel.: +81-99-285-7743

Abstract: In this study, we investigate the Pickering emulsion polymerization of styrene using scaled-down chitin nanofibers (SD-ChNFs) as stabilizers to produce nanochitin/polystyrene composite particles. Prior to emulsion polymerization, an SD-ChNF aqueous dispersion was prepared by disintegrating bundles of the parent ChNFs with an upper hierarchical scale in aqueous acetic acid through ultrasonication. After styrene was added to the resulting dispersions, the mixtures at the desired weight ratios (SD-ChNFs to styrene = 0.1:1–1.4:1) were ultrasonicated to produce Pickering emulsions. Radical polymerization was then conducted in the presence of potassium persulfate as an initiator in the resulting emulsions to fabricate the composite particles. The results show that their average diameters decreased to a minimum of 84 nm as the weight ratios of SD-ChNFs to styrene increased. The IR and ^{1}H-NMR spectra of the composite particle supported the presence of both chitin and polystyrene in the material.

Keywords: chitin nanofiber; composite particle; Pickering emulsion polymerization; polystyrene; scaled-down

Citation: Watanabe, R.; Izaki, K.; Yamamoto, K.; Kadokawa, J.-i. Preparation of Nanochitin/ Polystyrene Composite Particles by Pickering Emulsion Polymerization Using Scaled-Down Chitin Nanofibers. *Coatings* **2021**, *11*, 672. https://doi.org/10.3390/ coatings11060672

Academic Editor: Philippe Evon

Received: 20 May 2021
Accepted: 31 May 2021
Published: 1 June 2021

Publisher's Note: MDPI stays neutral with regard to jurisdictional claims in published maps and institutional affiliations.

1. Introduction

Chitin is one of the most abundant polysaccharides and acts as a structural material in biological systems. It has a regular structure consisting of β(1→4)-linked repeating units of *N*-acetyl D-glucosamine [1–3]. It remains an unutilized biomass resource, principally because of its intractable bulk structure and insolubility in water and common organic solvents. An efficient approach to the functional materialization of chitin is nanofibrillation, that is, the fabrication of nanocrystals and nanofibers [4–10] because of the exceptional properties of bio-based nanomaterials, such as their lightweight character, high tensile strength, low thermal expansion coefficient, biocompatibility, and nanosheet formability for sensing and electronic devices [11–17]. We previously developed a facile method for fabricating chitin nanofibers (ChNFs) with widths of approximately 20–60 nm based on a bottom-up approach where self-assembling regeneration from a chitin/ionic liquid (1-allyl-3-methylimidazolium bromide (AMIMBr)) ion gel using methanol was achieved [18,19]. This was based on our finding that AMIMBr efficiently dissolves and swells chitin [20]. Filtration of the resulting ChNF/methanol dispersion resulted in a film with a highly entangled nanofiber morphology.

We also previously reported styrene-in-water Pickering emulsion polymerization using self-assembled ChNFs as stabilizers to fabricate ChNF/polystyrene composite particles [21,22]. The stabilizers were additionally modified by anionic maleyl groups to provide amphiphilicity and simultaneously enhance dispersibility in aqueous media. Pickering emulsions are emulsions of any type, either oil-in-water, water-in-oil, or even multiple types which are stabilized by solid particles or other solid materials. These replace surfactants in general emulsions [23,24]. Pickering emulsion polymerization using polymerizable

137

hydrophobic substrates has proven to be a fascinating method for fabricating composite particles with well-defined morphologies [25]. In our system using maleylated ChNFs as stabilizers, particle sizes were controllable, where the sizes gradually decreased with increasing weight ratios of ChNFs to styrene from 0.02:1 to 0.2:1; however, a further increase in the ChNF ratio could not be obtained to produce smaller particles because of the insufficient dispersibility of the ChNF stabilizer under these higher content conditions in aqueous media.

Nevertheless, we found that the self-assembled ChNFs had a bundle-like structure and were constructed through the assembly of thinner fibrils [26]. Accordingly, the treatment of a ChNF film with aqueous sodium hydroxide for partial generation of amino groups on the chitin chains resulted in disentanglement of the bundles by cationization and electrostatic repulsion in 1.0 mol/L aqueous acetic acid with ultrasonication. This enabled individual thin fibril materials called scaled-down ChNFs (SD-ChNFs) to be obtained [27,28].

In the present study, we use SD-ChNFs as stabilizers for Pickering emulsion polymerization of styrene to fabricate composite particles (Figure 1). Due to their better dispersibility than the aforementioned maleylated ChNFs in aqueous media, nanochitin/polystyrene composite particles with smaller sizes could be obtained with a higher stabilizer content. As previous Pickering emulsion approaches for the fabrication of polystyrene particles have been conducted mostly using inorganic stabilizers such as SiO_2 and Fe_2O_3 substrates [29–32], the present materials have an advantageous potential to be used in biological and medicinal applications owing to the biocompatibility of ChNFs.

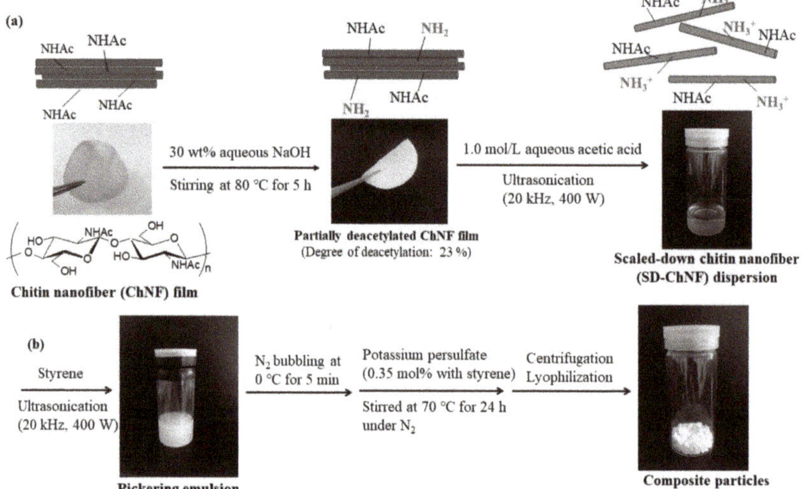

Figure 1. Procedures for the preparation of (**a**) a scaled-down chitin nanofiber (SD-ChNF) dispersion and (**b**) Pickering emulsion polymerization of styrene using SD-ChNFs as stabilizers.

2. Materials and Methods

2.1. Materials

Chitin powder from crab shells was purchased from Wako Pure Chemicals (Tokyo, Japan). An ionic liquid, AMIMBr, was prepared by the reaction of 1-methylimidazole with 3-bromo-1-propene according to a previously described method [33]. Other reagents and solvents were commercially available and used without further purification.

2.2. Preparation of SD-ChNFs

A mixture of chitin (0.120 g, 0.590 mmol) with AMIMBr (1.00 g, 4.92 mmol) was allowed to stand at room temperature for 24 h and subsequently heated with stirring at

80 °C for 24 h to obtain a chitin ion gel (10 wt%). The gel was then soaked in methanol (30 mL) at room temperature for a week for regeneration, followed by ultrasonication (Branson 1510, 42 kHz, 70 W, Branson Ultrasonics Corporation, Brookfield, CT, USA) for 4 h to yield a self-assembled ChNF dispersion with methanol. The dispersion was subjected to filtration to isolate ChNFs, and the isolated ChNFs were dried to obtain a ChNF film (0.117 g).

A mixture of the resulting film (0.117 g) with 30 wt% aqueous sodium hydroxide (30.0 mL) was heated at 80 °C for 5 h. The produced film was separated by filtration, washed with water and methanol, and dried to yield a partially deacetylated ChNF film (0.0824 g). From the integrated ratio of the signals assignable to acetamido protons to the signals assignable to anomeric (H1) protons in the ^{1}H-NMR spectrum of the sample hydrolyzed from the produced partially deacetylated (PDA-)ChNF film in DCl/D_2O, the degree of deacetylation (DDA) value was estimated to be 23.0%.

A mixture of the partially deacetylated ChNF film with 1.0 mol/L aqueous acetic acid (10.0 mL) was then ultrasonicated using a homogenizer (Branson Advanced-Digital Sonifier 450 (20 kHz, 400 W), Branson Ultrasonics Corporation, Brookfield, CT, USA) at room temperature for 30 min to yield an aqueous dispersion of SD-ChNFs.

2.3. Preparation of Nanochitin/Polystyrene Composite Particles

The typical experimental procedure was the following (run 5 in Table 1): Styrene (0.356 g, 3.42 mmol) was added to the aqueous dispersion of SD-ChNFs (0.501 g) and the mixture was ultrasonicated (Branson Advanced-Digital Sonifier 450 (20 kHz, 400 W), Branson Ultrasonics Corporation, Brookfield, CT, USA) for 30 min to produce a Pickering emulsion. After N_2 gas was bubbled into the emulsion for 5 min at 0 °C, potassium persulfate (2.80 mg, 0.0104 mmol, 0.35 mol% with styrene) was added. The resulting emulsified mixture was then kept at 70 °C for 24 h under stirring in a nitrogen atmosphere. The reaction mixture was centrifuged for 45 min at 3500 rpm, and the residual product was lyophilized to yield composite particles (0.0657 g). For scanning electron microscopy (SEM) measurement, a mixture of the product (1.00 mg) and water (10.0 mL) was ultrasonicated for 1 h to yield a re-dispersion.

Table 1. Average diameters and yields of composite particles.

Run	Weight Ratio (SD-ChNFs to Styrene)	Average Diameter (nm) [a]	Standard Deviation [a]	Yield (mg)
1	0.1:1	259	24.1	32.5
2	0.2:1	167	19.0	38.4
3	0.4:1	130	8.40	45.2
4	1:1	100	23.5	50.5
5	1.4:1	84	11.8	65.7

[a] Determined by SEM images.

2.4. Measurements

^{1}H-NMR spectra were recorded using a JEOL ECX 400 spectrometer (JEOL, Akishima, Tokyo, Japan). Microscopic laser images were obtained using a Keyence VK-8500 laser microscope (Keyence, Osaka, Japan). SEM images were obtained using a Hitachi S-4100H electron microscope (Hitachi High-Technologies Corporation, Tokyo, Japan) with an accelerating voltage of 5 kV. IR spectra were recorded on a PerkinElmer Spectrum Two spectrometer (PerkinElmer Japan Co., Ltd., Yokohama, Japan).

3. Results and Discussion

Prior to performing Pickering emulsion polymerization, we prepared SD-ChNF dispersions in aqueous acetic acid according to a previously reported procedure as follows (Figure 1a) [27,28]. The self-assembled ChNF film was first fabricated by regeneration from the chitin/AMIMBr ion gel using methanol to produce the ChNF dispersion and subsequent filtration [18]. In Figure 2a, the SEM image of the ChNF/methanol dispersion shows the same nanoscale morphology as that reported in a previous study [18]. The film was then subjected to treatment with 30 wt% aqueous sodium hydroxide at 80 °C for 5 h to achieve partial deacetylation. A DDA value of 23% was estimated by ^{1}H-NMR analysis after the acidic hydrolysis and dissolution in DCl/D_2O. The produced film was mixed with 1.0 mol/L aqueous acetic acid through ultrasonication using a homogenizer (20 kHz, 400 W) to produce the SD-ChNF/aqueous acetic acid dispersion. The sizes of the resulting nanochitin observed in the SEM image of the dispersion (Figure 2b) were clearly reduced as compared with those of the parent ChNF/methanol dispersion (Figure 2a), indicating the successful preparation of the desired SD-ChNF dispersion.

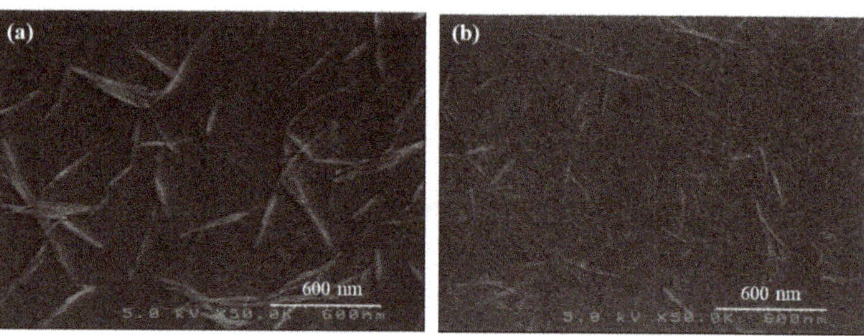

Figure 2. SEM images of the (**a**) parent ChNF/methanol dispersion and (**b**) SD-ChNF/aqueous acetic acid dispersion.

The Pickering emulsions were prepared by mixing styrene with the resulting dispersion (weight ratios of SD-ChNFs to styrene = 0.1:1–1.4:1, runs 1–5 in Table 1)) (Figure 1b). The laser microscopic images of the obtained mixtures (images of Pickering emulsions of runs 1 and 5 are representatively shown in Figure 3) supported the adequate formation of the emulsions at all weight ratios, even at higher ChNF contents (runs 3–5) than those that did not induce the formation of clear emulsions using the maleylated ChNFs [21]. These results suggest the practice of scaling down the parent ChNFs into SD-ChNFs for improving dispersibility in aqueous media in order to effectively act as stabilizers in Pickering emulsions.

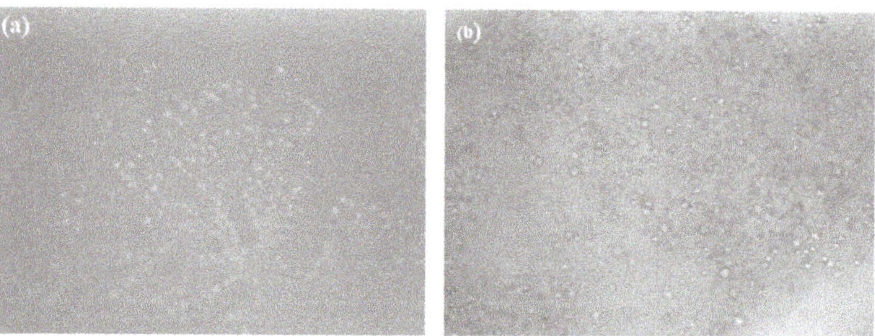

Figure 3. Microscopic laser images of Pickering emulsions for (**a**) run 1 and (**b**) run 5.

Pickering emulsion polymerization of styrene was then performed using the same procedure as that used for the maleylated ChNFs reported in our previous studies [21,22]. After nitrogen gas was bubbled into the emulsions, radical polymerization of styrene was conducted in the presence of potassium persulfate (0.35 mol% with styrene) at 70 °C for 24 h while stirring in a nitrogen atmosphere (Figure 1b). The products were isolated by centrifugation and lyophilized to obtain SD-ChNF/polystyrene composite particles. The IR spectrum of the product (run 2, Figure 4c) exhibited characteristic absorptions from chitin at 3437 cm^{-1} (O–H), 1656 and 1618 cm^{-1} (C=O of amide I), 1552 cm^{-1} (C=O of amide II), and 1072 cm^{-1} (C–O–C), as well as at 697 cm^{-1} for polystyrene (C–H of aromatic ring), respectively (Figure 4a,b). The ^{1}H-NMR spectrum of a solubilized fraction of the product of run 4 in CDCl$_3$ is shown in Figure 5, where the observed signals are assignable to polystyrene, but the chitin component was not analyzed by NMR measurement due to its insolubility in common NMR solvents. These data indicate that the particles consisted of both chitin and polystyrene. After the products were re-dispersed in water, SEM images of the spin-coated samples from the re-dispersions were captured to evaluate the morphologies and sizes of the composite particles.

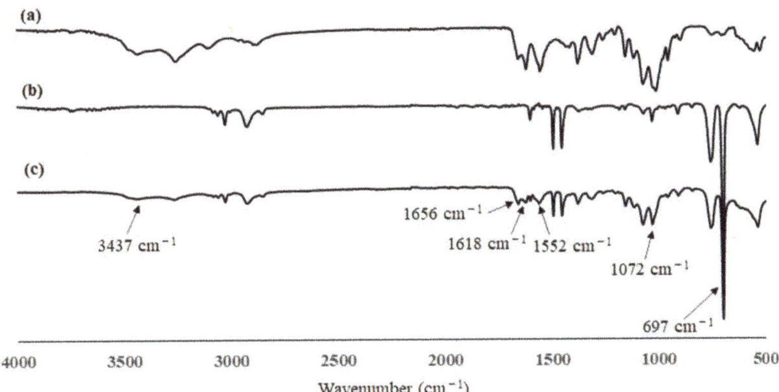

Figure 4. IR spectra of (**a**) the partially deacetylated (PDA-)ChNF film, (**b**) polystyrene, and (**c**) composite particles for run 2.

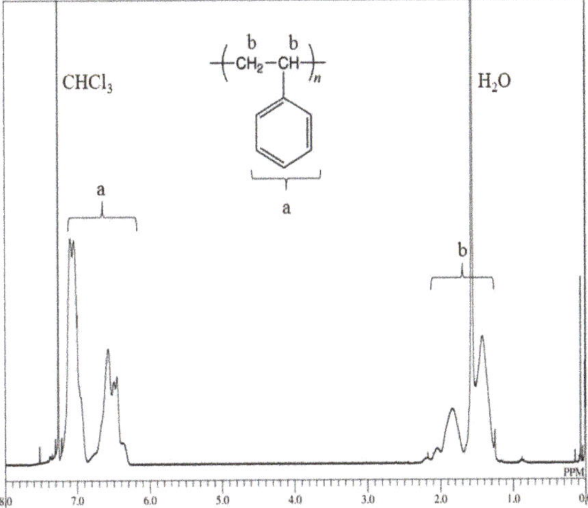

Figure 5. ^{1}H-NMR spectrum of solubilized fraction of composite particles for run 4 in CDCl$_3$.

The SEM images of the samples obtained using all the SD-ChNF/styrene weight ratios clearly show the particle morphologies (Figure 6). The particle sizes (average diameters) were calculated based on the vertical and horizontal lengths of 50 particles in each SEM image (Table 1). The standard deviation values for all the average diameters were relatively small, indicating the formation of composite particles with uniform sizes in all cases. The average diameter of the particles obtained by the SD-ChNF/styrene weight ratio of 0.1:1 (run 1) was comparable to that obtained by the same weight ratio of the maleylated ChNFs with an upper hierarchical scale as reported in our previous study (259 and 266 nm, respectively) [21]; however, at a higher weight ratio (0.2:1, run 2), the average diameter of the particles using the SD-ChNFs was smaller than that using the maleylated ChNFs (167 and 199 nm, respectively). Moreover, the average diameters were smaller with increasing SD-ChNF/styrene weight ratios and reached 84 nm with an SD-ChNF/styrene weight ratio of 1.4:1. These results strongly indicate that the SD-ChNFs efficiently acted as stabilizers for the present Pickering emulsion polymerization to fabricate composite particles that were smaller than 100 nm. The weight yields of composite particles increased when increasing the SD-ChNF/styrene weigh ratios, indicating the formation of more stable emulsions in accordance with increased SD-ChNF ratios.

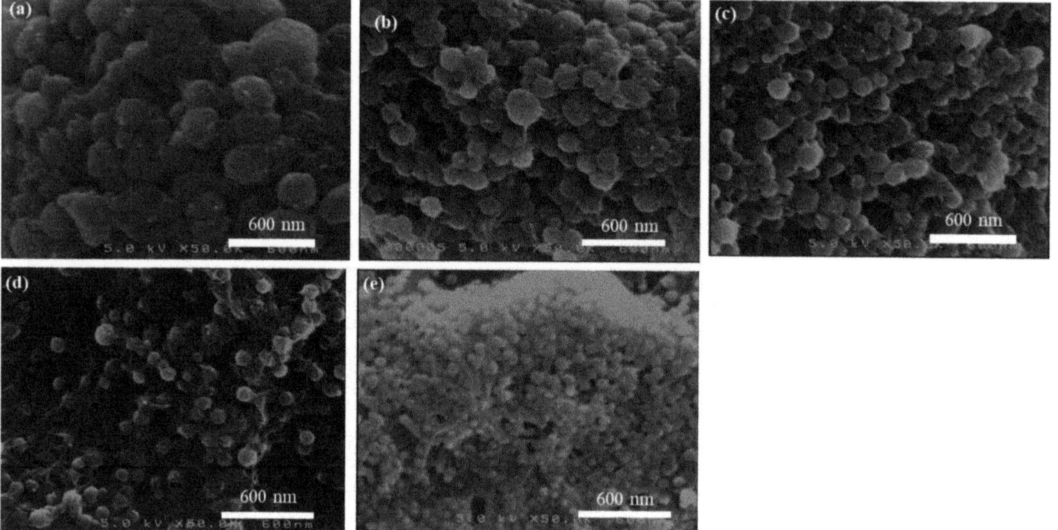

Figure 6. SEM images of SD-ChNF/polystyrene composite particles for (**a**–**e**) runs 1–5.

4. Conclusions

This study has investigated the Pickering emulsion polymerization of styrene using SD-ChNFs as stabilizers to fabricate nanochitin/polystyrene composite particles. The SD-ChNFs used as stabilizers efficiently formed Pickering emulsions with styrene in aqueous acetic acid, even under higher weight ratios of SD-ChNFs to styrene (e.g., 1.4:1). Radical polymerization of styrene in the resulting emulsions was then conducted in the presence of a radical initiator to fabricate the desired composite particles. Their average diameters decreased with increasing weight ratios of SD-ChNFs to styrene to a minimum of 84 nm. The resulting nanochitin/polystyrene composite particles can be converted into hollow particles composed of nanochitin/polystyrene shells by dissolving the inner polystyrene. The obtained materials have potential to be used as carriers for the controlled release of hydrophobic medical drugs owing to their biocompatibility and hydrophobicity, respectively. In the future, the approach described in this study will be employed in Pickering emulsions using other monomers to obtain nanochitin-based composite particles with functional polymers such as biodegradable and biocompatible polymers.

Author Contributions: J.-i.K. and K.Y. contributed to the methodology and writing of the paper. R.W. and K.I. performed the experiments and analysis. All authors have read and agreed to the published version of the manuscript.

Funding: This research received no external funding.

Institutional Review Board Statement: Not applicable.

Informed Consent Statement: Not applicable.

Data Availability Statement: Not applicable.

Acknowledgments: Not applicable.

Conflicts of Interest: The authors declare no conflict of interest.

References

1. Pillai, C.K.S.; Paul, W.; Sharma, C.P. Chitin and chitosan polymers: Chemistry, solubility and fiber formation. *Prog. Polym. Sci.* **2009**, *34*, 641–678. [CrossRef]
2. Rinaudo, M. Chitin and chitosan: Properties and applications. *Prog. Polym. Sci.* **2006**, *31*, 603–632. [CrossRef]
3. Kurita, K. Chitin and chitosan: Functional biopolymers from marine crustaceans. *Mar. Biotechnol.* **2006**, *8*, 203–226. [CrossRef] [PubMed]
4. Ifuku, S.; Nogi, M.; Abe, K.; Yoshioka, M.; Morimoto, M.; Saimoto, H.; Yano, H. Preparation of chitin nanofibers with a uniform width as α-chitin from crab shells. *Biomacromolecules* **2009**, *10*, 1584–1588. [CrossRef]
5. Ifuku, S. Preparation of chitin nanofibers from crab shell and their applications. *Kobunshi Ronbunshu* **2012**, *69*, 460–467. [CrossRef]
6. Ifuku, S.; Saimoto, H. Chitin nanofibers: Preparations, modifications, and applications. *Nanoscale* **2012**, *4*, 3308–3318. [CrossRef]
7. Ifuku, S. Chitin and chitosan nanofibers: Preparation and chemical modifications. *Molecules* **2014**, *19*, 18367–18380. [CrossRef]
8. Muzzarelli, R.A.A.; El Mehtedi, M.; Mattioli-Belmonte, M. Emerging biomedical applications of nano-chitins and nano-chitosans obtained via advanced eco-friendly technologies from marine resources. *Mar. Drugs* **2014**, *12*, 5468–5502. [CrossRef]
9. Rolandi, M.; Rolandi, R. Self-assembled chitin nanofibers and applications. *Adv. Colloid Interface sci.* **2014**, *207*, 216–222. [CrossRef]
10. Zhang, X.; Rolandi, M. Engineering strategies for chitin nanofibers. *J. Mater. Chem. B* **2017**, *5*, 2547–2559. [CrossRef] [PubMed]
11. You, J.; Li, M.; Ding, B.; Wu, X.; Li, C. Crab chitin-based 2D soft nanomaterials for fully biobased electric devices. *Adv. Mater.* **2017**, *29*, 1606895. [CrossRef] [PubMed]
12. Anraku, M.; Tabuchi, R.; Ifuku, S.; Nagae, T.; Iohara, D.; Tomida, H.; Uekama, K.; Maruyama, T.; Miyamura, S.; Hirayama, F.; et al. An oral absorbent, surface-deacetylated chitin nano-fiber ameliorates renal injury and oxidative stress in 5/6 nephrectomized rats. *Carbohydr. Polym.* **2017**, *161*, 21–25. [CrossRef] [PubMed]
13. Koizumi, R.; Azuma, K.; Izawa, H.; Morimoto, M.; Ochi, K.; Tsuka, T.; Imagawa, T.; Osaki, T.; Ito, N.; Okamoto, Y.; et al. Oral administration of surface-deacetylated chitin nanofibers and chitosan inhibit 5-fluorouracil-induced intestinal mucositis in mice. *Int. J. Mol. Sci.* **2017**, *18*, 11. [CrossRef]
14. Satam, C.C.; Irvin, C.W.; Lang, A.W.; Jallorina, J.C.R.; Shofner, M.L.; Reynolds, J.R.; Meredith, J.C. Spray-Coated Multilayer Cellulose Nanocrystal—Chitin Nanofiber Films for Barrier Applications. *ACS Sustain. Chem. Eng.* **2018**, *6*, 10637–10644. [CrossRef]
15. Mushi, N.E.; Nishino, T.; Berglund, L.A.; Zhou, Q. Strong and Tough Chitin Film from α-Chitin Nanofibers Prepared by High Pressure Homogenization and Chitosan Addition. *ACS Sustain. Chem. Eng.* **2019**, *7*, 1692–1697. [CrossRef]
16. Naghdi, T.; Golmohammadi, H.; Yousefi, H.; Hosseinifard, M.; Kostiv, U.; Horák, D.; Merkoçi, A. Chitin Nanofiber Paper toward Optical (Bio)sensing Applications. *ACS Appl. Mater. Interfaces* **2020**, *12*, 15538–15552. [CrossRef]
17. Sharma, P.R.; Sharma, S.K.; Lindström, T.; Hsiao, B.S. Water Purification: Nanocellulose-Enabled Membranes for Water Purification: Perspectives (Adv. Sustainable Syst. 5/2020). *Adv. Sustain. Syst.* **2020**, *4*, 2070009. [CrossRef]
18. Kadokawa, J.; Takegawa, A.; Mine, S.; Prasad, K. Preparation of chitin nanowhiskers using an ionic liquid and their composite materials with poly(vinyl alcohol). *Carbohydr. Polym.* **2011**, *84*, 1408–1412. [CrossRef]
19. Tajiri, R.; Setoguchi, T.; Wakizono, S.; Yamamoto, K.; Kadokawa, J. Preparation of self-assembled chitin nanofibers by regeneration from ion gels using calcium halide · dihydrate/methanol solutions. *J. Biobased Mater. Bioenergy* **2013**, *7*, 655–659. [CrossRef]
20. Prasad, K.; Murakami, M.; Kaneko, Y.; Takada, A.; Nakamura, Y.; Kadokawa, J. Weak gel of chitin with ionic liquid, 1-allyl-3-methylimidazolium bromide. *Int. J. Biol. Macromol.* **2009**, *45*, 221–225. [CrossRef]
21. Noguchi, S.; Sato, K.; Yamamoto, K.; Kadokawa, J.I. Preparation of composite and hollow particles from self-assembled chitin nanofibers by Pickering emulsion polymerization. *Int. J. Biol. Macromol.* **2019**, *126*, 187–192. [CrossRef] [PubMed]
22. Noguchi, S.; Yamamoto, K.; Kadokawa, J.I. Preparation of chitin-based fluorescent hollow particles by Pickering emulsion polymerization using functional chitin nanofibers. *Int. J. Biol. Macromol.* **2020**, *157*, 680–686. [CrossRef] [PubMed]
23. Chevalier, Y.; Bolzinger, M.-A. Emulsions stabilized with solid nanoparticles: Pickering emulsions. *Colloid. Surf. A* **2013**, *439*, 23–34. [CrossRef]
24. Yang, Y.; Fang, Z.; Chen, X.; Zhang, W.; Xie, Y.; Chen, Y.; Liu, Z.; Yuan, W. An Overview of Pickering Emulsions: Solid-Particle Materials, Classification, Morphology, and Applications. *Front. Pharm.* **2017**, *8*, 287. [CrossRef]
25. Bon, S.A.F. Pickering Emulsion Polymerization. In *Encyclopedia of Polymeric Nanomaterials*; Kobayashi, S., Müllen, K., Eds.; Springer: Berlin/Heidelberg, Germany, 2014; pp. 1–6.
26. Kadokawa, J.; Kawano, A.; Yamamoto, K. Fabrication of semi-crystalline film by hexanoylation on self-assembled chitin nanofibers. *ChemistrySelect* **2019**, *4*, 797–801. [CrossRef]
27. Hashiguchi, T.; Yamamoto, K.; Kadokawa, J. Preparation of soft materials by scaling down self-assembled chitin nanofibers. In Proceedings of the 101st CSJ Annual Meeting, Online, 19–22 March 2021.
28. Hashiguchi, T.; Yamamoto, K.; Kadokawa, J. Fabrication of highly flexible nanochitin film and its composite film with anionic polysaccharide. *Carbohydr. Polym.* **2021**. submitted.

29. Zhang, K.; Wu, W.; Meng, H.; Guo, K.; Chen, J.F. Pickering emulsion polymerization: Preparation of polystyrene/nano-SiO2 composite microspheres with core-shell structure. *Powder Technol.* **2009**, *190*, 393–400. [CrossRef]

30. Zhou, H.; Shi, T.; Zhou, X. Preparation of polystyrene/SiO_2 microsphere via Pickering emulsion polymerization: Synergistic effect of SiO2 concentrations and initiator sorts. *Appl. Surf. Sci.* **2013**, *266*, 33–38. [CrossRef]

31. Kim, Y.J.; Liu, Y.D.; Seo, Y.; Choi, H.J. Pickering-Emulsion-Polymerized Polystyrene/Fe_2O_3 Composite Particles and Their Magnetoresponsive Characteristics. *Langmuir* **2013**, *29*, 4959–4965. [CrossRef] [PubMed]

32. Kim, S.D.; Zhang, W.L.; Choi, H.J. Pickering emulsion-fabricated polystyrene–graphene oxide microspheres and their electrorheology. *J. Mater. Chem. C* **2014**, *2*, 7541–7546. [CrossRef]

33. Zhao, D.B.; Fei, Z.F.; Geldbach, T.J.; Scopelliti, R.; Laurenczy, G.; Dyson, P.J. Allyl-functionalised ionic liquids: Synthesis, characterisation, and reactivity. *Helv. Chim. Acta* **2005**, *88*, 665–675. [CrossRef]

Article

Study on the Grafting of Chitosan-Essential Oil Microcapsules onto Cellulosic Fibers to Obtain Bio Functional Material

Aicha Bouaziz [1,2,†], Dorra Dridi [3,†], Sondes Gargoubi [4,*], Abir Zouari [5], Hatem Majdoub [6], Chedly Boudokhane [5] and Aghleb Bartegi [1]

1   Bio-Resources, Integrative Biology & Valorization (BIOLIVAL, LR14ES06),
    Higher Institute of Biotechnology of Monastir, University of Monastir, 5000 Monastir, Tunisia;
    bouaziz.aicha@gmail.com (A.B.); a.bartegi@gmail.com (A.B.)
2   Higher School of Health Sciences and Techniques of Sousse, University of Sousse, 4054 Sousse, Tunisia
3   Unit of Analysis and Process Applied to the Environment (UR17ES32) Issat Mahdia, University of Monastir,
    5000 Monastir, Tunisia; dorra.dridi.jeddi@gmail.com
4   Textile Engineering Laboratory—LGTex, University of Monastir, 5000 Monastir, Tunisia
5   Research Unity of Applied Chemistry and Environment, Faculty of Science of Monastir,
    5000 Monastir, Tunisia; zouariabir2@gmail.com (A.Z.); chimi.tex@planet.tn (C.B.)
6   Laboratory of Advanced Materials and Interfaces, Faculty of Sciences of Monastir, University of Monastir,
    5000 Monastir, Tunisia; hatemmajdoub.fsm@gmail.com
*   Correspondence: gargoubisondes@yahoo.fr
†   A. Bouaziz and D. Dridi contributed equally to this work as first authors.

Citation: Bouaziz, A.; Dridi, D.; Gargoubi, S.; Zouari, A.; Majdoub, H.; Boudokhane, C.; Bartegi, A. Study on the Grafting of Chitosan-Essential Oil Microcapsules onto Cellulosic Fibers to Obtain Bio Functional Material. *Coatings* **2021**, *11*, 637. https://doi.org/10.3390/coatings11060637

Academic Editor: Philippe Evon

Received: 15 April 2021
Accepted: 21 May 2021
Published: 27 May 2021

Abstract: The purpose of this work was to prepare chitosan–essential oil microcapsules using the simple coacervation method and to graft them onto cellulosic fibers to obtain bio functional textile. The microcapsules morphology was characterized by optical microscopy. The 2D dimethyloldihydroxyethylene urea resin (DMDHEU) was used as a binding agent to graft microcapsules on the surface of cellulosic fibers. Scanning Electron Microscopy (SEM) photographs and Attenuated Total Reflectance-Fourier Transformed Infrared (ATR-FTIR) analyses were performed to prove the interaction between cellulosic fibers and microcapsules. Furthermore, the properties of the different fabrics such as mechanical strength and air permeability were investigated. Furthermore, washing durability was evaluated. Finally, the antibacterial activity of the finished fibers against the strains *Escherichia coli* (*E. coli*) and *Staphylococcus aureus* (*S. aureus*) was evaluated. The results evidence the ability of treated fabrics to induce bacteria growth inhibition. The coacervation method is a simple process to incorporate cinnamon essential oil on the cellulosic fiber's surface. The use of essential oils as active agents seems to be a promising tool for many protective textile substrates such as antimicrobial masks, bacteriostatic fabrics and healthcare textiles.

Keywords: cellulosic; fiber; microcapsules; chitosan; essential oil; bio functional material

1. Introduction

In recent years, the incidence of viruses, bacteria, parasites and fungi has risen considerably, especially among immune-compromised patients, geriatric and pediatric [1]. The uncontrolled growth of infectious diseases, which are becoming harder to treat, is a silent threat with long-term consequences for global public health and economy [2]. The COVID-19 crisis has brought into sharp focus the impact of infectious diseases. There are vital lessons to learn from this pandemic. Now more than ever, we need to make robust and comprehensive investments into the way we prepare and respond to health emergencies. This has the potential to avoid into long-term, global health threats and economic crises. A major challenge facing textile industries is to enhance the production of protective textiles. Natural fibers such as cotton are sensitive to microorganisms attack since their structure is hydrophilic and porous, allowing the retention of oxygen, moisture and nutrients required for microorganism development.

The healthcare sector consumes large amounts of textiles such as staff uniforms, patient clothing, operation room gowns and bed linen. Placing advanced properties and green criteria during production can increase the demand for antimicrobial finishing preventing microorganisms from adhering to the textile surface.

During the last decade, a number of antimicrobial textiles have been developed using techniques like the incorporation of antimicrobial agents into the synthetic fibers during extrusion, the surface grafting of active compounds or the fiber modification by physical or chemical treatments [3].

Several antimicrobial agents were imparted to textiles like triclosan, quaternary ammonium compounds, metal salts, polybiguanides and *N*-halamines [4].

However, the main disadvantage remains that these active components present drawbacks such as leaching of textiles during washing causing environmental problems and toxicity when exposed to sunlight such as for Triclosan, an efficient antimicrobial agent that has been banned by many countries [4].

Textile antimicrobial finishing needs to be effective against microorganisms, but the more important issue is the increasing demand of non-toxicity to the environment and the user [5]. Therefore, medical textile industries need to use new active agents that are safe and which present a broad spectrum of activity to ensure the control of microbes' multiplicity. In addition, optimum efficiency must be reached while providing the properties of textiles.

Recently, essential oils have been emerging as effective antimicrobial agents [6]. They are highly concentrated substances extracted from plant organs such as flowers, leaves, bark and seeds [7].

They contain various aromatic compounds [8]. These natural products have the advantage of providing a broad spectrum of biological activities [9]. This property makes them promising active agents to fight against the spread of multi-resistant microorganisms, which is one of the greatest challenges medical textile industries need to face.

Despite their antimicrobial activity, essential oils are not widely exploited in the textile sector since they do not dissolve in water [10]. To overcome this problem, Chitosan could be used to encapsulate these active compounds [11]. The target of the present work is to achieve cinnamon oil encapsulation using Chitosan and to graft the obtained microcapsules onto the cellulosic fiber's surface. The physio-chemical characteristics and antibacterial activity of finished samples were investigated.

2. Materials and Methods

2.1. Materials

Woven cotton fabric (areal mass density 280 g/m^2, threads per cm: warp 48 $\pm$ 2; weft 37 $\pm$ 1, and CIE whiteness 80) was purchased from the local market. The fabric was desized, bleached and mercerized.

Chitosan (deacetylation degree of 70%), acetic acid and NaOH were purchased from Sigma Aldrich. Tween 20 (surfactant) and dimethyloldihydroxyethylene urea DMDHEU (binding agent) were donated by Chimitex Plus company (Sousse, Tunisia). Barks of Ceylon cinnamon were obtained from shops. This species is not planted in Tunisia. However, the material is easy to obtain since cinnamon is commonly used as a spice in Tunisian cooking.

2.2. Methods

2.2.1. Essential Oil Extraction

The hydro-distillation was carried out for 3 h: 30 min, according to the following experimental protocol. An Amount of 250 g of mashed cinnamon sticks was laid out in a distillation flask containing water, then the system was brought to the boil. A condenser was placed between the distillation flask and the receiving container. The vapor produced in the flask drains the volatile components and then passes to the condenser, where it is condensed. After condensation, the volatile compounds are collected, cooled at room temperature and separated using a separatory funnel. The oil is recovered after decantation.

2.2.2. Microcapsules Preparation

Chitosan-essential oil microcapsules were prepared using the simple coacervation method [12]. Cinnamon essential oil (16 mL) was mixed with tween 20 (0.1 mL) and 1% Chitosan solution (50 mL) in 1% acetic acid (pH = 4). The solution was emulsified using an Ultra-Turrax (IKA, Staufen, Germany) operating at a speed of 900 rpm. The obtained emulsion was then slowly added to 250 mL of NaOH (0.2 M)/ethanol (4/1). The mixture was gently stirred to induce microcapsules formation and reinforcement of their walls.

2.2.3. Grafting of Microcapsules onto Fabric Surface

The cotton samples were submerged in the finishing bath, which contains the microcapsules emulsion (80 g/L) and the binding agent (40 g/L). The samples were then padded to obtain a wet pick up of 80%, pre-dried at 90 °C for 15 min, cured at 150 °C for 5 min, and stored at room temperature.

2.3. Characterization and Measurements

2.3.1. Optical Microscopy

The aspects of the microcapsules solution and the treated fibers were observed using an optical microscope (Leica DM 500, Leica Microsystems, Heerbrugg, Switzerland) connected to a Colorview camera and controlled by analysis software. The microscope is equipped with different objectives, allowing imaging with different magnifications. A slide and a cover glass were used to place a droplet of the solution or a treated fiber between them in order to observe under microscope. Photographs were taken in transmission mode under ×100 magnification.

2.3.2. Scanning Electron Microscopy

The untreated and treated cotton fabrics were characterized by a scanning electron microscope (JEOL JSM5400, JEOL Ltd., Tokyo, Japan). Sputter coating of the samples with conductive gold was achieved and imaging parameters were adjusted to obtain high quality images and to protect the original state of cotton fabrics.

2.3.3. ATR-FTIR Spectroscopy

An attenuated Total Reflectance-Fourier Transformed Infrared (ATR-FTIR, Perkin Elmer, Waltham, MA, USA) was used to examine the chemical changes of chitosan after essential oil loading and fabric surface after microcapsules grafting. The spectra were obtained via Spectrum Two™ FTIR (Perkin Elmer) and recorded with a spectral resolution of 4 cm^{-1} in the range 4000–400 cm^{-1}.

2.3.4. Encapsulation Efficiency and In Vitro Release of Essential Oil

The encapsulation efficiency of essential oil was evaluated using a UV/VIS spectrophotometer (Specord 210 plus, Analytik Jena AG, Jena, Germany). To determine residual oil, the microcapsules solution (1 mL) was mixed with hexane (9 mL). The mixture was allowed to settle for 10 min. The extracted residual oil in hexane was determined using spectrophotometer at 285 nm. The encapsulation efficiency (%) was calculated using a calibration curve constructed from the values of different concentrations of the cinnamon essential oil. The wavelength of 285 nm was considered for all the measurements. This value corresponds to the characteristic wavelength of cinnamaldheyde the main component of cinnamon oil.

The UV-Visible spectrum of cinnamon essential oil is shown in Figure 1. The Encapsulation efficiency (percentage) was then obtained as a percentage from the following equation:

$$\text{Encapsulation efficiency } (\%) = \left(\frac{\text{Total amount of essential oil} - \text{residual amount of essential oil}}{\text{Total amount of essential oil}} \right) \times 100 \qquad (1)$$

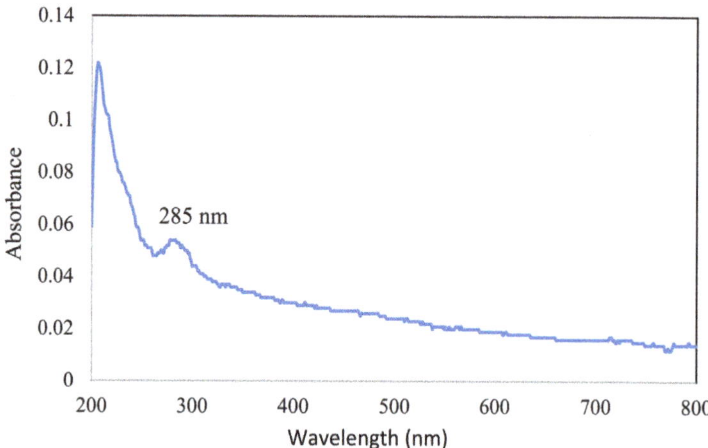

Figure 1. UV-visible spectra of cinnamon essential oil.

To determine the essential oil release, samples of 1 g of the finished fabric were submerged in hexane under agitation for 1 h at ambient temperature. The amount of oil released into the bath was quantified by UV/Vis Spectrophotometer at 285 nm. The Beer-Lambert law was used to measure the concentration of released oil based on the calibration curve of cinnamon oil.

2.3.5. Mechanical Properties

Mechanical tests were conducted according to NF EN ISO 13934-2 standards (in warp direction) via a testing machine (Lloyd LR 5 k, Lloyd Instruments Ltd., Largo, FL, USA). The load cell was 5 kN and the extension speed was 50 mm/min. Test samples had a gauge length of 75 mm and a thickness of 1 mm. Tests were run in triplicate.

2.3.6. Air Permeability

Air permeability of the specimens was determined according to the standard method ISO 9237 via an air permeability tester FX 3300 (Karl Schröder KG, Weinheim, Germany) at a constant pressure drop of 200 Pa.

2.3.7. Laundering Durability

Laundering durability was evaluated via a standard domestic washing process performed at 40 °C for 30 min using a detergent (anionic). After the washing process, samples were rinsed with water and dried at ambient temperature. To quantitatively evaluate the microcapsules attachment durability, microcapsules were counted on fibers surface using the optical microscope. A lot of fibers were separated from the treated fabric, the treated fabric after 5 washing cycles and the treated fabric after 10 washing cycles. The mean number and the standard deviation (SD) were calculated.

2.4. Antibacterial Activity

Antibacterial activity of treated fabrics was studied using the diffusion assay method. The tests were realized using the strains *Escherichia coli* (ATCC 8739), chosen because 80% of all infections are caused by gram-negative bacteria [13], and *Staphylococcus aureus* (ATCC 25923), used as a representative strain of gram-positive bacteria [14].

Prior to performing the experiments, all cotton samples were sterilized in the following conditions: Temperature 120 °C, pressure 100 kPa and time 20 min. The autoclaving process is fast and does not result in a loss of encapsulated essential oil. The microcapsules are able to protect the oil from volatilisation [15,16].

The microorganisms were cultured in petri dishes containing a nutrient agar and were incubated at 37 °C for 24 h. To adjust cell density to 108 colony forming units per milliliter (CFU/mL), the cultured microorganisms were diluted using a sterile saline physiologic solution. The optical density of the diluted suspension was considered to estimate the cell density. A total of 100 µL of the diluted suspension was spread with a sterile cotton swab on the surface of the nutrient agar contained in the petri dish. Then, the cotton samples (1×1 cm^2) were placed into a cultured nutrient agar petri dish and slightly pressed on the surface (transversely). The petri dishes with the cotton samples were incubated for 24 h at 37 °C. It is important to mention that the samples were lightly rubbed before the test.

Tests were conducted in duplicate. Images showing bacteria growth in the presence of cotton fabrics were saved and then treated with the software ImageJ. This software allows having quantifiable data from images. The number of pixels of the area covered with bacteria (near the edges of the fabric) is calculated and compared to that of the total uncovered area (the total area excluding the sample area). Therefore, the percentage of the area covered with bacteria is calculated.

3. Results and Discussion

3.1. Morphological Structure

The morphology of Chitosan-cinnamon oil microcapsules was investigated based on their optical micrograph. (Figure 2a). Optical micrograph shows that most of the microcapsules are spherical with no visible aggregation in them. The observed microcapsules present non-uniform size (the diameter varied between 20–50 µm) with clear differentiation between wall and core material. In previous studies, it has been reported that the non-uniformity of size results from a partial agglomeration of oil suspension before encapsulation [11].

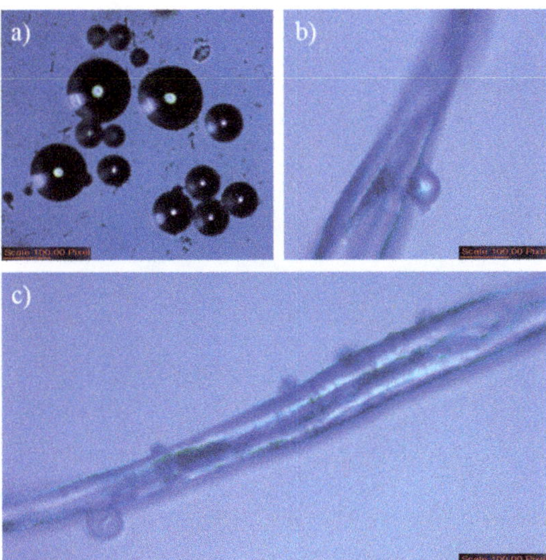

Figure 2. Optical photographs of (**a**) microcapsules solution (**b**) microcapsule on the cellulosic fiber surface and (**c**) a number of microcapsules throughout the length of the cellulosic fiber.

Optical microscopy was also used to observe the cellulosic fibers after treatment. As shown in Figure 2b,c micrographs of the treated cotton fibers confirmed that microcapsules have been successfully deposited on their surface. The microcapsules have conserved their spherical shape (Figure 2b) and their distribution was observed throughout the length of the fiber (Figure 2c).

The morphology of the treated sample was studied by SEM. Figure 3 shows the known spiral structure of cotton fiber. The fiber's surface appears relatively smooth with the presence of spherical entities, which are the microcapsules.

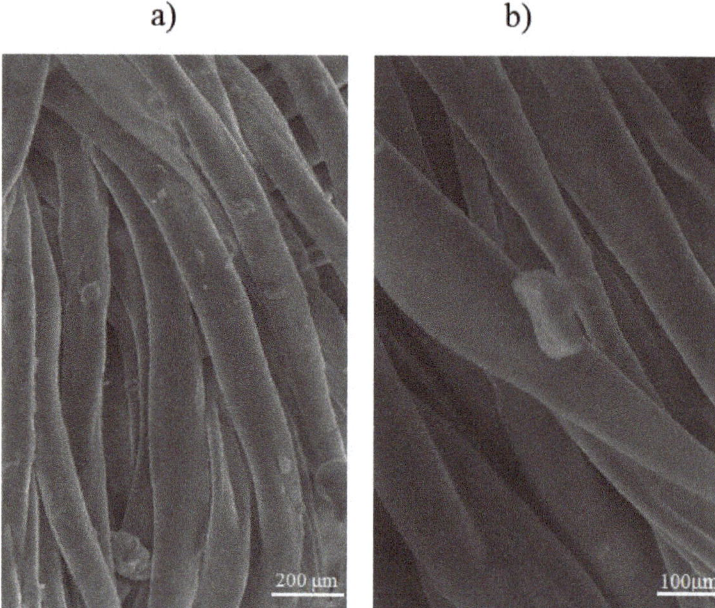

Figure 3. SEM images of treated fabric: (**a**) under ×200 magnification and (**b**) under ×400 magnification.

The treatment does not appear to damage the state of the fibers as evidenced by the SEM images. The fibers do not show any deterioration or degradation aspects in fibers.

3.2. ATR-FTIR Results

The solution of Chitosan microcapsules loaded with essential oil was analyzed by ATR FTIR. The spectrum (Figure 4) shows changes compared to the spectrum of pure Chitosan such as the fusion of the peak at 1650 cm^{-1} (amide I) with the peak at 1550 cm^{-1} (amide II). The two peaks appear with a wide absorption at 1640 cm^{-1}. In addition, the spectrum of microcapsules solution shows a characteristic large band around 3329 cm^{-1}. This band reveals the presence of OH groups (microcapsules aqueous solution).

The FTIR-ATR spectra of control fabric and treated fabric are shown in Figure 5. As expected, the FTIR spectra show characteristic peaks of cellulose. The stretching vibration of the hydroxyl group gives a broad peak at 3339 cm^{-1}. The band at 2897 cm^{-1} is characteristic of CH stretching vibration. Typical bands observed in the region of 1630–900 cm^{-1} are assigned to cellulose. The stretching and bending vibrations of –CH$_2$ and –CH, –OH and C–O bonds in cellulose gives the bands at 1428, 1368, 1317, 1061 cm^{-1} and 895 cm^{-1}. The bands at 895 cm^{-1} and around 1420–1430 cm^{-1} are associated to the amorphous and crystalline structure of the cellulose, respectively [17].

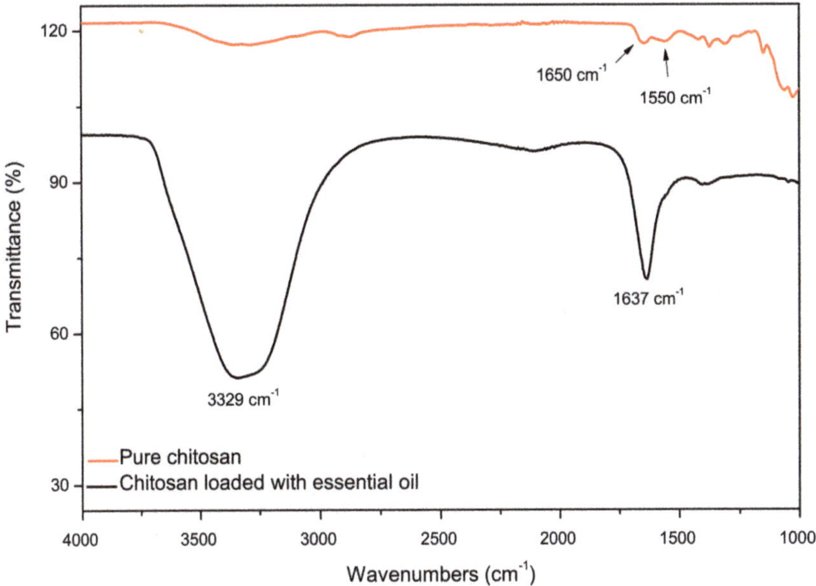

Figure 4. ATR-FTIR spectra of pure chitosan and chitosan loaded with essential oil.

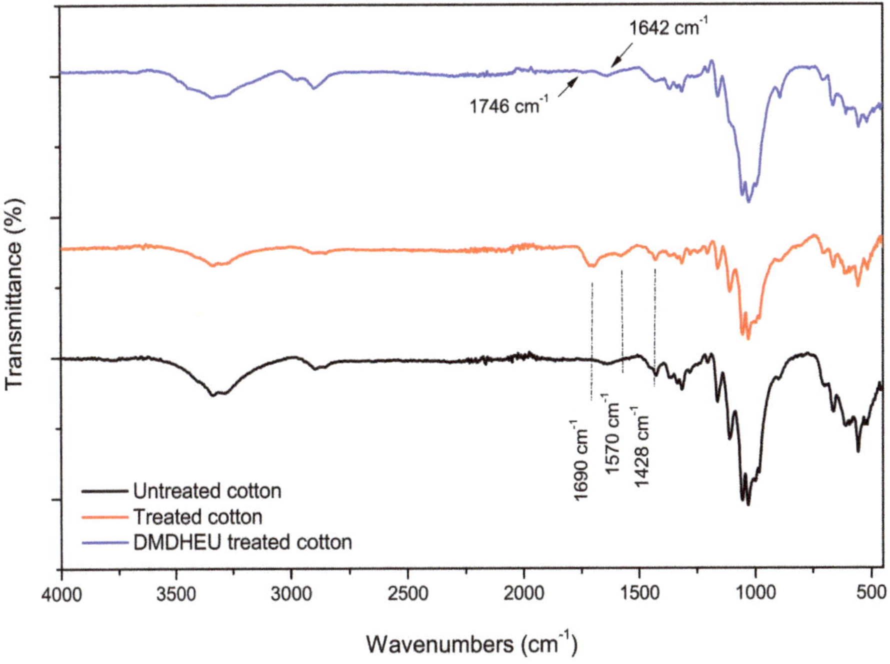

Figure 5. ATR-FTIR spectra of the treated, DMDHEU treated and untreated cotton fabrics.

The ATR-FTIR spectrum of DMDHEU treated cotton (without microcapsules) shows additional absorption peaks at 1746 and 1642 cm^{-1}. These peaks correspond to the C=O bond of ester groups resulting from crosslinking reaction (Figure 6) [18,19].

Figure 6. Chemical reaction between DMDHEU resin and Cotton/Microcapsules.

The peak around 3200–3400 cm^{-1} was broadened after the grafting of microcapsules. This funding is probably due to the overlapping of hydroxyl groups with the N-H vibration from Chitosan structure. Furthermore, the intensity of bands at 1690, 1570 and 1428 cm^{-1} was increased. These peaks belonged to the functional groups of amide I and amide II originated from the Chitosan [20]. This finding is in agreement with SEM results and confirmed the successful attachment of Chitosan-cinnamon oil microcapsules.

3.3. Encapsulation Efficiency and In Vitro Release of Essential Oil

The essential oil loading efficiency is an important detail for cotton functionalization purposes using microcapsules. It gives an idea of the amount of active component entrapped into them. The microcapsules loading efficiency was found to be 78.95 ± 3.64%. This funding is in agreement with previous results in studies dealing with chitosan-essential oil microcapsules preparation [21,22].

The essential oil release presents the fraction of cinnamon oil incorporated into cotton fabric grafted with microcapsules. The release of the essential oil was found to be 6.16 ± 0.71 µL/g, which was close to the values reported in the literature [16].

3.4. Mechanical Properties

Figure 7 shows the load–extension curves of the control and treated cotton fabrics. Both fabrics show a brittle failure with a sudden load drop when reaching a maximum load.

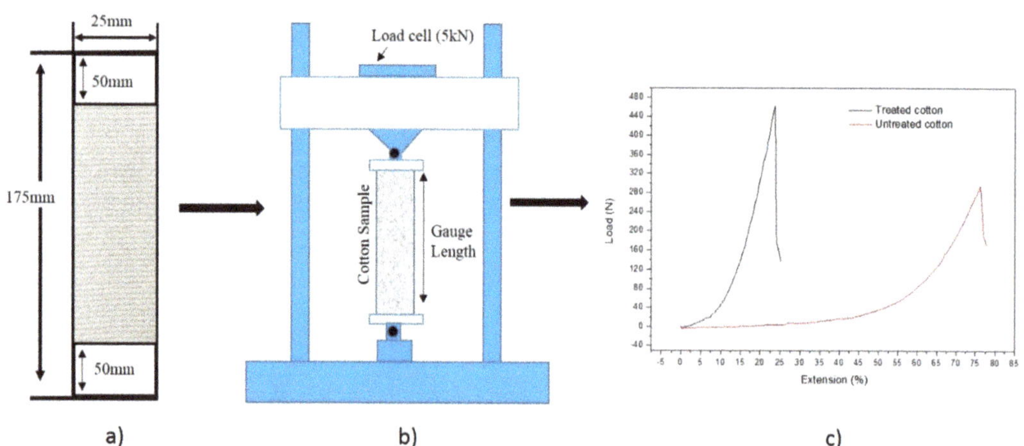

Figure 7. Tensile strength test: (**a**) Cotton sample dimensions, (**b**) The Apparatus for tensile test and (**c**) Load-extension curves of control and treated cotton fabrics.

As presented in Figure 7, the untreated sample showed the lower maximum breaking strength. After the microcapsules grafting, the value of maximum breaking strength increased. This result could be explained by the change of the surface topography. The roughening effect creates additional contact points within the fibers, resulting in enhanced interfiber friction. The low friction induces a weak cohesion between fibers and yarns [23].

Elongation provides an indication of the stretch rate of the fabric. Treated fabric was found to exhibit a reduced elongation rate; this behavior is due to less mobility of yarns and fibers and within the fabric. The binding agent, used for enhancing microcapsules attachment, results in an increase in the interaction force between yarns and fibers, leading to the reduction in their movement during extension.

3.5. Air Permeability

Air permeability is an important property for the final use of the treated textile. It provides an idea of the breathability of fabrics. Figure 8 shows that air permeability had slightly decreased after treatment. During the grafting process, yarn hair is generated and the space between the yarns is reduced. Fabric porosity is affected, resulting in low air permeability [24].

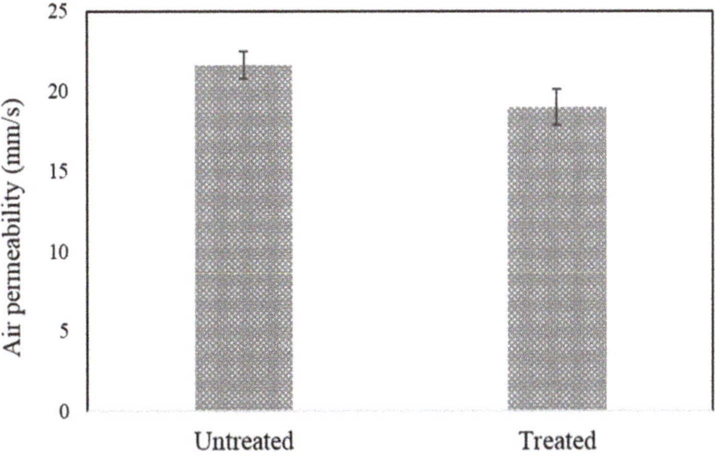

Figure 8. Air permeability of cotton fabrics.

3.6. Laundering Durability

Durability of microcapsules adhesion after the washing process is an interesting detail for farther functional application of treated fabrics.

Washing resistance is a big concern for the textile finishing industry, so it is necessary to control the washing durability of finished fabrics. Results in the literature showed that fabrics grafted with microcapsules exhibit a washing resistance after 5 to 40 times [16].

Figure 9 shows that an important number of microcapsules was still remaining on the fibers after five washing cycles. However, a decrease in the number of fixed microcapsules was noticed after ten washing cycles.

The mean number was considered to determine the washing durability of the treatment. Table 1 presents the results of microcapsules counting for different treated fabrics. The average number is 7.2 for the treated fabric before wash, 5 after five cycles and 0.9 after ten cycles.

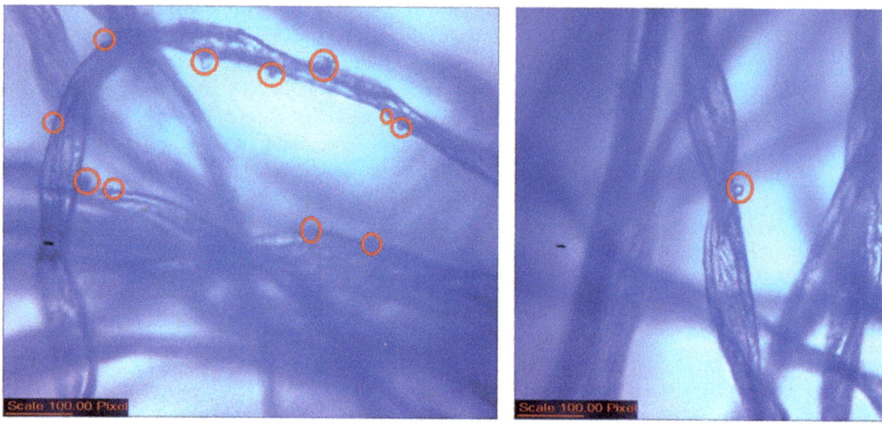

Figure 9. Optical photographs of treated fabric after washing cycles.

Table 1. Microcapsules number before and after washing cycles.

Tested Sample		Microcapsules Number	
Lot	**Number of Fibers**	**Mean**	**SD**
Before wash	12	7.2	1.77
After 5 cycles	9	5	1.12
After 10 cycles	7	0.9	0.72

The content of microcapsules on the fibers has gradually decreased as the number of washing cycles increased. The washing resistance of the treated fabrics after five washing cycles was due to the penetration of microcapsules into the voids between the fibers. In addition, the Chitosan, forming the microcapsules walls, is sticky by nature. This property makes the microcapsules resistant to wash.

Furthermore, the use of DMDHEU during the finishing process enhances the microcapsules' attachment to the surface of the fibers. Strong chemical bonds are formed. These bonds prevent the microcapsules from falling during the washing stage.

3.7. Antibacterial Activity

Figure 10 shows the qualitative and quantitative results of antibacterial activity. Untreated cotton did not show an inhibiting effect of bacteria. The bacterial growth was expanded around the samples. However, a peripheral inhibition was observed at the edges of treated samples. The microcapsules grafting on cotton fibers has led to a reduction in the bacteria growth of around 90% for the strain *S. aureus* and around 97% for the strain *E. coli*.

The treated fabrics were able to inhibit the growth of the tested strains. According to previous reports, the resulting antibacterial activity is mainly due to the cinnamaldehyde, the major compound of cinnamon oil [25] after the oil release through the microcapsules wall and not from the Chitosan itself.

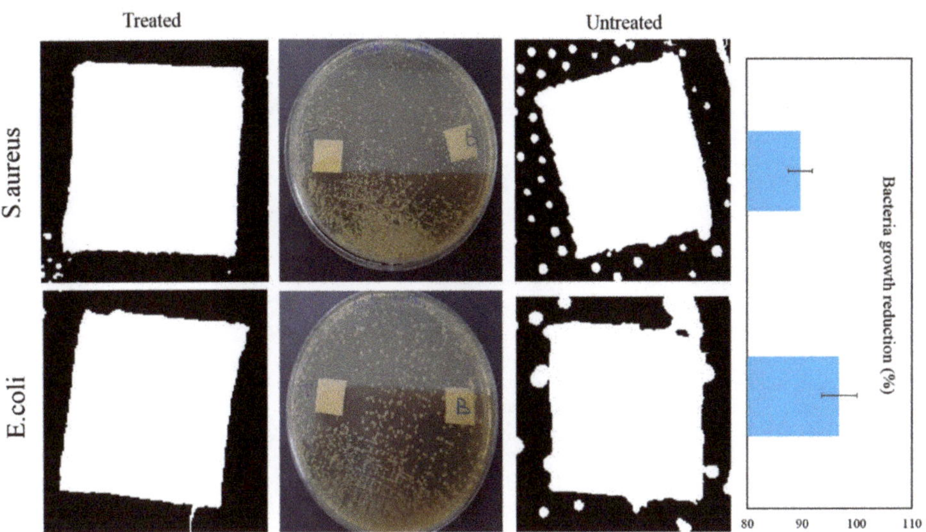

Figure 10. Antibacterial activity results of treated and untreated fabrics.

4. Conclusions

During this study, cinnamon oil was encapsulated using the coacervation technique and immobilized on the cotton fabrics.

Various characterization methods have allowed the evaluation of the microcapsules morphology and the efficiency of their grafting on the cellulosic fibers. Furthermore, mechanical properties and air permeability of different samples were tested. Washing durability was also evaluated. Finally, the resistance of treated fabric to bacteria growth was assessed using a qualitative method on gram-positive and gram-negative bacteria strains. Cotton grafting with chitosan-essential oil microcapsules was found to be a promising finishing process for the development of bio functional textile materials. Besides, essential oil appears as an interesting safe antibacterial agent.

Author Contributions: Conceptualization, A.B. (Aicha Bouaziz), D.D. and S.G.; methodology, A.B. (Aicha Bouaziz), D.D. and S.G.; experiments, A.B. (Aicha Bouaziz), D.D., S.G. and A.Z.; validation, H.M., C.B. and A.B. (Aghleb Bartegi); analysis, A.B. (Aicha Bouaziz), D.D., S.G., A.Z. and H.M.; investigation, A.B. (Aicha Bouaziz), D.D. and S.G.; resources, C.B. and H.M.; writing original draft preparation, A.B. (Aicha Bouaziz), D.D., S.G. and H.M.; writing review and editing A.B. (Aicha Bouaziz), D.D., S.G. and H.M.; supervision, A.B. (Aghleb Bartegi), H.M. and C.B.; project administration, S.G. and C.B. All authors have read and agreed to the published version of the manuscript.

Funding: This research received no external funding.

Institutional Review Board Statement: Not applicable.

Informed Consent Statement: Not applicable.

Data Availability Statement: Data sharing is not applicable to this article.

Conflicts of Interest: The authors declare no conflict of interest.

References

1. Lee, P.P.; Lau, Y.-L. Cellular and molecular defects underlying invasive fungal infections—revelations from endemic mycoses. *Front. Immunol.* **2017**, *8*, 735. [CrossRef] [PubMed]
2. Prestinaci, F.; Pezzotti, P.; Pantosti, A. Antimicrobial resistance: A global multifaceted phenomenon. *Pathog. Glob. Health* **2015**, *109*, 309–318. [CrossRef] [PubMed]

3.   Morais, D.S.; Guedes, R.M.; Lopes, M.A. Antimicrobial approaches for textiles: From research to market. *Materials* **2016**, *9*, 498. [CrossRef] [PubMed]

4.   Shabbir, M.; Yusuf, M.; Mohammad, F. Insights into Functional Finishing Agents for Textile Applications. In *Hanbook of Textile Coloration and Finishing: Studium Press*; USA LLC: New York, NY, USA, 2017; pp. 97–115.

5.   Joshi, M.; Purwar, R.; Ali, S.W.; Rajendran, S. Antimicrobial textiles for health and hygiene applications based on eco-friendly natural products. *Med. Healthc. Text.* **2010**, 84–92. [CrossRef]

6.   Mittal, R.P.; Rana, A.; Jaitak, V. Essential oils: An impending substitute of synthetic antimicrobial agents to overcome antimicrobial resistance. *Curr. Drug Targets* **2019**, *20*, 605–624. [CrossRef]

7.   Tongnuanchan, P.; Benjakul, S. Essential oils: Extraction, bioactivities, and their uses for food preservation. *J. Food Sci.* **2014**, *79*, R1231–R1249. [CrossRef]

8.   Tassoul, M.; Shaver, R. Effect of a mixture of supplemental dietary plant essential oils on performance of periparturient and early lactation dairy cows. *J. Dairy Sci.* **2009**, *92*, 1734–1740. [CrossRef]

9.   Ali, B.; Al-Wabel, N.A.; Shams, S.; Ahamad, A.; Khan, S.A.; Anwar, F. Essential oils used in aromatherapy: A systemic review. *Asian Pac. J. Trop. Biomed.* **2015**, *5*, 601–611. [CrossRef]

10.   de Souza, H.J.B.; Dessimoni, A.L.d.A.; Ferreira, M.L.A.; Botrel, D.A.; Borges, S.V.; Viana, L.C.; Oliveira, C.R.d.; Lago, A.M.T.; Fernandes, R.V.d.B. Microparticles obtained by spray-drying technique containing ginger essential oil with the addition of cellulose nanofibrils extracted from the ginger vegetable fiber. *Dry. Technol.* **2020**, 1–15. [CrossRef]

11.   Javid, A.; Raza, Z.A.; Hussain, T.; Rehman, A. Chitosan microencapsulation of various essential oils to enhance the functional properties of cotton fabric. *J. Microencapsul.* **2014**, *31*, 461–468. [CrossRef]

12.   Bustos, R.; Romo, L.; Yáñez, K.; Díaz, G.; Romo, C. Oxidative stability of carotenoid pigments and polyunsaturated fatty acids in microparticulate diets containing krill oil for nutrition of marine fish larvae. *J. Food Eng.* **2003**, *56*, 289–293. [CrossRef]

13.   Savaloni, H.; Haydari-Nasab, F.; Abbas-Rohollahi, A. Antibacterial effect, structural characterization, and some applications of silver chiral nano-flower sculptured thin films. *J. Theor. Appl. Phys.* **2015**, *9*, 193–200. [CrossRef]

14.   Davis, C.; Wagle, N.; Anderson, M.; Warren, M. Bacterial and fungal killing by iontophoresis with long-lived electrodes. *Antimicrob. Agents Chemother.* **1991**, *35*, 2131–2134. [CrossRef]

15.   Hsieh, W.-C.; Chang, C.-P.; Gao, Y.-L. Controlled release properties of chitosan encapsulated volatile citronella oil microcapsules by thermal treatments. *Colloids Surf. B Biointerfaces* **2006**, *53*, 209–214. [CrossRef]

16.   Zhang, T.; Luo, Y.; Wang, M.; Chen, F.; Liu, J.; Meng, K.; Zhao, H. Double-layered microcapsules significantly improve the long-term effectiveness of essential oil. *Polymers* **2020**, *12*, 1651. [CrossRef]

17.   Hospodarova, V.; Singovszka, E.; Stevulova, N. Characterization of cellulosic fibers by FTIR spectroscopy for their further implementation to building materials. *Am. J. Anal. Chem.* **2018**, *9*, 303–310. [CrossRef]

18.   Kongdee, A.; Bechtold, T. Cellulose-Protein Textiles: Utilisation of Sericin in Textile Finishing. In *Ecotextiles*; Elsevier: Amsterdam, The Netherlands, 2007; pp. 216–221.

19.   İlleez, A.A.; Dalbaşı, E.S.; Özçelik Kayseri, G. Seam Properties and Sewability of Crease Resistant Shirt Fabrics. *AATCC J. Res.* **2017**, *4*, 28–34. [CrossRef]

20.   Haque, A.N.M.A.; Remadevi, R.; Wang, X.; Naebe, M. Adsorption of anionic Acid Blue 25 on chitosan-modified cotton gin trash film. *Cellulose* **2020**, *27*, 9437–9456. [CrossRef]

21.   Wijesirigunawardana, P.B.; Perera, B.G.K. Development of a cotton smart textile with medicinal properties using lime oil microcapsules. *Acta Chim. Slov.* **2018**, *65*, 150–159. [CrossRef]

22.   Hussain, M.R.; Iman, M.; Maji, T.K. Determination of degree of deacetylation of chitosan and their effect on the release behavior of essential oil from chitosan and chitosan-gelatin complex microcapsules. *Int. J. Adv. Eng. Appl.* **2013**, *6*, 4–12.

23.   Lam, Y.L.; Kan, C.W.; Yuen, C.W.; Au, C.H. Objective measurement of fabric properties of the plasma-treated cotton fabrics subjected to cocatalyzed wrinkle-resistant finishing. *J. Appl. Polym. Sci.* **2011**, *119*, 2875–2884. [CrossRef]

24.   Du, W.; Zuo, D.; Gan, H.; Yi, C. Comparative study on the effects of laser bleaching and conventional bleaching on the physical properties of indigo kapok/cotton denim fabrics. *Appl. Sci.* **2019**, *9*, 4662. [CrossRef]

25.   El Atki, Y.; Aouam, I.; El Kamari, F.; Taroq, A.; Nayme, K.; Timinouni, M.; Lyoussi, B.; Abdellaoui, A. Antibacterial activity of cinnamon essential oils and their synergistic potential with antibiotics. *J. Adv. Pharm. Technol. Res.* **2019**, *10*, 63. [CrossRef]

Article

Electrospun Sericin/PNIPAM-Based Nano-Modified Cotton Fabric with Multi-Function Responsiveness

Jia Li [1,2,3], Bo-Xiang Wang [2,3], De-Hong Cheng [2,3], Zhi-Mei Liu [3], Li-Hua Lv [1], Jing Guo [1,4,*] and Yan-Hua Lu [2,3,*]

1   School of Textile and Material Engineering, Dalian Polytechnic University, Dalian 116034, China; lj18840597623@163.com (J.L.); lvlh@dlpu.edu.cn (L.-H.L.)
2   Liaoning Provincial Key Laboratory of Functional Textile Materials, Eastern Liaoning University, Dandong 118000, China; bxwang0411@163.com (B.-X.W.); chengdehongldxy1@163.com (D.-H.C.)
3   School of Chemical Engineering, Eastern Liaoning University, Dandong 118003, China; 13842507319@126.com
4   Liaoning Engineering Technology Research Center of Function Fiber and Its Composites, Dalian Polytechnic University, Dalian 116034, China
*   Correspondence: guojing8161@163.com (J.G.); yanhualu@aliyun.com (Y.-H.L.); Tel.: +86-137-0409-1879 (J.G.); +86-159-4253-2087 (Y.-H.L.)

Abstract: There is a significant interest in developing environmentally responsive or stimuli-responsive smart materials. The purpose of this study was to investigate multi-function responsive cotton fabrics with surface modification on the nanoscale. Three technologies including electrospinning technology, interpenetrating polymer network technology, and cross-linking technology were applied to prepare the multi-function sericin/poly(N-isopropylacrylamide)/Poly(ethylene oxide) nanofibers, which were then grafted onto the surfaces of cotton textiles to endow the cotton textiles with outstanding stimuli-responsive functionalities. The multi-function responsive properties were evaluated via SEM, DSC, the pH-responsive swelling behavior test and contact angle measurements. The results demonstrate that with this method, multi-function responsive, including thermo- and pH-responsiveness, cotton fabrics were fast formed, and the stimuli-responsiveness of the materials was well controlled. In addition, the antimicrobial testing reveals efficient activity of cotton fabrics with the sericin/PNIPAM/PEO nanofiber treatments against Gram-positive bacteria and Gram-negative bacteria such as *Staphylococcus aureus* and *Escherichia coli*. The research shows that the presented strategy demonstrated the great potential of multi-function responsive cotton fabrics fabricated using our method.

Keywords: sericin; poly(N-isopropylacrylamide); cotton fabrics; electrospinning

Citation: Li, J.; Wang, B.-X.; Cheng, D.-H.; Liu, Z.-M.; Lv, L.-H.; Guo, J.; Lu, Y.-H. Electrospun Sericin/PNIPAM-Based Nano-Modified Cotton Fabric with Multi-Function Responsiveness. *Coatings* **2021**, *11*, 632. https://doi.org/10.3390/coatings11060632

Academic Editor: Esther Rebollar

Received: 23 March 2021
Accepted: 12 May 2021
Published: 25 May 2021

Publisher's Note: MDPI stays neutral with regard to jurisdictional claims in published maps and institutional affiliations.

1. Introduction

Textiles are versatile materials composed of natural or synthetic fibers, with a wide range of applications [1,2]. With the global economy and technological advancements, smart textiles [3–5] are one of the research hotspots in the field of textiles and garments. Intelligent textiles are a sort of smart fabric or material which can be responsive to the external environment or an outside stimulus in behavior, including electrical, chemical, biological agents or physical temperature [6,7]. Currently, in order to introduce some kind of advanced function or special performance, smart textile materials are often modified by direct coating technology such as roll coating, the spraying method or deposition and surface treatment approaches, such as plasma treatment technology, ultraviolet irradiation [8], etc. These technologies are very useful and applied for smart textiles in improving or introducing some kind of certain performance based on the concrete requirements in actual applications. However, it should be mentioned that the utilization of traditional coating technology often inevitably hides the original capabilities of fabrics. Additionally, surface process techniques often concern a complicated process or environmental pollution along with surface structure destruction. This inevitably leads to the degeneration of some other characteristics of fabrics, such as tenderness, breathability and hygroscopicity.

157

As a comparison, electrospinning technology [9] has shown enormous potential such as filtration, separation [10,11], drug delivery of biomedical materials [12,13] and intelligent fabric modification because of great advantages such as the smaller fiber diameter, the higher porosity and permeability rate, large specific surface area, and high mechanical capacity [14,15], as well as its effective approaches of synergizing the required functions of materials and outstanding performance of textiles. However, the fabrics of intelligent modification with electrospinning technology are generally e-textiles and pressure-sensitive fabrics [16,17]. Furthermore, the fabrics of intelligent modification with electrospinning technology are generally e-textiles [18] and pressure-sensitive fabrics [19]. Though coupling electrospinning technology to thermo-responsive polymers such as poly(N-isopropylacrylamide) can effectively produce smart responsive nanofibers [20–23], few studies use them to modify textiles. In addition, most thermo-responsive textiles are conducted by coating smart polymers in industrial technologies. Currently, the coating smart polymers are mainly prepared and applied in the form of three-dimensional hydrogels with the shortcoming of smaller poriness and lower surface area and response rate.

In our previous work [24], multifunctional mulberry silk fabrics were successfully prepared with PNIPAM/chitosan/PEO nanofibers. The modified mulberry silk fabrics represented brilliant temperature- and pH-susceptivity and antibacterial capabilities. This prompted us to investigate other fabric applications such as cotton fabrics. Based on this, we chose the sericin of the Antheraea pernyi silks as the main raw materials. This took full advantage and improved the application of Antheraea pernyi silk sericin.

Here, we present a strategy for the fabrics, which brings some new ideas of modified fabrics. First, we prepared a kind of thermo-responsive polymer hydrogel by blending sericin/PEO solutions with N-isopropylacrylamide using an interpenetrating polymer network technology. Second, we prepared thermo-responsive nanofiber materials through electrospinning technology. Third, the modified cotton textiles that possessed temperature- and pH-sensitivity were produced by using the nanofiber network. The properties of the smart textiles were investigated and contrasted with those of the other samples. In addition, the biological property of smart textiles was examined with an antimicrobial activity test. This study suggests the potential of the temperature- and pH-responsive smart textiles to tissue engineer support materials, medicine slow release materials and other responsive materials.

2. Experimental Procedure

2.1. Materials

The raw materials (*Antheraea pernyi* silk cocoons, Liaoning Tussah Silk Institute Co., Ltd., Dandong, China) were applied to prepare the regenerated sericin (SS). N-isopropylacrylamide (NIPAM, 98%, Aladdin Reagent, Shanghai, China), N,N,N',N'-tetramethylethylenediamine (TEMED, 99%, Aladdin Reagent, Shanghai, China), ammonium peroxodisulfate (APS, Sinopharm Chemical Reagent Co., Ltd., Shanghai, China) and glutaric dialdehyde (GA, 25%, Kermel Reagent, Tianjin, China) were all analytical grade and used without further purification. Poly(ethylene oxide) (PEO, Mw = 400,000 Da, Shanghai, China) was used without further purification.

2.2. Preparation of the Sericin/PNIPAM/PEO Composite Spinning Solution

Antheraea pernyi is a wild nonmulberry silkworm species, which is a protein fiber and composed of fibroin and sericin [25]. Antheraea pernyi silks were considered increasingly for the textile and apparel industries [26] and medical applications [27] in recent years for their excellent mechanical properties and biocompatibility. As previously reported, regenerated sericin (SS) was extracted from *Antheraea pernyi* raw silk cocoons by using LiBr extractive technique [28]. Then, the regenerated sericin and PEO were dissolved in water under continuous magnetic stirring at 80 °C to prepare a mixed solution. The concentrations of sericin and PEO were 6 wt.% and 15 wt.%, respectively. The monomer solution (6.5 wt.%) was produced by putting the NIPAM monomer in distilled water and the initiator agent (APS) was all mixed with the above sericin/PEO solution, respectively. Then, the sericin/PNIPAM/PEO interpenetrating network blending spinning solution was prepared by adding TEMED in an as-prepared mixed solution and stirred at 40 °C for 1 h. The concentration values of the sericin, PEO and NIPAM were chosen by our early investigator's literature [24,29]. The formulations, conductivity and surface tension of blending solution is shown in Tables 1 and 2, respectively. The HAAKE RheoStress 1 rheometer (25 mm diameter plate with 1 mm gap) was used to estimate the blending spinning solution viscosity at room temperature under the increasing shear rate from 2 to 100 s^{-1} (see Figure 1).

Table 1. The formulations of the interpenetrating network blending spinning solution.

Code	Sericin: NIPAM: PEO (Volume Ratio)	Sericin (mL)	NIPAM (mL)	PEO (mL)	APS (mg)	5%TEMED (uL)
1	25/45/30	5	9	6	1.36	27.2
2	40/30/30	4	3	3	0.45	9.1
3	55/15/30	11	3	6	0.45	9.1

Table 2. The conductivity and surface tension of the interpenetrating network blending spinning solution.

Code	Sericin: NIPAM: PEO (Volume Ratio)	Conductivity (mS/cm)	Surface Tension (mN/m)
1	25/45/30	0.64 ± 0.15	43.5 ± 0.15
2	40/30/30	0.84 ± 0.15	47.7 ± 0.17
3	55/15/30	0.96 ± 0.12	46.4 ± 0.20

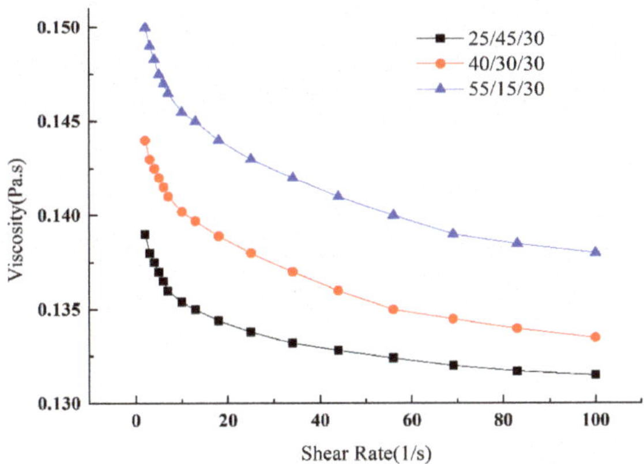

Figure 1. Viscosity of the interpenetrating network blending spinning solution.

2.3. Fabrication of Electrospun Sericin/PNIPAM/PEO Nanofibers and Functionalization of Cotton Fabric

Cotton fabric was pretreated by putting it into ultrasonic cleaners which contain distilled water for 0.5 h at room temperature and then dried in the air. The primary sericin/PNIPAM/PEO spinning solutions were loaded into a syringe equipped with a 0.5 mm diameter metal needle and electrospun on the prepared cotton fabric using an electrospinning machine. The functional cotton fabric was prepared by putting the above cotton fabric with nanofibers on the airtight container with glutaraldehyde vapor at room temperature for 24 h for the further cross-linking. The glutaraldehyde showed adequate safety, having been well applied in the field of biological medicinal materials [30,31]. The illustration of the functional cotton fabric is shown in Figure 2.

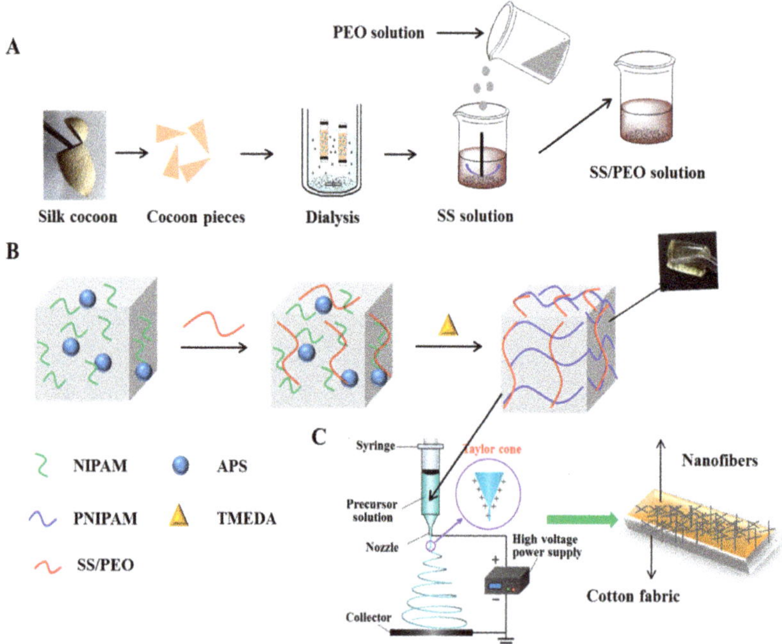

Figure 2. Schematic illustration of the (**A**) fabrication and (**B**) preparation of multifunctional cotton fabric.

The machine (KH08, Jinan, China) was purchased from Jinan Liangrui Technology Co., which had a high voltage direct current (HVDC) and a syringe pump. It was used to prepare the electrospun nanofibers. In this work, all of the electrospinning process was performed at 65% relative humidity (RH), room temperature. The electrospinning parameters are shown as follows: applied voltage 30 kV, working distance 15 cm and flow rate 0.03 mL/h. The parameters for the electrospinning experiments were chosen by our early investigator's literature [24,29]. The nanofibers were deposited at about 1 h and obtained the mats.

2.4. Characterization

2.4.1. Scanning Electron Microscopy

The microstructure of the nanofiber mats and cotton fabric was studied by scanning electron microscope (JSM-IT100, JEOL Ltd., Tokyo, Japan) with acceleration of voltage 20 KV. Before the imaging process, all the samples were sputter-coated with gold.

2.4.2. Fourier Transform Infrared Spectroscopy

Fourier transform infrared spectroscopy (FT-IR, Tensor-37, Bruker, Berlin, Germany) was used to determine the characteristics of the microstructure of NIPAM, sericin, PNIPAM, sericin/PNIPAM/PEO nanofibers and the functionalized cotton fabric with the KBr technique at wavelengths ranging between 400 cm^{-1} and 4000 cm^{-1}. The samples were obtained in a vacuum dryer for 4 h before acquiring the spectra.

2.4.3. Differential Scanning Calorimetry

Differential scanning calorimeter (DSC 7, Perkin Elmer, Waltham, MA, USA) was applied to evaluate thermal behaviors of untreated cotton fabric and functionalized cotton fabric under nitrogen with the flow rate at about 20 mL/min. The swollen sample weight was about 10 mg. The above samples were put in the aluminum sample holder and frozen below $-20\,°C$. The samples were operated at 0–50 °C, 2 °C min^{-1}. The samples performed heating/cooling cycles three times.

2.4.4. Contact Angle Measurements

Static water contact angle tests were accomplished by using a drop shape analyzer (PT-705, Shanghai, China) with the needle method at flow rate of 1.0 μLs^{-1} in 16% of relative humidity and temperature ranging from 15 °C to 45 °C. The Young Laplace drop profile fitting method was applied to evaluate the static contact angle results. The needle specification is shown as follows: diameter 0.5 mm and the water droplet volume ~5 μL. The standard deviation of the measurement series might create the error bars.

2.4.5. pH-Responsive Swelling

Sericin is an inherently weak amphipathic polyelectrolyte with the acidic side and alkaline side base, which makes it sensitive to pH [32,33]. The swelling behaviors of sericin/PMIPAM/PEO hydrogels were conducted within a variety of pH values (1.0~11.0), in view of the crucial aspect of the parameters in responsiveness.

In order to determine the swelling ratio, the dried samples were immersed in various pH buffers for 24 h in the temperature 15 °C and 37 °C that completes with the hydrogels' low critical solution temperature range. Before swollen sample mass testing, it was necessary to remove the surface water of specimens. The average of five samples was used to obtain the final results. The swelling ratio (SR) was calculated as follows [34]:

$$SR(\%) = [(W_t - W_d)/W_d] \times 100 \tag{1}$$

where W_t and W_d are the masses of the swollen sample and dried sample, respectively.

2.4.6. Antimicrobial Activity Measurements

The antibacterial activity of samples was investigated against Gram-positive bacteria and Gram-negative bacteria such as *Staphylococcus aureus* and *Escherichia coli* on the basis of the AATCC 100 test method and GB/T 31713-2015. Before the assay, all samples were made into small pieces at about 0.5×0.5 cm^2. Additionally, UV exposure was applied to sterilize the above samples at about 30 min. The above samples were put, respectively, in the bacterial suspension by using sterile forceps and then immersed in a flask including phosphate-buffered saline at 0.3 mM/70 mL. The phosphate-buffered saline consisted of monopotassium phosphate and the cell culture solution. The cell concentration was about 1×10^5–4×10^5 (CFU)/mL. We shook the above flask on a rotary shaker at 150 rpm, 24 °C, 1 h. The 0.5 mL of culture solution was obtained, respectively, from incubated samples before and after shaking for 1 h, and then diluted and placed on the agar plates that were cultivated at 37 °C, 1 h. The inhibitory rate (%) was decided as follows [35]:

$$R(\%) = [(B - A)/B] \times 100 \tag{2}$$

where R, B and A were the percentage of bacterial reduction and the bacterial colonies before and after shaking for 24 h, separately.

3. Results and Discussion

3.1. SEM

The surface morphology of the sericin/PNIPAM/PEO nanofibers and functionalized cotton fabric under optimal conditions was decided by SEM (see Figure 3). The final images of the sericin/PNIPAM/PEO nanofibers indicated that the fibers were uniform and continuous without any bead formation (Figure 3a). As shown in Figure 3a, there was breakage in nanofibers when the spinning solution had more NIPAM. The cause might be that the increasing NIPAM caused the bad mechanical properties of nanofibers. This indicates that sericin, NIPAM and PEO were well blended. Based on the morphological analysis of the functionalized cotton fabrics (Figure 3b,c), the sericin/PNIPAM/PEO nanofibers were overlapped and fixed on the surface of the cotton fabric formation bilayer structure. The cross-linked reaction mechanism between sericin and cotton fabric is shown in Figure 3d. The FTIR spectra analysis also confirms the chemical cross-linking reaction of the cotton fabrics and sericin in the presence of glutaraldehyde [36]. The results of SEM showed that the functionalized cotton fabric was also successfully prepared by a combination of electrospinning technology and an interpenetrating polymer network technology.

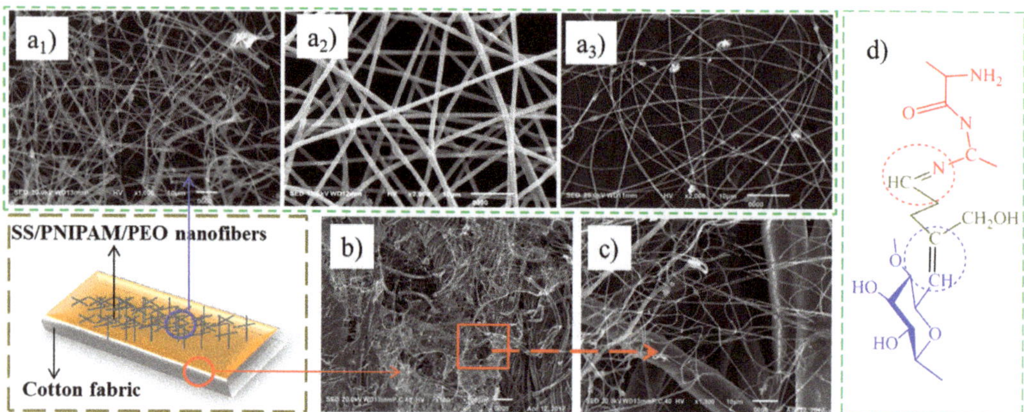

Figure 3. The morphology of (**a**) Sericin/PNIPAM/PEO nanofibers: ((**a₁**) 25/45/30; (**a₂**) 40/30/30; (**a₃**) 55/15/30); (**b,c**) Functionalized cotton fabric; (**d**) Reaction mechanism between sericin and cotton.

3.2. FT-IR

Figure 4A presents the Fourier transform infrared spectroscopy of the NIPAM and PNIPAM powders ranging from 2000 cm^{-1} to 1000 cm^{-1}. The disappearance of the 1622 cm^{-1} (–C=C–) peak showed the success of the synthesis of PNIPAM between NIPAM monomers (Figure 4A(a)). The 1652 cm^{-1} connected with the C=O, the 1549 cm^{-1} was related to the C–N stretching vibration and N–H flexural vibration, and the 1245 cm^{-1} corresponded to the C–N–H of PNIPAM [37]. FT-IR was employed to investigate the relationship between sericin/PNIPAM/PEO nanofibers and cotton fabric. The H–C=N– and H–C=C were formed by the crosslinking reaction between GA and sericin and cotton fabric, respectively.

From Figure 4B(e), it could be observed that the characteristic absorption bands at 3084 cm^{-1} could be contributing to the H–C=C group, which proved the cross-linking reaction between the –OH group on cotton and GA. The characteristic band at 1635 cm^{-1} corresponded to the H–C=N– stretching vibration of the modified cotton fabric. This suggested that the cross-linking reaction between the sericin and GA occurred. The –CHO produced a characteristic band at 1713 cm^{-1}. As we know, there was no extra GA in the resulting modified cotton fabric. Thus, the GA could make the reaction with the cotton textiles and sericin, separably. It was indicated that the glutaraldehyde could be a "bridge" that can crosslink the cotton fabric and the nanofibers [33,38].

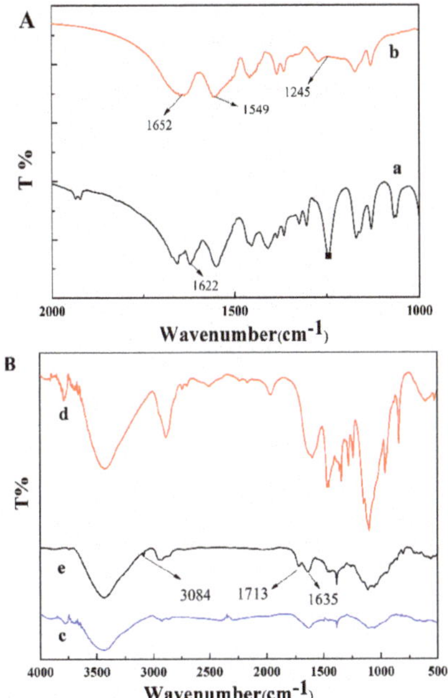

Figure 4. (A): The FT-IR spectra (2000–1000 cm^{-1}) of (**a**) NIPAM powders; (**b**) PNIPAM powders; (**B**): The FT-IR spectra (4000–500 cm^{-1}) of (**c**) Sericin/PNIPAM/PEO nanofibers; (**d**) Cotton fabric; (**e**) Functionalized cotton fabric.

3.3. Thermosensitive Behavior

Figure 5A shows the schematic illustration of PNIPAM involving the hydrophobic/hydrophilic transformation. PNIPAM interacted preferentially with water molecules and formed a hydrogen bond in the temperature range below LCST [39]. When increasing the temperature above LCST, PNIPAM molecules gradually turn into hydrogen bonds between polymer molecules. Thus, it was severely restricted for each molecule's mobility because of the hydrogen-bonded polymer network.

The surface wettability of functionalized cotton fabric was measured with different temperatures by using the static contact angle of water droplets (5 μL). Figure 5B exhibits that the static water contact angle of functionalized cotton fabrics increased apparently with the increasing temperature of the fabrics, demonstrating a hydrophilicity/hydrophobicity change of functionalized cotton fabrics. The wettability of functionalized cotton fabrics became a hydrophobic state and the contact angle increased up to the maximum at about 95 ± 1.5°, with the temperature improved.

Figure 5C shows the differential scanning calorimetry (DSC) thermogram curves of modified cotton fabric and cotton fabric. The initial temperature was considered as the lower critical solution temperature (LCST) value that was estimated with the intersection points of the two tangents (see Figure 5). It could be seen that the LCST of nanofibers and modified cotton fabric is 32.03 °C and 34.47 °C, respectively. Additionally, there was no phase transition peak in common cotton fabric. It was shown that the LCST of modified cotton fabric is between 30 °C and 36 °C, close to human body temperature.

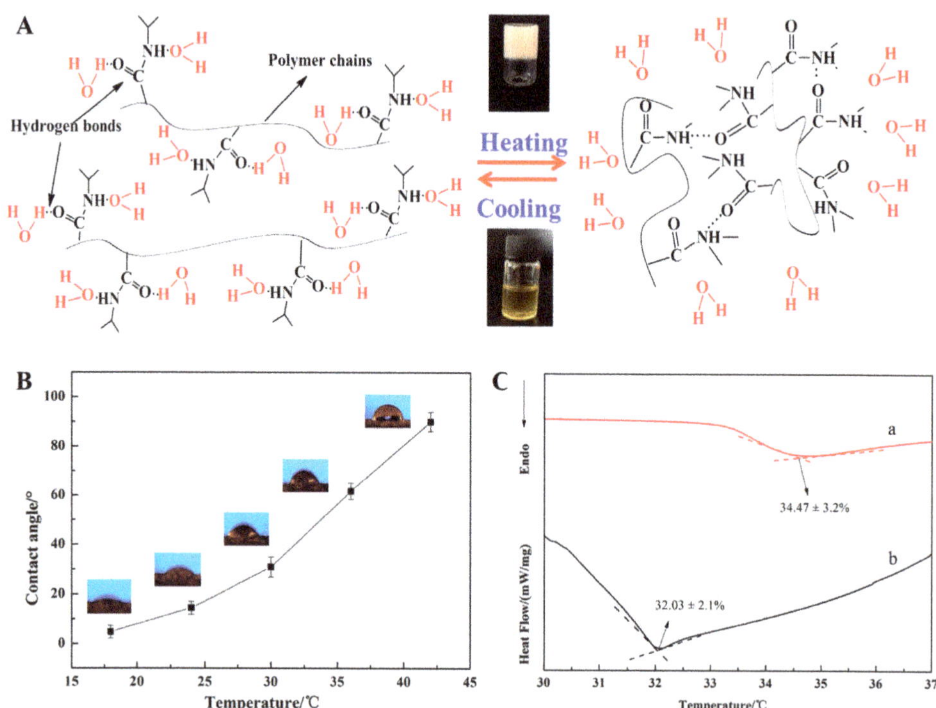

Figure 5. Representation of (**A**) the hydrophobic–hydrophilic phase transition of PNIPAM hydrogels; (**B**) Contact angle measurements of multifunctional cotton fabric at different temperatures; (**C**) Differential scanning calorimetry thermogram curves of modified cotton fabric and cotton fabric. (**a**) Cotton fabric; (**b**) Functionalized cotton fabric. (The final results of data are shown as mean ± standard deviation).

3.4. pH-Responsive Swelling Behaviors

In Figure 6, the images show that sericin/PNIPAM/PEO hydrogels had different swelling ratio changes according to the pH swelling medium at temperatures of 15 °C and 37 °C. From Figure 6, it was shown that all the hydrogels presented the same trends and temperature-sensitive property under the pH test range. The swelling ratio of hydrogels was much higher at 15 °C than at 37 °C, which showed a temperature-sensitive characteristic. This was largely due to the network shrinking of PNIPAM when the temperature of the solutions was above LCST (37 °C). The swelling ratio of sericin/PNIPAM/PEO (55/15/30) hydrogel decreased a little with a rise in temperature. The minimum swelling ratio of hydrogel appeared at about pH 4.0. This is due to the non-charged and hydrophobic form of sericin side chains. The increasing swelling ratios of sericin/PNIPAM/PEO (55/15/30) hydrogel appeared with reducing the pH buffer solutions down to 4.0 or increasing it above 4.0. This was because the –NH$_2$ of sericin was protonated into –NH$_3^+$ and –COOH of sericin charged to –COO$^-$ and then enhanced the electrostatic repulsion of interpenetrating network hydrogels.

The results showed that the swelling ratio was lower in all the specimens at pH 4.0 and pH 8.0, no matter what the composition of the hydrogels was. Additionally, there were better swelling ratios under acid or alkaline conditions.

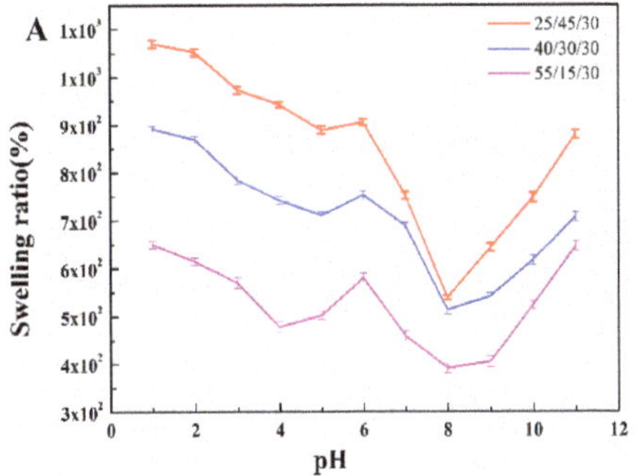 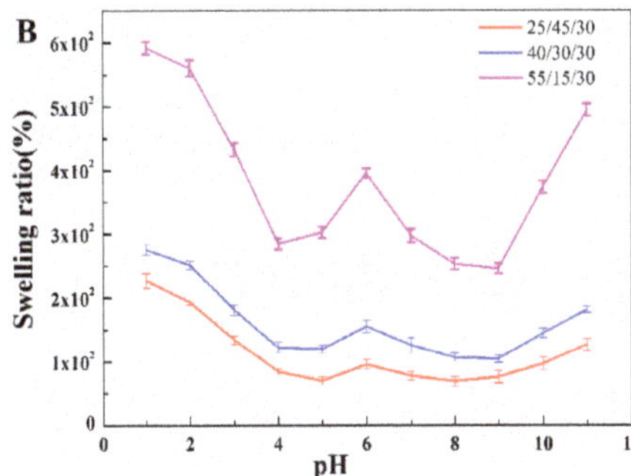

Figure 6. Swelling ratio of sericin/PNIPAM/PEO hydrogels in pH buffer solutions at (**A**) 15 °C and (**B**) 37 °C, the final results of data are shown as mean ± standard deviation.

With the increasing sericin concentration, the swelling ratios increased at 15 °C (Figure 6A), and the order of swelling ratios was a complete transformation at 37 °C (Figure 6B). Along with the increase in sericin content, the number of –COOH and –NH$_2$ was increased and resulted in the enhancement of the electrostatic repulsion of hydrogel. As can be seen, the sericin/PNIPAM/PEO hydrogels showed simultaneous pH and temperature sensitivity. It was an effective way to adjust the equilibrium swelling ratio and the swelling property of the hydrogels by changing the mixed proportion of hydrogels.

3.5. Antimicrobial Activity

The antibacterial activity of samples was evaluated against Gram-positive bacteria and Gram-negative bacteria such as *Staphylococcus aureus* (ATCC 6538) and *Escherichia coli* (ATCC 8739). The antimicrobial capability of the untreated cotton fabric and functionalized cotton fabric is shown in Table 3. The functionalized cotton fabrics exhibited antimicrobial property against the above microorganisms. The sericin/PNIPAM/PEO nanofibers could inhibit the growth of bacteria. All functionalized cotton fabrics showed lower antibacterial efficiency against *E. coli* and higher antibacterial property against *S. aureus*. It was revealed that the antibacterial efficiency of the functionalized cotton fabrics exhibited the highest antibacterial efficiency up to 85% of *E. coli* and 90% of *S. aureus*. This may be due to the fact that the sericin could control the free movement of the bacteria, inhibited respiration and finally rose to the death of bacteria. The sericin molecules had the electrostatic attraction with cell membranes of bacteria. The electrostatic attraction would decrease bacterial conductivity, reduce cell membrane permeability and partially inhibit metabolism. As a result, the intracellular composition containing water and protein was discharged and finally led to the death of bacteria [40].

Table 3. Antibacterial characteristic of cotton fabrics against *S. aureus* and *E. coli* bacteria.

Sample	E. coli					S. aureus				
	Surviving Cells (CFU/mL)					Surviving Cells (CFU/mL)				
	1	2	3	4	5	1	2	3	4	5
Untreated cotton fabric	251,000	232,000	243,000	235,000	250,000	241,000	234,000	228,000	221,000	230,000
Functionalized cotton fabric	27,000	36,000	26,000	18,000	19,000	5000	7000	9000	14,000	12,000
Reduction(%)	89.24	84.48	89.30	92.34	92.40	97.93	97.01	96.05	93.67	94.78
Mean ± SD	89.55 ± 3.2%					95.89 ± 1.7%				

4. Conclusions

In conclusion, a simple and effective strategy has been designed and proved for the surface modification of cotton textiles with multi-function responsiveness. The multi-function modified cotton fabric showed obvious temperature- and pH responsiveness behavior and antibacterial activity. The swelling ratio decreased significantly under the increasing temperature to the LCST of the hydrogels. The low values of equilibrium swelling ratios appeared at about pH 4.0 and 8.0. This is due to the salting-out effect and the isoelectric point (PI) of SS. The bacterial decline of multi-function modified cotton textiles against *E. coli* and *S. aureus* was all above 89%. Through the different combinations of textiles and smart polymers via electrospinning technology, the strategy shows robust advantages for the preparation of textiles with multi-function responsive properties. Moreover, this is also a simple, rapid and green environmental protection method for the fabrication of multi-functional fabrics. This study may provide the strategy of modified fabrics that can possess environmentally responsive or stimuli-responsive properties and that are supposed to be prospective materials, especially in smart and functional textile applications.

Here, this study is just a first step for the preparation of smart fabrics with nanofibers as a finishing technology. However, we would perform comprehensive functional testing for the real applications. Other applications of sericin/PNIPAM-based nano-modified cotton textiles such as mechanical properties, comfort capabilities and biocompatibility testing, and cell cytotoxicity tests will be shown in the following manuscripts.

Author Contributions: Conceptualization, J.G. and Y.-H.L.; methodology, J.L.; experimentation, J.L., B.-X.W., D.-H.C., and Z.-M.L.; date analysis, J.L. and L.-H.L. supervision, J.L., J.G. and Y.-H.L. All authors have read and agreed to the published version of the manuscript.

Funding: This work was supported by the National Natural Science Foundation of China (grant numbers: 51773024 and 51873084), the Liaoning Province Department of Education fund item (grant numbers: LNSJYT201901 and LNSJYT202006 and LNSJYT202022), the Natural Science Foundation of Liaoning (grant numbers: 2019JH8/10100047 and 2019-ZD-0533), and the Science and Technology Innovation Fund Project of Dalian (2019J12SN71). The Innovation Team Foundation of Liaoning (grant number: LT2017017).

Institutional Review Board Statement: Not Applicable.

Informed Consent Statement: Not applicable.

Data Availability Statement: Data is contained within the article.

Conflicts of Interest: The authors declare no conflict of interest.

References

1. Gong, J.; Xu, B.; Tao, X. Three-Dimensionally Conformal Porous Microstructured Fabrics via Breath Fig-ures: A Nature-Inspired Approach for Novel Surface Modification of Textiles. *Sci. Rep.* **2017**, *7*, 1–9.
2. Trovato, V.; Teblum, E.; Kostikov, Y.; Pedrana, A.; Re, V.; Nessim, G.D.; Rosace, G. Electrically conductive cotton fabric coatings developed by silica sol-gel precursors doped with surfactant-aided dispersion of vertically aligned carbon nanotubes fillers in organic solvent-free aqueous solution-ScienceDirect. *J. Colloid Interface Sci.* **2020**, *586*, 120–134. [CrossRef]
3. Zhu, S.; Wang, M.; Qiang, Z.; Song, J.; Wang, Y.; Fan, Y.; You, Z.; Liao, Y.; Zhu, M.; Ye, C. Multi-functional and Highly Conductive Textiles with Ultra-high Durability through 'Green' Fabrication Process. *Chem. Eng. J.* **2020**, *406*, 127140. [CrossRef]
4. Ivanoska-Dacikj, A.; Stachewicz, U. Smart textiles and wearable technologies—Opportunities offered in the fight against pandemics in relation to current COVID-19 state. *Rev. Adv. Mater. Sci.* **2020**, *59*, 487–505. [CrossRef]
5. Martínez-Pérez, C.A. Electrospinning: A promising technique for drug delivery systems. *Rev. Adv. Mater. Sci.* **2020**, *59*, 441–454. [CrossRef]
6. Huang, Z.S.; Shiu, J.W.; Way, T.F.; Rwei, S.P. A Thermo-Responsive Random Copolymer of Poly(NIPAm-co-FMA) for Smart Textile Applications. *Polymers* **2019**, *184*, 121917. [CrossRef]
7. Kreuzer, L.P.; Widmann, T.; Hohn, N.; Wang, K.; Bießmann, L.; Peis, L.; Moulin, J.F.; Hildebrand, V.; Laschewsky, A.; Papadakis, C.M. Swelling and Exchange Behavior of Poly(sulfobetaine)-Based Block Copolymer Thin Films. *Macromolecules* **2019**, *52*, 3486–3498. [CrossRef]
8. Tao, X. *Handbook of Smart Textiles*; CRC Press: Boca Raton, FL, USA, 2016.

9. Sun, B.; Long, Y.Z.; Zhang, H.D.; Li, M.M.; Duvail, J.L.; Jiang, X.Y.; Yin, H.L. Advances in three-dimensional nanofibrous macrostructures via electrospinning. *Prog. Polym. Sci.* **2014**, *39*, 862–890. [CrossRef]

10. Cui, J.; Lu, T.; Li, F.; Wang, Y.; Lei, J.; Ma, W.; Zou, Y.; Huang, C. Flexible and Transparent Composite Nanofibre Membrane that was Fabricated via a "Green" Electrospinning Method for Efficient Particulate Matter 2.5 Capture. *J. Colloid Interface Sci.* **2020**, *582*, 506–614. [CrossRef]

11. Barani, M.; Bazgir, S.; Hosseini, M.K.; Hosseini, P.K. Eco-facile application of electrospun nanofibers to the oil-water emulsion separation via coalescing filtration in pilot- scale and beyond. *Process Saf. Environ. Prot.* **2020**, *148*, 342–357. [CrossRef]

12. Morais, M.; Coimbra, P.; Pina, M. Comparative Analysis of Morphological and Release Profiles in Ocular Implants of Acetazolamide Prepared by Electrospinning. *Pharmaceutics* **2021**, *13*, 260. [CrossRef] [PubMed]

13. Luraghi, A.; Peri, F.; Moroni, L. Electrospinning for drug delivery applications: A review. *J. Control. Release* **2021**, *33*, 1–37.

14. Zhu, J.; Jiang, S.; Hou, H.; Agarwal, S.; Greiner, A. Low Density, Thermally Stable, and Intrinsic Flame Retardant Poly(bis(benzimidazo)Benzophenanthroline-dione) Sponge. *Macromol. Mater. Eng.* **2018**, *303*. [CrossRef]

15. Park, S.M.; Lee, S.J.; Lim, J.; Kim, B.C.; Han, S.J.; Kim, D.S. Versatile fabrication of size and shape-controllable nanofibrous concave mi-crowells for cell spheroid formation. *ACS Appl. Mater. Interfaces* **2018**, *10*, 37878–37885. [CrossRef]

16. Yang, W.; Gong, W.; Hou, C.; Su, Y.; Guo, Y.; Zhang, W.; Li, Y.; Zhang, Q.; Wang, H. All-fiber tribo-ferroelectric synergistic electronics with high thermal-moisture stability and comfortability. *Nat. Commun.* **2019**, *10*, 1–10. [CrossRef] [PubMed]

17. Zhou, Y.; He, J.; Wang, H.; Qi, K.; Nan, N.; You, X.; Shao, W.; Wang, L.; Ding, B.; Cui, S. Highly sensitive, self-powered and wearable electronic skin based on pressure-sensitive nanofiber woven fabric sensor. *Sci. Rep.* **2017**, *7*, 1–9. [CrossRef]

18. Pawłowska, S.; Rinoldi, C.; Nakielski, P.; Ziai, Y.; Urbanek, O.; Li, X.; Kowalewski, T.A.; Ding, B.; Pierini, F. Ultraviolet Light-Assisted Electrospinning of Core–Shell Fully Cross-Linked P(NIPAAm-co-NIPMAAm) Hydrogel-Based Nanofibers for Thermally Induced Drug Delivery Self-Regulation. *Adv. Mater. Interfaces* **2020**, *7*, 2000247. [CrossRef]

19. Jin, S.; Liu, M.; Zhang, F.; Chen, S.; Niu, A. Synthesis and characterization of pH-sensitivity semi-IPN hydrogel based on hydrogen bond between poly(N-vinylpyrrolidone) and poly(acrylic acid). *Polymer* **2006**, *47*, 1526–1532. [CrossRef]

20. Sengor, M.; Ozgun, A.; Gunduz, O.; Altintas, S. Aqueous electrospun core/shell nanofibers of PVA/microbial transglutaminase cross-linked gelatin composite scaffolds. *Mater. Lett.* **2019**, *263*, 127233. [CrossRef]

21. Nakielski, P.; Pawłowska, S.; Rinoldi, C.; Ziai, Y.; De Sio, L.; Urbanek, O.; Zembrzycki, K.; Pruchniewski, M.; Lanzi, M.; Salatelli, E.; et al. Multifunctional Platform Based on Electrospun Nanofibers and Plasmonic Hydrogel: A Smart Nanostructured Pillow for Near-Infrared Light-Driven Biomedical Applications. *ACS Appl. Mater. Interfaces* **2020**, *12*, 54328–54342. [CrossRef]

22. Guo, J.W.; Wang, C.F.; Lai, J.Y.; Lu, C.H.; Chen, J.K. Poly(N-isopropylacrylamide)-gelatin hydrogel membranes with thermo-tunable pores for water flux gating and protein separation. *J. Membr. Sci.* **2020**, *618*, 118732. [CrossRef]

23. Zheng, X.; Liu, X.; Zha, L. Fabrication of ultrafast temperature-responsive nanofibrous hydrogel with superelasticity and its 'on–off' switchable drug releasing capacity. *J. Appl. Polym. Sci.* **2021**, *138*. [CrossRef]

24. Li, J.; Wang, B.; Lin, J.; Cheng, D.; Lu, Y. Multifunctional Surface Modification of Mulberry Silk Fabric via PNIPAAm/Chitosan/PEO Nanofibers Coating and Cross-Linking Technology. *Coatings* **2018**, *8*, 68. [CrossRef]

25. Malay, A.D.; Sato, R.; Yazawa, K.; Watanabe, H.; Ifuku, N.; Masunaga, H.; Hikima, T.; Guan, J.; Mandal, B.B.; Damrongsakkul, S.; et al. Relationships between physical properties and sequence in silkworm silks. *Sci. Rep.* **2016**, *6*, 27573. [CrossRef] [PubMed]

26. Lu, Y.; Cheng, D.; Lu, S.; Huang, F.; Li, G. Preparation of quaternary ammonium salt of chitosan nanoparticles and their textile properties on Antheraea pernyi silk modification. *Text. Res. J.* **2014**, *84*, 2115–2124. [CrossRef]

27. Varone, A.; Knight, D.; Lesage, S.; Vollrath, F.; Rajnicek, A.M.; Huang, W. The potential of Antheraea pernyi silk for spinal cord repair. *Sci. Rep.* **2017**, *7*, 1–10. [CrossRef]

28. Wang, B.; Zhang, S.; Wang, Y.; Si, B.; Cheng, D.; Liu, L.; Lu, Y. Regenerated Antheraea pernyi Silk Fibro-in/Poly(N-isopropylacrylamide) Thermosensitive Composite Hydrogel with Improved Mechanical Strength. *Polymers* **2019**, *11*, 302. [CrossRef] [PubMed]

29. Li, J.; Wang, B.X.; Cui, Y.F.; Yu, Z.C.; Hao, X.; Huang, F.Y.; Cheng, D.H.; Lu, Y.H. Fabrication and Super-Antibacterial Property of Nanosil-ver/Sericin/Poly(ethylene oxide) Nanofibers through Electrospinning-Combined Postdeposition Method. *J. Nanomater.* **2016**, *2016*, 45. [CrossRef]

30. Pinto, R.V.; Gomes, P.S.; Fernandes, M.H.; Costa, M.E.; Almeida, M.M. Glutaraldehyde-crosslinking chitosan scaffolds reinforced with calcium phosphate spray-dried granules for bone tissue applications. *Mater. Sci. Eng.* **2020**, *109*, 110557. [CrossRef]

31. Valizadeh, S.; Naseri, M.; Babaei, S.; Hosseini, S.M.; Imani, A. Development of bioactive composite films from chitosan and carboxymethyl cellulose using glutaraldehyde, cinnamon essential oil and oleic acid. *Int. J. Biol. Macromol.* **2019**, *134*, 604–612. [CrossRef]

32. Ai, L.; He, H.; Wang, P.; Cai, R.; Tao, G.; Yang, M.; Liu, L.; Zuo, H.; Zhao, P.; Wang, Y. Rational Design and Fabrication of ZnONPs Functionalized Sericin/PVA Antimicrobial Sponge. *Int. J. Mol. Sci.* **2019**, *20*, 4796. [CrossRef]

33. Jahanshahi, M.; Kowsari, E.; Haddadi-Asl, V.; Khoobi, M.; Lee, J.H.; Kadumudi, F.B.; Talebian, S.; Kamaly, N.; Mehrali, M. Sericin grafted multifunctional curcumin loaded fluorinated graphene oxide nanomedicines with charge switching properties for effective cancer cell targeting. *Int. J. Pharm.* **2019**, *572*, 118791. [CrossRef]

34. Wu, W.; Li, W.; Wang, L.Q.; Tu, K.; Sun, W. Synthesis and characterization of pH- and temperature-sensitive silk sericin/poly(N-isopropylacrylamide) interpenetrating polymer networks. *Polym. Int.* **2006**, *55*, 513–519. [CrossRef]

35. Deng, X.; Nikiforov, A.Y.; Coenye, T.; Cools, P.; Aziz, G.; Morent, R.; De Geyter, N.; Leys, C. Antimicrobial nano-silver non-woven polyethylene terephthalate fabric via an atmospheric pressure plasma deposition process. *Sci. Rep.* **2015**, *5*, 10138. [CrossRef] [PubMed]
36. Purwar, R.; Verma, A.; Batra, R. Antimicrobial gelatin/sericin/clay films for packaging of hygiene products. *J. Polym. Eng.* **2019**, *39*, 744–751. [CrossRef]
37. Wang, B.; Wu, X.; Li, J.; Hao, X.; Lin, J.; Cheng, D.; Lu, Y. Thermosensitive Behavior and Antibacterial Activity of Cotton Fabric Modified with a Chitosan-poly(N-isopropylacrylamide) Interpenetrating Polymer Network Hydrogel. *Polymers* **2016**, *8*, 110. [CrossRef] [PubMed]
38. Cui, Y.; Xing, Z.; Yan, J.; Lu, Y.; Xiong, X.; Zheng, L. Thermosensitive Behavior and Super-Antibacterial Properties of Cotton Fabrics Modified with a Sercin-NIPAAm-AgNPs Interpenetrating Polymer Network Hydrogel. *Polymers* **2018**, *10*, 818. [CrossRef]
39. Minier, S.; Kim, H.J.; Zaugg, J.; Mallapragada, S.K.; Vaknin, D.; Wang, W. Poly(N-isopropylacrylamide)-grafted gold nanoparticles at the vapor/water interface. *J. Colloid Interface Sci.* **2021**, *585*, 312–319. [CrossRef]
40. Xue, R.; Liu, Y.; Zhang, Q.; Liang, C.; Qin, H.; Liu, P.; Wang, K.; Zhang, X.; Chen, L.; Wei, Y. Shape Changes and Interaction Mechanism of Escherichia coli Cells Treated with Sericin and Use of a Sericin-Based Hydrogel for Wound Healing. *Appl. Environ. Microbiol.* **2016**, *82*, 4663–4672. [CrossRef]

Article

Life Cycle Assessment of Olive Pomace as a Reinforcement in Polypropylene and Polyethylene Biocomposite Materials: A New Perspective for the Valorization of This Agricultural By-Product

Gabriela Espadas-Aldana [1,*], Priscila Guaygua-Amaguaña [2], Claire Vialle [1], Jean-Pierre Belaud [2], Philippe Evon [1] and Caroline Sablayrolles [1]

[1] Laboratoire de Chimie Agro-Industrielle, Université de Toulouse, INRA, ENSIACET, 4 Allée Emile Monso, CS 44362, 31030 Toulouse, France; claire.vialle@inp-toulouse.fr (C.V.); philippe.evon@inp-toulouse.fr (P.E.); caroline.sablayrolles@inp-toulouse.fr (C.S.)

[2] Laboratoire de Génie Chimique, Université de Toulouse, CNRS, ENSIACET, 4 allée Emile Monso, CS 84234, 31432 Toulouse, France; priscila.guayguaamaguana@etu.toulouse-inp.fr (P.G.-A.); jeanpierre.belaud@inp-toulouse.fr (J.-P.B.)

* Correspondence: gabriela.espadasaldana@inp-toulouse.fr

Citation: Espadas-Aldana, G.; Guaygua-Amaguaña, P.; Vialle, C.; Belaud, J.-P.; Evon, P.; Sablayrolles, C. Life Cycle Assessment of Olive Pomace as a Reinforcement in Polypropylene and Polyethylene Biocomposite Materials: A New Perspective for the Valorization of This Agricultural By-Product. *Coatings* **2021**, *11*, 525. https://doi.org/10.3390/coatings11050525

Academic Editor: Devis Bellucci

Received: 21 April 2021
Accepted: 28 April 2021
Published: 29 April 2021

Abstract: The main environmental impact of olive oil production is the disposal of residues such as pomace and water vegetation. During the olive oil extraction process, the olive stone is milled and discharged within the olive pomace. However, olive stone flour can be valorized as filler for polymeric composites. A life cycle assessment of the olive pomace valorization was carried out by focusing on the manufacturing process of a biocomposite made of two different thermoplastic matrices, i.e., polyethylene and polypropylene. The functional unit is the production of 1 m² of a lath made of an olive pomace-based biocomposite. The analysis was carried out with the SimaPro PhD 9.1.1.1 software, and the database used for the modeling was Ecoinvent 3.6. The obtained results reveal that the hotspot of the whole process is the twin-screw compounding of the olive stone fraction, with the polymeric matrix and coupling agent, and that human health is the most affected damage category. It represents 89% for both scenarios studied: olive stone fraction/polypropylene (OSF/PP) and olive stone fraction/polyethylene (OSF/PE). Further research directions include the use of biosourced polymer matrices, which could reduce the impact of olive pomace-based composite manufacturing.

Keywords: olive stone; biocomposite; LCA; circular economy; filler

1. Introduction

Olive oil is one of the main agricultural products in the Mediterranean countries. The organoleptic and healthy properties of olive oil associated with its high consumption level have made the cultivation of olive trees expand worldwide, along with its consumption trends.

The production of olive oil includes several phases. First, the olives are washed in order to remove the impurities collected during the harvesting (e.g., leaves, twigs, stones, etc.). After washing, the olives are crushed to facilitate the release of the oil from the vacuoles. Then, the crushed olives are subjected to a malaxation process, where the paste is mixed, allowing small oil droplets to combine into bigger ones. The last step is the separation of the oil from the rest of the olive components. The oil can be extracted by mechanical pressing (i.e., a discontinuous process) or centrifugation (i.e., a continuous process) [1,2].

It is worth noting that the pressing method is the oldest system for olive oil extraction, known as the traditional method. Nevertheless, it is still used in some small mills and has a processing extraction yield of around 86–90% [3].

The centrifugation method covers the need for a continuous extraction process. This method works on the basis of centrifugal force, where the less dense liquid phase forms a concentric inner layer, whereas the denser solid particles are pushed against the wall of the rotating bowl [3]. This extraction process presents two operation alternatives: the three-phase and the two-phase horizontal centrifugation methods.

The difference between the two-phase and three-phase horizontal centrifugation is not whether or not water is added, as it is often mistaken. The difference lies in the number of output streams that the decanter has:

- On the one hand, the three-phase centrifuge has as an output for olive oil, vegetable water, also known as olive mill wastewater (OMW) (alpechin in Spanish) and pomace (orujo in Spanish).
- On the other hand, the two-phase centrifuge has as an output for olive oil, and other for wet pomace (alperujo, contraction of alpechin and orujo in Spanish).

It should be noted that the two-phase decanter requires a minimum moisture content of the olive paste (about 50%) to facilitate the separation process. If the olive paste is too dry before being introduced into the decanter, it will be necessary to incorporate a certain amount of water until the moisture level required for a proper operation of the decanter is reached [4]. For this reason, water can be added to the two-phase centrifuge system as well. Compared to the traditional method, the three-phase centrifugation device increases water utilization (from 1.25 to 1.75 times more) [3]. Moreover, valuable components, especially natural antioxidants, can be lost in the water phase (OMW), thus reducing the olive oil quality.

The rising popularity of olive oil has increased the generation of its by-products: the olive pomace (OP), a general term used to refer to the pomace obtained from all the different olive oil extraction processes, and an effluent known as OMW, derived from traditional pressing and from the three-phase system, as mentioned before [5,6].

This OP is a mixture of residual skin, pulp, and fragments of the crushed olive stone [7]. The main components of this solid residue are cellulose, hemicelluloses, and lignins. Residual fat and proteins are also present in noteworthy quantities. The moisture content of the solid residues is 22–25% for traditional pressed olive pomace, 65–74% for pomace from a two-phase system, and 40–50% for that from a three-phase one [1,5]. On the other hand, OMW is a red-to-black colored acidic liquid, with 83–92% content of water, its main components being phenolic compounds, sugars, and organic acids. OMW also reveals an important quantity of potassium [5].

On average, olive fruit contains 20 wt.% of oil, and the remaining 80 wt.% together with the added water form OP [8]. Olive oil processing is considered inefficient due to the high volume of waste generated [9]. This particular industry has a seasonal production, which generates a high amount of waste in a short period of time. The olive oil industry causes many environmental impacts in terms of resource depletion, land degradation, air emissions, and waste generation. Moreover, the management of olive oil residues is an economic burden to producers [10–14]. In Europe, the production of OP reaches approximately 6.8 million tons per year [15].

Currently, olive oil by-products are discharged on agricultural land by controlled spreading [5,10,16]. Due to its high content in phenolic and lipidic constituents, organic acids, low pH, and salinity, OP should not be used as agricultural spreading [17]. Moreover, OP is resistant to bacterial degradation, which makes it a significant source of environmental pollution [17]. Another use of OP has been in animal nutrition. As an example, it is used in Tunisia in mixture with bran or even cactus to feed dromedaries or sheep. In countries such as Italy and Greece, cows are fed with OP. However, it can cause digestive problems in animals due to its high degree of lignification [18]. On the other hand, OP has also been used for composting, or to produce a non-phytotoxic product through biological conversion (bioremediation), which can be used as a fertilizer [5,19,20].

Over the years, many other methods have been proposed for olive oil waste disposal and valorization. These techniques include thermo-chemical processes, anaerobic digestion, fermentation, blending, and chemical extraction of bioactive compounds [1,21–23]. Valorization routes also include the production of activated carbons, cosmetic applications, the production of polyols, and the improvement of the thermal properties of cement mortar [24–26].

In the most recent years, research has focused on the valorization of OP on the biocomposite field [7,17,23,27–33]. Due to their natural richness in lignocellulosic fibers, this sector uses agricultural wastes (or by-products) as renewable fillers for polymeric matrices. The obtained results from olive stone flour have confirmed its viability as a cheap reinforcing filler for the polypropylene matrix, thus opening new perspectives for the use of this by-product [32]. The developed composites could find applications in buildings, in the automotive industry, and as outdoor products, e.g., deck floors, furniture, park benches, etc. Specific examples of this application are those from the GO-OLIVA project (Spain), which developed Olipast, a new sustainable packaging material from olive pits [34], and from the Biolive company, which commercializes the Bio-Pura product, used for the manufacture of television components (Turkey). The Biolive company is also working to produce shrink wrap for beer can from this material; and other end products for various applications, including consumer electronic casings, automotive interiors, toys, and packaging. Approximately 3.5 tons of bioplastics can be transformed from 5 tons of locally sourced olive seeds [35,36].

Since 2015, French authorities have developed "the Energy Transition Law for Green Growth". This law has focused on waste management as an essential pillar to ensure the transition to a circular economy model [37]. Moreover, the use of the circular economy is becoming increasingly important, especially in the field of agriculture, one of the main suppliers of waste. In particular, much research is being carried out to transform agricultural residues by sustainable processes.

For all these reasons, it is crucial to solve the waste management issues generated by OP and to explore the alternatives to convert this by-product into a co-product. Previous publications, like the one by Espadas-Aldana et al. [12] compiled the studies of environmental life cycle assessment of olive oil and some waste management techniques. Several authors have focused on the life cycle assessment of olive oil extraction waste treatment [23,38–42]. Here, the current case study is on the valorization of the olive pomace and its life cycle assessment, focusing on the manufacturing process of a biocomposite. To the authors' knowledge, this is the first environmental life cycle assessment study made on biocomposites from olive pomace in order to innovate on the by-product valorization from olive oil production.

The aim of this article is to evaluate the environmental performance of a biocomposite composed of olive pomace reinforcement and a polymeric matrix. Data for the compounding process come from a pilot scale experiment to produce lath for terraces. Two scenarios were investigated which are differentiated by the polymeric matrix used: i.e., one made from polyethylene and the other from polypropylene.

2. Materials and Methods

Life cycle assessment (LCA) is defined as "a tool to assess the potential environmental impacts and resources used throughout a product's life cycle" [43]. The ISO 14040 [44] standard states that the LCA framework includes four phases: Goal and scope definition, Inventory analysis, Impact assessment, and Interpretation.

2.1. Goal and Scope of the Study

The following LCA focuses on the manufacturing process of a biocomposite made of olive stone fraction (OSF), which is part of the olive pomace (OP), and two different polymeric matrices: polypropylene (PP) and polyethylene (PE).

The goal of this LCA is to analyze and compare the environmental impacts of the different scenarios and to identify the unit process with the strongest environmental impacts, in order to improve the current valorization of olive pomace.

2.1.1. Functional Unit

In order to build the production inventory and set the scope of the study, the functional unit is defined. Based on similar works [45–47] and EN 15804 [48], the functional unit chosen is the production of 1 m^2 of lath (used as building material) made from the olive pomace-based composite. Figure 1a,b show a diagram of the lath and its dimensions.

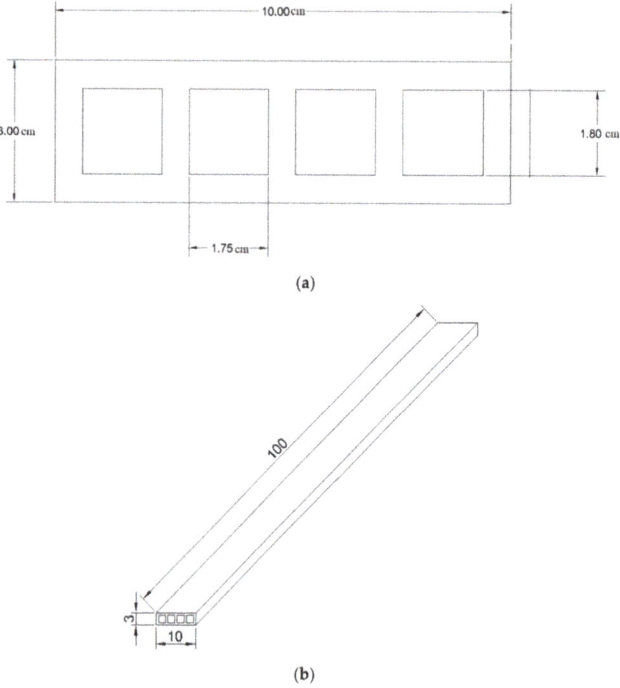

Figure 1. (**a**) Profile section of the lath made from the olive pomace-based composite. (**b**) Overview of the lath made from the olive pomace-based composite (measurements are in centimeters).

2.1.2. System Boundary

The ISO 14044 [49] states that the system boundary is a "set of criteria specifying which unit processes are part of a product system". In order to focus on the impacts related to the development of a new biocomposite made from OP as filler, and PP or PE, respectively, as thermoplastic matrix, a "cradle-to-gate" approach life cycle assessment was carried out. This study is centered on the generation of the raw materials and manufacturing of the biocomposites. In Figure 2, system A shows the completed life cycle of the biocomposite; system B comprehends the generation of the raw materials, and the manufacturing process of the composite; system C1 shows the production of the raw materials used in the process; system C2 shows the manufacturing process of the product and the necessary pretreatment (i.e., drying, milling, etc.) to prepare the OP fraction. Finally, system C3 includes the use and end of life of the product. It is important to mention that system B comprehends the complete system boundary of the LCA herein. Namely, the system studied takes into account the production of all the raw materials needed (i.e., OP, PP, PE, and the coupling agents added to the compound to reinforce the matrix/filler interface), and the production of both composites.

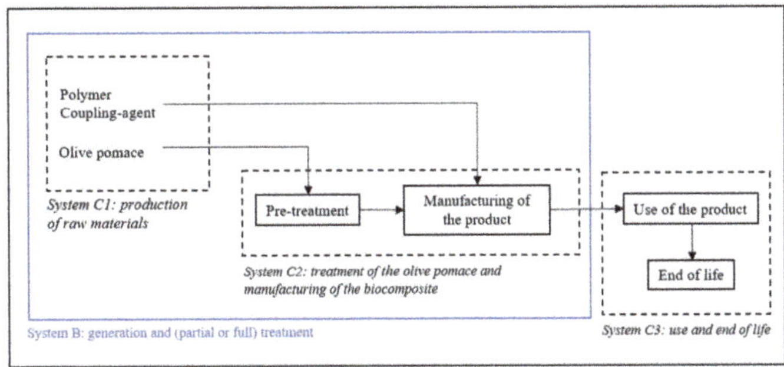

Figure 2. System boundary of the study.

Several hypotheses must be considered in the actual approach in order to avoid overlaps in the decision-making process:

- The necessary infrastructure is not taken into account, consequently excluding their manufacture as well as their dismantling,
- The electricity is considered to come from the mixed French energy supply,
- The cleaning of the devices used in the process is neglected,
- The transportation of the olive pomace is not taken into account.

The current study presents two different scenarios, which correspond to the mixture of the olive stone and the two different polymeric matrices (Table 1). The polymeric matrices studied are PP and PE. The OSF acts as a filler in the polymeric matrix [27,28].

Table 1. Scenarios studied.

Filler	Polymeric Matrix	Scenario
Olive stone fraction	Polypropylene	OSF/PP
Olive stone fraction	Polyethylene	OSF/PE

2.2. Life Cycle Inventory Analysis

The life cycle inventory (LCI) collects and compiles all data on elementary flows from all processes in the studied product system(s) drawing on a combination of different sources. The output is a compiled inventory of elementary flows that is used as the basis of the subsequent life cycle impact assessment phase.

2.2.1. Process Tree

Based on the analysis of various publications [7,17,28,30–32,50], all the manufacturing processes of olive pomace-based composites presented similar unit operations. First, crude olive pomace is dried in an oven at 60 °C for 24 h. Then, crude olive pomace is milled in a ball mill device at ambient temperature, 86 rpm for 30 min [7].

The powder passes into an electric sieving machine (RITEC, model 400, Signes, France) through a 1.25 mm mesh for 10 min. Two fractions are obtained, the fine fraction corresponds to the pulp-rich fraction (PF) and is recovered at the bottom of the sieve; whereas, the coarse fraction is retained in the sieve [31].

The coarse fraction is further ground in a knife mill (SM 300, Retch, Haan, Germany), with a speed of 1500 rpm and a grid size of 1 mm [7]. Then, the ground powder is sieved through a 0.4 mm mesh to separate the OSF from the intermediate fraction [7]. Polymer granules and the OP-based filler were then blended into a twin-screw extrusion compounding device. The coupling agents used are PE-g-MA (polyethylene-grafted-maleic anhydride) and PP-g-MA (polypropylene-grafted-maleic anhydride) agents, for the PE and PP thermoplastic matrices, respectively.

The presence of dust during the grinding process is very common. The main particle size present is PM10 (particles with the size smaller than 10 μm); nevertheless there are also particles with smaller size, such as PM2.5 (particles with the size smaller than 2.5 μm). Regardless of the milling method, organic dust is always produced when lignocellulosic materials are ground [51]. Dust is harmful to the working environment. Therefore, the particulate matter (i.e., the smallest particles) have to be recovered by a cyclone, and the content conveyed to a dry storage bin [52]. The equipment used for grinding includes a cyclone that recovers all the particulates, and collects them directly on a container [53].

The development of the biobased lath is part of a confidential work carried out by the Laboratoire de Chimie Agro-industrielle (Université de Toulouse, Toulouse, France) on behalf of a biosourced plastics industrialist [54]. Composites had been produced by mixing the OP (60% filler) in two polymeric matrices: PE and PP. The addition of a coupling agent has also been carried out to reinforce the matrix/fiber interface. As a result, adding OP to the polymeric matrices showed a mechanical reinforcement of the material, which was illustrated by the increase of the elastic modulus simultaneously with the decrease of the elongation at break, both in tensile and in bending.

Figure 3 itemizes the process tree of the manufacturing of the olive stone composite, with the different stages of the process and their flows.

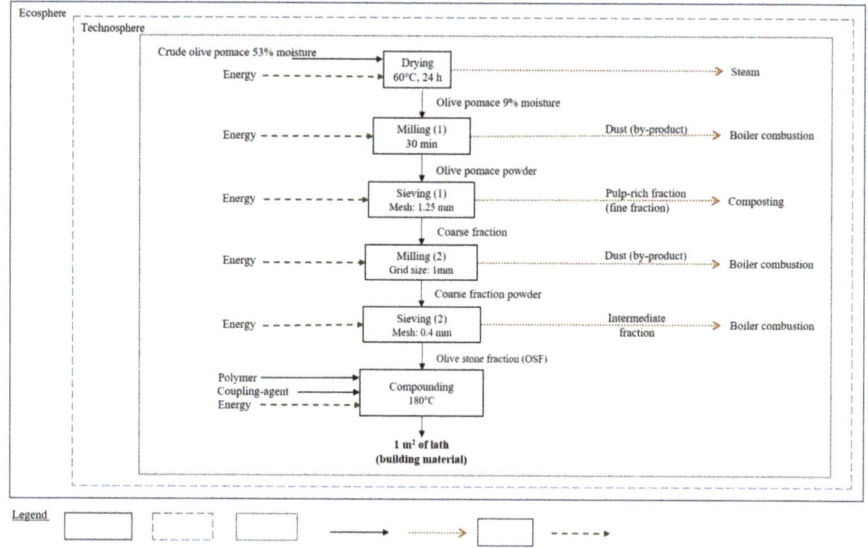

Figure 3. Process tree of the olive pomace-based composite.

2.2.2. Data Collection

Inputs: Raw Materials

- Olive pomace

OP is a mixture of different fractions, the pulp-rich fraction (PF), stone-rich fraction (OSF) and intermediate fraction, whose percentage weights are 31.3, 56.4, and 11.7%, respectively [7]. The moisture of the crude olive pomace considered for the process is 53%, which is close to a three-phase system pomace. The moisture of the material is then decreased to 9% thanks to a drying process [7]. The energy consumption linked to the drying, crushing, and sieving of the olive pomace comes from the technical data of the machinery used and literature data [53,55–57]. The grinding of the olive husk gave 80% of the weight of the sample [58]. This value was used in the milling of the coarse fraction due to missing data. For the modeling of "Crude olive pomace 53% moisture", the olive pomace dataset was adapted from the AGRIBALYSE v3.0 database according to Avadí [59].

The above-mentioned database considers that the impact of olive cultivation is attributed only to olive oil. Nevertheless, a part of the impact of olive oil production is attributed to the virgin olive pomace at an economic allocation of 2.32%. These characteristics belong to the "Olive pomace" file presented in Table 2. Then, this virgin olive pomace follows a new extraction process to obtain pomace oil and de-oiled pomace. There again, a part of the impact is attributed to the de-oiled pomace with an economic allocation of 9.47%, named "Olive pomace, processed" (Table 2).

Table 2. Background data for the modeling of olive pomace, processed. Adapted from [59].

Process	Input/Output	Amount	Comment
	Input	-	-
	Substrate	4 kg	No impacts from olive agricultural production were included (empty process)
Olive pomace	Water	0.16 kg	Economic allocation key for wet pomace (70%
	Energy	0.005 kW·h	moisture) by the PEFCR *: 2.32%
	Output	-	-
	Olive pomace	1.60 kg	Virgin olive pomace
	Input	-	-
	Substrate	2.00 kg	"Olive pomace" (previous process)
	Water	0 kg	
Olive pomace, processed	Energy	0.01 kW·h	Economic allocation key for dry pomace by the PEFCR *: 9.47%
	Energy	0.70 MJ	
	Output	-	-
	Olive pomace, processed	0.93 kg	De-oiled olive pomace
	Water	1.07 kg	Calculate by mass difference

* PEFCR: Product Environmental Footprint Category Rules for olive oil [60].

- Polymeric matrices

The word "composite" indicates that two or more separate materials are combined on a macroscopic scale to form a structural unit for various engineering applications. The composite is constituted by the reinforcement (olive stone) and the matrix. Polymers with thermoplastic behavior are usually used as matrix materials in composites [61].

In the present study, the polymers considered for the matrices are PP and PE, as they are very common thermoplastic polymers used for many applications. Being able to originate from petrol just as from renewable resources, PE is produced through radical polymerization, anionic polymerization, and cationic polymerization, while PP is obtained from high temperature cracking of petroleum hydrocarbons and propane. The properties of PP are almost similar to those of PE. However, PP does not present stress-cracking problems, and it offers electrical and chemical resistance at high temperatures. Besides, it has a little lower density, and its structure is hard and more rigid [62].

The product studied is 1 m^2 of lath made of olive pomace-based composite. The dimensions of the profile are 10 cm $\times$ 3 cm on the outside, and 1.72 cm $\times$ 1.8 cm for the 4 interior spaces, and 1 m in length, as shown in Figures 1 and 2. For a superficial area of 1 m^2, the total quantity of profiles used is 10. The total area is 0.0174 m^2, and the volume of 1 m^2 of lath (building material) is 0.0174 m^3. For the latter calculation, the hypothesis of lath contiguous (no space between them) was supposed.

Table 3 shows the density of each composite and the mass needed for the production of 1 m^2 of lath. The density of the composites was taken from the extrapolation of the results of Uitterhaegen et al. [54].

Table 3. Density and mass for the production of 1 m^2 of lath made of olive pomace-based composite. Extrapolated from Uitterhaegen et al. [54].

Composite	Density (kg/m^3)	Mass (kg)
OSF/PP	1060.9	18.458
OSF/PE	1068.0	18.584

- Coupling agents

The coupling agent improves the compatibility between the natural fiber and the polymeric matrix. The main incompatibility cause of natural fibers and polymer inside composites is due to the hydrophilic properties of natural fibers and the hydrophobic ones of the thermoplastic matrices. To improve the reinforcement effect of the filler, and especially to ensure efficient load transfer from the matrix to the filler, some authors [27,28,32] used a polymer-based coupling agent to improve the mechanical properties, especially the maximal strengths, of lignocellulosic-plastic composites.

In the present study, polypropylene-grafted-maleic anhydride (PP-g-MA) and polyethylene-grafted-maleic anhydride (PE-g-MA) were used as coupling agents. The percentage used in the formulation of the composite was 6 wt.%

Table 4 shows the data used in the modelling of the coupling agents. In particular, the percentages of maleic anhydride of each compound were taken from the formulation of the commercial compound.

Table 4. Data for the modeling of the coupling agents (PP-g-MA, PE-g-MA).

Process	Input/Output	Amount	Comment	Ref.
	Input	-	-	-
PP-g-MA	Maleic anhydride	0.09 kg	8–10%wt, an average between the two values was taken as reference.	[63]
	Polypropylene	0.91 kg	-	-
	Output	-	-	-
	PP-g-MA	1.00 kg	The data considered for the modeling is the formulation of the commercial compound.	-
	Input	-	-	-
PE-g-MA	Maleic anhydride	0.005 kg	~0.5 wt.%	[64]
	Polyethylene	0.995 kg	-	-
	Output	-	-	-
	PE-g-MA	1.00 kg	The data considered for the modeling is the formulation of the commercial compound.	-

Outputs: By-Products of the Olive Stone Composite Manufacturing

The by-products obtained during the processing of the composite (Figure 3), such as olive stone dust and intermediate and fine fractions, can be used as biofuels. The most common waste management approach for the by-products is incineration [28]. As with other fuels, the heating value depends on the moisture content and ranges from around 17 MJ/kg (moisture 10%) to 20 MJ/kg (moisture 6%) [65]. Incineration of natural fibers results in recovery of energy and carbon credits [66]. Another common waste management approach is composting [23,40].

- Boiler combustion of olive stone dust

Olive mills used combustion to obtain thermal or electric energy, due to the high calorific power of the dried olive husk (4000 kcal/kg) [67]. However, the energy obtained by combustion is used for dryness of the fresh two-phase olive waste mill, which decreases the total energy recovery [10,68].

Wood lower heating values (LHVs) range between 10.5 MJ/kg for wet wood and 18.6 MJ/kg for dry wood [69].

The properties of the olive pomace present more advantages than other biomass. Indeed, it has a low sulfur content between 0.12% and 0.26%, and an LHV in the range of 16.4–18.6 MJ/kg [10]. Moreover, the LHV of olive stone presents similar values in the range of 16.2–19.2 MJ/kg. Table 5 shows some LHVs of olive stone found in the literature. It is worth mentioning that olive pit is another term used to refer to the olive stone.

Table 5. Lower heating value of olive stone.

Type of Waste	Lower Heating Value (MJ/kg)	Reference
Olive pit	19.0	[70]
Olive pit	16.2	[71]
Olive pit	17.3	[72]
Olive pit	19.2	[73]
Olive stone	16.3	[74]
Olive stone	17.0	[75]

For the modeling of the "boiler combustion" on SimaPro (PhD 9.1.1.1, PRé Sustainability, Amersfoort, The Netherlands), the combustion of natural wood chips from forest (Heat production, untreated waste wood, at a furnace 1000–5000 kW CH) was chosen as process reference. Included activities start from the delivery of waste wood to the combustion of untreated waste wood chips. It comprises the infrastructure (dust collector, furnace), the wood requirements (LHV), the emissions to air, the electricity needed for its operation, and the disposal of the ashes. The LHV of wood used in the file was 14.0 MJ/kg.

On the other hand, the LHV of olive stone is bigger than that of wood. Therefore, a corrective factor between the olive stone and wood was used in order to adjust the modeling.

The following data were used for comparing the olive stone per kg of wood.

The LHV of wood is 14.0 MJ/kg and the average LHV of an olive pit is 17.5 MJ/kg. After comparing these values, the amount of olive pit that would replace the wood in the boiler is 0.80 kg olive pit per kg of wood.

An example is the milling process, where the quantity of olive pit dust produced is 0.20 kg. This value was replaced by 0.25 kg of wood as a factor of 0.80 kg olive pit/kg wood was used for this calculation.

- Composting of pulp-rich fraction

The main by-product of the "Sieving 1" process is the olive pulp. For the treatment of this waste, the industrial composting of biowaste process was chosen (Biowaste {CH} | treatment of biowaste, industrial composting | Cut-off, U). The composting treatment is a process of controlled decomposition and humidification of biodegradable materials under managed conditions, which is aerobic and which allows the development of temperatures suitable for mesophilic and thermophilic bacteria as a result of biologically produced heat. The inventory refers to 1 kg of fresh weight of biogenic waste.

The activities of the process include energy demand for operating a compost plant as well as process emissions, infrastructure of the compost plant, and transports related to the collection of the biogenic waste.

2.2.3. Inventory Tables

Pretreatment data were obtained from the literature. Pre-industrial scale trials carried out in the Laboratoire de Chimie Agro-industrielle (Université de Toulouse, Toulouse, France) have provided the data of the compounding process [54].

The tables below show the data used for the completed modeling of the process. Table 6 shows the Global LCI data corresponding to the foreground system for the production of the olive pomace-based composite. The data of each process are reported per 1 kg of output. Table 7 shows the description of the main background processes from Ecoinvent 3.6 considered in this study, and Table 8 shows the LCI data corresponding to the production of an olive pomace-based composite reported per functional unit.

Table 6. Global life cycle inventory data corresponding to the foreground system for the production of the olive-based composite. Data of each process are reported per 1 kg of output.

Process	Flow	Amount	Comments	Ref.
Drying	**Input**	-	-	-
	Crude olive pomace (53% moisture)	1.09 kg	-	-
	Energy	0.06 kW·h	-	[56]
	Output	-	-	-
	Water (steam)	0.09 kg	Emissions to air	-
	Olive pomace (9% moisture)	1.00 kg	-	-
Milling (1)	**Input**	-	-	-
	Energy	0.11 kW·h	Ball mill DECO-PM-2X10L/15L, Electrical detail: 220 VAC, 50 Hz, 1.5 kW	[55]
	Output	-	-	-
	Olive pomace powder	0.80 kg	Olive husk sample is ground before use, giving 80% of the weight of the sample	[58]
	Dust (by-product)	0.20 kg	-	-
Sieving (1)	**Input**	-	-	-
	Energy	0.04 kW·h	Sieving machine RITEC, power: 0.48 kW	[57]
	Output	-	-	-
	Coarse fraction	0.687 kg	Fractions of OP:OSF 56.4%, intermediate fraction 11.7%	[7]
	Pulp-rich fraction	0.313 kg	Fractions of OP:PF 31.3%,	[7]
Milling (2)	**Input**	-	-	-
	Energy	0.1 kW·h	Cutting mill SM 300, Retsch. Input: 30 kg/h. Power: 3 kW	[53]
	Output	-	-	-
	Dust (by-product)	0.20 kg	All milling processes are considered 80%	-
	Coarse fraction powder	0.80 kg	-	-
Sieving (2)	**Input**	-	-	-
	Energy	0.04 kW·h	Sieve RITEC, power: 0.48 kW	[57]
	Output	-	-	-
	Intermediate fraction	0.17 kg	Fractions of OP: intermediate fraction 11.7%	[7]
	Olive stone-rich fraction	0.83 kg	Fractions of OP:PF 31.3%, OSF 56.4%	[7]
Compounding	**Input**	-	-	-
	Polymeric matrix	0.34 kg	34 wt.% polymer	-
	PE-g-MA/PP-g-MA	0.06 kg	6 wt.% coupling agent	-
	Energy (OSF/PP)	0.3566 kW·h	The information available was obtained by extrapolating the results of Uitterhaegen et al.	[54]
	Energy (OSF/PE)	0.2836 kW·h	The information available was obtained by extrapolating the results of Uitterhaegen et al.	[54]
	Output	-	-	-
	Composite OSF/PP or OSF/PE	1 kg	-	-

Table 7. Description of the main Ecoinvent 3.6 database processes considered in this study for the background processes.

Input	Ecoinvent Database v.3.6
Combustion of dust	Waste wood, untreated {CH} I heat production, untreated waste wood, at furnace 1000–5000 kW I Cut-off, U
Composting	Biowaste {CH} I treatment of biowaste, industrial composting I Cut-off, U
Energy	Electricity, medium voltage {FR} I market for I Cut-off, U
Maleic anhydride	Maleic anhydride {GLO} I market for maleic anhydride I Cut-off, U
Polyethylene	Polyethylene, high density, granulate {GLO} I market for I Cut-off, U
Polypropylene	Polypropylene, granulate {GLO} I market for I Cut-off, U
Steam	Water (Emissions to air)

Table 8. LCI data corresponding to production of the olive pomace-based composite. Data are reported per functional unit.

Name of the Process	Amount for 1 m² Profile Construction of OSF/PP	Unit	Amount for 1 m² Profile Construction of OSF/PE	Unit
Drying	30.42	kg	30.63	kg
Milling (1)	24.34	kg	24.50	kg
Sieving (1)	16.72	kg	16.83	kg
Milling (2)	13.37	kg	13.47	kg
Sieving (2)	11.07	kg	11.15	kg
Compounding	18.46	kg	18.58	kg

2.3. Life Cycle Impact Assessment

The system scenarios were developed and analyzed with SimaPro PhD 9.1.1.1 software (PRé Sustainability, Amersfoort, The Netherlands). The life cycle impact assessment (LCIA) results were assessed by the ReCiPe 2016 Endpoint v1.04 (Hierarchist; H) method, normalized and weighted based on an average world environmental impact for the year 2000 (Word ReCiPe H/A,2000). The methodology takes into account the midpoint indicators from CML, and the endpoint indicators from Ecoindicator [76]. The database used for the modeling was Ecoinvent 3.6.

Modeling elements are used to link midpoint indicators to one or more endpoint indicators, which are representative of different topics or "areas of protection" (AoP) that "defend" our interests as a society with regards to human health, ecosystems or ecosystem services and resources [43]. The ReCiPe 2016 methodology includes 18 midpoint impact categories, and three areas of protection or endpoints [77]. The different midpoint indicators contribute to a small set of endpoint indicators as can be observed in Figure 4.

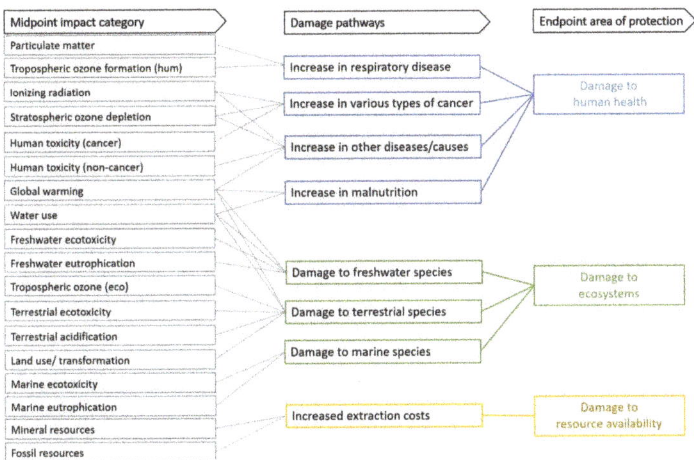

Figure 4. Summary of the impact categories that are covered in the ReCiPe 2016 method and their relation to the areas of protection (endpoint). The dotted line means that there is no constant mid-to-endpoint factor for fossil resources. Adapted from Huijbregts et al. [77].

3. Results

Figures 5 and 6 show the result of the LCA of 1 m² of lath (building material) made of OSF/PE and OSF/PP composites, respectively. As we can observe in both production processes, the main hotspot is the compounding process, which affects all the impact categories, the most affected ones being global warming (88%), freshwater eutrophication (87%), and fossil resource scarcity (95%). Inside the compounding process, the main hotspot is the production of PE and PP. Fossil resource scarcity is the most affected midpoint indicator, due to the production of ethylene and propylene (in the petrochemical industry). Another impact category that is highly affected by compounding is human carcinogenic toxicity, which reaches values of 86% and 82% for OSF/PE and OSF/PP composites, respectively.

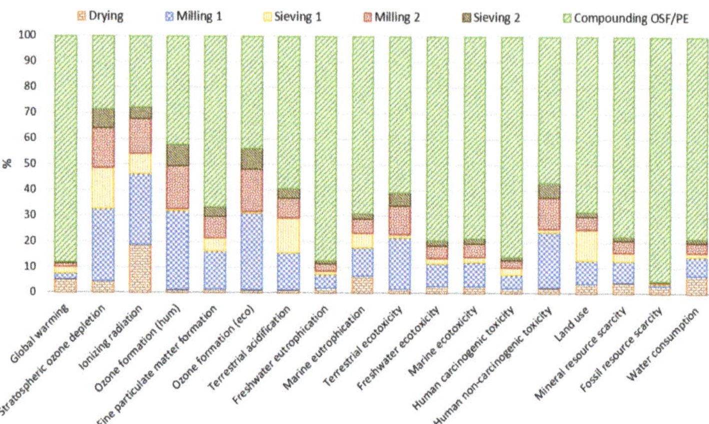

Figure 5. Contribution of each process to the potential environmental impact of the OSF/PE scenario. Characterization, ReCiPe 2016 Midpoint (H).

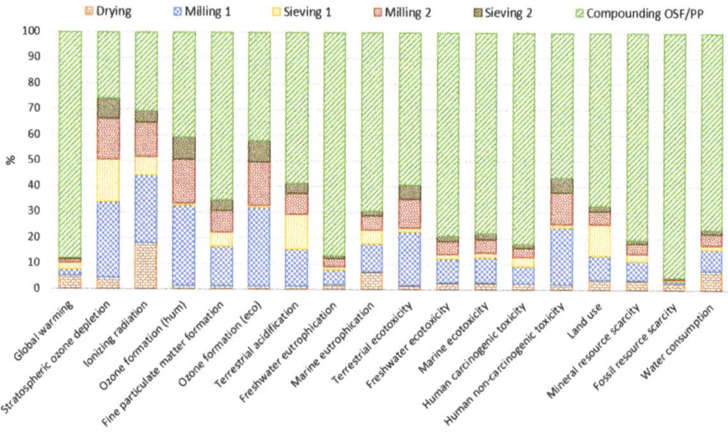

Figure 6. Contribution of each process to the potential environmental impact of the OSF/PP scenario. Characterization, ReCiPe 2016 Midpoint (H).

Figure 7 shows the pie chart of the OSF/PE and OSF/PP compounding processes. From the three damage categories, the most influenced is human health, with a value of 88%, follow by ecosystems and resources availability, with 6% in both categories. In the human health category, the main contributor is the polymeric matrix (PE and PP) (83%), whereas the coupling agent (PP-g-MA and PE-g-MA) contributed 15% of the impact in the mentioned category of each scenario. The electricity need for the process has the lowest contribution, i.e., only 2%.

From Tables 9 and 10, it is possible to observe the main input and substance that contribute to the pollution of each process. The contribution percentages of each input are expressed based on 100% of each midpoint indicator. The substance percentages are first calculated from the total of all substances of each process, and the percentage expressed in the table is the percentage that corresponds only to the input mentioned (e.g., in the drying process, 85% of the global warming is due to the input of "Olive pomace, processed" and the remaining 15% is due to "Electricity"). A total of 92% of the substances are from "Carbon dioxide, fossil" and this breaks down to 79% from "Olive pomace, processed" and 13% from "Electricity". Table 9 details the pollutants corresponding to drying, milling and sieving processes, which are similar for both scenarios. Table 10 details the pollutants corresponding to the compounding process in both scenarios.

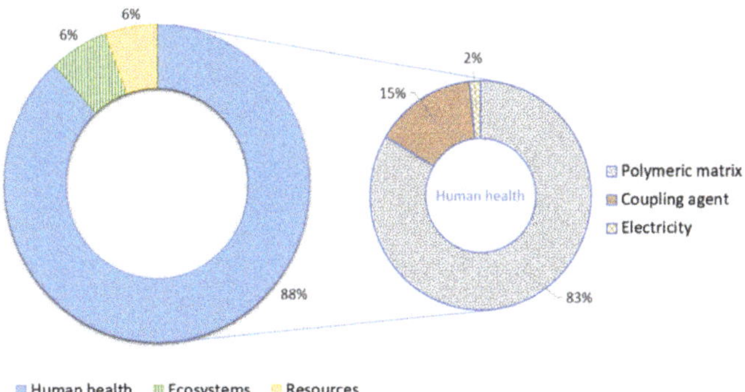

Figure 7. Contribution of endpoint indicators for the compounding process of OSF/PE and OSF/PP. ReCiPe 2016 Endpoint (H).

As mentioned before, freshwater eutrophication is highly impacted by the compounding process. This indicator shows that along the process the main impacts are produced by consumption of energy (electricity) from the machines, the combustion and composting process of the waste, and the impacts coming from the polymeric matrices (PP, PE) in both cases. Tables 9 and 10 show that the main pollutant in human carcinogenic toxicity is Chromium VI in water. The values of this midpoint indicator were 721 g 1,4-DCB for the OSF/PE composite, and 574 g 1,4-DCB for the OSF/PP one. Terrestrial ecotoxicity is one of the main contributors to pollution in the case study. The production of 1 FU of the OSF/PE composite released 61.4 kg 1,4-DCB, while the production of the OSF/PP one released 58.8 kg 1,4-DCB. The two main metals emitted during the life cycle of the product were zinc and copper, which generated damages on the ecosystems, especially the soil [78]. The results show that fossil resource scarcity indicator is similar in both cases, with 13.2 kg oil eq for the OSF/PE composite and 13.3 kg oil eq for the OSF/PP one. The mineral resources scarcity indicator (Table 10) shows that the polymeric matrices (PE and PP) depleted the environment with different substances: the OSF/PE composite mostly drains gold, while the OSF/PP one mainly drains titanium. All the results for each impact category for both biocomposites are presented in Table 11.

Table 9. Midpoint indicators and their main pollutants of the analysis of 1 FU of the composites production with the ReCiPe 2016 (H) method.

Midpoint Indicator	Drying		Milling 1		Sieving 1		Milling 2		Sieving 2	
	Input	Substance	Input	Substance	Input	Substance	Input	Substance	Input	Substance
Global warming	Olive pomace, processed (85%)	Carbon dioxide, fossil (79%)	Electricity (58%)	Carbon dioxide, fossil (52%)	Biowaste, composting (84%)	Methane, biogenic (50.3%)	Electricity (56%)	Carbon dioxide, fossil (50%)	Heat production, waste wood (55%)	Carbon dioxide, fossil (40%)
Stratospheric ozone depletion	Olive pomace, processed (66%)	Dinitrogen monoxide (33%)	Heat production (Waste wood) (90%)	Dinitrogen monoxide (90%)	Biowaste, composting (95%)	Dinitrogen monoxide (94%)	Heat production (Waste wood) (91%)	Dinitrogen monoxide (91%)	Heat production, waste wood (94%)	Dinitrogen monoxide (94%)

Table 9. *Cont.*

Midpoint Indicator	Drying		Milling 1		Sieving 1		Milling 2		Sieving 2	
	Input	Substance	Input	Substance	Input	Substance	Input	Substance	Input	Substance
Ionizing radiation	Electricity (74%)	Radon-222 (73%)	Electricity (94%)	Radon-222 (92%)	Electricity (96%)	Radon-222 (94%)	Electricity (94%)	Radon-222 (91%)	Electricity (90%)	Radon-222 (88%)
Ozone formation, Human health	Olive pomace, processed (76%)	Nitrogen oxides (73%)	Heat production (Waste wood) (98%)	Nitrogen oxides (97%)	Biowaste, composting (73%)	Nitrogen oxides (70%)	Heat production (Waste wood) (98%)	Nitrogen oxides (98%)	Heat production, waste wood (99%)	Nitrogen oxides (98%)
Fine particulate matter formation	Olive pomace, processed (58%)	Sulfur dioxide (23%)	Heat production (Waste wood) (92%)	Nitrogen oxides (68%)	Biowaste, composting (94%)	Ammonia (83%)	Heat production (Waste wood) (93%)	Nitrogen oxides (69%)	Heat production, waste wood (95%)	Nitrogen oxides (70%)
Ozone formation, Terrestrial ecosystems	Olive pomace, processed (76%)	Nitrogen oxides (71%)	Heat production (Waste wood) (98%)	Nitrogen oxides (97%)	Biowaste, composting (74%)	Nitrogen oxides (68%)	Heat production (Waste wood) (98%)	Nitrogen oxides (97%)	Heat production, waste wood (99%)	Nitrogen oxides (98%)
Terrestrial acidification	Olive pomace (61%)	Sulfur dioxide (31%)	Heat production (Waste wood) (93%)	Nitrogen oxides (83%)	Biowaste, composting (98%)	Ammonia (95%)	Heat production (Waste wood) (94%)	Nitrogen oxides (83%)	Heat production, waste wood (96%)	Nitrogen oxides (85%)
Freshwater eutrophication	Electricity (63%)	Phosphate, water (63%)	Heat production (Waste wood) (58%)	Phosphate, water (42%)	Biowaste, composting (51%)	Phosphate, water (51%)	Heat production, waste wood (61%)	Phosphate, water (43%)	Heat production, waste wood (71%)	Phosphate, water (50%)
Marine eutrophication	Electricity (73%)	Nitrate (70%)	Electricity (85%)	Nitrate (81%)	Biowaste, composting (51%)	Nitrate (47%)	Electricity (83%)	Nitrate (80%)	Electricity (76%)	Nitrate (73%)
Terrestrial ecotoxicity	Electricity (65%)	Copper (44%)	Heat production (Waste wood) (90%)	Zinc (48%)	Biowaste, composting (59%)	Copper (38%)	Heat production (Waste wood) (91%)	Zinc (49%)	Heat production, waste wood (94%)	Zinc (51%)
Freshwater ecotoxicity	Electricity (75%)	Copper (58%)	Heat production (Waste wood) (59%)	Copper (30%)	Electricity (55%)	Copper (43%)	Heat production, waste wood (62%)	Copper (31%)	Heat production, waste wood (71%)	Copper (36%)
Marine ecotoxicity	Electricity (69%)	Copper (51%)	Heat production (Waste wood) (63%)	Copper (25%)	Electricity (54%)	Copper (40%)	Heat production, waste wood (65%)	Copper (26%)	Heat production, waste wood (74%)	Copper (30%)
Human carcinogenic toxicity	Electricity (54%)	Chromium VI, water (38%)	Heat production (Waste wood) (64%)	Chromium VI, water (49%)	Biowaste, composting (81%)	Chromium VI, water (77%)	Heat production (Waste wood) (66%)	Chromium VI, water (51%)	Heat production, waste wood (75%)	Chromium VI, water (58%)
Human non-carcinogenic toxicity	Electricity (62%)	Arsenic, water (29%)	Heat production (Waste wood) (89%)	Zinc, soil (56%)	Electricity (57%)	Zinc (20%)	Heat production (Waste wood) (90%)	Zinc, soil (57%)	Heat production, waste wood (93%)	Zinc, soil (59%)
Land use	Electricity (78%)	Occupation, forest, intensive (65%)	Electricity (60%)	Occupation, forest, intensive (50%)	Biowaste, composting (87%)	Occupation, forest, intensive (68%)	Electricity (58%)	Occupation, forest, intensive (48%)	Heat production, waste wood (53%)	Occupation, forest, intensive (15%)
Mineral resource scarcity	Electricity (63%)	Uranium (42%)	Electricity (59%)	Uranium (39%)	Biowaste, composting (55%)	Iron (30%)	Electricity (56%)	Uranium (38%)	Heat production, waste wood (55%)	Uranium (3%)
Fossil resource scarcity	Olive pomace, processed (89%)	Gas, natural/m^3 (86%)	Electricity (77%)	Gas, natural/m^3 (49%)	Biowaste, composting (64%)	Oil, crude (51%)	Electricity (76%)	Gas, natural/m^3 (48%)	Electricity (67%)	Gas, natural/m^3 (42%)
Water consumption	Olive pomace, processed (63%)	Water, turbine use, unspecified natural origin, FR (25%)	Electricity (64%)	Water, turbine use, unspecified natural origin, FR (80%)	Electricity (88%)	Water, turbine use, unspecified natural origin, FR (88%)	Electricity (62%)	Water, turbine use, unspecified natural origin, FR (78%)	Electricity (51%)	Water, turbine use, unspecified natural origin, FR (70%)

Table 10. Midpoint indicators and their main pollutants in the analysis of 1 FU on the compounding process of the composites with the ReCiPe 2016 (H) method.

Midpoint Indicator	Compounding OSF/PE		Compounding OSF/PP	
	Input	Substance	Input	Substance
Global warming	Polyethylene, high density (84%)	Carbon dioxide, fossil (67%)	Polypropylene (83%)	Carbon dioxide, fossil (66%)
Stratospheric ozone depletion	Polyethylene, high density (78%)	Dinitrogen monoxide (68%)	Polypropylene (74%)	Dinitrogen monoxide (65%)
Ionizing radiation	Electricity (79%)	Radon-222 (77%)	Electricity (85%)	Radon-222 (83%)
Ozone formation, Human health	Polyethylene, high density (84%)	Nitrogen oxides (74%)	Polypropylene (84%)	Nitrogen oxides (75%)
Fine particulate matter formation	Polyethylene, high density (84%)	Sulfur dioxide (44%)	Polypropylene (83%)	Sulfur dioxide (46%)
Ozone formation, Terrestrial ecosystems	Polyethylene, high density (84%)	Nitrogen oxides (69%)	Polypropylene (84%)	Nitrogen oxides (71%)
Terrestrial acidification	Polyethylene, high density (84%)	Sulfur dioxide (62%)	Polypropylene (83%)	Sulfur dioxide (62%)
Freshwater eutrophication	Polyethylene, high density (83%)	Phosphate, water (83%)	Polypropylene (82%)	Phosphate, water (82%)
Marine eutrophication	Polyethylene, high density (75%)	Nitrate (58%)	Polypropylene (73%)	Nitrate (54%)
Terrestrial ecotoxicity	Polyethylene, high density (83%)	Copper (56%)	Polypropylene (82%)	Copper (59%)
Freshwater ecotoxicity	Polyethylene, high density (82%)	Copper (49%)	Polypropylene (81%)	Copper (49%)
Marine ecotoxicity	Polyethylene, high density (82%)	Copper (45%)	Polypropylene (81%)	Copper (45%)
Human carcinogenic toxicity	Polyethylene, high density (83%)	Chromium VI, water (78%)	Polypropylene (82%)	Chromium VI, water (76%)
Human non-carcinogenic toxicity	Polyethylene, high density (82%)	Zinc (50%)	Polypropylene (81%)	Zinc (49%)
Land use	Polyethylene, high density (79%)	Occupation, forest, intensive (37%)	Polypropylene (77%)	Occupation, forest, intensive (35%)
Mineral resource scarcity	Polyethylene, high density (80%)	Gold (24%)	Polypropylene (81%)	Titanium (22%)
Fossil resource scarcity	Polyethylene, high density (85%)	Oil, crude (50%)	Polypropylene (85%)	Oil, crude (50%)
Water consumption	Polyethylene, high density (81%)	Water, turbine use, unspecified natural origin, FR (6%)	Polypropylene (79%)	Water, turbine use, unspecified natural origin, FR (4%)

Table 11. Results of the environmental impact categories from a cradle-to-gate perspective for the production of 1 m^2 olive pomace-based composite lath.

Impact Category	Unit	OSF/PE	OSF/PP
Global warming	kg CO_2 eq	20.52	20.25
Stratospheric ozone depletion	kg CFC11 eq	1.46×10^{-5}	1.39×10^{-5}
Ionizing radiation	kBq Co-60 eq	7.38	7.67
Ozone formation, human health	kg NO_x eq	0.09	0.09
Fine particulate matter formation	kg PM2.5 eq	0.03	0.03
Ozone formation, Terrestrial ecosystems	kg NO_x eq	0.09	0.09
Terrestrial acidification	kg SO_2 eq	0.08	0.08
Freshwater eutrophication	kg P eq	3.87×10^{-3}	3.72×10^{-3}
Marine eutrophication	kg N eq	4.30×10^{-4}	4.30×10^{-4}
Terrestrial ecotoxicity	kg 1,4-DCB	61.36	58.76
Freshwater ecotoxicity	kg 1,4-DCB	0.70	0.68
Marine ecotoxicity	kg 1,4-DCB	0.92	0.90
Human carcinogenic toxicity	kg 1,4-DCB	0.72	0.57
Human non-carcinogenic toxicity	kg 1,4-DCB	18.27	17.89
Land use	m^{2}a crop eq	0.24	0.23
Mineral resource scarcity	kg Cu eq	0.05	0.06
Fossil resource scarcity	kg oil eq	13.18	13.32
Water consumption	m^3	0.24	0.21

Figure 8 represents the single score result of the LCA production of 1 m^2 of the OSF/PE and OSF/PP laths, respectively. It is worth mentioning that the score used in the charts is called eco-points (Pt). One eco-point can be interpreted as one thousandth of the annual environmental load of one average European inhabitant [79].

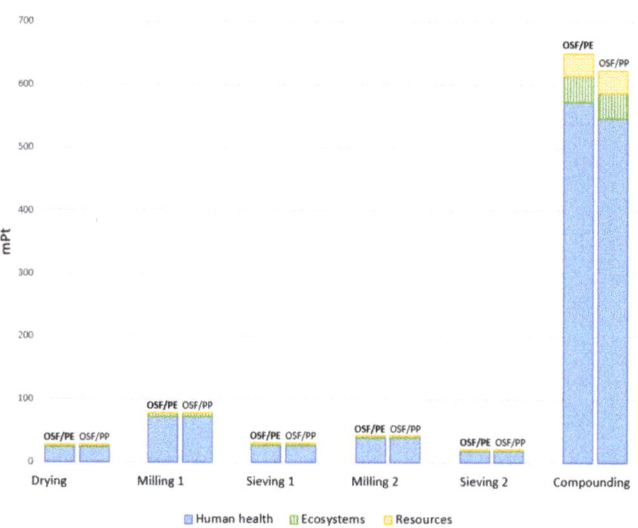

Figure 8. Single score results of OSF/PE and OSF/PP scenarios. ReCiPe 2016 Endpoint (H).

In both scenarios, the most impactful process was compounding, where human health was the more affected area of protection. The results show similar values for the compounding of OSF/PE and OSF/PP composites, 574 and 548 mPt, respectively. The human health category repeated along the different processes. Therefore, it can be identified as an area of concern. The contribution for this endpoint category was 89% for both scenarios. Ecosystems and resources availability had a small impact in comparison with human health, their total values being 53.7 mPt (6%) and 36.5 mPt (4%), respectively, for the OSF/PE composite, and 52.5 (6%) mPt and 37.1 mPt (5%), respectively, for the OSF/PP composite. As mention before, the main contributor to these damage categories was the compounding process. For ecosystems, the contribution of this process to the total damage category was 76% for both scenarios, while compounding represented 96% of the resources availability category.

4. Discussion and Perspectives

4.1. Comparison with a Conventional Lath

For comparing the performance of the herein biocomposites made of olive stone flour and its application as a building material, it is important to compare it with a business-as-usual scenario for this type of application. A common material used for this is "PVC decking".

This term is used to refer to plastic decking that uses cellular polyvinyl chloride (PVC) as the building material of the lath. Cellular PVC is lighter than standard PVC due to the addition of a foaming agent throughout the manufacturing process [80]. The authors reported that the production of 9.3 m^2 of installed PVC decking in service for 25 years produces 426 kg CO_2 eq. That project considered the full life cycle of the decking product, starting from the raw material extraction to the final disposal in a municipal solid waste landfill [80]. The preliminary results (cradle-to-gate) of this project reported the emission of 368 kg CO_2 eq. This translates into 39.6 kg CO_2 eq for 1 m^2 of PVC decking. In contrast, the herein work results show that the production of 1 m^2 of OSF/PE and OSF/PP lath releases 20.5 kg CO_2 eq and 20.3 kg CO_2 eq, respectively. This may indicate that the pollution caused by PVC decking is almost two times bigger than that caused by the olive pomace-based composites.

What is certain is that PVC is considered carcinogenic and environmentally hazardous due to the presence of organochlorines. In contrast, PP and PE are chlorine-free plastics [81].

4.2. Biosourcing the Matrix

The most common polymeric matrices used for biocomposites are produced from petrochemicals and are not biodegradable. This type of polymer generates long-term negative impacts on the environment and human health [17]. It should nevertheless be noted here that biosourced polyolefins exist on the market, especially BioPE, which is produced from sugar cane waste, and, to a much lesser extent, BioPP.

Biodegradable and biosourced polymers are raising great interest, due to the growing environmental concerns and the decline of the fossil resources [30]. Among the mentioned polymers, PLA (poly(lactic acid)) and polyhydroxyalkanoates (PHAs) (e.g., PHB (polyhydroxybutyrate), PHV (polyhydroxyvalerate), PHBV (polyhydroxy-butyrate-co-valerate)) are commercially available polymers, produced in large scale and able to compete with more traditional petroleum-based plastics [17,30,82]. The environmental assessment of this biodegradable composite reinforced with OP can be the objective for further studies.

4.3. Impact of the Diversion of the Olive Pomace from Its Original Use

It is important to contemplate the consequences of diverting the olive pomace from its original use.

The conventional uses of olive oil by-products explained earlier (1. Introduction) are their controlled spreading, their use as a mixture in animal nutrition, and composting.

Olive wastes are frequently discharged on soil. The production of 1 L of extra virgin olive oil (EVOO) leads to the emission of around 57 g CO_2 eq. This practice is widespread around the European Community as it is supported by the law in many countries [6,83].

Duman et al. [23] studied the composting scenario in Turkey, and found that 2.25 kg of OP in a mixture with 0.34 kg of wheat straw and 0.67 kg of poultry manure can produce 2.09 kg of compost and release to the atmosphere 6.82 kg CO_2 eq, with nitrous oxide as the main emission [23].

For the production of the herein functional unit (1 m^2 of lath (building material) made of olive pomace-based composite), an average of 46.7 kg OP is used. Compared with the results of Duman et al. [23], the composting of the same quantity of OP would liberate to the environment 141.5 kg CO_2 eq, which is almost seven times higher than the pollution caused by the olive pomace-based composites.

The controlled spreading of those 46.7 kg OP would liberate to the environment 0.665 kg CO_2 eq. This is taking into account that, on average, olive fruit contains 20 wt.% of oil, and the remaining 80 wt.% together with the water added during the olive oil extraction process form OP. The production of 1 L of EVOO generates 4 kg of OP. The controlled spreading of OP is about thirty times less polluting than the production of 1 m^2 of olive pomace-based composites. Nevertheless, the direct application of this olive oil by-product can cause negative effects on soil due to its high mineral salt content, low pH and presence of polyphenols [6].

4.4. Carbon Sequestration

To deepen the information obtained on this work, the importance of an expected benefit of biocomposites can be discussed, namely long-term carbon sequestration.

The U.S. Environmental Protection Agency defines biogenic CO_2 emissions as CO_2 emissions associated with the natural carbon cycle, together with the emissions resulting from the combustion, harvest, digestion, fermentation, decomposition, or processing of biologically based materials [84]. Products and residues of the olive orchard cultivation contain biogenic carbon derived from the uptake of CO_2 by the crop [85].

When talking about olive oil production, the biogenic carbon from the agricultural phase (olive tree permanent components, especially taking into account the longevity of olive groves) does not become part of the product (olive oil). Therefore, in accordance with PAS 2050, the assessment of biogenic carbon should be taken into account when performing an LCA that considers this phase within its system boundaries [12]. Storing carbon during a long lifespan (composite decking can last between 25 and 30 years) can mitigate climate change because of the delay in the carbon emissions into the technosphere. This advantage, however, could not be quantified with the static LCA approach used in this study. A rough estimation of biogenic carbon, based on the work herein presented, shows that 1 FU of the olive pomace based-composite could contribute to delaying the emission of over 32.5 kg CO_2 (considering a humidity content of OP of 53% and a percentage of total carbon for a three-phase centrifugation of 29%).

5. Conclusions

In this study, the valorization of an olive pomace-based biocomposite was evaluated through the environmental criteria, with the LCA method. The functional unit is the production of 1 m^2 of a lath (building material) made of an olive pomace-based biocomposite. Two scenarios with different thermoplastic matrices (PE, PP) were assessed. In both cases, the twin-screw compounding process contributed the major burden in most of the midpoint impact categories. Compounding is mainly affected by the production of the respective polymers. When comparing the OSF/PE and OSF/PP materials, the impacts are relatively similar. Therefore, a further study at the end of life of the material should be carried out to conclude which of the proposed biocomposites is the less polluting one.

The results obtained showed that human health is the most affected area of protection; it represented 89% for both scenarios, i.e., OSF/PP and OSF/PE. The main contributors to this damage category are energy, carbon dioxide and sulfur dioxide, used and produced in the manufacturing process. Ecosystems and resources availability represented a lower contribution to the total impact. Scenarios presented values of 6% and 4.5% for ecosystems and resources availability, respectively.

The comparison of the olive pomace-based-composite with the business-as-usual scenario shows that the biocomposite released half of the pollution produced by the PVC decking, when considering the same FU. Besides, given the long lifespan of the biocomposites, the CO_2 stored on it can mitigate climate change because of the delay in the carbon emissions into the technosphere.

This case study evaluates a new path for olive pomace-based composites as an alternative eco-material for the building sector, based on environmental criteria. Future works can include the use of biosourced polymer matrices, which could reduce the impact of the production of olive pomace-based composites.

Author Contributions: Methodology, G.E.-A.; investigation, G.E.-A., P.G.-A. and P.E.; writing—original draft, G.E.-A. and P.G.-A.; writing—review and editing, C.V., J.-P.B., P.E. and C.S.; validation, C.V., J.-P.B., P.E. and C.S.; supervision, C.V., J.-P.B. and C.S.; visualization, G.E.-A. All authors have read and agreed to the published version of the manuscript.

Funding: The work was carried out as part of the doctoral thesis of G. Espadas-Aldana, her work being financed by the Mexican National Council for Science and Technology (CONACYT Mexico) under the scholarship No. 471707.

Institutional Review Board Statement: Not applicable.

Informed Consent Statement: Not applicable.

Data Availability Statement: Data is contained within the article.

Conflicts of Interest: The authors declare no conflict of interest.

References

1. Del Mar Contreras, M.; Romero, I.; Moya, M.; Castro, E. Olive-derived biomass as a renewable source of value-added products. *Process Biochem.* **2020**, *97*, 43–56. [CrossRef]
2. Notarnicola, B.; Salomone, R.; Petti, L.; Renzulli, P.A.; Roma, R.; Cerutti, A.K. *Life Cycle Assessment in the Agri-Food Sector: Case Studies, Methodological Issues and Best Practices*; Springer International Publishing: Berlin/Heidelberg, Germany, 2015; ISBN 978-3-319-11939-7.
3. Boskou, D. *Olive Oil: Chemistry and Technology*, 2nd ed.; AOCS Publishing: Urbana, IL, USA, 2006; ISBN 978-1-893997-88-2.
4. Vidal, A.M. Mejora de Las Características Organolépticas, Funcionales y Nutricionales de Aceites de Oliva Vírgenes. Ph.D. Thesis, Universidad de Jaén, Jaén, Spain, 2019.
5. Dermeche, S.; Nadour, M.; Larroche, C.; Moulti-Mati, F.; Michaud, P. Olive mill wastes: Biochemical characterizations and valorization strategies. *Process Biochem.* **2013**, *48*, 1532–1552. [CrossRef]
6. Salomone, R.; Cappelletti, G.M.; Malandrino, O.; Mistretta, M.; Neri, E.; Nicoletti, G.M.; Notarnicola, B.; Pattara, C.; Russo, C.; Saija, G. Life cycle assessment in the olive oil sector. In *Life Cycle Assessment in the Agri-food Sector: Case Studies, Methodological Issues and Best Practices*; Notarnicola, B., Salomone, R., Petti, L., Renzulli, P.A., Roma, R., Cerutti, A.K., Eds.; Springer International Publishing: Cham, Germany, 2015; pp. 57–121. ISBN 978-3-319-11940-3.
7. Lammi, S.; Le Moigne, N.; Djenane, D.; Gontard, N.; Angellier-Coussy, H. Dry fractionation of olive pomace for the development of food packaging biocomposites. *Ind. Crops Prod.* **2018**, *120*, 250–261. [CrossRef]
8. De la Casa, J.A.; Castro, E. Recycling of washed olive pomace ash for fired clay brick manufacturing. *Constr. Build. Mater.* **2014**, *61*, 320–326. [CrossRef]
9. Ravindran, R.; Jaiswal, A.K. Exploitation of food industry waste for high-value products. *Trends Biotechnol.* **2016**, *34*, 58–69. [CrossRef] [PubMed]
10. Azbar, N.; Bayram, A.; Filibeli, A.; Muezzinoglu, A.; Sengul, F.; Ozer, A. A review of waste management options in olive oil production. *Crit. Rev. Environ. Sci. Technol.* **2004**, *34*, 209–247. [CrossRef]
11. Bhatnagar, A.; Kaczala, F.; Hogland, W.; Marques, M.; Paraskeva, C.A.; Papadakis, V.G.; Sillanpää, M. Valorization of solid waste products from olive oil industry as potential adsorbents for water pollution control—A review. *Environ. Sci. Pollut. Res.* **2014**, *21*, 268–298. [CrossRef] [PubMed]
12. Espadas-Aldana, G.; Vialle, C.; Belaud, J.-P.; Vaca-Garcia, C.; Sablayrolles, C. Analysis and trends for life cycle assessment of olive oil production. *Sustain. Prod. Consum.* **2019**, *19*, 216–230. [CrossRef]
13. Nunes, M.A.; Pimentel, F.B.; Costa, A.S.G.; Alves, R.C.; Oliveira, M.B.P.P. Olive by-products for functional and food applications: Challenging opportunities to face environmental constraints. *Innov. Food Sci. Emerg. Technol.* **2016**, *35*, 139–148. [CrossRef]
14. Salomone, R.; Ioppolo, G. Environmental impacts of olive oil production: A life cycle assessment case study in the province of Messina (Sicily). *J. Clean. Prod.* **2012**, *28*, 88–100. [CrossRef]
15. Les Coproduits de L'olivier, Déchets ou Matières Premières D'avenir? Available online: https://www.olivardeche.fr/articles.php?lng=fr&pg=81&fbclid=IwAR27WcF_zf0qqNu-r869l13AP2zWz8Z9a_E-6F2Dmh14RgcvsAvpOyI3r7I (accessed on 27 January 2021).
16. Aparicio-Ruiz, R.; Harwood, J. *Handbook of Olive Oil: Analysis and Properties*, 2nd ed.; Springer: New York City, NY, USA, 2013; ISBN 978-1-4614-7776-1.
17. Lammi, S.; Gastaldi, E.; Gaubiac, F.; Angellier-Coussy, H. How olive pomace can be valorized as fillers to tune the biodegradation of PHBV based composites. *Polym. Degrad. Stab.* **2019**, *166*, 325–333. [CrossRef]
18. Alibes, X.; Berge; Martilotti, F.; Nefzaoui, A.; Zoïopoulos, P.E. *Utilisation Des Sous-Produits de l'Olivier En Alimentation Animale Dans Le Bassin Méditerranéen*; FAO: Rome, Italy, 1983.
19. Haddadin, M.S.Y.; Haddadin, J.; Arabiyat, O.I.; Hattar, B. Biological conversion of olive pomace into compost by using trichoderma harzianum and phanerochaete chrysosporium. *Bioresour. Technol.* **2009**, *100*, 4773–4782. [CrossRef]
20. Morillo, J.A.; Antizar-Ladislao, B.; Monteoliva-Sánchez, M.; Ramos-Cormenzana, A.; Russell, N.J. Bioremediation and biovalorisation of olive-mill wastes. *Appl. Microbiol. Biotechnol.* **2009**, *82*, 25–39. [CrossRef]
21. Christoforou, E.; Fokaides, P.A. A review of olive mill solid wastes to energy utilization techniques. *Waste Manag.* **2016**, *49*, 346–363. [CrossRef]
22. Cossu, A.; Degl'Innocenti, S.; Agnolucci, M.; Cristani, C.; Bedini, S.; Nuti, M. Assessment of the life cycle environmental impact of the olive oil extraction solid wastes in the European Union. *Open Waste Manag. J.* **2013**, *6*, 12–20. [CrossRef]
23. Duman, A.K.; Özgen, G.Ö.; Üçtuğ, F.G. Environmental life cycle assessment of olive pomace utilization in Turkey. *Sustain. Prod. Consum.* **2020**, *22*, 126–137. [CrossRef]
24. Barreca, F.; Fichera, C.R. Use of olive stone as an additive in cement lime mortar to improve thermal insulation. *Energy Build.* **2013**, *62*, 507–513. [CrossRef]
25. El-Sheikh, A.H.; Newman, A.P.; Al-Daffaee, H.K.; Phull, S.; Cresswell, N. Characterization of activated carbon prepared from a single cultivar of Jordanian olive stones by chemical and physicochemical techniques. *J. Anal. Appl. Pyrolysis* **2004**, *1*, 151–164. [CrossRef]
26. Matos, M.; Barreiro, M.F.; Gandini, A. Olive stone as a renewable source of biopolyols. *Ind. Crops Prod.* **2010**, *32*, 7–12. [CrossRef]
27. Banat, R. Olive pomace flour as potential organic filler in composite materials: A brief review. *Am. J. Polym. Sci.* **2019**, *9*, 10–15.

28. Boufi, S. Biocomposites from olive-stone flour. In *Lignocellulosic Fibre and Biomass-Based Composite Materials*; Elsevier: Amsterdam, The Netherlands, 2017; pp. 387–408. ISBN 978-0-08-100959-8.

29. Kaya, N.; Atagur, M.; Akyuz, O.; Seki, Y.; Sarikanat, M.; Sutcu, M.; Seydibeyoglu, M.O.; Sever, K. Fabrication and characterization of olive pomace filled PP Composites. *Compos. Part B Eng.* **2018**, *150*, 277–283. [CrossRef]

30. Koutsomitopoulou, A.F.; Bénézet, J.C.; Bergeret, A.; Papanicolaou, G.C. Preparation and characterization of olive pit powder as a filler to PLA-matrix bio-composites. *Powder Technol.* **2014**, *255*, 10–16. [CrossRef]

31. Lammi, S.; Barakat, A.; Mayer-Laigle, C.; Djenane, D.; Gontard, N.; Angellier-Coussy, H. Dry fractionation of olive pomace as a sustainable process to produce fillers for biocomposites. *Powder Technol.* **2018**, *326*, 44–53. [CrossRef]

32. Naghmouchi, I.; Mutjé, P.; Boufi, S. Olive stones flour as reinforcement in polypropylene composites: A step forward in the valorization of the solid waste from the olive oil industry. *Ind. Crops Prod.* **2015**, *72*, 183–191. [CrossRef]

33. Aouat, H.; Hammiche, D.; Boukerrou, A.; Djidjelli, H.; Grohens, Y.; Pillin, I. Effects of interface modification on composites based on olive husk flour. *Mater. Today Proc.* **2021**, *36*, 94–100. [CrossRef]

34. Researchers Develop Compostable Plastic Packaging from Olive Waste. Available online: https://www.oliveoiltimes.com/world/researchers-develop-compostable-plastic-packaging-from-olive-waste/81373 (accessed on 27 January 2021).

35. Biolive Biological and Chemical Technologies. Available online: http://www.biolivearge.com/indexen.html (accessed on 27 January 2021).

36. Material Insight: Bioplastic Made from Olives. Available online: https://www.materialconnexion.com/material-insight-bioplastic/ (accessed on 27 January 2021).

37. Belaud, J.-P.; Adoue, C.; Vialle, C.; Chorro, A.; Sablayrolles, C. A Circular economy and industrial ecology toolbox for developing an eco-industrial park: Perspectives from french policy. *Clean Technol. Environ. Policy* **2019**, *21*, 967–985. [CrossRef]

38. Chatzisymeon, E.; Foteinis, S.; Mantzavinos, D.; Tsoutsos, T. Life cycle assessment of advanced oxidation processes for olive mill wastewater treatment. *J. Clean. Prod.* **2013**, *54*, 229–234. [CrossRef]

39. Christoforou, E.A.; Fokaides, P.A. Life cycle assessment (LCA) of olive husk torrefaction. *Renew. Energy* **2016**, *90*, 257–266. [CrossRef]

40. El Hanandeh, A. Energy recovery alternatives for the sustainable management of olive oil industry waste in Australia: Life cycle assessment. *J. Clean. Prod.* **2015**, *91*, 78–88. [CrossRef]

41. Parascanu, M.M.; Sánchez, P.; Soreanu, G.; Valverde, J.L.; Sanchez-Silva, L. Environmental assessment of olive pomace valorization through two different thermochemical processes for energy production. *J. Clean. Prod.* **2018**, *186*, 771–781. [CrossRef]

42. Parascanu, M.M.; Puig Gamero, M.; Sánchez, P.; Soreanu, G.; Valverde, J.L.; Sanchez-Silva, L. Life cycle assessment of olive pomace valorisation through pyrolysis. *Renew. Energy* **2018**, *122*, 589–601. [CrossRef]

43. Hauschild, M.; Rosenbaum, R.K.; Olsen, S. *Life Cycle Assessment: Theory and Practice*; Springer International Publishing: Berlin/Heidelberg, Germany, 2018; ISBN 978-3-319-56474-6.

44. *ISO 14040 Environmental Management—Life Cycle Assessment—Principles and Framework*; International Organization for Standardization (ISO): Geneve, Switzerland, 2006.

45. Al-Ma'adeed, M.; Ozerkan, G.; Kahraman, R.; Rajendran, S.; Hodzic, A. Life cycle assessment of particulate recycled low density polyethylene and recycled polypropylene reinforced with talc and fiberglass. *Key Eng. Mater.* **2011**, *471–472*, 999–1004. [CrossRef]

46. Sommerhuber, P.F.; Wenker, J.L.; Rüter, S.; Krause, A. Life cycle assessment of wood-plastic composites: Analysing alternative materials and identifying an environmental sound end-of-life option. *Resour. Conserv. Recycl.* **2017**, *117*, 235–248. [CrossRef]

47. Vidal, R.; Martínez, P.; Garraín, D. Life cycle assessment of composite materials made of recycled thermoplastics combined with rice husks and cotton linters. *Int. J. Life Cycle Assess.* **2009**, *14*, 73–82. [CrossRef]

48. *EN15804:2012+A1:2013 European Committee for Standardization Sustainability of Construction Works—Environmental Product Declarations—Core Rules for the Product Category of Construction Products*; European Standard: Brussels, Belgium, 2013.

49. *ISO 14044: Environmental Management—Life Cycle Assessment—Requirements and Guidelines*; International Organization for Standardization (ISO): Geneve, Switzerland, 2006.

50. Naghmouchi, I.; Mutjé, P.; Boufi, S. Polyvinyl chloride composites filled with olive stone flour: Mechanical, thermal, and water absorption properties. *J. Appl. Polym. Sci.* **2014**, *131*. [CrossRef]

51. Sobczak, P.; Mazur, J.; Zawiślak, K.; Panasiewicz, M.; Żukiewicz-Sobczak, W.; Królczyk, J.; Lechowski, J. Evaluation of dust concentration during grinding grain in sustainable agriculture. *Sustainability* **2019**, *11*, 4572. [CrossRef]

52. Zhao, X.; Roberts, G.; Roberts, E. Bio-Composite and Bioplastic Materials and Method. U.S. Patent Application No. 16/345,668, 12 September 2019.

53. RETSCH Cutting Mill SM 300—Retsch. Available online: https://www.retsch.com/products/milling/cutting-mills/sm-300/function-features/ (accessed on 10 July 2020).

54. Uitterhaegen, E.; Parinet, J.; Labonne, L.; Mérian, T.; Ballas, S.; Véronèse, T.; Merah, O.; Talou, T.; Stevens, C.V.; Chabert, F.; et al. Performance, durability and recycling of thermoplastic biocomposites reinforced with coriander straw. *Compos. Part Appl. Sci. Manuf.* **2018**, *113*, 254–263. [CrossRef]

55. DECO 20L Ball Mill Lab Experiment. Available online: http://m.deco-ballmill.com/laboratory-ball-mill/lab-roller-ball-mill/20l-ball-mill-lab-experiment.html (accessed on 9 July 2020).

56. Kylili, A.; Christoforou, E.; Fokaides, P.A. Environmental evaluation of biomass pelleting using life cycle assessment. *Biomass Bioenergy* **2016**, *84*, 107–117. [CrossRef]

57. RITEC Tamiseurs MC-RITEC. Available online: http://ritec.fr/separer/tamiseurs/MC (accessed on 10 July 2020).

58. Aslan, S.S. Olive oil recovery methods. *Terr. Aquat. Environ. Toxicol.* **2010**, *4*, 118–119.

59. Avadí, A. Screening LCA of French organic amendments and fertilisers. *Int. J. Life Cycle Assess.* **2020**, *25*, 698–718. [CrossRef]

60. Schau, E.M.; Michalopoulos, G.; Russo, C. *Product Enviromental Footprint Category Rules for Olive Oil*; European Commission: Brussels, Belgium, 2016.

61. Qin, Q.H. Introduction to the composite and its toughening mechanisms. In *Toughening Mechanisms in Composite Materials*; Qin, Q., Ye, J., Eds.; Woodhead Publishing: Cambridge, UK, 2015; pp. 1–32. ISBN 978-1-78242-279-2.

62. Koerner, G.R.; Koerner, R.M. Polymeric Geomembrane Components in Landfill Liners. In *Solid Waste Landfilling*; Cossu, R., Stegmann, R., Eds.; Elsevier: Amsterdam, The Netherlands, 2018; pp. 313–341. ISBN 978-0-12-818336-6.

63. Sigma-Aldrich Polypropylene-Graft-Maleic Anhydride. Available online: https://www.sigmaaldrich.com/catalog/product/aldrich/427845 (accessed on 3 September 2020).

64. Sigma-Aldrich Polyethylene-Graft-Maleic Anhydride. Available online: https://www.sigmaaldrich.com/catalog/product/aldrich/456624 (accessed on 3 September 2020).

65. Pattara, C.; Cappelletti, G.M.; Cichelli, A. Recovery and use of olive stones: Commodity, environmental and economic assessment. *Renew. Sustain. Energy Rev.* **2010**, *14*, 1484–1489. [CrossRef]

66. Joshi, S.V.; Drzal, L.T.; Mohanty, A.K.; Arora, S. Are natural fiber composites environmentally superior to glass fiber reinforced composites? *Compos. Part Appl. Sci. Manuf.* **2004**, *35*, 371–376. [CrossRef]

67. Intini, F.; Kühtz, S.; Rospi, G. Energy recovery of the solid waste of the olive oil industries—LCA analysis and carbon footprint assessment. *J. Sustain. Energy Environ.* **2011**, *2*, 157–166.

68. Roig, A.; Cayuela, M.L.; Sánchez-Monedero, M.A. An overview on olive mill wastes and their valorisation methods. *Waste Manag.* **2006**, *26*, 960–969. [CrossRef]

69. Morris, J. Recycle, bury, or burn wood waste biomass?: LCA answer depends on carbon accounting, emissions controls, displaced fuels, and impact costs. *J. Ind. Ecol.* **2017**, *21*, 844–856. [CrossRef]

70. Miranda, T.; Esteban, A.; Rojas, S.; Montero, I.; Ruiz, A. Combustion analysis of different olive residues. *Int. J. Mol. Sci.* **2008**, *9*, 512–525. [CrossRef]

71. Dogru, M. Experimental results of olive pits gasification in a fixed bed downdraft gasifier system. *Int. J. Green Energy* **2013**, *10*, 348–361. [CrossRef]

72. Mami, M.A.; Mätzing, H.; Gehrmann, H.-J.; Stapf, D.; Bolduan, R.; Lajili, M. Investigation of the olive mill solid wastes pellets combustion in a counter-current fixed bed reactor. *Energies* **2018**, *11*, 1965. [CrossRef]

73. Mata-Sánchez, J.; Pérez-Jiménez, J.A.; Díaz-Villanueva, M.J.; Serrano, A.; Núñez-Sánchez, N.; López-Giménez, F.J. Development of olive stone quality system based on biofuel energetic parameters study. *Renew. Energy* **2014**, *66*, 251–256. [CrossRef]

74. Iglesias Loredo, R. Valorización Energética del Orujillo de Aceituna. Master's Thesis, Universidad del Pais Vasco, Bilbao, Spain, 2019.

75. Rodríguez, G.; Lama, A.; Rodríguez, R.; Jiménez, A.; Guillén, R.; Fernández-Bolaños, J. Olive stone an attractive source of bioactive and valuable compounds. *Bioresour. Technol.* **2008**, *99*, 5261–5269. [CrossRef]

76. PRé Sustainability. *SimaPro Database Manual: Methods Library*; SimaPro: Utrecht, The Netherlands, 2020.

77. Huijbregts, M.A.J.; Steinmann, Z.J.N.; Elshout, P.M.F.; Stam, G.; Verones, F.; Vieira, M.; Zijp, M.; Hollander, A.; van Zelm, R. ReCiPe2016: A harmonised life cycle impact assessment method at midpoint and endpoint level. *Int. J. Life Cycle Assess.* **2017**, *22*, 138–147. [CrossRef]

78. Haye, S.; Slaveykova, V.I.; Payet, J. Terrestrial ecotoxicity and effect factors of metals in life cycle assessment (LCA). *Chemosphere* **2007**, *68*, 1489–1496. [CrossRef] [PubMed]

79. Vervaeke, M. Life cycle assessment software for product and process sustainability analysis. *J. Chem. Educ.* **2012**, *89*, 884–890. [CrossRef]

80. Bergman, R.; Suo-Han, H.; Oneil, E.; Eastin, I. *The Consortium for Research on Renewable Industrial Materials*; CORRIM: Seattle, WA, USA, 2013; p. 101.

81. Thorton, J.; Healthy Building Network; Institute for Local Self-Reliance. *Environmental Impacts of Polyvinyl Chloride Building Materials*; Healthy Building Network: Washington, DC, USA, 2002; ISBN 978-0-9724632-0-1.

82. Gamon, G.; Evon, P.; Rigal, L. Twin-screw extrusion impact on natural fibre morphology and material properties in poly(lactic acid) based biocomposites. *Ind. Crops Prod.* **2013**, *46*, 173–185. [CrossRef]

83. Batuecas, E.; Tommasi, T.; Battista, F.; Negro, V.; Sonetti, G.; Viotti, P.; Fino, D.; Mancini, G. Life cycle assessment of waste disposal from olive oil production: Anaerobic digestion and conventional disposal on soil. *J. Environ. Manag.* **2019**, *237*, 94–102. [CrossRef]

84. Carbon Dioxide Emissions Associated with Bioenergy and Other Biogenic Sources. Available online: /climatechange/carbon-dioxide-emissions-associated-bioenergy-and-other-biogenic-sources (accessed on 13 January 2021).

85. Rinaldi, S.; Barbanera, M.; Lascaro, E. Assessment of carbon footprint and energy performance of the extra virgin olive oil chain in Umbria, Italy. *Sci. Total Environ.* **2014**, *482–483*, 71–79. [CrossRef]

Article

Printing with Natural Dye Extracted from *Impatiens glandulifera* Royle

Maja Klančnik

Department of Textiles, Graphic Arts and Design, Faculty of Natural Sciences and Engineering, University of Ljubljana, Snežniška 5, SI-1000 Ljubljana, Slovenia; maja.klancnik@ntf.uni-lj.si; Tel.: +386-1-20-03-200

Abstract: Invasive alien plants that damagingly overgrow native ecosystems can be beneficially used to produce natural dyes. Natural dyes are healthier and more environmentally friendly than synthetic dyes, so their use on textiles and other products that come into contact with humans is desirable. In this study, the possibility of using a natural dye extracted from the purple petals of the invasive plant *Impatiens glandulifera* Royle (Himalayan balsam) for screen printing on various substrates; woven fabrics and different papers made from virgin fibers, recycled fibers, and from fibers of Japanese knotweed, was investigated. The prints were evaluated by color measurements and fastness properties. With the violet dye extract, purple-brown prints were obtained on papers made from Japanese knotweed, and more brown prints on other substrates. They had excellent rub fastness but faded significantly when exposed to light. The wash fastness of the prints on cotton fabrics was moderate and poor on polyester fabrics, but the prints had good resistance to wet ironing. The addition of acid to the printing paste resulted in a lighter violet color, the addition of alkali caused a drastic color change to green, both additives increased the light fastness of the prints but reduced the fastness on fabrics to wet treatments.

Keywords: natural dye; Himalayan balsam; invasive plant; printing; textile; paper

Citation: Klančnik, M. Printing with Natural Dye Extracted from *Impatiens glandulifera* Royle. *Coatings* **2021**, *11*, 445. https://doi.org/10.3390/coatings11040445

Academic Editor: Philippe Evon

Received: 9 March 2021
Accepted: 9 April 2021
Published: 13 April 2021

Publisher's Note: MDPI stays neutral with regard to jurisdictional claims in published maps and institutional affiliations.

1. Introduction

Ecological awareness among the public has led to a renewed investigation of natural dyes derived from plants for dyeing and printing textiles. Natural dyes can be extracted from any part of the plant like roots, leaves, fruits, seeds, petals [1]. They have certain advantages over synthetic dyes such as non-toxicity, medicinal properties (e.g., antibacterial activity), UV protective effects, biodegradability, and natural renewability of the plant source [1,2]. Disadvantages of natural dyes are their low wash and light fastnesses and that they can only achieve limited hues, mainly yellow, reddish, and brown [2,3].

Natural dyes from various plant sources, such as alkanet [4–6], rhubarb [4,5,7], manjistha [6], turmeric [5,8–10], marigold [8], chrysanthemum seed [9], locust bean seed [11], madder, buckthorn, walnut bark [12], red poppy [1], *Butea monosperma* flower [13], golden dock, cutch [14], pomegranate peel [14,15], nutshell, orange tree leaves, dyer's chamomile [15], and annatto [6,14,16] have already been investigated for textile printing.

The aim of this study was to investigate the possible use of an extract from the flower of *Impatiens glandulifera* Royle for printing, which as a natural dye, has not previously been studied for textile dyeing or printing.

Impatiens glandulifera Royle (of the Balsaminaceae family), commonly known as Himalayan balsam, originally native to Southern Asia (Himalayas), also known as policeman's helmet because of its hat-shaped flower, is an invasive species that is spreading rapidly across Europe and North America, causing harmful effects on native plant species by competing for space, nutrients, and monopolizing pollinators [17].

Due to the harmful spread of invasive alien plants, their removal from the natural ecosystem should be planned, including finding their new beneficial uses. They can be a source for the production of paper [18] or the extraction of natural dyes used for wood coating [19] or textile dyeing and printing. A yellow dye from the rhizome of Japanese knotweed has been successfully used for dyeing plasma-pretreated cellulosic fabrics [20], and natural dyes from the leaves of Japanese knotweed [21] and the petals of Goldenrod [22] for dyeing cationic pretreated cotton fabrics in shades of brown and light yellow. The extract of the rhizome of Japanese knotweed has also been successfully used for screen printing on papers and fabrics, where an interesting phenomenon of darkening of the prints from yellow to orange-brown on exposure to light has been observed [23].

In the present study, an extract of the purple petals of *Impatiens glandulifera* Royle was used to prepare printing inks for screen printing on fabrics and various papers. Anti-pruritic, anti-anaphylactic, and anti-nociceptive effects of the flowers have already been reported [24]. The use of the flower extract as a dye for human contact products may be desirable due to its medicinal properties. For printing with the low-substantive natural dye, a pigment printing paste with a binder was used as a simple alternative to today's ecologically unacceptable and mostly toxic metal salt mordants [12,15]. In pigment printing, the molecules of the natural dye are held on the fabric surface in a film of polymeric binder that adheres to the fibers [25].

The ability to print was tested on different substrates, such as cotton and polyester fabrics, two commercial papers; one made from virgin cellulose fibers and the other from recycled cellulose fibers, and two innovative papers made from different contents of the stem fibers of the invasive alien plant Japanese knotweed. The violet extract from the petals of *Impatiens glandulifera* Royle is a mixture of different types of flavonoids [24], the color of which can be altered by changing the pH. Therefore, the influence of alkali and acid additives in the printing paste on the color change of the ink, prints, and their fastness properties on different substrates was also observed. The quality of prints was determined by color measurements and fastness tests to rubbing, light exposure, and on fabrics as well as to wet ironing and washing.

2. Materials and Methods

2.1. Printing Materials

The following materials were used for printing:

- cotton fabric, woven in linen (manufacturer: Tekstina Ltd., Ajdovščina, Slovenia) with a surface mass of 122.73 g/m^2, hereinafter referred to as CO;
- polyester fabric, woven in linen (supplier: Luna Ltd., Ljubljana, Slovenia) with a surface mass of 175.99 g/m^2, hereinafter referred to as PES;
- papers made from Japanese knotweed in two grammages; 200 and 240 g/m^2, manufactured by Pulp and Paper Institute (Ljubljana, Slovenia), hereinafter referred to as J.k. 200 and J.k. 240, where J.k. 200 contains about 30% fibers from the stem of Japanese knotweed and J.k. 240 contains about 40% fibers from the stem of Japanese knotweed, the remainder being commercial cellulose fibers in a ratio of conifers: deciduous (eucalyptus) 1:1;
- paper made from virgin cellulose fibers with the trade name 'IQ premium' (produced by Mondi Group Ltd., Vienna, Austria), grammage of 200 g/m^2, referred to as Paper 1;
- recycled paper unknown producer, grammage of 200 g/m^2, referred to as Paper 2.

2.2. Printing Ink Preparation

The natural violet dye was prepared by the National Institute of Chemistry (Ljubljana, Slovenia) by extraction from the purple flowers of *Impatiens glandulifera* Royle with 1M HCl in methanol.

For the screen printing inks, the printing paste for pigment printing was prepared according to the following recipe with ingredients from AchitexMinerva Ltd. (Vaiano Cremasco, Italy); 150 g of a self-crosslinking acrylic binder Binder SECONC, 18 g of acrylic thickener Clear MCS, and up to 1000 g of demineralized water.

The violet dye was added in six different concentrations; 0.5, 1, 2, 3, 4, and 5 g per 100 g of the prepared printing paste with pH 7.43. The violet inks from light to dark shades were obtained. To prepare an alkaline printing paste, 1 g of Na_2CO_3 calc. (Sigma-Aldrich, St. Louis, MO, USA) was added to 100 g of the prepared printing paste, thus raising the pH value to 8.99. To this alkaline paste, 3 g of the violet dye was then added and a bluish-green printing ink was obtained. To prepare an acidic printing paste, 2 mL of CH_3COOH 60% (Carl Roth, Karlsruhe, Germany) was added to 100 g of the originally prepared paste to obtain a pH of 5.38, and then 3 g of a violet dye was mixed into the paste and a lighter violet printing ink was obtained.

2.3. Printing

The printing inks were applied to the substrates using the semi-automatic screen printing machine SD 05 (RokuPrint Ltd., Dornstadt, Germany) with the flat printing screen made of polyester fabric with 77 threads/cm and a thread diameter of 55 μm with three strokes of a squeegee.

The prints were dried at room temperature overnight and then cured at 150 °C for 5 min.

2.4. Color Measurements

Color measurements of the prints were made two weeks after printing using the spectrophotometer Eye-One i1 Pro (X-Rite, Grand Rapids, MI, USA) with 45/0 plane geometry, illuminant D65, 10° standard observer, and measurement aperture of 4.5 mm diameter. The CIELAB color coordinates: L^* (lightness), a^* (red-green value), b^* (yellow-blue value) were measured, an average of three measurements was taken for each sample, and h_{ab} (hue) and C^*_{ab} (chroma) were calculated.

2.5. Fastness Tests

The abrasion resistance test of the ink layer on printed papers was carried out according to ASTM D 5264 using the digital ink rub tester RT-01 (Labthink Ltd., Neu-Isenburg, Germany) at a rubbing pressure of 2 kg and at a rubbing frequency of 1.8 s^{-1} 500 times [26]. The transfer of ink from the prints to the white paper (Paper 1) used as a receptor and the inked surface of the prints after rubbing were visually assessed.

The fastness of printed ink on fabrics to dry and wet rubbing was carried out using the electronic Crocmeter rub tester (SDL Atlas, Rock Hill, SC, USA) according to ISO 105-X12: 2016 [27]. The resistance of printing ink on fabrics to wet ironing was tested according to ISO 105-X11: 1994 [28]. The fastness of the prints to domestic and commercial laundering was tested according to ISO 105-C06: 2012- test A1S at 40 °C for 30 min in the standard washing machine Launder-O-meter (SDL Atlas, Rock Hill, SC, USA) [29]. The color change of the prints and the staining of the white adjacent fabrics after the tests were visually assessed with the greyscales according to ISO 105-A02 [30] and ISO 105-A03 [31], respectively, with ratings from 1 to 5, 5 being the best value.

The fastness of the color of the prints on fabrics and papers was tested according to the standard ISO 105-B02: 2014 to artificial light (Xenon arc fading lamp) in the testing instrument Xenotest Alpha (Atlas, Rancho Cucamonga, CA, USA) under the following conditions: 35 °C, 35% RH, and 72 h, alongside with the standard blue reference scale [32]. After the light fastness test, the color change of the exposed prints was evaluated visually by the blue wool references with ratings from 1 to 8, 8 being the best value, and colorimetrically by calculating the overall color difference (ΔE^*_{ab}) between exposed and non-exposed prints according to the formula CIE 1976 [33].

For the presentation of fastness test results for rubbing, wet ironing, and washing, the prints made with inks containing concentrations of 1, 3, and 5 g of dye per 100 g of initial printing paste were selected as representative prints for light, medium and dark shades and for comparison with the prints made with 3 g of dye per 100 g of alkaline and acidic printing pastes. All prints were considered for the light fastness results.

3. Results and Discussion

3.1. Color of Prints

The extract of the flower of *Impatiens glandulifera* Royle is a complex mixture of different colorants. The petals contain various types of flavonoids such as flavones, flavanone and flavonol monoglucosides and diglucosides, anthocyanins as well as phenolic acids, coumarins, and quinone pigments [24].

For the violet color of the extract, quinones and various types of flavonoids, among the latter mostly anthocyanins, are responsible. All the colorants are pH-sensitive. However, anthocyanins are the most strongly colored of all the flavonoids, with a predominant influence on the color of the extract. The color of anthocyanins can change from red to blue depending on the pH value of the surrounding solution [34]. Thus, in the initial printing paste with a pH of 7.43, the dye extract was violet, where anthocyanins were predominantly in the form of the quinoidal base (A) and some in the anionic form (A^-) [34,35]. In the alkaline printing paste of pH of 8.99, further, deprotonation took place and more anionic quinoidal base (A^-) was formed, consequently the extract became bluish-green. In the acidic printing paste of pH 5.38, the color of the ink was lighter violet, where more colorless carbinol (pseudo-base) and slightly yellow chalcone structures were formed. If the pH was lowered even further, a more reddish color would appear due to the increased formation of a red flavylium cation (AH^+) of the anthocyanins. However, further increasing the acidity of the printing paste was not practical in our case. The thickening agent used was polyacrylate, which swells due to the repulsive action of its anionic carboxylate groups [25]. The acid causes the carboxylate groups of the thickener to convert to carboxyl groups, so the printing paste loses the high viscosity necessary for screen printing.

The highest concentration of the extract (i.e., 5 g/100 g) also reduced the viscosity of the printing paste, which caused printing difficulties, especially on more absorbent cotton fabrics where the edges of the prints were smeared. The extract was slightly acidic and therefore impaired the performance of the acrylate-based thickener.

After immediate printing with the violet-colored printing inks, the prints looked purple, but after a day, when the prints were dry, they were browner. This can be explained by the fact that the final color of the print is also influenced by other flavonoids (derivatives of quercetin, kaempferol, etc.) [24,36], naphthoquinones [36], and tannins [37], which are present in the extract.

The spectrophotometric measurements of the prints from light to dark colors are summarized in Table 1. Table 2 contains excerpts of photographs of some prints to show the visual color difference between prints with different dye concentrations on different substrates. The color of the prints is influenced by the base color of the substrate, especially at lower dye concentrations (Tables 1 and 2). The spectrophotometric measurements of the base color of these substrates have already been published [23].

Table 1. Color values of prints with different dye concentrations on papers and fabrics.

Substrate	L^*	a^*	b^*	C^*_{ab}	h_{ab} (°)
		Prints of 0.5 g dye/100 g			
Paper 1	90.78	1.85	−5.41	5.72	288.88
Paper 2	91.02	−0.51	3.69	3.72	97.83
J.k. 200	82.08	2.07	14.00	14.16	81.60
J.k. 240	74.42	4.53	21.35	21.83	78.01
CO	87.65	1.28	6.26	6.39	78.42
PES	84.67	2.74	0.93	2.90	18.73

Table 1. *Cont.*

Substrate	L^*	a^*	b^*	C^*_{ab}	h_{ab} (°)
Prints of 1 g dye/100 g					
Paper 1	88.01	1.76	−2.43	3.00	305.97
Paper 2	90.01	−0.83	4.67	4.74	100.05
J.k. 200	81.20	2.36	14.58	14.77	80.82
J.k. 240	73.33	4.72	20.75	21.28	77.18
CO	84.29	2.87	8.23	8.72	70.75
PES	76.12	5.73	7.22	9.22	51.58
Prints of 2 g dye/100 g					
Paper 1	84.96	2.89	−0.17	2.89	356.70
Paper 2	88.15	−0.74	5.31	5.36	97.89
J.k. 200	78.80	3.42	15.10	15.48	77.24
J.k. 240	71.62	5.21	20.00	20.67	75.40
CO	77.93	5.40	11.11	12.35	64.07
PES	74.47	5.64	7.59	9.46	53.41
Prints of 3 g dye/100 g					
Paper 1	82.04	3.48	2.62	4.36	36.98
Paper 2	86.90	−0.42	6.75	6.76	93.56
J.k. 200	75.15	4.78	15.31	16.04	72.66
J.k. 240	68.87	5.95	18.09	19.04	71.80
CO	70.38	8.61	13.78	16.25	58.01
PES	66.12	7.15	9.54	11.92	53.14
Prints of 4 g dye/100 g					
Paper 1	79.04	2.69	1.81	3.24	33.85
Paper 2	85.54	−0.33	5.28	5.29	93.61
J.k. 200	73.33	5.26	15.40	16.28	71.13
J.k. 240	68.41	5.91	18.62	19.53	72.38
CO	68.98	8.11	12.99	15.31	58.01
PES	64.13	7.06	8.97	11.42	51.79
Prints of 5 g dye/100 g					
Paper 1	77.94	4.41	1.60	4.70	19.97
Paper 2	84.08	0.09	6.99	6.99	89.24
J.k. 200	72.69	6.02	15.58	16.70	68.86
J.k. 240	67.73	6.49	16.94	18.14	69.04
CO	65.56	9.49	14.26	17.13	56.36
PES	57.62	7.28	8.33	11.06	48.87
Prints of 3 g dye/100 g (alkaline)					
Paper 1	80.54	−1.84	12.23	12.36	98.54
Paper 2	85.24	−2.47	11.22	11.49	102.40
J.k. 200	75.15	1.24	19.64	19.68	86.39
J.k. 240	66.36	2.60	22.68	22.83	83.47
CO	73.66	0.58	17.95	17.96	88.16
PES	70.67	−1.12	12.35	12.40	95.18
Prints of 3 g dye/100 g (acidic)					
Paper 1	84.99	0.66	−0.17	0.68	345.28
Paper 2	87.23	−1.86	5.50	5.80	108.73
J.k. 200	78.06	3.80	15.92	16.36	76.57
J.k. 240	71.84	5.71	20.56	21.34	74.48
CO	75.30	6.36	11.65	13.27	61.39
PES	74.75	6.43	7.06	9.55	47.66

Table 2. Photographs of substrates and prints.

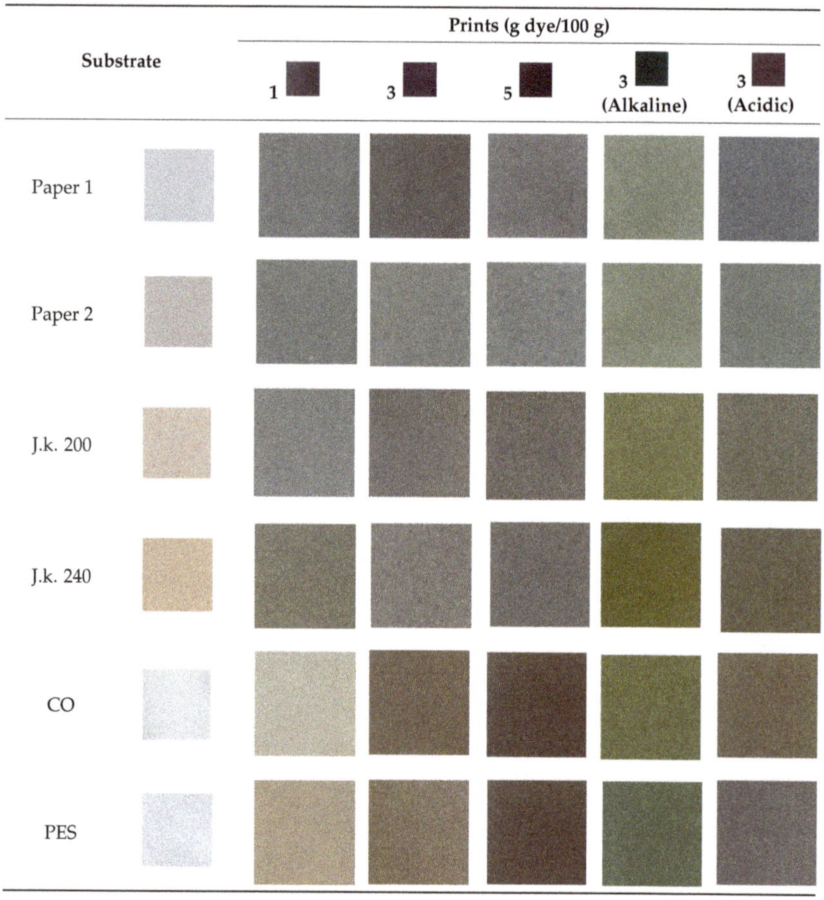

Substrate		1	3	5	3 (Alkaline)	3 (Acidic)
		Prints (g dye/100 g)				
Paper 1						
Paper 2						
J.k. 200						
J.k. 240						
CO						
PES						

The paper made from virgin fibers (Paper 1) was the whitest of all printing materials, with a violet-blue hue. The prints on Paper 1 obtained with the lower dye concentrations of 0.5 g to 2 g/100 g were in the red-blue range of the CIELAB color space with a violet hue, and with higher dye concentrations were in the red-yellow range with a reddish-brown hue. They looked brown.

The prints on recycled paper (Paper 2) were in the green-yellow range and only at the highest dye concentration in the red-yellow range. They had a greenish-yellow hue as the recycled paper itself. The prints looked grey-brown.

The Japanese knotweed papers were the darkest among all the printing materials with the highest chroma values. Because J.k. 200 contained fewer stem fibers of Japanese knotweed, its color was slightly lighter and less saturated than that of J.k. 240. The prints on Japanese knotweed papers were in the red-yellow range with a brownish-yellow hue, as was the hue of the papers themselves. The prints of higher dye concentrations on Japanese knotweed papers looked more purple-brown than on other materials.

The prints on fabrics were in the red-yellow range with an orange-brown hue. The colorimetric measurements showed a greater increase in darkness and saturation of the prints on the fabrics with an increase in dye concentration than on papers. Due to the higher absorption of the ink, darker and more saturated colors of the brown prints were obtained.

The prints produced with the bluish-green alkaline printing ink were green. The spectrophotometric measurements showed that the prints on Paper 1, Paper 2, and polyester fabric were greener than the prints on J.k. papers and cotton fabric, which had a yellowish hue.

The prints made with the acidic printing paste (3 g dye/100 g) were slightly lighter, duller, and with lower chromatic coordinates than prints made with the initial printing paste. They had a similar lightness and hue as the prints obtained with 2 g of dye per 100 g of the initial printing paste.

3.2. Fastness of Prints

The results of visual assessments of the resistance of the prints on paper to dry rubbing are summarized in Table 3. The visual assessments of the resistance of the prints on textiles to rubbing, ironing, washing according to the grayscale are summarized in Table 4. The results of the lightfastness of the prints according to the blue reference scale and ΔE^*_{ab} between exposed and unexposed prints are summarized in Table 5.

Table 3. Colorfastness of prints on papers to dry rubbing.

Examined Parameters	Paper 1	Paper 2	J.k. 200	J.k. 240
Prints of 1 g dye/100 g				
Printed surface	unchanged	unchanged	unchanged	unchanged
Ink transfer to white paper	no	no	no	no
Prints of 3 g dye/100 g				
Printed surface	unchanged	unchanged	unchanged	slight transfer of white paper to the printed surface
Ink transfer to white paper	no	no	no	no
Prints of 3 g dye/100 g (alkaline)				
Printed surface	unchanged	unchanged	unchanged	slight transfer of white paper to the printed surface
Ink transfer to white paper	no	no	no	no
Prints of 3 g dye/100 g (acidic)				
Printed surface	unchanged	unchanged	unchanged	unchanged
Ink transfer to white paper	no	no	no	no
Prints of 5 g dye/100 g				
Printed surface	unchanged	unchanged	unchanged	slight transfer of white paper to the printed surface
Ink transfer to white paper	no	no	no	no

Table 4. Colorfastness of prints on fabrics to dry and wet rubbing, wet hot pressing, laundering at 40 °C.

Fabrics	Rubbing		Hot Pressing			Washing	
	Staining of Dry Cloth	Staining of Wet Cloth	Staining of Wet Cloth	Staining of Dry Cloth	Color Change	Staining of Cloth	Color Change
Prints of 1 g dye/100 g							
CO	5	5	5	5	5	5	4
PES	5	5	5	5	5	5	3–4
Prints of 3 g dye/100 g							
CO	5	5	4–5	4–5	4	5	3
PES	5	5	4–5	4–5	4	5	2–3
Prints of 3 g dye/100 g (alkaline)							
CO	5	4–5	4–5	5	4	5	2–3
PES	5	4	4	3–4	3	5	1
Prints of 3 g dye/100 g (acidic)							
CO	5	5	4–5	5	4	5	2
PES	5	5	4	3–4	3	5	1–2
Prints of 5 g dye/100 g							
CO	5	5	4	4–5	4	5	2–3
PES	5	4–5	3–4	4	3–4	5	2–3

Table 5. Colorfastness of prints to light.

Evaluated Parameters	Paper 1	Paper 2	J.k. 200	J.k. 240	CO	PES
Prints of 0.5 g dye/100 g						
Blue scale assessment	2	3	4	4	5	4
ΔE^*_{ab}	15.71	5.96	3.22	3.26	2.79	4.79
Prints of 1 g dye/100 g						
Blue scale assessment	2–3	3–4	4	3–4	5	3
ΔE^*_{ab}	12.43	5.97	2.97	5.28	4.95	10.65
Prints of 2 g dye/100 g						
Blue scale assessment	2–3	4	4	4	3	3
ΔE^*_{ab}	12.06	5.32	5.96	6.16	9.51	12.46
Prints of 3 g dye/100 g						
Blue scale assessment	2	3–4	3	2	3	2
ΔE^*_{ab}	15.10	8.30	8.65	9.96	13.34	16.29
Prints of 4 g dye/100 g						
Blue scale assessment	2	4	2	2	3	2
ΔE^*_{ab}	15.15	5.80	10.10	9.07	12.30	16.40

Table 5. *Cont.*

Evaluated Parameters	Paper 1	Paper 2	J.k. 200	J.k. 240	CO	PES
Prints of 5 g dye/100 g						
Blue scale assessment	2	4	2	2	4	3
ΔE^{*}_{ab}	15.22	6.10	11.79	11.10	12.69	16.82
Prints of 3 g dye/100 g (alkaline)						
Blue scale assessment	3	4	4	3	3	2
ΔE^{*}_{ab}	8.85	8.32	8.52	9.50	10.28	16.01
Prints of 3 g dye/100 g (acidic)						
Blue scale assessment	3	4	4	4	3	3
ΔE^{*}_{ab}	8.62	7.83	8.01	8.90	13.01	15.21

The prints on papers made with the initial (pH 7.43), alkaline (pH 8.99), and acidic (pH 5.38) printing pastes exhibited good abrasion resistance when rubbed 500 times under a high pressure of 2 kg (Table 3). There was no ink transfer from the printing surface to the white paper. The test method of dry rubbing on papers according to ASTM D 5264, which simulates the effects of storage, shipment, and handling of printed products, was carried out under more severe conditions than on fabrics according to ISO 105-X12, where only 10 rubs under a pressure of 1 kg were used. The printing inks made from *Impatiens glandulifera* Royle extract were found to have excellent resistance to dry rubbing and can be used for printing on packaging or other paper products that are constantly rubbed with other papers.

The prints on fabrics also had excellent fastness to dry and wet rubbing (grade 5) with no color transfer, except at the highest dye concentration on polyester fabric, where slight staining of the wet cloth was seen (grade 4–5, in Table 4). The good resistance of prints to rubbing is the result of the good strength of the bond between the polymeric binder film and the fibers for adhesion [38], which was well observed on cellulose fibers; in the case of cotton fabric and papers.

The prints on fabrics had good fastness to wet ironing; the average grade was 4, except for the highest dye concentration on polyester fabric, where the average grade was 3–4 (Table 4). The results showed weaker adhesion of the binder film to hydrophobic polyester fibers, which was expected due to the lack of available polar groups on polyester for stronger bonding [25].

The fastness of the prints on fabrics to a single commercial or domestic laundering at 40 °C was fair; slightly better on cotton, where the grade of color loss was 3, however, the color changed from brown to green. The loss of color on polyester was slightly higher, the grade was 2–3 (Table 4), indicating not-so-stable cross-links of the binder to hydrolysis during washing and weaker interactions of the binder film with polyester fibers [25]. There was no staining of the white adjacent fabric in the washing bath, which is due to a low substantivity of the washed natural dye to the fibers.

The prints made with alkaline paste on cotton and polyester had excellent fastness to dry rubbing, but lower fastness to wet rubbing for half a grade on cotton and one grade on polyester, compared to the initial paste (Table 4). The prints on polyester with alkaline and acidic printing pastes had lower fastness to wet ironing for one grade (grade 3–4). Alkali and acid in the printing paste also impaired the colorfastness to washing. The alkali reduced the wash fastness on cotton for half a grade and polyester for one and a half grade. The acid reduced the wash fastness on both fabrics for one grade compared to the initial printing paste.

The poorer resistance of alkaline and acidic inks to wet treatments indicates weaker adhesion of the polymeric binder with the fibers. A non-optimal pH (lower or higher than 7.5–8) of the printing paste results in the polyacrylate thickener not being present in the appropriate form of carboxylates ($-COO^-$) [25] or at least not in a sufficiently large amount, which is why it cannot convert into the appropriate amount of polyacrylic acid during thermal fixation to accelerate the cross-linking of the binder and its interaction with the fibers.

The prints on papers and fabrics faded significantly when exposed to light, becoming lighter and yellower (Table 5). It has already been reported that light exposure of anthocyanin pigments accelerates their destruction [39]. A large color difference between exposed and unexposed prints was observed at higher dye concentrations. The average grade of fading according to the blue reference scale was 2 for prints on Papers 1,3 on the Japanese knotweed papers and polyester fabric, 4 on Paper 2, and cotton fabric, indicating poor to fair lightfastness of the printing ink [40].

The alkaline and acidic printing pastes increased the light fastness of the prints on papers for one grade compared to the neutral printing paste. The alkaline or acidic printing pastes, which improved lightfastness on paper, had no visual effect on the lightfastness of the prints on cotton fabrics; though spectrophotometric measurements showed slightly improved lightfastness of the prints. On polyester fabric, the acidic paste increased the light fastness of the prints for one grade.

The addition of acetic acid or soda ash in the prints could neutralize the destructive effect of UV light on the dye molecules, which has already been demonstrated for ascorbic acid and gallic acid [41,42], which have been effectively used to improve the lightfastness of cotton dyed with natural dyes in the post-treatment process. However, the effects of acid and alkali on the lightfastness properties of the prints need further investigation.

4. Conclusions

A violet dye extracted from the purple-pink petals of the invasive alien plant *Impatiens glandulifera* Royle can be used as a natural dye with a pigment printing paste to produce screen printing inks up to a concentration of 5 g of the dye per 100 g of printing paste. Higher concentrations of the dye extract impaired the performance of the acrylate-based thickener and consequently lowered the viscosity of the printing ink.

The prints were purple-brown on Japanese knotweed papers and browner on other substrates. Greater absorption of the inks and darker colors were obtained on the fabrics. All prints had very good rubbing fastness, but faded considerably when exposed to light (average grade 3). The wash fastness of the prints on cotton fabric was fair (grade 3) and poor on polyester fabric (grade 2–3), but the prints had good resistance to wet ironing (grade 4).

The addition of acid to the printing paste caused a lighter violet color and the addition of alkali drastically changed the color of the ink to green. Both additives increased the light fastness of the prints, especially on papers for one grade; however, they decreased the wet fastness of the prints on fabrics, on average for one grade.

Due to the excellent adhesion resistance of the printing ink to extreme numbers of rubbing repetitions, the natural dye can be successfully used for printing on paper products that are not directly exposed to light. However, its use on textiles is limited due to its lower wash fastness.

Funding: This research was funded by the European research project UIA02-228 APPLAUSE.

Institutional Review Board Statement: Not applicable.

Informed Consent Statement: Not applicable.

Data Availability Statement: Not applicable.

Acknowledgments: The author thanks the project partners; the National Institute of Chemistry (Ljubljana, Slovenia) for the preparation of dye extract and Pulp and Paper Institute (Ljubljana, Slovenia) for the production of Japanese knotweed papers.

Conflicts of Interest: The author declares no conflict of interest. The funder had no role in the design of the study; in the collection, analyses, or interpretation of data; in the writing of the manuscript, or in the decision to publish the results.

References

1. Hebeish, A.; Shahin, A.A.; Rekaby, M.; Ragheb, A.A. New environment-friendly approach for textile printing using natural dye loaded chitosan nanoparticles. *Egypt. J. Chem.* **2015**, *58*, 659–670.
2. Rungruangkitkrai, N.; Mongkholrattanasit, R. Eco-Friendly of Textiles Dyeing and Printing with Natural Dyes. In Proceedings of the International Conference: Textiles & Fashion 2012, Bangkok, Thailand, 3–4 July 2012; pp. 1–16.
3. Devi, S.; Karuppan, P. Reddish brown pigments from *Alternaria alternata* for textile dyeing and printing. *Indian J. Fibre Text. Res.* **2015**, *40*, 315–319.
4. Rekaby, M.; Salem, A.A.; Nassar, S.H. Eco-friendly printing of natural fabrics using natural dyes from alkanet and rhubarb. *J. Text. Inst.* **2009**, *100*, 486–495. [CrossRef]
5. Salem, A.A.; Shahin, M.F.; El Sayad, H.S.; El Halwagy, A.A. Transfer printing of polyester fabrics with natural dyes. *Res. J. Text. Appar.* **2013**, *17*, 61–67. [CrossRef]
6. Chattopadhyay, S.N.; Pan, N.C.; Khan, A. Printing of jute fabric with natural dyes extracted from manjistha, annatto and ratanjot. *Indian J. Fibre Text. Res.* **2018**, *43*, 352–356.
7. Osman, H. Eco-friendly printing of textile substrates with rhubarb natural dye nanoparticles. *World Appl. Sci. J.* **2014**, *29*, 592–599. [CrossRef]
8. Teli, M.D.; Sheikh, J.; Shastrakar, P. Exploratory investigation of chitosan as mordant for eco-friendly antibacterial printing of cotton with natural dyes. *J. Text.* **2013**, *2013*, 320510. [CrossRef]
9. Pratoomtong, T. The property of screen ink from natural mordant, colorant, and additive for art. *IJBSS* **2015**, *6*, 68–76.
10. Ragheb, A.A.; Tawfik, S.; Abd-El Thalouth, J.I.; Mosaad, M.M. Development of printing natural fabrics with curcuma natural dye via nanotechnology. *IJPSR* **2017**, *8*, 611–620. [CrossRef]
11. Abd-El-Thalouth, J.I. Synthesis and application of eco-friendly natural-printing paste for textile coloration. *J. Am. Sci.* **2011**, *7*, 632–640. [CrossRef]
12. Bahtyari, M.I.; Benli, H.; Yavas, A. Printing of wool and cotton fabrics with natural dyes. *Asian J. Chem.* **2013**, *25*, 3220–3224. [CrossRef]
13. Babel, S.; Gupta, R. Screen printing on silk fabric using natural dye and natural thickening agent. *J. Text. Sci. Eng.* **2016**, *6*, 1000230. [CrossRef]
14. Savvidis, G.; Karanikas, E.; Nikolaidis, N.; Eleftheriadis, I.; Tsatsaroni, E. Ink-jet printing of cotton with natural dyes. *Color. Technol.* **2013**, *130*, 200–204. [CrossRef]
15. Bahtyari, M.I.; Benli, H.; Yavas, A.; Akça, A.C. Use of different natural dye sources for printing of cotton fabrics. *Text. Appar.* **2017**, *27*, 259–265.
16. Savvidis, G.; Zarkogianni, M.; Karanikas, E.; Lazaridis, N.; Nikolaidis, N.; Tsatsaroni, E. Digital and conventional printing and dyeing with the natural dye annatto: Optimization and standardization process to meet future demands. *Color. Technol.* **2012**, *129*, 55–63. [CrossRef]
17. Cafa, G.; Baroncelli, R.; Ellison, C.A.; Kurose, D. *Impatiens glandulifera* (Himalayan balsam) chloroplast genome sequence as a promising target for populations studies. *PeerJ* **2020**, *8*, e8739. [CrossRef] [PubMed]
18. Karlovits, I.; Kavčič, U.; Lavrič, G.; Šinkovec, A.; Torić, V. Digital printability of papers made from invasive plants and agro-industrial residues. *Cellul. Chem. Technol.* **2020**, *54*, 523–529. [CrossRef]
19. Horvat, M.; Iskra, J.; Pavlič, M.; Žigon, J.; Merela, M. Wood dyes from invasive alien plants. *Les Wood* **2020**, *69*, 37–48. [CrossRef]
20. Gorjanc, M.; Savić, A.; Topalić-Trivunović, L.; Mozetič, M.; Zaplotnik, R.; Vesel, A.; Grujić, D. Dyeing of plasma treated cotton and bamboo rayon with *Fallopia Japonica* extract. *Cellulose* **2016**, *23*, 2221–2228. [CrossRef]
21. Gorjanc, M.; Kert, M.; Mujadžić, A.; Simončič, B.; Forte-Tavčer, P.; Tomšič, B.; Kostanjšek, K. Cationic pretreatment of cotton and dyeing with *Fallopia Japonica* leaves. *Tekstilec* **2019**, *62*, 181–186. [CrossRef]
22. Topič, T.; Gorjanc, M.; Kert, M. The influence of the treatment process on the dyeability of cotton fabric using goldenrod dye. *Tekstilec* **2018**, *61*, 192–200. [CrossRef]
23. Klančnik, M. Screen printing with natural dye extract from Japanese knotweed rhizome. *Fiber Polym.* **2021**, in press.
24. Viera, M.N.; Winterhalter, P.; Jerz, G. Flavonoids from the flowers of *Impatiens glandulifera* Royle isolated by high performance countercurrent chromatography. *Photochem. Anal.* **2016**, *27*, 116–125. [CrossRef]
25. Broadbent, A.D. Printing. In *Basic Principles of Textile Coloration*; Society of Dyers and Colourists: Bradford, UK, 2001; pp. 509–526.
26. ASTM. *Standard Practice for Abrasion Resistance of Printed Materials by the Sutherland Rub Tester*; ASTM D 5264-98; ASTM International: West Conshohocken, PA, USA, 2019.
27. ISO. *Textiles—Tests for Colour Fastness—Part X12: Colour Fastness to Rubbing*; ISO 105-X12; ISO: Geneva, Switzerland, 2016.
28. ISO. *Textiles—Tests for Colour Fastness—Part X11: Colour Fastness to Hot Pressing*; ISO 105-X11; ISO: Geneva, Switzerland, 1994.

29. ISO. *Textiles—Tests for Colour Fastness—Part C06: Colour Fastness to Domestic and Commercial Laundering*; ISO 105-C06; ISO: Geneva, Switzerland, 2012.

30. ISO. *Textiles—Tests for Colour Fastness—Part A02: Grey Scale for Assessing Change in Colour*; ISO 105-A02; ISO: Geneva, Switzerland, 1993.

31. ISO. *Textiles—Tests for Colour Fastness—Part A03: Grey Scale for Assessing Staining*; ISO 105-A03; ISO: Geneva, Switzerland, 2019.

32. ISO. *Textiles—Tests for Colour Fastness—Part B02: Colour Fastness to Artificial Light: Xenon Arc Fading Lamp Test*; ISO 105-B02; ISO: Geneva, Switzerland, 2014.

33. Broadbent, A.D. Colour measurement. In *Basic Principles of Textile Coloration*; Society of Dyers and Colourists: Bradford, UK, 2001; pp. 465–468.

34. Khoo, H.E.; Azlan, A.; Tang, S.T.; Lim, S.M. Anthocyanidins and anthocyanins: Colored pigments as food, pharmaceutical ingredients, and the potential health benefits. *Food Nutr. Res.* **2017**, *61*, 1361779. [CrossRef]

35. Rose, P.M.; Cantrill, V.; Benohoud, M.; Tidder, A.; Rayner, C.M.; Blackburn, R.S. Application of anthocyanins from blackcurrant (*Ribes nigrum* L.) fruit waste as renewable hair dyes. *J. Agric. Food Chem.* **2018**, *66*, 6790–6798. [CrossRef] [PubMed]

36. Szewczyk, K.; Cicek, S.S.; Zidorn, C.; Granica, S. Phenolic constituents of the aerial parts of *Impatiens glandulifera* Royle (Balsaminaceae) and their antioxidant activities. *Nat. Prod. Res.* **2019**, *33*, 2851–2855. [CrossRef] [PubMed]

37. Szewczyk, K.; Kalemba, D.; Komsta, L.; Nowak, R. Comparison of essential oil composition of selected *Impatiens* species and its antioxidant activities. *Molecules* **2016**, *21*, 1162. [CrossRef] [PubMed]

38. Yaman, N.; Ozdogan, E.; Seventekin, N. Improvement Fastnesses and Color Strength of Pigment Printed Textile Fabric. *J. Eng. Fibers Fabr.* **2012**, *7*, 40–46. [CrossRef]

39. Amogne, N.Y.; Ayele, D.W.; Tsigie, Y.A. Recent advances in anthocyanin dyes extracted from plants for dye sensitized solar cell. *Mater. Renew. Sustain. Energy* **2020**, *9*, 23. [CrossRef]

40. Broadbent, A.D. Testing of dyes and dyeings. In *Basic Principles of Textile Coloration*; Society of Dyers and Colourists: Bradford, UK, 2001; pp. 527–540.

41. Criesta, D.; Vilarem, G. Improving light fastness of natural dyes on cotton yarn. *Dyes Pigment.* **2006**, *70*, 238–245. [CrossRef]

42. Thiagarajan, P.; Nalankilli, G. Improving light fastness of reactive dyed cotton fabric with antioxidant and UV absorbers. *Indian J. Fibre Text. Res.* **2013**, *38*, 161–164.

Article

Green Sound-Absorbing Composite Materials of Various Structure and Profiling

Eulalia Gliscinska [1,*], Javier Perez de Amezaga [1,2], Marina Michalak [1] and Izabella Krucinska [1]

[1] Faculty of Material Technologies and Textile Design, Institute of Material Science of Textiles and Polymer Composites, Lodz University of Technology, 116 Zeromskiego Street, 90-924 Lodz, Poland; javier.perez.tomas@gmail.com (J.P.d.A.); marina.michalak@p.lodz.pl (M.M.); izabella.krucinska@p.lodz.pl (I.K.)

[2] Barcelona School of Industrial Engineering (ETSEIB), Polytechnic University of Catalonia, C. Jordi Girona, 31, 08034 Barcelona, Spain

[*] Correspondence: klata@p.lodz.pl

Abstract: This article presents thermoplastic sound-absorbing composites manufactured on the basis of renewable raw materials. Both the reinforcing material and the matrix material were biodegradable and used in the form of fibers. In order to mix flax fibers with polylactide fibers, the fleece was fabricated with a mechanical system and then needle-punched. The sound absorption of composites obtained from a multilayer structure of nonwovens pressed at different conditions was investigated. The sound absorption coefficient in the frequency ranging from 500 Hz to 6400 Hz was determined using a Kundt tube. The tests were performed for flat composites with various structures, profiled composites, and composite/pre-pressed nonwoven systems. Profiling the composite plate by convexity/concavity has a positive effect on its sound absorption. It is also important to arrange the plate with the appropriate structure for the incident sound wave. For the composite layer with an added pre-pressed nonwoven layer, a greater increase in sound absorption occurs for the system when a rigid composite layer is located on the side of the incident sound wave. The addition of successive nonwoven layers not only increases the absorption but also extends the maximum absorption range from the highest frequencies towards the lower frequencies.

Keywords: green composite; nonwoven; sound absorption; structure; profiling

Citation: Gliscinska, E.; Perez de Amezaga, J.; Michalak, M.; Krucinska, I. Green Sound-Absorbing Composite Materials of Various Structure and Profiling. *Coatings* **2021**, *11*, 407. https://doi.org/10.3390/coatings11040407

Academic Editor: Philippe Evon

Received: 11 February 2021
Accepted: 29 March 2021
Published: 31 March 2021

Publisher's Note: MDPI stays neutral with regard to jurisdictional claims in published maps and institutional affiliations.

1. Introduction

Green composites have less environmental impact at the production, use, and post-use stages than in the case of composites based on chemical fibers [1]. As filling material for producing biocomposites, the natural fibers of wood, cork, horsehair, nettle, leaves, paper cut in a shredder, chicken feather calamus cut into small cubes, fine sawdust, and straw are usually used [2–7]. As the matrix material, among others, polylactide, poly-hydroxybutyrate, starch, chitosan, gum Arabic, and green epoxy resin are used [1,8]. Nowadays, different kinds of natural materials are investigated for the reinforcement of sound-absorbing composites because of their cheap production cost, eco-friendly composition, and their relevant properties related to the application of interest [9]. The results of the sound absorption of composites based on natural filling materials and biopolymers are promising, and present the high potential of such materials as sound absorbers. These composites exhibit different sound absorption depending not only on the type of filling/reinforcing material, but also on the sound frequency range. Usually, however, high sound absorption is observed at frequencies above 2000 Hz [1,2,10]. Sound absorption of selected frequency ranges depends on the structure, density, and thickness of the absorber. The results for pineapple leaf/epoxy composite show that for a given material density, by increasing its thickness, we extend the absorption range towards lower frequencies [11].

Among various forms of the filling component, the fibers are the most beneficial from the point of view of the mechanical properties of the composite [12]. The tensile strength of a composite based on flax fibers is much higher than those reinforced with particles. It can be about 20 times the strength of the cork-based composite, and about 3.6 times the strength of the straw-based composite [1]. Fibrous products are increasingly used as a sound-absorbing material. New structures with the participation of natural or synthetic fibers are still being developed, and the literature on the subject contains research results concerning the use of fibers that differ in terms of raw material, dimensions, arrangement, and specific surface area [13–15]. Mamtaz et al. stated that synthetic fibers, due to their thinner diameter and antifungal quality, are a better sound absorptive material than natural fibers; unfortunately, they also have a greater impact on the environment [16]. However, when talking about the acoustic properties of the fibers, many factors should be taken into account, such as their thinness, the shape of their cross-section, and the bulk density of the material. For example, the sound absorption of fabrics made from 3 denier polyester fibers is 5 times greater than that of the material from 15 denier fibers [17]. The polyester fibers with octalobal or trilobal cross-sections are better as sound insulators than round fibers because of their higher total surface area [17]. The absorption of glass fibers with a bulk density of 54 kg/m^3 is higher than that of kapok fibers, with a bulk density of 10 kg/m^3, but lower than those fibers with a density of 15 kg/m^3 [18].

The microscopic structure and surface morphology of natural fibers such as flax, bamboo, kenaf, kapok, coir, cotton, broom, giant reeds, cane, coconut, hemp, etc. are conducive to sound absorption [18–22]. Natural fibers, due to their unique hollow and multi scale structures, show better sound absorption compared to high-modulus fibers such as glass or carbon, especially at frequencies above 1000 Hz. The ramie, jute, and flax fibers are characterized by a noise reduction coefficient at the level of 0.6–0.65, while glass fibers are at the level of 0.35, and carbon fibers at 0.45 [23]. The literature reports indicate that, for example, natural kapok fibers show sound absorption comparable to the widely used reinforcing glass fibers [18,24]. They are also characterized by a much lower density, which is of great importance in the design of lightweight composites. Yang and Li stated that composites made of natural fibers, such as jute, ramie, flax, and epoxy resin can exhibit similar or even better sound absorption than composites based on high-modulus fibers such as glass or carbon, which is important for aeronautical applications [23]. Mohanty and Fatima presented natural rubber-based jute composites manufactured by a compression molding process as biodegradable soundproofing materials for noise control applications, e.g., in home appliances, building construction, and cars [25]. Ersoy and Kucuk proposed that biodegradable tea leaf-fibers, as a product of renewable bio-resources, can be used as a sound absorber. The backing of these fibers with a cotton fabric layer gives sound absorption comparable to nonwoven polypropylene [26]. According to Zulkifli et al., an even better effect can be obtained by using coconut fibers [27]. The sound absorption of composite materials is also positively influenced by the addition of ultra-short/ultra-fine fibers obtained from natural fibers, e.g., flax fibers. Such fibers obtained by an enzymatic process and mechanical treatment, due to the larger total fiber surface, give greater interaction with sound waves, and thus increase the sound absorption of the material [28,29]. Research has shown that urea formaldehyde resin fiber boards made of various fibers, such as bagasse, banana, bamboo, coir, and corn husk, are characterized by better sound absorption if they have a lower density [30].

Much attention is also paid to fibrous layered products, as increasing the thickness and density of the sound-absorbing layer promotes sound absorption at the mid-to-high frequency ranges [10]. Layer systems where the layers differ in the type of fibers and in the textile structure are described. The most commonly used structures are nonwovens, woven fabrics, knitted fabrics, or nanofiber membranes. Sometimes fibrous layers are combined with other materials, such as cork [31,32], foam [33], or a honeycomb grid [34]. The use of several fibrous layers differing in apparent density favors increased sound absorption and widened absorption bands. The same is true in the case of laminated composites based on textiles. The use of a system of textile layers differing in structure allows for obtaining a composite that has a lower thickness than the system of textile layers, and a comparable absorption [28,29,35]. By using layers that differ in structure, one can control the dependence of the sound absorption on the sound frequency [36]. Each material absorbs and reflects sound waves to some extent. The proportions between the energy of the reflected and absorbed waves can vary depending on the frequency of the sound. Soft, porous materials are good sound absorbing materials. Hard and smooth materials absorb sound waves very poorly. However, for some applications, it is necessary that at least the surfaces of the material should be hard and smooth. The structure of the material can be created in the process of its production [37].

Much attention is also paid to the importance of the material shape for its acoustic performance. Sharma et al. presented an analytical framework for a metasurface with a lattice of closely spaced spherical cavities embedded in a thin, soft medium. The strong resonance of the cavities was confirmed, and it has been shown that the high sound absorption of the metasurface is due to the strong multiple scattering of waves between the cavities and the conversion of longitudinal waves into shear waves dissipated then in the elastic medium [38]. Azad et al. investigated the effects of large-scale pyramidal and convex-shaped diffusers in an empty non-diffuse room on its acoustical parameters. The statistical analysis and measurement results showed that the influence of diffuser type on the room acoustic characteristics is significant, especially at high frequencies [39]. Recently, a growing interest in practical applications has been enjoyed by sonic crystals, i.e., finite arrays of periodically distributed scatterers for which very little sound transmission occurs in certain frequency bands. It is caused by the destructive interference of scattered waves in the latttice structure. These bands depend on the shape of the scatterers, the distance between the scatterers as a lattice constant, and the filling fraction. The center frequencies of these bands can be predicted from Bragg's law, and, for a greater lattice constant, they are at lower frequencies. The literature gives results for rigid diffusers in the form of steel cylinders or trees. Measurement results indicated that if the diffusers are arranged in a lattice configuration, they can more effectively attenuate certain low frequency bands. The attenuation level depends on the filling fraction, and the attenuation frequencies depend on the type of lattice pattern and the angle of sound incidence on the barrier [40,41].

In the case of thermoplastic composites, both the reinforcing component and the thermoplastic polymer can be used in the form of fibers. Having fibers, they can be used in the form of a hybrid structure, e.g., a needle-punched nonwoven. This structure allows for a high degree of mixing of both components, necessary to ensure good wettability of the re-inforcement by the matrix. Nonwovens provide a number of functional benefits, including thermal and acoustic insulation. The most important advantages of the needle-punching technology include high production efficiency, the possibility of obtaining structures of high thickness, and the possibility of joining layers with different fiber orientation. The literature reports present the results of the research on sound absorption by nonwovens in terms of nonwoven technology and the aspect of web orientation angle. The effect of the web orientation angle on the sound absorption properties for thermally bonded nonwovens of multiangle layered webs was tested by Lee et al. [29].

The present work concerns research on the possibilities of producing green composite materials of various layered structures and different profiling from nonwovens, and determining the impact of these factors on the sound absorption by the composite [23,42].

2. Materials and Methods

To obtain green composites, flax fibers (LI), from Safilin Ltd., Milakowo, Poland, were used as a reinforcement. The dimensions and quality of waste short flax fibers, the so-called noils, were very diverse, Figure 1. The length of the flax fibers ranged from a few mm up to 115 mm, and the transverse dimensions were from 16 μm to 560 μm. As a matrix material, biodegradable polylactide fibers (PLA) were used [43,44]. These commercial thermoplastic fibers, Ingeo Fiber type SLN2660D (linear density 6.7 dtex, length 64 mm), with a melting point in the range of 165–170 °C and finished with polylactide resin without any hazardous substances, were delivered by Far Eastern Textile Ltd., Taipei, Taiwan. In order to obtain a homogeneous composite material, it was necessary to perfectly blend the fibers in the nonwoven fabric. The mixing process consisted of passing the fibers twice through the carding machine. The fleece with a parallel system of fiber arrangement was obtained. Then, the needle-punching process of the fleece layer was carried out on an Asselin needle-punching machine (France). The following technological parameters were used: type of needles—$15 \times 18 \times 40 \times 3^1/_2$ RB (Groz-Beckert®, Albstadt, Germany); number of needles punching—$40/cm^2$; depth of needle-punching—12 mm. The needle-punched nonwoven was thus obtained.

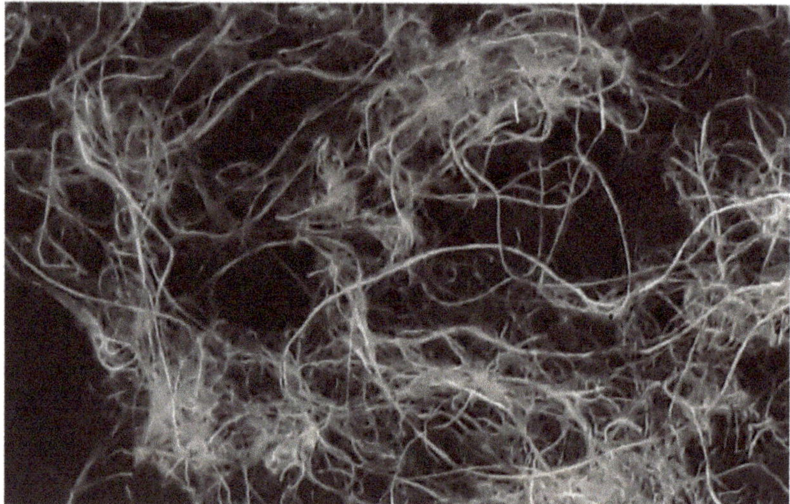

Figure 1. View of waste flax fibers.

Composites were formed from textile multilayer structures in a hydraulic press machine, Hydromega, Gdynia, Poland, with heated top and bottom plates and a water cooling system. A multilayer structure of nonwovens, sandwiched between two layers of Teflon foil, was put into the press mold. The mold was then closed and the heating was turned on. After reaching the pressing temperature, i.e., the melting point of the thermoplastic fibers, the consolidation stage was carried out under a pressure of 0.58 MPa for 5 min. Finally, the heating was turned off and the water cooling system was turned on to bring the temperature down to room temperature. For profiling the composites, a Teflon plate with holes was used (during pressing it was placed under the multilayer structure of nonwovens), and a tool with a spherical tip was used to obtain concavities in the composite plates, Figure 2.

The acoustic properties of the composites were determined by means of a small-sized impedance tube, type 4206 (Bruel&Kjaer, Denmark) using two $\frac{1}{4}$-inch condenser microphones, type 4187, Figure 3. The physical sound absorption coefficient (a quotient of acoustic energy absorbed by the given material to the energy of the acoustic incident wave) was determined for each sample by the method using the coefficient of a standing wave, according to the standard procedure: PN-EN ISO 10534-2 in the frequency range of 500–6400 Hz. This range of sound frequencies is adequate and sufficient to observe the sound absorbing behavior of the tested materials [2,12,28]. In this method, using a Kundt tube, the sound wave coming from sound source is directed perpendicularly at the surface of tested material. The acoustic pressure is measured by microphones at two locations on the wall of the tube. Then, the signals are transferred to analyzer. In this method, the impedance tube is connected to the sound source on one side, and on the opposite side the test sample is placed. The noise source generates plane waves in the tube directed perpendicular to the sample surface. The sound pressure is measured thanks to the two microphones in fixed positions in the tube wall. Then, the signals are transferred to the analyzer and the interference distribution of the field is determined. Based on this, the sound absorption coefficient is calculated. Before the measurement, the instrument was calibrated each time the sample material was changed. The samples with a diameter of 29 mm were cut with a punch. Three samples were tested for each variant.

Figure 2. View of the composites and the Teflon plate (in white color).

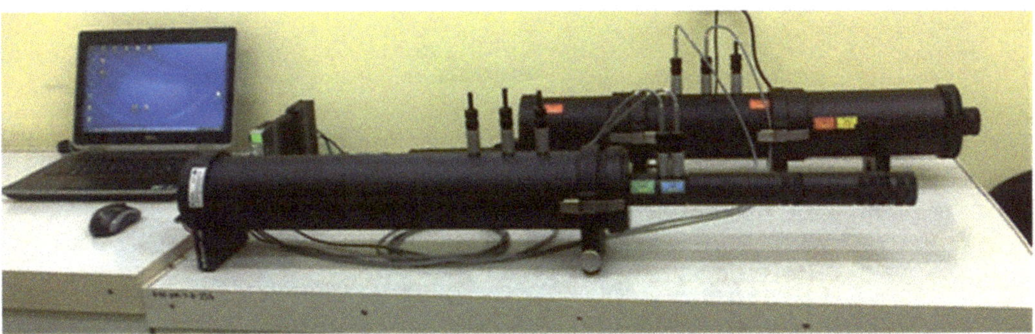

Figure 3. The Kundt device.

3. Results

Comparatively, the pre-pressed nonwovens, the composites made of nonwovens, and the layered systems composed of pre-pressed nonwovens and composites were used for the acoustic measurements.

3.1. Nonwovens

Nonwovens were obtained with an area weight of $180 \, \mathrm{g/m^2}$, differing in fiber composition: fiber composition I—20% LI fibers and 80% PLA fibers, mixed together—"LI/PLA", fiber composition II—100% PLA fibers—"PLA".

3.2. Pre-Pressed Nonwovens

The obtained LI/PLA nonwoven was pre-pressed on the nonwoven press machine at different conditions. The parameters of the pre-pressing process and the characteristics of the nonwoven sound-absorbing materials obtained after the pre-pressing process are presented in Table 1.

Table 1. Variants and characterization of the pre-pressed nonwovens.

	Process Parameters			Characteristics of the Pre-Pressed Nonwovens	
	Temperature, °C	Pressure, MPa	Time, s	Thickness, mm	Surface Description
1N (1 layer of nonwoven)	130	5	30	1.55	Nonwoven with low compression, with a very fibrous structure without hard or plastic zones.
2N (2 layers of nonwoven)	130	10	30	1.75	Nonwoven similar to 1N, but with a more compact structure and more rigid because double the pressure was applied.
3N (1 layer of nonwoven)	140	15	60	1.08	Nonwoven more compact than the other two. It presents an almost plastic structure.
4N (1 layer of nonwoven)	140	15 (pressure with metal mesh)	60	0.8	Structurally similar to 3N, but thinner and with a drilled surface due to the metal mesh.
5N (1 layer of nonwoven)	140	10	120	1.24	It could be described as a step between nonwovens 2N and 3N. Quite compact but more fibrous than 3N

The dependence of the sound absorption coefficient on the sound frequency for individual variants of the pre-pressed nonwovens is shown in Figure 4. For single, very thin layers of pre-pressed nonwovens, 0.8 to 1.75 mm thick, an increase in sound absorption is observed with increasing sound frequency. In the case of sounds with frequencies up to 4800 Hz, the highest similar values of sound absorption coefficient were obtained for the 2N and 3N pre-pressed nonwovens. Higher frequency sounds are better absorbed by the 1.08 mm thick 3N pre-pressed nonwoven, presenting an almost plastic structure, than by the 1.75 mm thick 2N pre-pressed nonwoven with a compact fibrous and rigid structure. Nonwoven 3N shows the highest value of the sound absorption coefficient, i.e., 0.44 at the sound frequency of 6400 Hz. The other three pre-pressed nonwovens show a lower absorption. In all frequency ranges, the dependence of the sound absorption coefficient on the sound frequency is similar for these three nonwovens. The maximum value of the sound absorption coefficient, obtained at 6400 Hz, is 0.23 for 1.55 mm thick 1N nonwoven with a fibrous structure, 0.20 for 1.24 mm thick 5N nonwoven with a compact/fibrous structure, and 0.17 for 0.8 mm thick 4N nonwoven with an almost plastic structure but with a mesh surface.

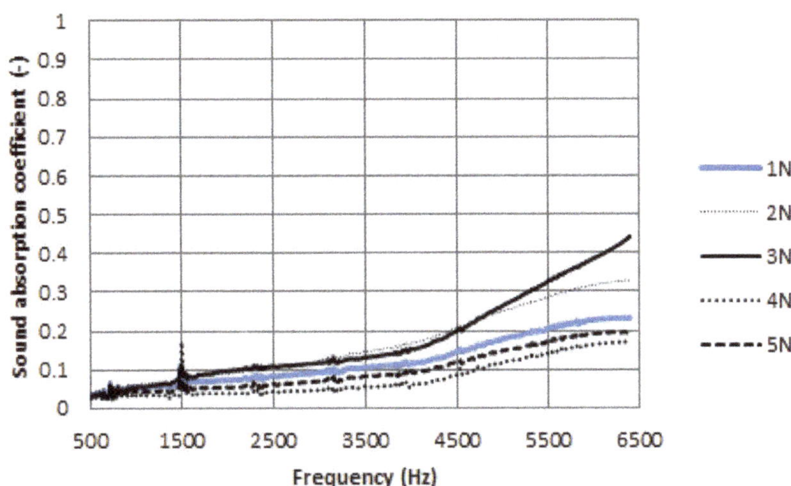

Figure 4. Sound absorption coefficient of individual pre-pressed nonwovens.

The combination of successive nonwovens with each other causes the thickness of the resulting absorbent systems to be greater than that of the individual layers, but does not mean adding up their individual sound absorption. It can be seen from Figure 5 that adding three nonwovens, i.e., 2N, 3N, and 4N, to the nonwoven 1N successively increases the value of the absorption coefficient by a value corresponding to the individual nonwovens in a given frequency range. The addition of another layer, i.e., a 5N nonwoven, no longer increases the sound absorption by a value corresponding to this nonwoven, but rather only slightly. The contribution of the next added layer with specific sound absorption characteristics to the increase of a system's sound absorption coefficient depends on the resulting system structure and sound frequency. The same layer can show a different absorption as a separate layer, and a different one to the arrangement with another layer, because then, a new structure is created, which constitutes different conditions for attenuating the energy of the acoustic wave. As a consequence, the sound absorption of the system is different from that resulting from adding up the absorption of both layers. Table 2 shows that the tested pre-pressed nonwovens, very thin and with low absorption, can be combined into multilayer systems in order to increase sound absorption in the frequency range of 2500–5500 Hz in relation to the total absorption of the individual layers. The share of each next added nonwoven layer characterized by a specific sound absorption in the increase in the system's sound absorption coefficient depends on the sound frequency and on the sound absorption of the system without this layer.

If both previous graphs, Figures 4 and 5, are compared, it is possible to observe that the shapes of the curves for individual nonwovens are completely different when the nonwovens are together. For individual nonwovens, the curves are practically flat, with an important increase for high frequencies. However, for the nonwovens together, the curves are concave, more similar than a composite curve, which have higher values for a wider range of frequencies.

In the case of sound-absorbing porous materials, a low-frequency sound absorption is higher if the material is thicker. A homogeneous material of high thickness or a layered material with layers of different structure can be used. Homogeneous material can be used with a thick material layer or a different layer structure.

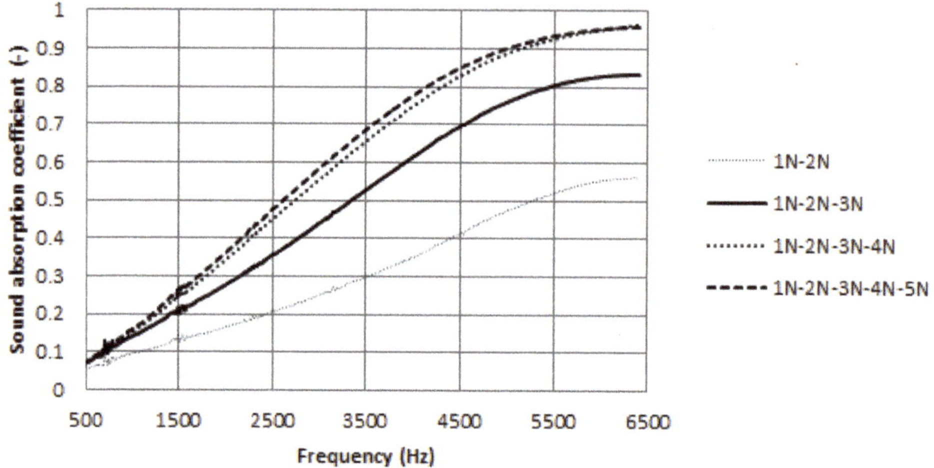

Figure 5. Sound absorption coefficient of nonwoven layered systems.

Table 2. Comparison between sound absorption values for layered systems of pre-pressed nonwovens—sound absorption coefficient calculated (in black color) and measured (in blue color).

Layered System of Pre-Pressed Nonwovens	Sound Frequency, Hz						
	500	1500	2500	3500	4500	5500	6400
1N + 2N	0.06	0.14	0.20	0.25	0.36	0.39	0.56
	0.06	0.14	0.20	0.30	0.41	0.52	0.56
1N + 2N + 3N	0.09	0.22	0.31	0.38	0.57	0.73	1.00
	0.07	0.22	0.36	0.53	0.70	0.80	0.83
1N + 2N + 3N + 4N	0.12	0.27	0.36	0.44	0.66	0.89	1.17
	0.07	0.25	0.46	0.66	0.84	0.94	0.96
1N + 2N + 3N + 4N + 5N	0.15	0.32	0.43	0.52	0.78	1.06	1.37
	0.07	0.26	0.48	0.69	0.85	0.94	0.96

However, joining the layers of nonwovens, the thickness of the material can increase up to several cm, which is why, in this work, pre-pressed nonwovens were used. However, combining layers of nonwovens causes an increase in the material thickness of up to several cm, so, in this research work, the pre-pressed nonwovens were proposed.

The multilayer structures consist of several pre-pressed nonwoven layers with different acoustic characteristics, and are a promising material for noise reduction. Thanks to such acoustic systems, it is possible to increase the level of absorption and extend the frequency range of high absorption.

3.3. Composites

Variants of the composites obtained from LI/PLA and PLA nonwovens are presented in Table 3.

Composites obtained according to variants 1, 2, and 3 were made of a nonwoven under conditions differing in system temperature, and this factor influenced the structure of the composite. By changing the temperature of the pressing process in the range from 165 to 180 °C, we changed the structure of the composite from fibrous to plastic. Differences in the sound absorption of the composites produced resulted from their structure. The values of the sound absorption coefficient for the individual sound frequencies are presented in Figures 6–8.

Table 3. Variants and structure characterization of the composites.

No.	Component Layers	Temperature, °C	Thickness, mm	Description of Composite Structure up/down
1	8xLI/PLA	165	6.0	fibrous/plastic
2	8xLI/PLA	170	5.0	porous plastic/more plastic
3	8xLI/PLA	180	5.5	porous plastic/porous plastic
4	1xPLA8xLI/PLA1xPLA	170	min 4.0 max 6.8	plastic/more fibrous/plastic, profiled

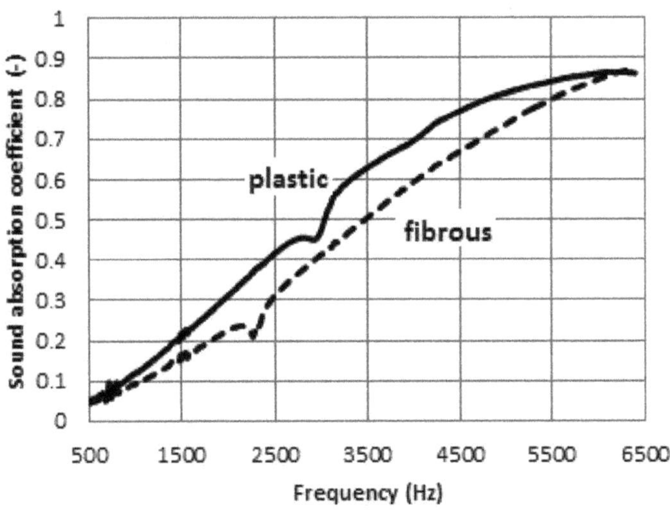

Figure 6. Sound absorption coefficient of composite 1.

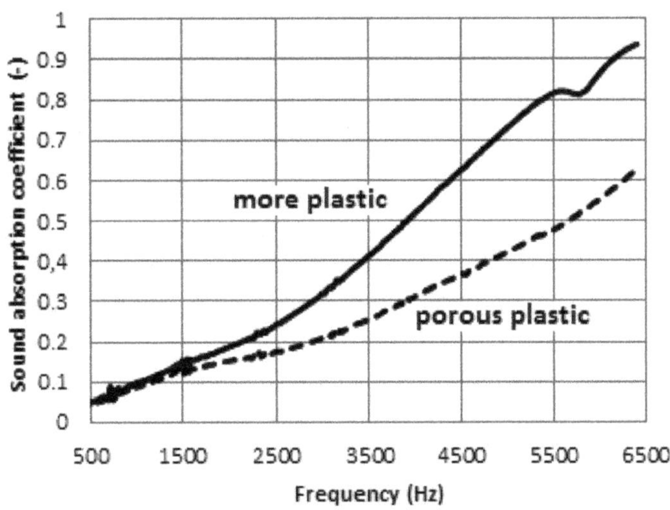

Figure 7. Sound absorption coefficient of composite 2.

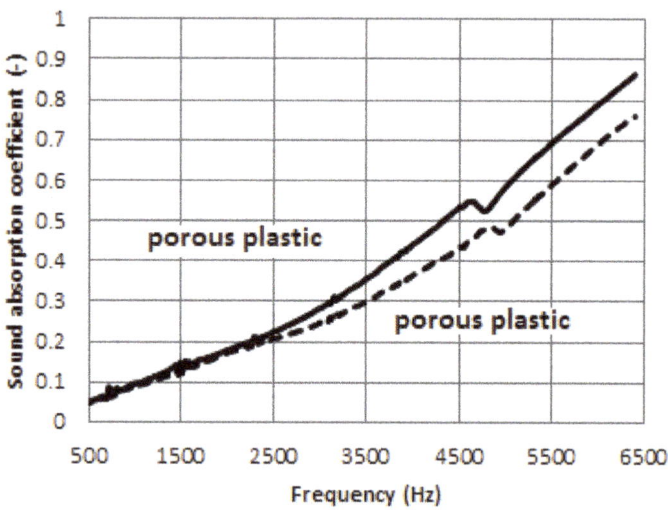

Figure 8. Sound absorption coefficient of composite 3.

For each sample, two different curves, depending on the receptor surface of the sound waves (two composite sides), are presented. For composite 1, Figure 6, the sound absorption coefficient is higher for middle and low frequencies than for the composites 2 and 3. It is more advantageous if the sound wave strikes the composite from the plastic side than the fibrous side. The highest values of sound absorption coefficient are above 0.9 at 6000–6400 Hz. The results for composite 2, Figure 7, show a clear difference between the sound absorption of surface 1 and 2. Surface 1, more plastic, presents a better structure for sound absorption at higher frequencies than surface 2, porous plastic. In low frequencies the difference is not so big. This sample is interesting because the maximum coefficient value of the plastic surface receptor is very high at 6400 Hz, between 0.9 and 1. For composite 3, Figure 8, the values of the sound absorption coefficient are similar, regardless of which the side of the composite faces the sound wave. This fact results from the similar structure of both surfaces of the composite. The maxim coefficient is lower than for sample 2's maxim coefficient. Generally, composites with a different structure for both surfaces give better sound absorption, and if the plastic surface is directed to the sound, the values of the sound absorption coefficient are higher.

Another important parameter of the composite is the profiling. Composite 4 was made on the basis of a blended nonwoven, but located between layers of PLA nonwoven, which, after compression, gives rigid, smooth plastic surfaces. Then, the modifications were used to profile the composite 4 plate, as shown in Figure 9. The results below show the influence of the surface profiling on the sound-absorption properties. Four samples were analyzed. The schematic view of the samples is shown below. The dependence of the sound absorption coefficient on the sound frequency for the profiled composite is shown in the diagrams in Figures 10–13. The composite plates were tested with the left side and right side facing the sound wave, respectively. From assessing the effects of profiling the composite plate, it can be seen that the proposed modifications have a beneficial effect on sound absorption. A two-sided flat plate shows the lowest values of the sound absorption coefficient in the entire tested frequency range, Figure 10. However, profiling a concave in the plate and directing it with its front face to the sound wave causes an increase in sound absorption, Figure 11.

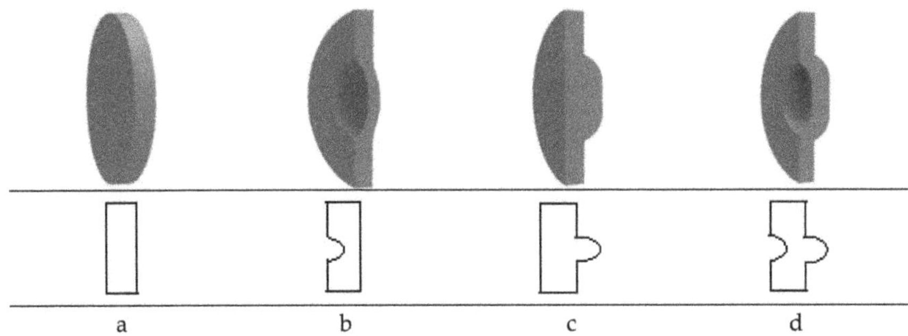

Figure 9. Profiled composite 4 (**a–d**).

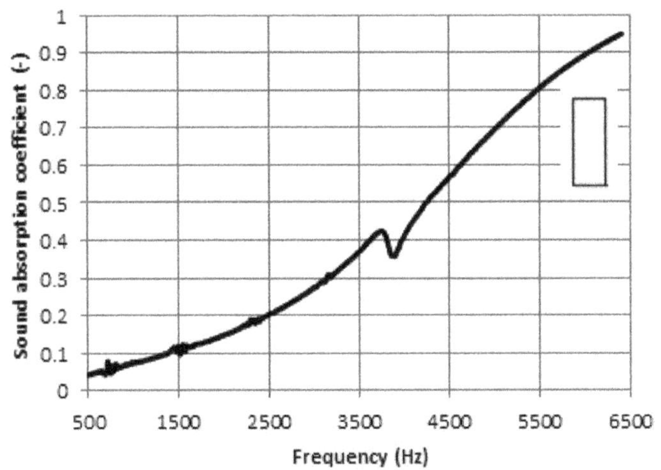

Figure 10. Sound absorption coefficient of composite 4a.

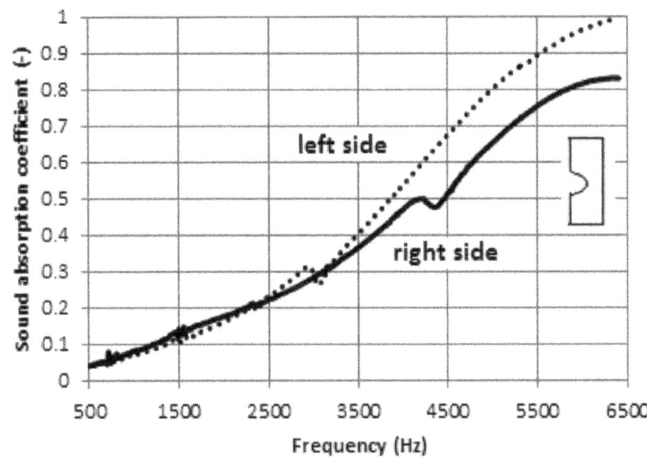

Figure 11. Sound absorption coefficient of composite 4b.

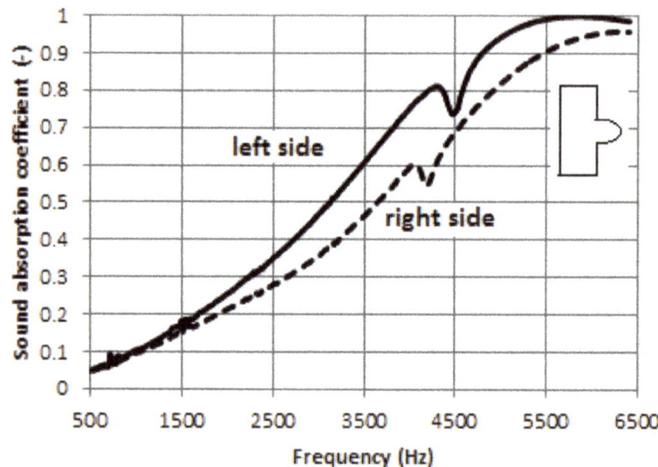

Figure 12. Sound absorption coefficient of composite 4c.

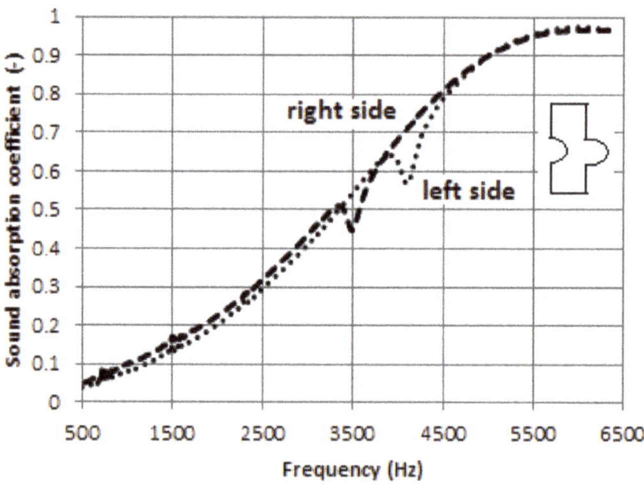

Figure 13. Sound absorption coefficient of composite 4d.

Taking into account the published results of previous studies [21], which indicated the validity of the use of convexity on the back of the composite plate, which was confirmed in Figure 12, the absorption of the plate with concavity was measured comparatively. The results showed that convexity on the plate is more advantageous than concavity, Figures 11 and 12. In the case of a plate with convexity, the flat side facing the sound wave achieves the greatest sound absorption among the tested profiled plates, and the value of the absorption coefficient is equal to 1 for sounds with a frequency above 5500 Hz, Figure 12, Table 4. Among the tested profiled samples, a plate with convexity on one side has the greatest thickness over the entire area. This favors an increase in sound absorption. In addition, when such a sample is positioned with the flat side to the sound wave, and there is a convexity on the other side, an air space is created between the sample and the wall of the measuring tube in which resonance may occur. Under the conditions of use, such an orientation of the sample with a convexity in relation to the wall will be most advantageous. In the case of a concave–convex plate (Figure 13), where it does not matter which side it faces the sound wave, the values of the sound absorption coefficient are higher than for a concave–flat, flat–concave plate (Figure 11), or convex–flat plate (Figure 12), and lower than for a flat–convex plate (Figure 12).

Table 4. Comparison between the sound absorption values for variants of composite 4.

Sound Frequency	Variant of Composite 4			
f (Hz)	4a	4b	4c	4d
1000	0.07	0.07	0.10	0.10
2000	0.15	0.17	0.25	0.22
3000	0.27	0.30	0.47	0.44
4000	0.41	0.53	0.75	0.70
5000	0.70	0.80	0.95	0.90
6000	0.90	0.96	1.00	0.96

The homogeneous porous material is not a good and practical low frequency sound absorber, as it would have to be extremely thick or very far from the back boundary surface. For example, at 500 Hz, the total wavelength is 0.688 m, so the porous material would need to be approximately 0.172 m from the back boundary surface to meet the 1/4 wavelength requirement for significant sound absorption. The lower the sound frequencies, the greater the material thickness/distance should be. In order to improve the sound absorption at low frequencies, the material thickness can be compensated for by air space in the rear. In studied porous composites, periodic inclusions have been employed to significantly enhance the sound wave manipulation abilities, Figures 2 and 9. The 4c and 4d profiled composites having the convexities at the rear provide just such an air space at the rear on the stiff wall side, and therefore exhibit better sound absorption towards lower frequencies than composites that are flat on both sides or flat–concave. Taking into account the front surface of the composite, when the back surface is flat, the best sound absorption is provided by the surface with convexities, then with concavities, and the worst by a flat surface. Sound absorption, apart from the thickness of the material, is also influenced by the surface area of interaction with the sound wave, the largest surface area is provided by convexity, then concavity, and the smallest one by a flat surface. The porosity of the material is also an important factor, and the convexity promotes increased porosity. A detailed presentation of the influence of individual factors on the obtained effect of sound absorption requires model considerations, which, in the case of fiber-based materials, and especially natural waste fibers with wide variation in thinness and length, and the sophisticated profiling of the composite plate, is an extremely complex issue [38,45,46]. It is known that resonant inclusions embedded in the soft matrix are conducive to the effective conversion of long longitudinal sound waves into short shear waves, which are subsequently absorbed. The literature reports that in the case of a periodically voided viscoelastic coating made of soft rubber embedded with a layer of cylindrical voids of infinite length, uncertainty in the geometric parameters has greater impact on the resonance frequency of the voids and sound transmission through the coating than the uncertainty in the material properties [47].

3.4. Pre-Pressed Nonwovens and Composites

In order to verify the possibility of obtaining an increase in sound absorption, systems consisting of pre-pressed nonwovens and composites in various combinations were tested. Composite 3 (described in Table 2 as "8xLI/PLA") was joined in a two-layer system successively with individual pre-pressed nonwovens. Successive material samples were stacked on top of each other without space between them, and put together into the measuring tube. The scheme of the arrangement of layers, where a nonwoven is situated on the side of the sound wave, and the composite on the back, is shown in Figure 14, and the sound absorption of such a system is shown in Figure 15. The scheme of the arrangement of layers, where the composite is situated on the side of the sound wave, and the pre-pressed nonwoven on the back, is shown in Figure 16, and the sound absorption of such a system is shown in Figure 17.

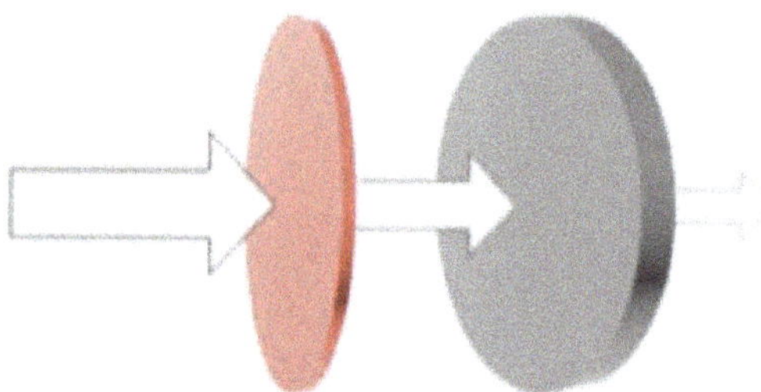

Figure 14. Schematic view of the sample's position inside the tube. From the left: one pre-pressed nonwoven—one composite. Both layers are together without space between them inside the tube.

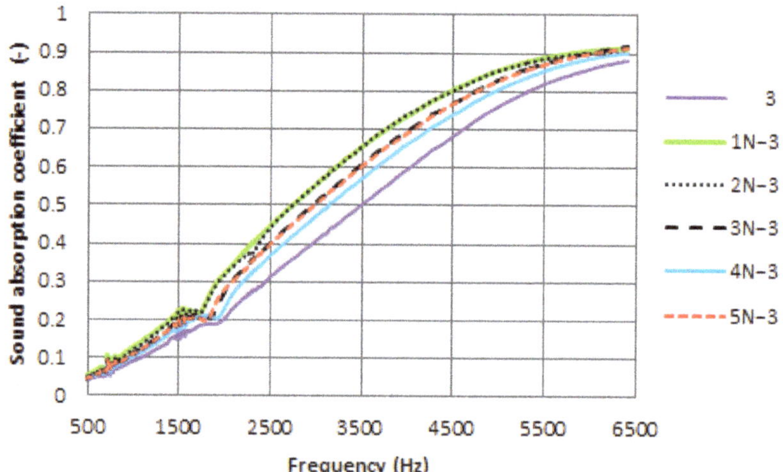

Figure 15. Sound absorption of composite 3 with one pre-pressed nonwoven layer in front.

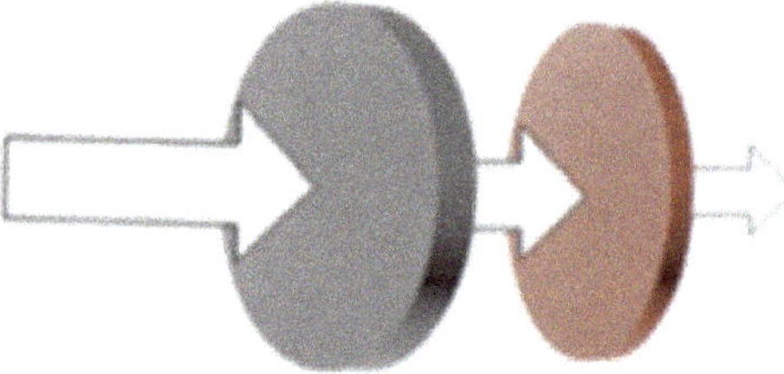

Figure 16. Schematic view of the sample's position inside the tube. From the left: one composite—one pre-pressed nonwoven. Both layers are together without space between them inside the tube.

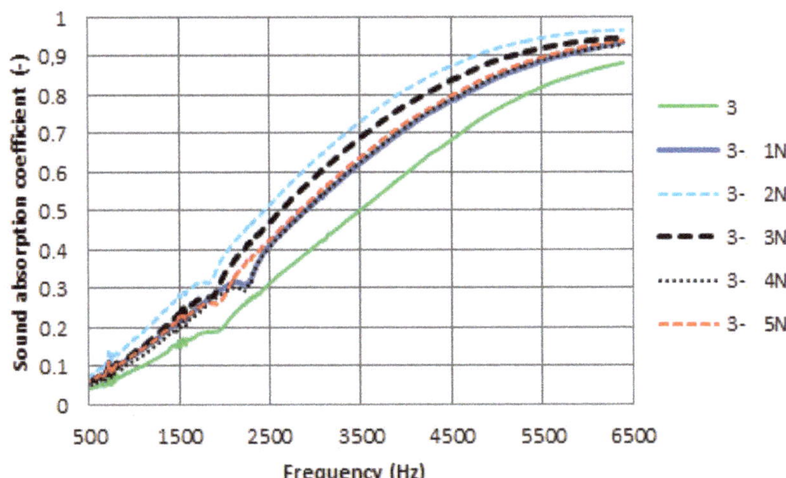

Figure 17. Sound absorption of composite 3 with one pre-pressed nonwoven layer behind.

The graphs of the dependence of the sound absorption coefficient on the sound frequency, Figures 4, 15 and 17, show that the combined systems show greater absorption than the individual layers separately, which results from the increase in the thickness of the final absorbing material and depends on the structure and, in consequence, on the absorption of the component layers. In the "pre-pressed nonwoven + composite" system, the 1N nonwoven and the 2N nonwoven provide the best effects. In such a system, the layer on the side of the sound wave is more fibrous, and on the other side, the layer is a porous composite. In the case of the "composite + pre-pressed nonwoven" system, the 2N nonwoven also provides the best effect. As shown in Figure 4, among the nonwovens tested, this nonwoven, next to the 3N nonwoven, shows the highest sound absorption in the sound frequency range up to 4800 Hz. The composite layer forms a better sound-absorbing system with a more fibrous and compact layer than with a more plastic layer. The comparison of the sound absorption coefficient for both systems, with the example of composite 3 and 2N pre-pressed nonwoven, i.e., 3-2N and 2N-3, is shown in Table 5. The results show that it is more advantageous to position the composite layer from the sound wave side, and the nonwoven as the back layer. Then, the absorptive surface of the system is plastic and the back is fibrous, and as shown in the earlier Figures 6 and 7, this structure, with more plastic surface, promotes sound absorption [48].

Table 5. Comparison between the sound absorption values for composite + pre-pressed nonwoven (3-2N) and pre-pressed nonwoven + composite (2N-3).

Sound Frequency	Variant of Sound-Absorbing System	
f (Hz)	3-2N	2N-3
1000	0.17	0.12
2000	0.39	0.32
3000	0.63	0.56
4000	0.81	0.74
5000	0.92	0.74
6000	0.96	0.90

Composite 3, with a thickness of 5.5 mm, shows a sound absorption very similar to that presented by Hao et.al. [49] for kenaf/polypropylene nonwoven composites with a thickness of 6 mm. The values of the sound absorption coefficient of these composites at high frequencies are high, about 0.9 at 6400 Hz. The addition of a pre-pressed nonwoven, with a thickness in the range from 0.8 mm to 1.75 mm, to the tested composite has the effect of increasing the sound absorption in a wide sound frequency range. For both two-layer systems, the values of the sound absorption coefficient increase with increasing sound frequency, and begin to stabilize at about 6000 Hz. The results of absorption are very good (above 0.7) at higher values of the tested frequency range, i.e., from 3750–4230 Hz to 6400 Hz for the "pre-pressed nonwoven + composite" systems, and from 3340–4000 Hz to 6400 Hz for the "composite + pre-pressed nonwoven" systems. Separate composite 3 reaches a value of sound absorption coefficient of 0.7 at frequencies from 4600 Hz to 6400 Hz. At the highest sound frequencies, the sound absorption coefficient for both kinds of systems is around 0.9 or even higher. At the lower sound frequency range, the absorption of the system increases by a value approximate to the absorption of the added layer; the higher the frequency, the smaller the increase in absorption.

An increase in the sound absorption level can be obtained by combining a fibrous material, e.g., a pre-pressed nonwoven, with a composite plate acting as a rigid membrane, which can also extend the range of high absorption. The absorber then consists of two types of material, and there are probably two mechanisms for damping the sound wave in different frequency ranges. A composite layering sequence in a multilayer structure also plays an important role, the most absorptive are the systems where the composite as a rigid material is on the side of the incident sound wave, and the more porous is on the back. All proposed layers differing only in pressing conditions are produced on the basis of the same LI/PLA nonwoven fabric, which simplifies the production process and makes it more economical.

In order to improve the results for low frequencies, the next tests were made with a combination of composite and pre-pressed nonwoven layers. The scheme of the arrangement of the layers, where the composite is located on the side of the sound wave and the individual nonwovens are placed on the back, is shown in Figure 18, and the sound absorption of such a system is shown in Figure 19. The values of the sound absorption coefficient increase as the frequency increases, and then stabilize. As more nonwoven layers are added, the results of absorption increase, but less and less. This is because the total thickness increases, and the values of the absorption coefficient of the final multilayer system result from the structure, and consequently from the absorption of the individual layers. Sound absorption of 0.9 for the system (3-1N-2N-3N-4N-5N) with a total thickness of only 11.92 mm occurs in the frequency range from 2700 Hz to 6400 Hz, and 0.7 from 1850 Hz to 6400 Hz, which is a very good result. The literature [50] states that, for bilayered nonwoven composite with the thickness of 12.02 mm, the absorption is 0.92 at a peak at 1500 Hz, and 0.7 in the range of about 1000–3500 Hz. The more layers the system has, the more its absorption is close to 1.0 and the range of highest absorption is extended towards lower frequencies. This means that this combination will have a good absorption for almost all frequencies. Then, this material could be used in a wide variety of applications. Adding the next layers to the system increases the maximum sound absorption and extends its frequency range. A nonwoven, more fibrous, and more open structure absorbs better low frequency sound, and a composite, more plastic, and rigid structure absorbs better high frequency sound, [51]. Combining such structures into one system extends the range of high sound absorption.

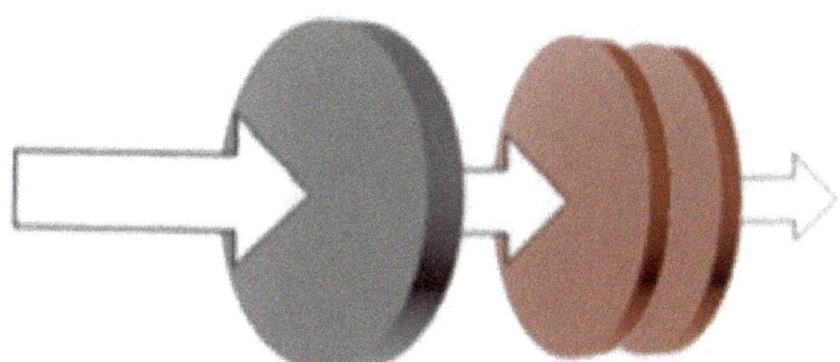

Figure 18. Schematic view of the sample's position inside the tube. From the left: one composite—pre-pressed nonwovens. All layers are together without space between them inside the tube.

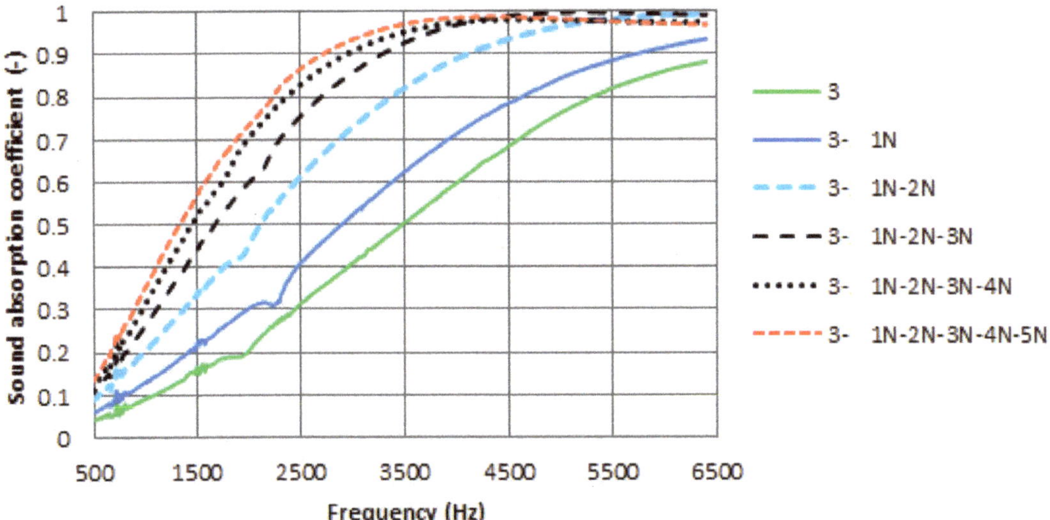

Figure 19. Sound absorption of composite 3 with the pre-pressed nonwoven layers behind.

4. Conclusions

The combination of successive pre-pressed nonwovens with each other causes the thickness of the resulting sound-absorbing systems to be greater than that of the individual layers. The contribution of the next added layer with specific sound absorption characteristics to the increase of a system's sound absorption coefficient depends on the resulting system structure and sound frequency.

Thermoplastic composites with different structures of both surfaces give better sound absorption than composites with the same surfaces. If the more plastic surface of the composite is directed to the sound, the values of the sound absorption coefficient are higher.

Shaping the composite plate has a positive effect on its sound absorption. In the case of a one-sided concave, it is better to position the plate with the concave side facing the sound and the flat side on the back. In the case of a one-sided convexity, it is better to place the plate with its flat side facing the sound, and the convex side at the back, then the absorption is the highest among the tested variants of profiled composites. For a composite plate with opposite profiles, i.e., concavity and convexity, no difference in absorption is related to the orientation of the plate relative to the sound.

The addition of a pre-pressed nonwoven to the composite, regardless of whether it is on the side of the incident sound wave or on the back, increases the sound absorption coefficient in the entire tested frequency range. However, it is preferable to have a composite on the sound side and a nonwoven on the back. Increasing the number of differentiated nonwoven layers in terms of structure means increasing the value of the sound absorption coefficient and broadening the frequency range of sounds most absorbed. The presented research is part of the work on sound absorbing composites on the basis of natural materials, and will be continued towards identifying the optimal composite structure. The proposed materials containing natural waste fibers will be an environmentally friendly, cheaper, and more sustainable alternative to traditional sound-absorbing materials. They could be used in the walls and partitions of vehicles, in buildings, or in door panels.

Author Contributions: Conceptualization, E.G.; formal analysis, J.P.d.A.; project administration, I.K.; supervision, E.G.; writing–original draft, J.P.d.A.; writing-review and editing, E.G. and M.M. All authors have read and agreed to the published version of the manuscript.

Funding: This research received no external funding.

Institutional Review Board Statement: Not applicable.

Informed Consent Statement: Not applicable.

Data Availability Statement: Not applicable.

Acknowledgments: This work was partially supported by statutory research fund of the Institute of Material Science of Textiles and Polymer Composites, Lodz University of Technology, Poland, no. I42/501/4-42-1-1.

Conflicts of Interest: The authors declare no conflict of interest.

References

1. Gliscinska, E.; Kaczor, M.; Milc, A.; Misztela, A.; Warczyk, P.; Krucinska, I. Biocomposites for Sound Absorption. *Compos. Theory Pract.* **2019**, *19*, 107–111.
2. Gliscinska, E.; Krucinska, I.; Michalak, M.; Puchalski, M.; Ciechanska, D.; Kazimierczak, J.; Bloda, A. Bio-Based Composites for Sound Absorption. *Compos. Renew. Sustain. Mater.* **2016**, *12*, 217–239. [CrossRef]
3. Rwawiire, S.; Tomkova, B.; Gliscinska, E.; Krucinska, I.; Michalak, M.; Militky, J.; Jabbar, A. Investigation of Sound Absorption Properties of Bark Cloth Nonwoven Fabric and Composites. *Autex Res. J.* **2015**, *15*, 173–180. [CrossRef]
4. Bledzki, A.K.; Reihmane, S.; Gassan, J. Thermoplastic Reinforced with Wood Fillers: A Literature Review. *Polymer. Plast. Tech. Eng.* **1998**, *37*, 451–468. [CrossRef]
5. Fernandes, E.M.; Correlo, V.M.; Chagas, J.A.M.; Mano, J.F.; Reis, R.L. Properties of New Cork-Polymer Composites: Advantages and Drawbacks as Compared with Commercially Available Fibreboard Materials. *Compos. Struct.* **2011**, *93*, 3120–3129. [CrossRef]
6. Gil, L. Cork Composites: A Review. *Materials* **2009**, *2*, 776–789. [CrossRef]
7. Tiuca, A.E.; Nemeşa, O.; Vermeşana, H.; Toma, A.C. New Sound Absorbent Composite Materials Based on Sawdust and Polyurethane Foam. *Compos. Part. B* **2019**, *165*, 120–130. [CrossRef]
8. Hassan, T.; Jamshaid, H.; Mishra, R.; Khan, M.Q.; Petru, P.; Novak, J.; Choteborsky, R. Hromasova, M. Acoustic, Mechanical and Thermal Properties of Green Composites Reinforced with Natural Fibers Waste. *Polymers* **2020**, *12*, 654. [CrossRef]
9. La Mantia, F.P.; Morreale, M. Green Composites: A Brief Review. *Compos. Part A* **2011**, *42*, 579–588. [CrossRef]
10. Kucuk, M.; Korkmaz, Y. The Effect of Physical Parameters on Sound Absorption Properties of Natural Fiber Mixed Nonwoven Composites. *Text. Res. J.* **2012**, *82*, 2043–2053. [CrossRef]
11. Adhika, D.R.; Prasetiyo, I.; Noeriman, A.; Hidayah, N. Widayani Sound Absorption Characteristics of Pineapple Leaf/Epoxy Composite. *Arch. Acoust.* **2020**, *45*, 233–240. [CrossRef]
12. Gliscinska, E.; Michalak, M.; Krucinska, I. Sound Absorption Property of Nonwoven Based Composites. *Autex Res. J.* **2013**, *13*, 150–155. [CrossRef]
13. Furstoss, M.; Thenail, D.; Galland, M.A. Surface Impedance Control for Sound Absorption: Direct and Hybrid Passive/Active Strategies. *J. Sound Vib.* **1997**, *203*, 219–236. [CrossRef]
14. Galland, M.A.; Mazeaud, B.; Sellen, N. Hybrid Passive/Active Absorbers for Flow Ducts. *Appl. Acoust.* **2005**, *66*, 691–708. [CrossRef]
15. Arenas, J.P.; Crocker, M.J. Recent Trends in Porous Sound-Absorbing Materials. *Sound Vib.* **2010**, *44*, 12–18.
16. Mamtaz, H.; Fouladi, M.H.; Al-Atabi, M.; Namasivayam, S.N. Acoustic Absorption of Natural Fiber Composites. *J. Eng.* **2016**, *42*, 5836107. [CrossRef]

17. Tascan, M.; Vaughn, E.A.; Stevens, K.A.; Brown, P.J. Effects of Total Surface Area and Fabric Density on the Acoustical Behavior of Traditional Thermal-Bonded Highloft Nonwoven Fabrics. *J. Text. Inst.* **2011**, *102*, 746–751. [CrossRef]
18. Xiang, H.F.; Wang, D.; Liua, H.C.; Zhao, N.; Xu, J. Investigation on Sound Absorption Properties of Kapok Fibers. *Chin. J. Polym. Sci.* **2013**, *31*, 521–529. [CrossRef]
19. Sreeja, R.; Premlet, B.; Prasanth, R. An Insight into the Composite Materials for Passive Sound Absorption. *J. Appl. Sci.* **2017**, *17*, 339–356. [CrossRef]
20. Gliscinska, E.; Michalak, M.; Krucinska, I. The Influence of Surface Asymmetry of Thermoplastic Composites on Their Sound Absorption. *Compos. Theory Pract.* **2014**, *14*, 150–154.
21. Ciaburro, G.; Iannace, G.; Puyana-Romero, V.; Trematerra, A. A Comparison Between Numerical Simulation Models for the Prediction of Acoustic Behavior of Giant Reeds Shredded. *Appl. Sci.* **2020**, *10*, 6881. [CrossRef]
22. Iannace, G.; Ciaburro, G.; Trematerra, A. Modelling Sound Absorption Properties of Broom Fibers Using Artificial Neural Networks. *Appl. Acoust.* **2020**, *163*, 107239. [CrossRef]
23. Yang, W.D.; Li, Y. Sound Absorption Performance of Natural Fibers and Their Composites. *Sci. China Tech. Sci.* **2012**, *55*, 2278–2283. [CrossRef]
24. Yang, T.; Hu, L.; Xiong, X.; Petru, M.; Noman, M.T.; Mishra, R.; Militky, J. Sound Absorption Properties of Natural Fibers: A Review. *Sustainability* **2020**, *12*, 8477. [CrossRef]
25. Mohanty, A.R.; Fatima, S. Noise Control Using Green Materials. *Sound Vib.* **2015**, *49*, 13–15.
26. Ersoy, S.; Kucuk, H. Investigation of Industrial Tea-Leaf-Fibre Waste Material for Its Sound Absorption Properties. *App. Acoust.* **2009**, *70*, 215–220. [CrossRef]
27. Zulkifli, R.Z.; Nor, M.J.M. Noise Control Using Coconut Coir Fiber Sound Absorber with Porous Layer Backing and Perforated Panel. *Am. J. Appl. Sci.* **2010**, *7*, 260–264. [CrossRef]
28. Krucinska, I.; Gliscinska, E.; Michalak, M.; Ciechanska, D.; Kazimierczak, J.; Bloda, A. Sound-Absorbing Green Composites Based on Cellulose Ultra-Short/Ultra-Fine Fibers. *Text. Res. J.* **2015**, *85*, 646–657. [CrossRef]
29. Lee, Y.E.; Joo, C.W. Sound Absorption Properties of Thermally Bonded Nonwovens on Composing Fibers and Production Parameters. *J. Appl. Polym. Sci.* **2004**, *92*, 2295–2302. [CrossRef]
30. Sharma, S.K.; Shukla, S.R.; Sethy, A.K. Acoustical Behaviour of Natural Fibres-Based Composite Boards as Sound-Absorbing Materials. *J. Indian Acad. Wood Sci.* **2020**, *17*, 66–72. [CrossRef]
31. Iasnicu, I.; Vasile, O.; Iatan, R. Sound Absorption Analysis for Layered Composite Made from Textile Waste and Cork. *Sisom Acoust.* **2015**, *78*, 292–299.
32. Iannace, G.; Ciaburro, G.; Guerriero, L.; Trematerra, A. Use of Cork Sheets for Room Acoustic Correction. *J. Green Build.* **2020**, *15*, 45–55. [CrossRef]
33. Lin, J.H.; Liao, Y.C.; Huang, C.C.; Lin, C.C.; Lin, C.M.; Lou, C.W. Manufacturing Process of Sound Absorption Composite Planks. *Adv. Mat. Res.* **2020**, *97-101*, 1801–1804. [CrossRef]
34. Tang, X.; Yan, X. Acoustic Energy Absorption Properties of Fibrous Materials: A review. *Compos.* **2017**, *101*, 360–380. [CrossRef]
35. Jiang, S.; Cai, Y.; Zhou, X.; Yan, X. Preparation and Properties of Composite Multilayer Sound Absorption Structures. *J. Text. Res.* **2012**, *33*, 20–25.
36. Zulkifli, R.; Nor, M.J.M.; Tahir, M.F.M.; Ismail, A.R.; Nuawi, M.Z. Acoustic Properties of Multi-Layer Coir Fibres Sound Absorption Panel. *J. Appl. Sci.* **2008**, *8*, 3709–3714. [CrossRef]
37. Chen, W.H.; Lee, F.C.; Chiang, D.M. On the Acoustic Absorption of Porous Materials with Different Surface Shapes and Perforated Plates. *J. Sound Vib.* **2000**, *237*, 337–355. [CrossRef]
38. Sharma, G.S.; Skvortsov, A.; MacGillivray, I.; Kessissoglou, N. Sound Scattering by a Bubble Metasurface. *Phys. Rev. B* **2020**, *102*, 214308. [CrossRef]
39. Azad, H.; Meyer, J.; Siebein, G.; Lokki, T. The Effects of Adding Pyramidal and Convex Diffusers on Room Acoustic Parameters in a Small Non-Diffuse Room. *Acoustics* **2019**, *1*, 618–643. [CrossRef]
40. Martinez-Sala, R.; Rubio, C.; Garcia-Raffi, L.M.; Sanchez-Perez, J.V.; Sanchez-Perez, E.A.; Llinares, J. Control of Noise by Trees Arranged like Sonic Crystals. *J. Sound Vib.* **2006**, *291*, 100–106. [CrossRef]
41. Sharma, G.S.; Eggler, D.; Peters, H.; Kessissoglou, N.; Skvortsov, A.; MacGillivray, I. Acoustic Performance of Periodic Composite Materials. In *Proceedings of the Australian Acoustical Society 2015*; Hunter Valley, Australia, 15–18 November 2015. Available online: http://dspace.nal.gov.au/xmlui/handle/123456789/403 (accessed on 20 January 2021).
42. Gliscinska, E.; Krucinska, I.; Michalak, M.; Puchalski, M.; Kazimierczak, J.; Bloda, A.; Ciechanska, D. Sound Absorption of Composites with Waste Cellulose Submicrofibres. In Proceedings of the Book of Abstracts 14th AUTEX Word Textile Conference, Bursa, Turkey, 26–28 May 2014; article on CD (poster presentations, poster 84). ISBN 978-605-63112-4-6.
43. Lee, S.H.; Kim, I.Y.; Song, W.S. Biodegradation of Polylactic Acid (PLA) Fibers Using Different Enzymes. *Macromol. Res.* **2014**, *22*, 657–663. [CrossRef]
44. Bax, B.; Mussig, J. Impact and Tensile Properties of PLA/Cordenka and PLA/flax Composites. *Compos. Sci. Technol.* **2008**, *68*, 1601–1607. [CrossRef]
45. Cao, L.; Fua, Q.; Sia, Y.; Bin, D.; Yua, J. Porous Materials for Sound Absorption. *Compos. Commun.* **2018**, *10*, 25–35. [CrossRef]
46. Qui, H.; Enhui, Y. Effect of Thickness, Density and Cavity Depth on the Sound Absorption Properties of Wool Boards. *Autex Res. J.* **2018**, *18*, 203–208. [CrossRef]

47.	Sharma, G.S.; Faverjon, B.; Dureisseix, D.; Skvortsov, A.; MacGillivray, I.; Audoly, C.; Kessissoglou, N. Acoustic Performance of a Periodically Voided Viscoelastic Medium with Uncertainty in Design Parameters. *J. Vib. Acoust.* **2020**, *142*, 61002. [CrossRef]
48.	Gliscinska, E.; Sankowski, D.; Krucinska, I.; Gocławski, J.; Michalak, M.; Rowinska, Z.; Sekulska-Nalewajko, J. Optical Coherence Tomography Image Analysis of Polymer Surface Layers in Sound-Absorbing Fibrous Composite Materials. *Polym. Test.* **2017**, *63*, 194–203. [CrossRef]
49.	Hao, A.; Zhao, H.; Chen, J.Y. Kenaf/Polypropylene Nonwoven Composites: The Influence of Manufacturing Conditions on Mechanical, Thermal, and Acoustical Performance. *Compos.* **2013**, *54*, 44–51. [CrossRef]
50.	Kucuk, M.; Korkmaz, Y. Sound Absorption Properties of Bilayered Nonwoven Composites. *Fibers Polym.* **2015**, *16*, 941–948. [CrossRef]
51.	Midha, V.K.; Chavhan, M.V. Nonwoven Sound Absorption Materials. *IJTFT* **2012**, *2*, 45–55.

Article

Low-Density Insulation Blocks and Hardboards from Amaranth (*Amaranthus cruentus*) Stems, a New Perspective for Building Applications

Philippe Evon [1,*], Guyonne de Langalerie [1,2], Laurent Labonne [1], Othmane Merah [1,3], Thierry Talou [1], Stéphane Ballas [2] and Thierry Véronèse [2]

[1] Laboratoire de Chimie Agro-Industrielle (LCA), Université de Toulouse, ENSIACET, INRAE, Toulouse INP, 4 allée Emile Monso, 31030 Toulouse, France; guyonne.delangalerie@gmail.com (G.d.L.); Laurent.Labonne@ensiacet.fr (L.L.); othmane.merah@ensiacet.fr (O.M.); Thierry.Talou@ensiacet.fr (T.T.)
[2] Ovalie Innovation, 2 rue Marguerite Duras, 32000 Auch, France; stephane.ballas@ovalie-innovation.com (S.B.); thierry.veronese@ovalie-innovation.com (T.V.)
[3] Département Génie Biologique, IUT A, Université Paul Sabatier, 24 rue d'Embaques, 32000 Auch, France
* Correspondence: Philippe.Evon@ensiacet.fr; Tel.: +33-(0)562-446080

Abstract: Nowadays, amaranth appears as a promising source of squalene of vegetable origin. Amaranth oil is indeed one of the most concentrated vegetable oils in squalene, i.e., up to 6% (w/w). This triterpene is highly appreciated in cosmetology, especially for the formulation of moisturizing creams. It is almost exclusively extracted from the liver of sharks, causing their overfishing. Thus, providing a squalene of renewable origin is a major challenge for the cosmetic industry. The amaranth plant has thus experienced renewed interest in recent years. In addition to the seeds, a stem is also produced during cultivation. Representing up to 80% (w/w) of the plant aerial part, it is composed of a ligneous fraction, the bark, on its periphery, and a pith in its middle. In this study, a fractionation process was developed to separate bark and pith. These two fractions were then used to produce renewable materials for building applications. On the one hand, the bark was used to produce hardboards, with the deoiled seeds acting as natural binder. Such boards are a viable alternative to commercial wood-based panels. On the other hand, the pith was transformed into cohesive and machinable low-density insulation blocks revealing a low thermal conductivity value.

Keywords: amaranth stem; bark; pith; insulation blocks; hardboards

Citation: Evon, P.; de Langalerie, G.; Labonne, L.; Merah, O.; Talou, T.; Ballas, S.; Véronèse, T. Low-Density Insulation Blocks and Hardboards from Amaranth (*Amaranthus cruentus*) Stems, a New Perspective for Building Applications. *Coatings* **2021**, *11*, 349. https://doi.org/10.3390/coatings11030349

Academic Editor: Yong X. Gan

Received: 23 February 2021
Accepted: 15 March 2021
Published: 18 March 2021

Publisher's Note: MDPI stays neutral with regard to jurisdictional claims in published maps and institutional affiliations.

1. Introduction

Amaranthus cruentus is an annual herb native to temperate and tropical regions. It is cultivated for the nutritional properties of its seeds [1–3]. It is a promising raw material as all parts are potentially usable for both food and non-food applications.

The amaranth seed contains a lipid fraction (7–8%), rich in squalene of vegetal origin (5–6%) [4]. Acting as an essential intermediate in the biosynthesis of cholesterol, steroid hormones, and vitamin D for humans, squalene is a triterpene largely used as food supplement or for cosmetics [5]. In particular, squalene is used in pharmaceutical and especially in cosmetic formulations for the treatment of skin disorders because of its moisturizing and protective activity against external agents, e.g., air, light, UV rays, environmental pollution, etc. [6]. It is a good emollient, and the cosmetic industry uses it especially for the manufacture of moisturizers, make-up, lipstick, and hair care products. Besides squalene, its hydrogenated product, squalane, is even often preferred to squalene as it has a greater oxidation stability, due to the absence of double bonds [7].

The amaranth seed also contains a substantial amount of proteins (15%). The latter could be used for bread flours as a protein reinforcement [8], and for their biocidal and antioxidant activities [9]. In addition, they would be potentially usable for their emulsifying capacity (surfactants for food or in cosmetic creams) [10] or for their adhesive properties in the panel industry [11–13].

Lastly, the amaranth seed is particularly rich in starch (up to 55%). After its plasticization/gelatinization (plus the denaturation of proteins) through a thermo-mechano-chemical pre-treatment using the twin-screw extrusion technology, amaranth seed (or cake) could thus be possibly transformed into thermoplastic granules for injection-molding applications [14,15].

During amaranth cultivation, a stem is also produced. Representing up to 80% (w/w) of the plant aerial part, it is composed of a pith fraction in its middle, and a bark (i.e., a woody part, or ligneous fraction) on its periphery. To our knowledge, the scientific literature does not refer to works that have already used the amaranth stem in the manufacture of bio-based materials, particularly for construction applications. Nevertheless, the use of natural fibers (or aggregates) in composites as a replacement for mineral fillers or glass or carbon fibers has many advantages. Especially, composites made from natural by-products are highly recyclable, environmentally friendly, cost-effective, and also safe for society. By way of an example, flax and jute fibers have already shown their full potential for the reinforcement of an acrylic resin, the obtained composites being usable in the automotive sector [16]. In addition, with a pith in its middle and a bark on its periphery, the structure of the amaranth stem is close to that of sunflower, which is already used in the building sector. For these different reasons, it is reasonable to assume that the pith and bark particles from the amaranth stem could also be used for the manufacture of bio-based building composite materials, thus providing real benefits for the near future.

Firstly, pith particles reveal an alveolar structure close to that of expanded polystyrene. The same was true for the pith fraction coming from sunflower stem [17], and this has recently allowed the development of sunflower-based low-density insulation materials for buildings [18–20]. The construction sector generates major environmental impacts, e.g., consumption of non-renewable raw materials, greenhouse gas (GHG) emissions, waste production, etc. [21]. Numerous research projects are therefore currently aimed at developing alternative materials with low environmental costs. In particular, the use of plant co-products in building materials is particularly interesting. Indeed, vegetable aggregates are renewable. They are mainly produced locally, and they also constitute a carbon sink.

In this topic, a low-density insulation block consisting solely of sunflower pith particles and a starch-based glue was recently developed [20]. Its good mechanical resistance allowed machining. In addition, its thermal insulating ability in dry state (32 mW/(m.K) at 25 °C) was equivalent to those of stone wool and expanded polystyrene, therefore corresponding to current standards for heat transfer. However, while the thermal conductivity of expanded polystyrene was not affected by ambient humidity, that of the sunflower-based samples increased by about 15% (i.e., reduced insulating properties), which was explained by the highly hygroscopic nature of these bio-based materials. Conversely, the latter revealed a high water vapor permeability (i.e., five to ten times higher than expanded polystyrene). Their power as water regulators could be useful in houses to regulate indoor humidity variation. For example, the risk of condensation at the interface between different layers of materials could be limited, and this could be useful in the case of the renovation of old buildings [20].

A more recent improvement has been the application of a surface treatment made of glycerol esters to make these sunflower-based insulation blocks more resistant to microbial growth [22]. Even in the presence of the glycerol esters, the potential of the light insulation made of sunflower pith was highlighted, with a 50 kg/m^3 density and a 35 mW/(m.K) thermal conductivity at 25 °C. Its high water vapor permeability was also preserved. The antimicrobial efficiency of glycerol esters was evidenced, contributing to better protection of these bio-based materials from microbial proliferation. At the same time, a non-flammability classification has been surprisingly assigned to the untreated insulation materials.

On its side, the amaranth bark reveals a high amount of lignocellulosic fibers. Such a characteristic could allow its use as a mechanical reinforcement inside biocomposites, in particular injection-moldable materials [23,24] or fiberboards (e.g., hardboards). For this second purpose, the bark pre-treatment through twin-screw extrusion-refining could make this reinforcement capability even more important, thanks to an improvement in the morphological characteristics of the fibers (i.e., increase in their mean aspect ratio, defined as the ratio of the length to the diameter, after refining).

The twin-screw extrusion technology has already proven itself in the processing of biomass, including mechanical pressing of vegetable oil from various oilseeds, continuous liquid/solid extraction of active biomolecules, starch plasticization, protein destructurization and denaturation, compounding of matrix/fiber type thermoplastic granules, production of second-generation bioethanol, etc. [25,26]. Especially, it has also recently been used as an alternative tool for the thermo-mechanical defibration of various agricultural by-products or those resulting from a first agro-industrial transformation such as rice straw [27,28], coriander straw [13,29], oleaginous flax shives [30,31], and sunflower bark [22], prior to hot pressing.

The function of defibration is to break down the lignins that protect cellulose and hemicelluloses. This pre-treatment contributes to increase the fiber-specific surface and to promote their self-adhesion [30]. The applied pre-treatments are most often mechanical, thermal, and/or chemical ones [32]. As a first non-expensive solution, a simple grinding enables the improvement of the particle size profile of micrometric fibers [30,33,34]. Defibration is however much more efficient when thermo-mechanical pulps are produced. Widely used in the paper industry, the thermo-mechanical defibration can be carried out from different pulping processes [35], including digestion plus defibration [36] and steam explosion [37–41]. More original pre-treatment methods involving enzymes have also been considered [42–44]. However, they have the disadvantage of being rather difficult to implement on a large scale.

Twin-screw extrusion, on the other hand, can be more easily deployed in the industry. Above all, it has recently shown its full potential for an efficient thermo-mechanical defibration of lignocellulosic by-products at a moderate cost. It was compared with a digestion plus defibration process from the paper industry for the pre-treatment of rice straw [27]. A significantly reduced cost (i.e., nine times less) was observed for the extrusion-refining process. In parallel, this technology was much more efficient, even with reduced amount of water added (i.e., 0.4–1.0 instead of at least 4.0 in the case of the pulping method), as the length of the refined fibers was better preserved using the extrusion-refining process: 21.2–22.6 for their average aspect ratio instead of 16.3–17.9 in the case of the digestion plus defibration process. With such significantly improved mechanical reinforcement capability, the extrusion-refined rice straw thus enabled the manufacture of fiberboards with particularly promising usage properties [28]. The optimal one was molded after the addition of 8.9% (w/w) commercial Biolignin™ (i.e., a pure and non-deteriorated *Organosolv* lignin extracted from wheat straw using a mixture of acetic acid and formic acid) as a binder. With a density of 1414 kg/m^3, the latter revealed very high bending properties (i.e., strength at break and elastic modulus of 50 MPa and 8.6 GPa, respectively), simultaneously with good water resistance (i.e., thickness swelling and water absorption of 24% and 18%, respectively, after 24 h soaking in water). Such a rice straw-based material could be used as a load-bearing board under high stress.

Another recent study focused on the extrusion-refining of coriander straw [13]. The results obtained during this study confirmed the interest of the twin-screw extrusion pre-treatment with regard to the average aspect ratio of the obtained fibres, which was evaluated at values of between 22.9 and 26.5 depending on the amount of water added during extrusion, as opposed to only 4.5 in the case of a straw that was simply crushed using a hammer mill. During this study, fiberboards combining the straw as a reinforcement and the cake resulting from the extraction of vegetable oil from the coriander seeds as a protein binder were manufactured through hot pressing. All biocomposites made of the extrusion-refined coriander fibers have shown better usage properties, and the optimal 100% coriander-based board was the one containing 40% (w/w) cake. The latter had the next characteristics: 1195 kg/m^3 density, 29.1 MPa flexural strength at break, 3.9 GPa elastic modulus, 24% thickness swelling, and 24% water absorption. Less dense and with lower bending properties than the optimal hardboard developed by Theng et al. from rice straw [28], it still had a water resistance after 24 h immersion of the same order of magnitude, thanks to a heat post-treatment applied after hot pressing. Its bending properties were nonetheless sufficient to use it in dry conditions, for general purposes (e.g., carcasses, domestic flooring, etc.) as well as furniture grade and load-bearing board (e.g., shelving). Additionally, the VOC and carbonyl compound emissions from this hardboard were studied, and no formaldehyde emission was detected [29]. In comparison with most of the wood-based materials available on the market (e.g., medium-density fiberboard (MDF) or chipboard, which have also been tested in that study), it was thus more environmentally and human-health friendly.

A single twin-screw extruder can even be used both as a continuous fiber-refining tool and as a tool for adding continuously an exogenous binder to the previously refined fibers, resulting at the extruder outlet in an homogeneous premix made of reinforcing fibers and a binder, ready to be transformed into cohesive fiberboards through hot pressing [13,31].

In respect with the biorefinery concept of whole plants, this study aimed to develop a continuous mechanical process allowing the fractionation of the amaranth stem into bark and pith, and to investigate the use of those two fractions for building material applications. Two families of innovative bio-based materials were investigated, i.e., bulk (loose fill) insulation and low-density blocks from the pith fraction on the one hand, and hardboards from the bark fraction on the other hand. The use as a natural binder of the solid residue (i.e., the cake) obtained after extraction of vegetable oil from the amaranth seeds using an apolar solvent was also investigated in this study, due to its natural richness in proteins and especially starch.

2. Materials and Methods

2.1. Raw Materials and Natural Binders

The amaranth (*Amaranthus cruentus*) stems and seeds used in this study were from the South-West part of France. They were cultivated near the city of Auch during the summer of 2018. They were supplied by Ovalie Innovation (Auch, France).

The starch-based binder used for producing the low-density insulation blocks had an 85% (w/w) starch content. Supplied by Bostik (Colombes, France) with reference number 28474, it is usually used as a glue for wallpapers.

The amaranth-based binder (AB) used for obtaining the hardboards was prepared from the seeds supplied by Ovalie Innovation. Firstly, the latter were grinded using a Foss (Hilleroed, Denmark) Cyclotec 1093 sample mill fitted with a 1 mm screen. Deoiling was then conducted using a Soxhlet extraction apparatus (Lenz Laborglas, Wertheim, Germany), and cyclohexane as extracting solvent.

2.2. Analytical Methods

Mineral content inside amaranth pith and bark were determined according to the ISO 749:1977 standard [45]. An estimation of the three parietal constituents, i.e., cellulose, hemicelluloses, and lignins, was made from the ADF-NDF (ADF, Acid Detergent Fiber; NDF, Neutral Detergent Fiber) method of Van Soest and Wine [46,47]. The water-soluble components were determined by measuring the mass loss of the test sample after 1 h in boiling water. All determinations were made in triplicate.

2.3. Grinding of Amaranth Stems and Separation between Bark and Pith Particles

The grinding of amaranth stems was done using an Electra (Poudenas, France) Goulu N hammer mill with no grid. The separation between bark and pith particles was then conducted using a vacuum cleaner. The suction step was repeated three times, first on a

Ritec 600 (Signes, France) vibrating sieve shaker and then twice on a rough and inclined conveyor belt.

During the initial sieving stage, the sieve shaker was fitted with two sieves with a mesh of 2 mm and 1 mm, respectively, from top to bottom. Three different fractions made of dense bark particles were thus generated, i.e., fines (<1 mm), and medium (from 1 to 2 mm) and large particles (>2 mm). Additionally, thanks to the vacuum cleaner positioned above the upper sieve, the light pith particles were sucked continuously.

The three bark fractions collected at the level of the sieve shaker were almost pure. On the opposite, the pith one still contained some bark impurities, which were sucked simultaneously with the pith particles. Two additional purification extra-steps were thus required. The latter were conducted using an inclined conveyor belt made of a rough rubber band. A ventilator with an air speed from 1.0 to 2.5 m/s was positioned at 1.3 m from the bottom of the conveyor belt, and the vacuum cleaner at its bottom. The movement of the conveyor belt was at constant speed, from bottom to top.

The pith fraction to be purified was fed at a constant mass flow rate onto the conveyor belt just in front of the ventilator. In doing so, the ventilation opposed the movement of the conveyor belt, and this allowed the pith particles, which were light and rather cylindrical in shape, to roll down to the bottom of the device where they were sucked. The bark particles to be removed, on the other hand, were much denser and had an elongated shape, so that they remained attached to the rough surface of the rubber band. They could then be picked up from the top of the device. In order to obtain a pith fraction with a purity of more than 90% (*w*/*w*), two successive passes of the pith on the conveyor belt were necessary, using an angle of inclination equal to 32° and 23°, respectively.

In Figure 1, a schematic diagram of the fractionation process of the amaranth stems, including their grinding and then their separation between bark and pith particles through three successive suction stages, is presented. This diagram also includes the building materials obtained from each of these two fractions and the equipment used to obtain them.

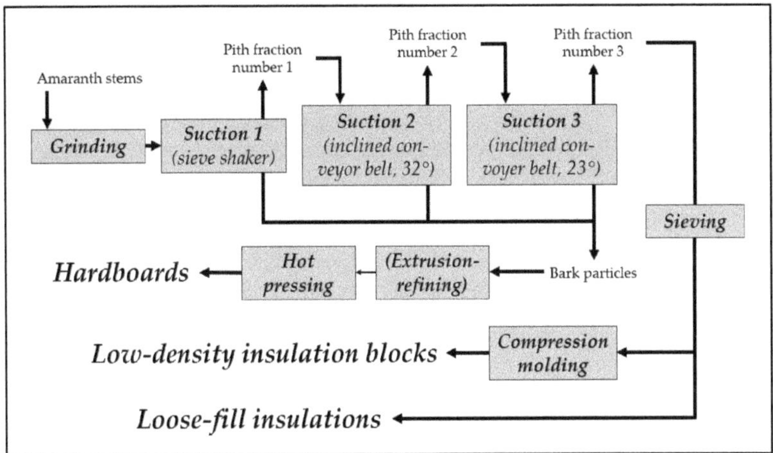

Figure 1. Schematic diagram of the fractionation process of the amaranth stems into bark and pith fractions, and building materials obtained from these fractions.

2.4. Sieving of Pith Particles

To determine the particle size distribution of the pith particles, a Retsch AS 300 (Haan, Germany) vibratory sieve shaker was used. The three implemented sieves had a mesh of 4 mm, 2 mm and 1 mm, respectively, from top to bottom. A sieve acceleration of $1.5\times g$ and a sieving time of 10 min were chosen for sieving.

2.5. Twin-Screw Extrusion-Refining of Fibers from Amaranth Bark

Amaranth bark particles with a diameter of more than 2 mm and a moisture content of $6.0 \pm 0.0\%$ were extrusion-refined in the presence of water using a Clextral (Firminy, France) Evolum HT 53 co-rotating and co-penetrating twin-screw extruder. The screws were 53 mm in diameter (D), and the barrel was 36 D in length with nine successive modules, each 4 D in length.

The thermo-mechanical fiber pre-treatment was inspired by a previous work developed from the coriander straw [13]. Amaranth bark was fed using a weight dosing device in module 1, and water was injected using a piston pump at the end of module 2. In particular, to prevent water from clogging the twin-screw extruder, a filtration module equipped with two filter grids made of 1 mm diameter holes was positioned at the end of the barrel, i.e., in penultimate position. The excess of water was thus removed (in the form of an aqueous filtrate).

The screw profile used was that of Uitterhaegen et al. [13]. In particular, two successive pairs (i.e., 2 D in length) of BB bilobe paddles mounted in a staggered pattern (i.e., with a 90° angle) were positioned in module 5 to ensure intimate mixing and homogenization of solid and liquid phases. In addition, four successive pairs (i.e., 1.5 D in length) of CF1C screws (i.e., conjugated cut-flight, single-flight screws with left-handed pitch) were positioned at the beginning of module 9. Thanks to the intense mechanical shear applied to the raw material in this location, this enabled an efficient extrusion-refining of fibers in amaranth bark.

The temperature profile used was as follows: 25 °C for module 1 (feeding module), 50 °C for module 2, 80 °C for module 3, 100 °C for modules 4 to 7, and 110 °C for module 9 (thermo-mechanical refining module).

The screw rotation speed was 250 rpm. The inlet flow rate of amaranth bark particles was 10 kg/h. Three different inlet flow rates of water were tested (i.e., 10 kg/h, 20 kg/h and 40 kg/h, respectively), corresponding to liquid/solid ratios of 1, 2, and 4, respectively.

2.6. Molding of Insulation Blocks and Hardboards

The compression molding of the insulation blocks was conducted at room temperature inside an aluminium mold with 15 cm sides using a conventional hydraulic press. A pressure of 9 kPa was applied at room temperature for 5 min to the mixture composed of pith particles (bulk test sample volume equal to 2700 cm^3), starch-based binder (10% (w/w) in proportion to the sum of the pith and binder masses), and water (3.7% (w/w) for the binder to water ratio). For the medium size (2–4 mm) of pith particles only, starch-based binder contents of 15%, 20% and 25% (w/w) were also tested.

Once molded, the insulation blocks were dried at 60 °C in a France Etuves (Chelles, France) XL2520 ventilated oven until constant weight (i.e., mass variation less than 0.1% after 24 h). The drying step was conducted with the objective to eliminate the water added to dissolve the binder. As the external starchy binder used was with physical curing, the adhesion was achieved as water evaporated.

The hot pressing of hardboards was conducted using 400 tons capacity Pinette Emidecau Industrie (Chalon-sur-Saône, France) heated hydraulic press. The conditions were standard ones [13,30], i.e., 200 °C for the mold temperature, 20 MPa for the applied pressure, and 5 min for the molding time.

A 150 mm × 150 mm aluminium mold was used to perform both compression molding and hot pressing. The material thickness was 4 cm and around 4.5 mm, respectively, for the insulation blocks and hardboards. All materials were then equilibrated in a climatic chamber at 60% relative humidity (RH) and 25 °C until constant weight (i.e., mass variation less than 0.1% after 24 h). The equilibration lasted around three weeks, after which the materials were analyzed.

2.7. Density Measurements

The bulk density of the pith and bark particles was measured using a 2 L test tube. Once filled with the particles, the latter were weighed in order to determine the bulk density. The tapped density of bark particles, before and after the extrusion-refining step, was determined using a Granuloshop (Chatou, France) Densitap ETD-20 volumenometer. All determinations were carried out in triplicate.

The density of the insulation blocks and hardboards was conducted from the bending test specimens. An electronic digital sliding caliper having a resolution of 0.01 mm was used to measure their thickness, width, and length, allowing the specimen volume to be calculated. In parallel, all the test specimens were also weighed, enabling their density to be determined. All determinations were carried out four times.

2.8. Physical Characterization of the Building Materials

The flexural properties of the insulation blocks and hardboards were measured using the three points bending technique according to the ISO 16978:2003 standard [48]. The universal testing machine used for those tests was an Instron (Norwood, MA, USA) 5500R machine fitted with a 500 N load cell. All test specimens were 3 cm wide, and 12 cm long. The grip separation was 8 cm, and the test speed was 2 mm/min. All determinations were conducted four times.

The thermal conductivity of loose pith particles and insulation blocks was measured at 25 °C using a Neotim (Albi, France) FP2C thermal conductivimeter with a hot wire. A 5% accuracy was obtained for these measurements. During the hot wire test, an electric current through the linear wire generated heat, which resulted in the increase of the temperature of the material tested. The wire length was considered as infinite, and its diameter as negligible. The wire and thermocouple were included in a Kapton probe. During the test, the probe was placed in the middle of the pith particles arranged in bulk, or between two samples of the insulation block. As the material tested was considered as semi-infinite, the heat conduction equation was solved in cylindrical coordinates. This allowed the determination of the thermal conductivity. Then, the thermal resistance was deduced from the thermal conductivity value by considering a 4 cm thickness for the loose pith particles as for the insulation blocks. Before the test, the pith particles were equilibrated in a climatic chamber under the same conditions as those used for the insulation blocks, i.e., 60% RH and 25 °C for three weeks. All determinations were conducted in triplicate.

The thickness swelling of hardboards was measured after 24 h immersion in water according to the ISO 16983:2003 standard [49]. Expressed in mass percentage, their water absorption level was determined at the same time. All determinations were conducted four times.

2.9. Statistical Analyses

Results are expressed as mean value $\pm$ standard deviation. One-way analysis of variance (ANOVA) was used in order to examine the significance of the effect of factors on studied traits. Duncan's multiple range test was used to compare individual means at a 5% probability level.

3. Results

3.1. Separation between Pith and Bark

Immediately after their harvesting, amaranth stems had to be dried (50 °C, 48 h) to facilitate their storage over time. Their structure was then manually studied from ten specimens. They were composed of a light pith fraction in their middle (24% (w/w)), and a bark one, fibrous and rigid, in their periphery (76% (w/w)) (Figure 2a).

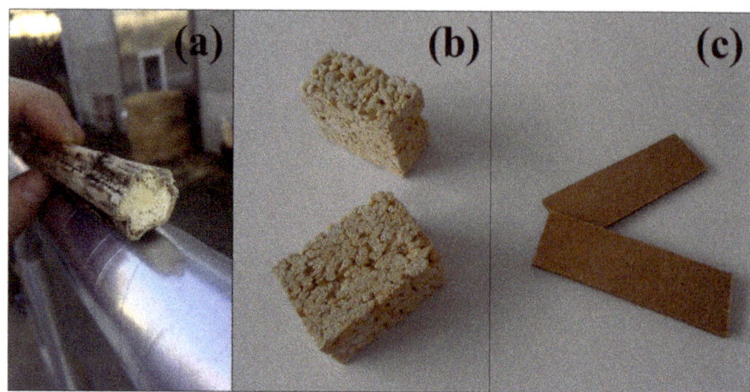

Figure 2. (**a**) Photograph of the amaranth stem (cross section); (**b**) Photograph of the optimal insulation block from the BP pith fraction; (**c**) Photograph of the optimal hardboard (HB6) from extrusion-refined bark.

The structure of the amaranth stem is reminiscent of sunflower or even corn. Nevertheless, the elements of the amaranth stem are not yet valorized in the field of renewable materials. Conversely, pith and bark fractions from sunflower stems have been the subject of recent work to transform them into low-density insulating blocks [22], or rigid [11] and semi-rigid [22] panels, respectively.

Based on the same principle as for sunflower, the amaranth stem fractions could therefore be used as basic elements for valorization in the materials sector. Indeed, the pith could be used as thermal insulation while the bark fraction could be used for the production of dense boards through hot pressing.

The large difference in density between bark and pith particles enabled their continuous mechanical separation through a two-stage process involving a grinding step of the stems, and then a suction one. On the one hand, the grinding step allowed the pith to be separated from the bark particles. On the other hand, during the suction step, the use of a vacuum cleaner resulted in the isolation of the lighter pith particles from those of bark. The suction operation was repeated three times, first on a vibrating sieve shaker and then twice on an inclined conveyor belt.

Table 1 shows the mass balance for each of the three suction stages based on 100 kg of crushed amaranth stem. For its part, Table 2 shows the evolution of the pith purity of the light fractions, sucked after the sieving step (P1) and after each passage on the conveyor belt (P2 and P3), respectively. The pith purity was measured twice by manually separating the residual bark particles from the pith ones. The pith purity was multiplied by more than four thanks to the two conveyor belt purification steps. It was greater than 90% (w/w) after the three consecutive suction steps, which was considered sufficient to produce low-density insulation blocks.

Table 1. Mass balance of the fractionation of 100 kg amaranth stem into bark particles and a sucked fraction containing the pith ones for the three successive suction stages, and mass of the two fractions collected at the end of the separation process.

Fraction	Suction Number 1	Suction Number 2	Suction Number 3	Mass Balance after Suction Number 3
Bark particles (kg)	80.9	11.7	2.6	95.2 [1]
Pith fraction sucked out (kg) [2]	19.1	7.4	4.8	4.8 [3]

[1] At the end of the fractionation process, the bark particles collected after each suction stage were mixed together to form a single fraction of bark particles. [2] The pith fractions sucked out after stages number 1 and 2 constituted the input materials for the following stages, i.e., stages number 2 and 3, respectively. [3] The pith fraction from stage number 3 was the fraction used in this study for the manufacture of the insulating materials.

Table 2. Purity in pith particles of the sucked fraction after each suction stage.

Suction Stage	Suction Number 1	Suction Number 2	Suction Number 3
Purity in pith particles (% (*w/w*))	22.9 ± 0.1	59.0 ± 0.3	91.0 ± 0.2

3.2. Chemical Composition of Pith and Bark, and Physical Characterization of Pith Particles

Table 3 shows the chemical composition of the pith and bark, both from the amaranth stem. While the pith contains a median amount of lignocellulosic fibers (47%), but a lot of minerals and water-soluble components (19% and 31%, respectively), the bark is much richer in lignocellulosic fibers (64%), and it has much lower mineral and especially water-soluble contents. Thus, the bark is indeed the woody part (or ligneous fraction) of the stem.

Table 3. Chemical composition of pith and bark from amaranth stem (% of the dry matter).

Components	Pith	Bark
Minerals	18.9 ± 0.2	9.3 ± 0.3
Cellulose	37.5 ± 0.5	43.6 ± 0.6
Hemicelluloses	12.1 ± 0.1	17.6 ± 0.4
Lignins	9.4 ± 0.2	20.3 ± 0.6
Water-soluble components	31.2 ± 0.2	14.8 ± 0.9

In addition to these previous comments, it should be pointed out here that the sum of the chemical compounds quantified in both pith and bark was greater than 100%. This is explained by the fact that minerals in ionic form as well as some hemicelluloses, especially those with the lowest molecular weights, may be soluble in water. These chemical constituents could then be quantified twice.

Due to their alveolar structure, the pith particles were light, with an estimated bulk density of only 32.2 kg/m^3 (Table 4). Their particle size distribution is also presented in Table 4. Considered as dust (i.e., fines), the particles smaller than 1 mm have not been kept for the manufacture of the low-density insulation blocks. The other three pith fractions generated after sieving (i.e., SP, MP and BP) were also characterized in bulk density (Table 4), and the results obtained revealed significant differences.

Table 4. Particle size distribution of the amaranth pith particles, and bulk density, thermal conductivity and thermal resistance in bulk at 25 °C of all pith particles (mix) and the SP, MP, and BP fractions generated after sieving [1].

Pith Fraction	Distribution (% (*w/w*))	Bulk Density (kg/m^3)	Thermal Conductivity (mW/(m.K))	Thermal Resistance [6] (m^2·K/W)
Mix [2]	100.0	32.2 ± 1.0 [b]	37.0 ± 0.0 [b]	1.081 ± 0.000 [a]
Fines (<1 mm)	15.0	n.d.	n.d.	n.d.
SP [3]	17.8	29.6 ± 0.5 [c]	37.0 ± 0.0 [b]	1.081 ± 0.000 [a]
MP [4]	41.7	32.0 ± 0.2 [b]	39.5 ± 0.7 [a]	1.013 ± 0.018 [b]
BP [5]	25.5	39.4 ± 0.7 [a]	41.0 ± 1.4 [a]	0.976 ± 0.034 [b]

[1] Means in the same column with the same superscript letter (a–c) are not significantly different at $p < 0.05$. [2] All pith fractions. [3] Small particles (1–2 mm). [4] Medium particles (2–4 mm). [5] Big particles (>4 mm). [6] Thermal resistance for a 4 cm thick bed of pith particles. n.d., non-determined.

For its part, Table 5 proposes a distribution in weight and another one in volume for the SP, MP, and BP fractions of pith particles, in proportion to the total weight or to the total volume, respectively, of the mixture made of these three pith fractions. As a reminder, these pith fractions were the three ones used to produce the low-density insulation blocks, on their own or mixed in the right weight proportions. The results in Table 5 show relatively close weight and volume distributions, with the intermediate pith fraction (MP particles) representing for both distributions a value close to 50%, followed by the coarser fraction (BP particles) and then the fraction made up of the smallest particles (SP particles).

Table 5. Distribution in weight (% (w/w)) and distribution in volume (% (v/v)) for the SP, MP, and BP fractions of pith particles, calculated in proportion to the total weight or to the total volume, respectively, of the mixture made of these three pith fractions.

Pith Fraction	Distribution in Weight (% (w/w))	Distribution in Volume (% (v/v))
SP	20.9	23.6
MP	49.1	51.1
BP	30.0	25.3

Thermal conductivity measurements were also performed on each bulk pith sample using the hot wire method. The thermal conductivity values obtained are shown in Table 4. The results revealed that they were of the same order of magnitude as those of a commercial cellulose wadding used as bulk insulation for which the thermal conductivity at 25 °C was between 38 and 44 mW/(m.K).

3.3. Low-Density Insulation Blocks from Pith

Low-density insulation blocks were produced from pith particles through compression molding using standard conditions already implemented in a previous study [20], i.e., 9 kPa for the applied pressure, 5 min for the molding time, and 25 °C for the mold temperature. Once molded, all blocks were placed in a ventilated oven to evaporate the water initially added to dissolve the starchy binder. The adhesion was achieved after complete evaporation, and all the resulting blocks were cohesive enough to be machined.

Blocks having a 10% (w/w) binder content were produced from the three sieved fractions and from the mix made of all pith particles. In addition to these four insulating materials, three other low-density blocks made from the medium pith particles (MP) were also molded using higher binder contents, i.e., 15%, 20%, and 25% (w/w), respectively. All the characteristics of these seven low-density insulation blocks are presented in Table 6. These include density, flexural properties, thermal conductivity, and thermal resistance.

Table 6. Characteristics of the low-density insulation blocks produced from pith particles through compression molding (9 kPa, 5 min, 25 °C) [1].

Pith Particles	Binder Content (%)	Density (kg/m^3)	Flexural Strength (kPa)	Elastic Modulus (kPa)	Thermal Conductivity (mW/(m.K))	Thermal Resistance [2] (m^2·K/W)
Mix	10	131.4 ± 2.4 [a]	59.3 ± 4.5 [b]	510 ± 29 [c]	60.3 ± 1.5 [a]	0.664 ± 0.017 [d]
SP	10	115.8 ± 2.2 [c]	15.8 ± 1.9 [f]	169 ± 21 [g]	57.5 ± 0.6 [b]	0.696 ± 0.007 [c]
MP	10	102.4 ± 5.9 [e]	23.0 ± 3.4 [e]	299 ± 44 [f]	54.8 ± 1.5 [d]	0.731 ± 0.020 [a]
MP	15	108.1 ± 1.9 [d]	29.5 ± 4.5 [d]	343 ± 40 [e]	54.8 ± 0.5 [d]	0.731 ± 0.007 [a]
MP	20	128.9 ± 3.9 [a]	45.3 ± 3.1 [c]	647 ± 28 [a]	56.5 ± 1.3 [bc]	0.708 ± 0.016 [bc]
MP	25	130.9 ± 1.5 [a]	44.7 ± 5.9 [c]	582 ± 15 [b]	60.3 ± 0.5 [a]	0.664 ± 0.006 [d]
BP	10	120.7 ± 1.7 [b]	84.5 ± 8.2 [a]	468 ± 17 [d]	56.0 ± 1.4 [c]	0.715 ± 0.018 [bc]

[1] Means in the same column with the same superscript letter (a–g) are not significantly different at $p < 0.05$. [2] Thermal resistance for a 4-cm thick low-density insulation block.

3.4. Twin-Screw Extrusion-Refining of Bark

During the twin-screw extrusion-refining pre-treatment, the amaranth bark undergoes defibring (i.e., destructuring) due to the mechanical stresses and temperature in the extruder. Three different inlet flow rates of water were tested in this study, i.e., 10 kg/h, 20 kg/h, and 40 kg/h, respectively, thus corresponding to liquid/solid ratios varying from

1.0 to 4.0. The results of the amaranth bark refining treatment in twin-screw extruder are shown in Table 7. As a first result of this treatment, the three extrudates obtained had the form of fluffy materials, with significantly reduced apparent and tapped densities compared to those of the simply ground bark particles (i.e., 147.7 ± 1.3 kg/m^3 and 148.5 ± 4.2 kg/m^3, respectively).

Table 7. Inlet flow rate of water, liquid/solid ratio, outlet flow rate of the filtrate, content of water-soluble components in the filtrate, extraction yield in water-soluble components, and apparent and tapped densities of the extrudate after drying for the three twin-screw extrusion-refining experiments applied to the amaranth bark particles (10 kg/h for their inlet flow rate) [1].

Experiment Number	1	2	3
Inlet flow rate of water (kg/h)	10	20	40
Liquid/solid ratio	1	2	4
Outlet flow rate of the extrudate (kg/h)	13.2	13.9	13.7
Moisture in the extrudate (%)	28.6 ± 0.6 [c]	37.6 ± 0.3 [b]	38.2 ± 0.1 [a]
Water-solubles in the extrudate (% of the dry matter)	13.1 ± 0.1 [a]	10.8 ± 0.6 [b]	7.5 ± 0.1 [c]
Outlet flow rate of the filtrate (kg/h)	0.1	12.7	29.7
Dry matter in the filtrate (%)	1.1 ± 0.0 [c]	5.9 ± 0.1 [a]	3.2 ± 0.3 [b]
Water evaporated (%) [2]	67.0	17.0	16.7
Extraction yield in water-solubles (%)	11.7	32.9	54.2
Apparent density of extrudate (kg/m^3)	63.3 ± 1.6 [b]	69.9 ± 2.0 [a]	63.5 ± 0.6 [b]
Tapped density of extrudate (kg/m^3)	70.6 ± 1.3 [b]	74.5 ± 1.9 [a]	71.7 ± 1.3 [b]

[1] Means in the same line with the same superscript letter (a–c) are not significantly different at $p < 0.05$. [2] The difference between the cumulative inlet flow rate (bark particles plus water) and the cumulative outlet one (extrudate plus filtrate) can be explained by the loss of part of the added water by evaporation, at the level of the filtration module and at the outlet of the barrel.

When the quantity of water increased, this facilitated the transport of the solid material whose residence time in the twin-screw extruder decreased. Thus, the extrudate obtained has undergone less shearing, and its fiber length was therefore better preserved. This was previously evidenced in the case of rice straw [27], and the same was true in this study. Nonetheless, the values obtained for both apparent and tapped densities of the extrusion-refined fibers remained in the same order of magnitude for the three tested liquid/solid ratios.

Looking at the content in water-solubles remaining in the extrudate, the values obtained were all three significantly different. With a higher liquid/solid ratio, the flow rate of the filtrate collected at the penultimate module increased, also bringing with it more water-soluble components initially present in the bark. The extrudate obtained from the higher liquid/solid ratio (i.e., 4.0) was therefore more depleted in water-soluble components. From the extrusion-refining data in Table 7, the flow rate of water-soluble compounds extracted from the bark and contained in the filtrate could be calculated. These extracted water-soluble compounds were then expressed as a percentage of the incoming water-solubles in bark, for the three liquid/solid ratios tested (i.e., 1.0, 2.0 and 4.0). The results were 11.7%, 32.9%, and 54.2%, respectively. It was therefore well observed that the extraction yield in water-soluble components increased with the amount of water added.

To conclude, although three liquid/solid ratios have been tested in the twin-screw extruder, only the extrudate produced using the liquid/solid ratio of 4.0 has been used in the production of hardboards through hot pressing. Indeed, due to the higher amount of water added during the thermo-mechanical defibration of the amaranth bark particles in the twin-screw extruder, the length of the fiber bundles was better preserved.

3.5. Hardboards from Bark

Hardboards were produced through hot pressing from ground (GB) or extrusion-refined (ERB) amaranth bark particles. The molding conditions used were as follows:

200 °C for the mold temperature, 20 MPa for the applied pressure, and 5 min for the molding time. In particular, the 200 °C mold temperature was used as optimal temperature to ensure an efficient mobilization of the internal binders inside bark, i.e., free sugars, hemicelluloses and lignins [13,30]. Thanks to the adhesive ability of these chemicals, cohesive hardboards were obtained without exogenous binder.

An improvement in the molding process consisted in adding to the bark particles an amaranth-based binder (AB). This exogenous binder had the form of grinded and deoiled amaranth seeds, and it was the starch and to a lesser extent, proteins contained in AB that have given it its aptitude for adhesion.

All the characteristics of the six hardboards produced (HB1 to HB6) are presented in Table 8. These include density, flexural properties and water resistance. In addition, these characteristics were also compared in Table 8 with those of two commercial wood-based materials, i.e., MDF and chipboard (CH).

Table 8. Density, flexural properties, and water resistance of hardboards (HB) produced from ground and extrusion-refined barks (GB and ERB, respectively) through hot pressing (200 °C mold temperature, 20 MPa applied pressure, and 5 min molding time) [1], and comparison with the properties of two commercial wood-based materials (i.e., MDF and chipboard (CH)).

Hardboard Number	Bark Form	AB (%)	Density (kg/m^3)	Flexural Strength (MPa)	Elastic Modulus (GPa)	Thickness Swelling (%)	Water Absorption (%)
HB1	GB	0	1234 ± 28 [a]	13.2 ± 2.4 [d]	2.1 ± 0.4 [d]	143 ± 11 [a]	124 ± 8 [a]
HB2	GB	20	1248 ± 34 [a]	18.8 ± 1.7 [c]	2.7 ± 0.4 [bc]	136 ± 10 [a]	124 ± 7 [a]
HB3	GB	40	1242 ± 31 [a]	18.5 ± 2.3 [c]	2.1 ± 0.4 [d]	132 ± 20 [a]	132 ± 11 [a]
HB4	ERB	0	1244 ± 35 [a]	19.8 ± 1.6 [c]	2.4 ± 0.2 [cd]	71 ± 4 [b]	56 ± 2 [b]
HB5	ERB	10	1279 ± 47 [a]	32.4 ± 2.6 [b]	2.9 ± 0.3 [b]	77 ± 4 [b]	60 ± 1 [b]
HB6	ERB	20	1318 ± 47 [a]	35.8 ± 2.9 [a]	4.2 ± 0.2 [a]	87 ± 5 [b]	69 ± 2 [b]
MDF [29]	–	–	589 ± 4	20.7 ± 2.1	1.5 ± 0.1	20 ± 1	51 ± 5
CH [29]	–	–	710 ± 7	10.2 ± 1.5	1.4 ± 0.1	25 ± 1	58 ± 3

[1] Means in the same column with the same superscript letter (a–d) are not significantly different at $p < 0.05$.

4. Discussion

4.1. Physical Characterization of Pith Particles

Once isolated, pith particles were sieved, resulting in the next particle size distribution: 18% for small particles (SP) (from 1 to 2 mm), 42% for medium particles (MP) (from 2 to 4 mm), and 25% for big particles (BP) (>4 mm) (Table 4). Fines (<1 mm) represented 15% (w/w) but they were not used in the present study. All pith particles revealed an alveolar structure, and this resulted in low bulk densities, e.g., only 30 kg/m^3 for the smaller particle size (SP).

Despite the fact that small pith particles had fewer inter-particle voids, due to a better arrangement (i.e., stacking) of the particles in relation to each other, it was these smaller particles (SP) that revealed the lowest bulk density value (Table 4). This could be explained by the fact that the larger ones (BP) and, to a lesser extent, the medium ones (MP) still contained some long fibers from the bark fraction after the sieving step of the pith particles. For the BP fraction, the decrease in bulk density that should have been observed due to more inter-particle voids was therefore largely compensated by the presence of these denser bark fibers, which contributed to increase the density of the BP pith fraction. Conversely, the smaller particles were too small to contain residual bark fibers after sieving.

As seen previously with bulk density, the thermal conductivity at 25 °C of the bulk pith therefore also increased with the particle size, with the residual bark particles present in the BP fraction being denser and above all more conductive than the pith particles (Table 4).

The amaranth pith thus appears as a promising bulk raw material for the thermal insulation of buildings, as previously observed for sunflower pith [20,22]. This makes it possible to position the pith or one of its sieved fractions, especially the smaller one (i.e., the SP fraction) for which the thermal conductivity and thermal resistance are significantly different from the two others, in the building insulation market. For example, it could be blown into the attic of houses or used as a filling for the interior partitions. However, it remains to be seen how the pith particles will behave over time. If the pith is compacted, it could become denser over the years and its thermal conductivity will increase. In the same way, for long-term use, it will also be necessary to judge the pith's behaviour to fire and its ability to resist fungi.

For future work, the coating of the pith particles by a hydrophobing agent (e.g., hydrogenated oils, vegetable oil derivatives, etc.) and a fireproofing product would improve their water and fire resistances, respectively. In the same way, glycerol esters could be a bio-based solution favourable to render the pith particles more resistant to microbial growth when coated at their surface [22].

4.2. Low-Density Insulation Blocks from Pith

Cohesive low-density insulation blocks were produced through compression molding thanks to the addition of the starchy binder. A progressive increase in the density of the blocks made of MP particles was observed with the binder content (Table 6). In parallel, the bending performance of the blocks improved since a larger binder quantity allowed the pith particles to be better impregnated with the starchy glue. Logically, the overall cohesion of the block was thus progressively improved. However, this densification resulted in a lower internal porosity inside the blocks, which then had a higher and higher thermal conductivity and therefore a lower and lower thermal resistance.

As the MP-based blocks remained machinable even with only 10% (w/w) binder, this content was retained for the other pith fractions (Figure 3) since it was expected to give to the insulating blocks a better thermal insulation performance. For such binder content added, the densest and most conductive insulating block was the one made from the mix of pith particles, which could be explained by the fact that this fraction still contained about 9% (w/w) of bark impurities (Table 2). For the blocks made from the sieved fractions, the values obtained for the density and thermal conductivity were significantly different. A significant reduction in the density and thermal conductivity was observed with increasing particle size when comparing the materials made from the SP and MP fractions, respectively. The most likely reason for such a result was the presence of more inter-particle voids in the case of the block made from the medium size particles. Conversely, and as it was already observed for the bulk pith particles, a significant increase in the density and, to a lesser extent, in thermal conductivity was obtained for the block made from the bigger particles. This was evidently due to the substantial amount of dense bark impurities remaining inside the BP fraction after sieving.

For this 10% (w/w) binder content, the flexural performance of the insulating blocks made from the sieved fractions revealed significant differences, and it increased with the size of the pith particles used. Indeed, the larger the particles were, the smaller was their cumulative surface area. Thus, with the pith particles from the BP fraction, more starch-based adhesive was present on their surface, which resulted in a better bonding of the particles to each other and therefore better mechanical performance in bending of the corresponding block. The flexural strength of the block originating from the mix of all the pith particles was also significantly different and logically median, i.e., situated between those of the MP and BP-based blocks.

For the optimal 10% (w/w) binder content, the best compromise between flexural and heat insulation properties was obtained using big particles (>4 mm) (Table 6). Light and insulating, this optimal BP-based block (Figures 2b and 3b) especially preserved very good machinability, and it could thus be positioned at all levels of the buildings, e.g., walls, floors, roofs, etc. However, its thermal conductivity at 25 °C (56 mW/(m.K)) was higher to those of sunflower-based insulation panels [20,22] and especially commercial expanded polystyrene [20]: 38–41 mW/(m.K) and only 32 mW/(m.K), respectively. These results made it a slightly less efficient insulator than the two other materials mentioned above. Indeed, based on the thermal resistance of these three types of insulating materials, defined as the ratio of the material thickness to its thermal conductivity, the thickness of the amaranth-based optimal block should be increased by 37–47% or by up to 75%, respectively, to reach the same thermal insulating capacity as sunflower-based low-density blocks or polystyrene.

Figure 3. (**a**) Photograph of the low-density insulation block made from the MP sieved fraction of pith particles; (**b**) Photograph of the low-density insulation block made from the BP sieved fraction of pith particles; (**c**) Photograph of the low-density insulation block made from the SP sieved fraction of pith particles; (**d**) Photograph of the low-density insulation block made from the mix of the SP, MP, and BP pith particles. All the insulation blocks have a 10% (*w*/*w*) starchy binder content.

However, these results are only preliminary, and some improvements are still possible. On the one hand, a better purification of the BP pith particles, which would then contain a lower proportion of bark impurities, should lead to a reduction in the thermal conductivity of the obtained insulating block. On the other hand, an adaptation of the compression molding conditions should also improve the insulating capacity of the amaranth-based block. For future work, its water vapor permeability will also have to be determined. As for the low-density blocks made from sunflower pith [20,22], it should be quite high, which could also make this insulating material a promising water regulator.

As for the pith particles used in bulk, additional developments will also be required to improve the durability of the optimal low-density insulation block before proposing it to the market. A reduction in its water sensitivity could be achieved thanks to the replacement of the starchy binder by another polysaccharide binder with physical curing, e.g., alginates and especially *Citrus* pectins and chitosan [50]. It could be also achieved thanks to its coating by hydrophobing agents even if its ability as a water regulator may be partially impaired after such post-treatment. The addition of a fireproofing product to the formulation would also render the amaranth-based block more resistant to fire. Lastly, glycerol esters appear as a promising renewable solution to protect it against microbial growth [22].

4.3. Hardboards from Bark

The high lignocellulose content for bark (43.6% cellulose, 17.6% hemicelluloses, and 20.3% lignins) contributed to its potential for producing hardboards (HB), i.e., dense fiberboards, through hot pressing. Cohesive hardboards were obtained with no external binder added (HB1 and HB4) (Table 8), due to the adhesive ability of some chemicals inside bark (i.e., free sugars, hemicelluloses, and lignins). Both HB1 and HB4 boards could thus be considered as promising binderless fiberboards.

Especially, the HB4 hardboard obtained from a previously extrusion-refined bark (ERB), using water at a 4 liquid/solid ratio, was much more mechanically resistant (+50% for flexural strength) than that from ground bark (GB). This was already observed for coriander straw [13], and for shives from oleaginous flax [30]. The extrusion-refining pre-treatment contributed to a much more favorable fiber morphology (large increase in their mean aspect ratio), just as to an efficient separation between cellulose, hemicelluloses, and lignins inside the extruded material, thus facilitating the mobilization of internal binders during hot pressing.

An additional improvement in the bending properties was obtained by adding to the bark grinded and deoiled amaranth seeds. Here, because of its richness in proteins and especially starch, this AB amaranth-based binder was used as a natural external binder, and its binding ability was evidenced for the ground bark, just as for the extrusion-refined one. In addition, especially for the ERB fibrous reinforcement, the more the amaranth-based binder added, the more the flexural strength and the more the elastic modulus. The best flexural properties were obtained when adding 20% AB to ERB (Table 8). Such amount of AB added allowed the ERB fibers to stick well together.

From the point of view of sensitivity to water of the hardboards produced, using the extrusion-refined bark divided the thickness swelling and water absorption values by up to 2 for a panel without exogenous binder, and by up to 1.5 for a panel hot pressed with 20% amaranth-based binder added (Table 8). The interest of the preliminary defibration of the amaranth bark in a twin-screw extruder regarding the water resistance of the obtained panels is thus indisputable.

The optimal hardboard (HB6) (Figure 2) has a 36 MPa flexural strength, which is significantly different from all the other strength values in the present study. This value is also 23% higher than that of another fiberboard recently developed from coriander (29 MPa) [13]. It is also much higher than the flexural strengths of two commercial wood-based materials tested in a previous study [29], i.e., MDF and chipboard: 73% and up to 251%, respectively (Table 8). HB6 board is thus undoubtedly a viable, sustainable alternative for replacing current building wood-based materials like plywood, chipboard, OSB, MDF, etc. According to the NF EN 312 standard [51], it is already usable for interior fittings (including furniture) in a dry environment (P2 type board). In addition to furniture, its possible uses in buildings include interior partitions, underlays for floors, panels for suspended ceilings, etc. However, its fire resistance will have to be characterized before using it inside houses. The addition of a flame retardant to the formulation prior to hot pressing may be considered in the event of poor fire resistance.

The optimal board nevertheless reveals a greater sensitivity to water after 24 h immersion than commercial MDF or chipboard type panels (Table 8). With much improved water resistance (10% max thickness swelling), it could also be used as a P7 type board, i.e., a board working under high stress, used in humid environment. For future work, multiple solutions could be investigated to improve the water resistance. Firstly, a reduction in thickness swelling could be attained thanks to a heat post-treatment. This was already evidenced for two types of renewable fiberboards, i.e., one made of coriander straw as reinforcement and press cake from seeds as binder [13], and another made of oleaginous flax shives as reinforcement and plasticized linseed cake as binder [31]. At the end of this post-treatment, an increase in the flexural strength was even observed at the same time as the reduction in thickness swelling for these two innovative materials.

In order to meet the same objective of increasing water resistance, other treatments may be tested after hot pressing, e.g., coating, chemical, or steam treatment [30]. In particular, the application of a coating based on vegetable oils or their derivatives could be a promising renewable solution to improve the stability in the board dimension after soaking in water. Among these bio-sourced additives, drying oils such as linseed oil or hydrogenated oils that have the property of being solid at room temperature appear to be good candidates.

For future industrial process intensification, one single extruder pass would allow the continuous production of a premix ready to be molded through hot pressing, and associating the refined fibers from amaranth bark as mechanical reinforcement, the amaranth-based binder and even water-repellent and/or flame-retardant additives [31]. Indeed, in addition to its ability evidenced in this study to refine plant fibers, the twin-screw extrusion technology is also known for its particularly intense mixing ability [25,26]. From the same machine, the amaranth bark could thus be extrusion-refined in the presence of water in the first half of the screw profile. Then, the deoiled amaranth seeds and any additives could be added in its middle thanks to a side feeder. Finally, the second half of the screw profile could be used for the intimate mixing of all these components with each other. Using the as-described combined process, the premix would be produced at a reduced cost before being transformed into hardboards.

5. Conclusions

In this study, the amaranth stem was used to develop innovative building materials. Representing up to 80% (w/w) of the plant aerial part, the stem is made of a pith in its middle and a bark on its periphery. Firstly, a fractionation process of the stem into these two different fractions was proposed. Secondly, bio-sourced composite materials were manufactured, i.e., thermal insulating materials made from pith (in bulk or in the form of low-density cohesive blocks), and hardboards from bark. All these materials are viable alternatives to some of the commercial materials nowadays used in buildings.

For future work, the scaling-up of the fractionation process of the amaranth stems into pith and bark fractions will be an important step for its subsequent use at an industrial scale. It should allow a better profitability of this crucial step for the process. Additional developments will also have to be conducted to improve the global performance of the amaranth-based buildings materials. In particular, various post-treatments (e.g., coating, cooking, etc.) should reduce their water sensitivity on the one hand and improve their resistance to microbial growth on the other hand. The evaluation of their fire resistance and, if necessary, the addition of fireproofing products should also be considered before they are used inside houses. Lastly, as it is expected that the low-density insulation blocks from pith will be promising water regulators, their water vapour permeability will also have to be determined. All these developments will undoubtedly facilitate the subsequent introduction of these different bio-sourced building materials on the market.

Author Contributions: Conceptualization, P.E.; methodology, P.E. and L.L.; validation, P.E., O.M., and T.T.; formal analysis, P.E. and O.M.; investigation, G.d.L. and L.L.; resources, S.B. and T.V.; writing—original draft preparation, P.E. and G.d.L.; writing—review and editing, P.E. and O.M.; supervision, P.E.; project administration, P.E. All authors have read and agreed to the published version of the manuscript.

Funding: This research received no external funding.

Institutional Review Board Statement: Not Applicable.

Informed Consent Statement: Not Applicable.

Data Availability Statement: Data sharing is not applicable.

Conflicts of Interest: The authors declare no conflict of interest.

References

1. Kachiguma, N.K.; Mwase, W.; Maliro, M.; Damaliphetsa, A. Chemical and mineral composition of amaranth (*Amaranthus* L.) species collected from central Malawi. *J. Food Res.* **2015**, *4*, 92–102. [CrossRef]
2. Ogrodowska, D.; Zadernowski, R.; Czaplicki, S.; Derewiaka, D.; Wronowska, B. Amaranth seeds and products—The source of bioactive compounds. *Pol. J. Food Nutr. Sci.* **2014**, *64*, 165–170. [CrossRef]
3. Rastogi, A.; Shukla, S. Amaranth: A new millennium crop of nutraceutical values. *Crit. Rev. Food Sci. Nutr.* **2013**, *53*, 109–125. [CrossRef] [PubMed]
4. León-Camacho, M.; Garcia-González, D.L.; Aparicio, R. A detailed and comprehensive study of amaranth (*Amaranthus cruentus* L.) oil fatty profile. *Eur. Food Res. Technol.* **2001**, *213*, 349–355. [CrossRef]
5. Ott, C.; Lacatusu, I.; Badea, G.; Grafu, I.A.; Istrati, D.; Babeanu, N.; Stan, R.; Badea, N.; Meghea, A. Exploitation of amaranth oil fractions enriched in squalene for dual delivery of hydrophilic and lipophilic actives. *Ind. Crop. Prod.* **2015**, *77*, 342–352. [CrossRef]
6. Huang, Z.R.; Lin, Y.K.; Fang, J.Y. Biological and pharmacological activities of squalene and related compounds: Potential uses in cosmetic dermatology. *Molecules* **2009**, *14*, 540–554. [CrossRef]
7. Lidgate, D.M.; Byars, N.E. Development of an emulsion-based muramyl dipeptide adjuvant formulation for vaccines. In *Vaccine Design, the Subunit and Adjuvant Approach*; Powell, M.F., Newman, M.J., Eds.; Springer Nature: Lausanne, Switzerland, 1995; pp. 313–324.
8. Sayed Ahmad, B. Étude de l'agroraffinage des graines d'Apiaceae, Lamiaceae et Chenopodiaceae pour la production de molécules bio-Sourcées en vue d'application en industrie cosmétique. Ph.D. Thesis, Institut National Polytechnique de Toulouse, Université de Toulouse, Toulouse, France, 2018.
9. Caselato-Sousa, V.M.; Amaya-Farfán, J. State of knowledge on amaranth grain: A comprehensive review. *J. Food Sci.* **2012**, *77*, 93–104. [CrossRef]
10. Evon, P.; Vandenbossche, V.; Labonne, L.; Vinet, J.; Pontalier, P.Y.; Rigal, L. The thermo-mechano-chemical twin-screw reactor, a new perspective for the biorefinery of sunflower whole plant: Aqueous extraction of oil and other biopolymers, and production of biodegradable fiberboards from solid raffinate. *Oilseeds Fats Crop. Lipids* **2016**, *23*, D505. [CrossRef]
11. Evon, P.; Amalia Kartika, I.; Rigal, L. New renewable and biodegradable particleboards from jatropha press cakes. *J. Renew. Mater.* **2014**, *2*, 52–65. [CrossRef]
12. Evon, P.; Vinet, J.; Labonne, L.; Rigal, L. Influence of thermo-pressing conditions on mechanical properties of biodegradable fiberboards made from a deoiled sunflower cake. *Ind. Crop. Prod.* **2015**, *65*, 117–126. [CrossRef]
13. Uitterhaegen, E.; Labonne, L.; Merah, O.; Talou, T.; Ballas, S.; Véronèse, T.; Evon, P. Impact of thermomechanical fiber pre-treatment using twin-screw extrusion on the production and properties of renewable binderless coriander fiberboards. *Int. J. Mol. Sci.* **2017**, *18*, 1539. [CrossRef] [PubMed]
14. Chabrat, E.; Abdillahi, H.; Rouilly, A.; Rigal, L. Influence of citric acid and water on thermoplastic wheat flour/poly(lactic acid) blends. I: Thermal, mechanical and morphological properties. *Ind. Crop. Prod.* **2012**, *37*, 238–246. [CrossRef]
15. Abdillahi, H.; Chabrat, E.; Rouilly, A.; Rigal, L. Influence of citric acid on thermoplastic wheat flour/poly(lactic acid) blends. II. Barrier properties and water vapor sorption isotherms. *Ind. Crop. Prod.* **2013**, *50*, 104–111. [CrossRef]
16. Samal, S.; Stuchlík, M.; Petrikova, I. Thermal behavior of flax and jute reinforced in matrix acrylic composite. *J. Anal. Calorim.* **2018**, *131*, 1035–1040. [CrossRef]
17. Maréchal, V.; Rigal, L. Characterization of by-products of sunflower culture: Commercial applications for stalks and heads. *Ind. Crop. Prod.* **1999**, *10*, 185–200. [CrossRef]
18. Vandenbossche, V.; Rigal, L.; Saiah, R.; Perrin, B. New agro-materials with thermal insulation properties. In Proceedings of the 18th International Sunflower Conference, Mar del Plata, Argentina, 27 February–1 March 2012; pp. 949–954.
19. Mati-Baouche, N.; de Baynast, H.; Lebert, A.; Sun, S.; Lopez-Mingo, C.J.S.; Leclaire, P.; Michaud, P. Mechanical, thermal and acoustical characterizations of an insulating bio-based composite made from sunflower stalks particles and chitosan. *Ind. Crop. Prod.* **2014**, *58*, 244–250. [CrossRef]
20. Sabathier, V.; Ahmed Maaloum, A.; Magniont, C.; Evon, P.; Labonne, L. Développement d'un panneau isolant 100% biosourcé à base de moelle de tournesol. In Proceedings of the National Congress of Research in IUT (CNRIUT'2017), Auxerre, France, 4–5 June 2017.
21. Buyle-Bodin, F.; Blanpain, O. Analyse du cycle de vie des ouvrages en béton. In *Annales du Bâtiment et des Travaux Publics*, 6th ed.; Editions ESKA: Paris, France, 2002; pp. 47–54.
22. Verdier, T.; Balthazard, L.; Montibus, M.; Magniont, C.; Evon, P.; Bertron, A. Using glycerol esters to prevent microbial growth on sunflower-based insulation panels. *Proc. Inst. Civ. Eng. Constr. Mater.* **2020**. [CrossRef]
23. Gamon, G.; Evon, P.; Rigal, L. Twin-screw extrusion impact on natural fibre morphology and material properties in poly(lactic acid) based biocomposites. *Ind. Crop. Prod.* **2013**, *46*, 173–185. [CrossRef]
24. Uitterhaegen, E.; Parinet, J.; Mérian, T.; Ballas, S.; Véronèse, T.; Merah, O.; Talou, T.; Labonne, L.; Stevens, C.V.; Chabert, F.; et al. Performance, durability and recycling of thermoplastic biocomposites reinforced with coriander straw. *Compos. Part A Appl. Sci. Manuf.* **2018**, *113*, 254–263. [CrossRef]

25. Evon, P.; Vandenbossche, V.; Candy, L.; Pontalier, P.Y.; Rouilly, A. Twin-screw extrusion: A key technology for the biorefinery. In *Biomass Extrusion and Reaction Technologies: Principles to Practices and Future Potential*, 1st ed.; Ayoub, A., Lucia, L., Eds.; ACS Symposium Series, eBooks; American Chemical Society: Washington, DC, USA, 2018; Volume 1304, pp. 25–44.

26. Vandenbossche, V.; Candy, L.; Evon, P.; Rouilly, A.; Pontalier, P.Y. The twin-screw extrusion technology, an innovative technique for the green extraction of food products. In *Green Food Processing Techniques: Preservation, Transformation and Extraction*, 1st ed.; Chemat, F., Vorobiev, E., Eds.; Academic Press, Elsevier: London, UK, 2019; Volume 10, pp. 289–314.

27. Theng, D.; Arbat, G.; Delgado-Aguilar, M.; Ngo, B.; Labonne, L.; Evon, P.; Mutjé, P. Comparison between two different pretreatment technologies of rice straw fibers prior to fiberboard manufacturing: Twin-screw extrusion and digestion plus defibration. *Ind. Crop. Prod.* **2017**, *107*, 184–197. [CrossRef]

28. Theng, D.; Arbat, G.; Delgado-Aguilar, M.; Ngo, B.; Labonne, L.; Mutjé, P.; Evon, P. Production of fiberboards from rice straw thermo-mechanical extrudates using thermopressing: Influence of fiber morphology, water addition and lignin content. *Eur. J. Wood Wood Prod.* **2019**, *77*, 15–32. [CrossRef]

29. Simon, V.; Uitterhaegen, E.; Robillard, A.; Ballas, S.; Véronèse, T.; Vilarem, G.; Merah, O.; Talou, T.; Evon, P. VOC and carbonyl compound emissions of a fiberboard resulting from a coriander biorefinery: Comparison with two commercial wood-based building materials. *Environ. Sci. Pollut. Res. Int.* **2020**, *27*, 16121–16133. [CrossRef] [PubMed]

30. Evon, P.; Barthod-Malat, B.; Grégoire, M.; Vaca-Medina, G.; Labonne, L.; Ballas, S.; Véronèse, T.; Ouagne, P. Production of fiberboards from shives collected after continuous fiber mechanical extraction from oleaginous flax. *J. Nat. Fibers* **2018**, *16*, 1–17. [CrossRef]

31. Evon, P.; Labonne, L.; Ullah Khan, S.; Ouagne, P.; Pontalier, P.Y.; Rouilly, A. Twin-screw extrusion process to produce renewable fiberboards. *J. Vis. Exp.* **2021**, *167*, e62072. [CrossRef]

32. Zhang, D.; Zhang, A.; Xue, L. A review of preparation of binderless fiberboards and its self-bonding mechanism. *Wood Sci. Technol.* **2015**, *49*, 661–679. [CrossRef]

33. Okuda, N.; Sato, M. Manufacture and mechanical properties of binderless boards from kenaf core. *J. Wood Sci.* **2004**, *50*, 53–61. [CrossRef]

34. Velásquez, J.A.; Ferrando, F.; Salvadó, J. Binderless fiberboard from steam exploded miscanthus sinensis: The effect of a grinding process. *Holz Roh. Werkst.* **2002**, *60*, 297–302. [CrossRef]

35. Migneault, S.; Koubaa, A.; Nadji, H.; Riedl, B.; Zhang, S.Y.T.; Deng, J. Medium-density fiberboard produced using pulp and paper sludge from different pulping processes. *Wood Fiber Sci.* **2010**, *42*, 292–303.

36. Theng, D.; Arbat, G.; Delgado-Aguilar, M.; Vilaseca, F.; Ngo, B.; Mutjé, P. All-lignocellulosic fiberboard from corn biomass and cellulose nanofibers. *Ind. Crop. Prod.* **2015**, *76*, 166–173. [CrossRef]

37. Velásquez, J.A.; Ferrando, F.; Farriol, X.; Salvadó, J. Binderless fiberboard from steam exploded miscanthus sinensis. *Wood Sci. Technol.* **2003**, *37*, 269–278. [CrossRef]

38. Xu, J.; Widyorini, R.; Yamauchi, H.; Kawai, S. Development of binderless fiberboard from kenaf core. *J. Wood Sci.* **2006**, *52*, 236–243. [CrossRef]

39. Quintana, G.; Velásquez, J.; Betancourt, S.; Gañán, P. Binderless fiberboard from steam exploded banana bunch. *Ind. Crop. Prod.* **2009**, *29*, 60–66. [CrossRef]

40. Mancera, C.; El Mansouri, N.E.; Vilaseca, F.; Ferrando, F.; Salvado, J. The effect of lignin as a natural adhesive on the physico-mechanical properties of *Vitis vinifera* fiberboards. *BioResources* **2011**, *6*, 2851–2860. [CrossRef]

41. Mancera, C.; El Mansouri, N.E.; Pelach, M.A.; Francesc, F.; Salvadó, J. Feasibility of incorporating treated lignins in fiberboards made from agricultural waste. *Waste Manag.* **2012**, *32*, 1962–1967. [CrossRef] [PubMed]

42. Felby, C.; Pedersen, L.S.; Nielsen, B.R. Enhanced auto adhesion of wood fibers using phenol oxidases. *Holzforschung* **1997**, *51*, 281–286. [CrossRef]

43. Felby, C.; Hassingboe, J.; Lund, M. Pilot-scale production of fiberboards made by laccase oxidized wood fibers: Board properties and evidence for cross-linking of lignin. *Enzym. Microb. Technol.* **2002**, *31*, 736–741. [CrossRef]

44. Felby, C.; Thygesen, L.G.; Sanadi, A.; Barsberg, S. Native lignin for bonding of fiber boards: Evaluation of bonding mechanisms in boards made from laccase-treated fibers of beech (*Fagus sylvatica*). *Ind. Crop. Prod.* **2004**, *20*, 181–189. [CrossRef]

45. ISO. *ISO 749:1977, Oilseed Residues—Determination of Total Ash*; International Organization for Standardization: Geneva, Switzerland, 1977.

46. Van Soest, P.J.; Wine, R.H. Use of detergents in the analysis of fibrous feeds. Part iv. Determination of plant cell-wall constituents. *J. Assoc. Off. Agric. Chem.* **1967**, *50*, 50–55.

47. Van Soest, P.J.; Wine, R.H. Determination of lignin and cellulose in acid-detergent fiber with permanganate. *J. Assoc. Off. Agric. Chem.* **1968**, *51*, 780–785. [CrossRef]

48. ISO. *ISO 16978:2003, Wood-Based Panels—Determination of Modulus of Elasticity in Bending and of Bending Strength*; International Organization for Standardization: Geneva, Switzerland, 2003.

49. ISO. *ISO 16983:2003, Wood-Based Panels—Determination of Swelling in Thickness after Immersion in Water*; International Organization for Standardization: Geneva, Switzerland, 2003.

50. Ampe, C.; Labonne, L.; Magniont, C.; Laborel-Préneron, A.; Evon, P. Thermal insulation blocks made of sunflower pith particles and polysaccharide-based binders: Influence of binder type and content on their characteristics. In Proceedings of the 4th International Conference on Bio-Based Building Materials (ICBBM 2021), Barcelona, Spain, 16–18 June 2021.
51. AFNOR. *NF EN 312 (2010-11), Particleboards—Specifications*; Association Française de Normalisation: La Plaine Saint-Denis, France, 2010.

Article

Contribution of Different Pretreatments to the Thermal Stability and UV Resistance Performance of Cellulose Nanofiber Films

Lianxin Luo [1,2], Xuchong Wang [1,2], Sheng Zhang [1,2], Xiaojun Yuan [1,2], Mingfu Li [3,4,*] and Shuangfei Wang [1,2,*]

1   College of Light Industry and Food Engineering, Guangxi University, Nanning 530004, China; luolianxin@gxu.edu.cn (L.L.); 1916301029@st.gxu.edu.cn (X.W.); 1916301042@st.gxu.edu.cn (S.Z.); 1816301034@st.gxu.edu.cn (X.Y.)
2   Guangxi Key Laboratory of Clean Pulp & Papermaking and Pollution Control, Guangxi University, Nanning 530004, China
3   University of Chinese Academy of Sciences, Beijing 100049, China
4   Ningbo Institute of Materials Technology & Engineering, Chinese Academy of Sciences, Ningbo 315201, China
*   Correspondence: limingfu@nimte.ac.cn (M.L.); wangsf@gxu.edu.cn (S.W.)

Abstract: Hot water (HW), green liquor (GL), and sodium chlorite (SC) pretreatments were used to pretreat sugarcane bagasse (SCB) and spruce (SP) and then to prepare cellulose nanofibers (CNFs) through high-pressure homogenization to explore the effect of physicochemical properties on the thermal stability and ultraviolet (UV) resistance performance of CNF films. The results indicated that the lignin content of HW-pretreated CNFs was higher than that of GL- and SC-pretreated CNFs, and the hemicellulose content of HW-pretreated CNFs was lower than that of GL- and SC-pretreated CNFs. The synergy of lignin and hemicellulose impacted the thermal stability of CNF films. The thermal stability of all the SP CNF films was higher than that of all the SCB CNF films. Hot water pretreatment improved the thermal stability of CNF films, and green liquor and sodium chlorite pretreatment decreased the thermal stability of CNF films. The highest thermal stability of SP-HW CNF films reached 392 °C, which was 5.4% higher than that of SP-SC CNF films. Furthermore, the ultraviolet resistance properties of different CNF films were as follows: SCB-HW > SCB-GL > SCB-SC and SP-HW > SP-GL > SP-SC. Green liquor pretreatment is an effective method to prepare CNFs. Conclusively, this research provides a basic theory for the preparation of CNFs and allows the improvement of CNF films in the application of thermal stability management and UV resistance fields.

Keywords: cellulose nanofiber; pretreatment; lignin; hemicellulose; physicochemical properties

Citation: Luo, L.; Wang, X.; Zhang, S.; Yuan, X.; Li, M.; Wang, S. Contribution of Different Pretreatments to the Thermal Stability and UV Resistance Performance of Cellulose Nanofiber Films. *Coatings* **2021**, *11*, 247. https://doi.org/10.3390/coatings11020247

Academic Editor: Alessandro Pezzella

Received: 5 January 2021
Accepted: 11 February 2021
Published: 19 February 2021

Publisher's Note: MDPI stays neutral with regard to jurisdictional claims in published maps and institutional affiliations.

1. Introduction

In recent years, the overexploitation and utilization of global fossil energy and fuels has led to environmental changes and energy shortages. Therefore, there is increasing interest for sustainable and environmentally friendly materials research [1,2]. Cellulose nanofibers (CNFs) have become a promising renewable material because of their unique structure, excellent performance, and natural abundance. It is widely found in plants and considered a nontoxic, degradable, and low-cost material, so it is a good substitute for synthesizing many products [3,4]. CNFs can be used to prepare pollution-free product components, such as membrane electrode assemblies, organic light-emitting diodes, writable touch screens, energy storage materials, and eco-friendly fabric softeners [5–8]. Thermal instability is the main problem of biomass materials because their components have a low melting temperature. Therefore, these electronic components are easily destroyed when they encounter high temperatures and need to be annealed or sterilized [9]. CNFs can be combined with other wood material components to produce a hybrid superlattice structure, reduce their thermal conductivity, and become a thermoelectric energy-harvesting material.

The thermal stability of CNFs is closely related to the lignin and hemicellulose in plant fibers [10]. Lignin and hemicellulose act as binders and fillers in lignocellulosic raw materials, and many cellulose molecular chains combine to form fiber bundles [11]. Lignin is one of the main components in lignocellulosic raw materials [12]. It is a natural high molecular weight polymer with a three-dimensional structure that is formed by connecting a phenylpropane structure through ether and carbon-carbon bonds [13]. Lignin can increase the water contact angle of cellulose fiber films, reduce their water absorption capacity, and improve the thermal stability and UV barrier properties of cellulose fiber-based films [14,15]. The degradation of wood fiber is a complex process involving a series of continuous reactions, and its thermal stability depends on the chemical composition, fiber size, crystal structure, and number of intermolecular and intramolecular hydrogen bonds. Nair et al. investigated the effect of lignin on the thermal stability of CNFs, indicating that the presence of lignin significantly improved the thermal stability of CNFs [16].

The preparation methods for CNFs also impact on their thermal stability. Pretreatments can change the cell wall structure and composition of lignocellulose, reduce energy consumption during the preparation of CNFs, and increase the yield of CNFs [17–19]. At present, the main pretreatments include acid, alkali, hot water, organosolv, and ionic liquid pretreatments [1,20,21]. Compared to gamma-valerolactone/water [22], catalyzed chemical oxidation [23] and organosolv pretreatment [24] were used to pretreat lignocellulose and prepare CNFs, and hot water pretreatment as a green and environmental method can save chemical reagent consumption [25]. Furthermore, green liquor as a kind of alkaline liquor mainly consists of sodium carbonate and sodium sulfide, which is the liquor recovered after the combustion of black liquor in the recovery boiler of the Kraft pulp mill. Green liquor was used to pretreat lignocellulosic biomass can decrease the overall cost for the pretreatment process [26], which could also alleviate the discharge of pulping wastewater. Therefore, using the green liquor pretreated lignocellulose to prepare CNFs can realize the reusability of the wastewater in pulp and papermaking. However, there is a lack of using green liquor pretreatment to pretreat lignocellulose and then prepare CNFs, and to further investigate the effect of pretreatment on the physicochemical properties of CNFs.

Therefore, in this work, hot water, green liquor and sodium chlorite were used to pretreat sugarcane bagasse and spruce and then prepare CNFs by high-pressure homogenization. Component analysis, transmission electron microscope (TEM), Fourier transform infrared spectroscopy (FTIR), thermogravimetric (TG), and ultraviolet (UV) spectroscopy were used to analyze the effect of different pretreatments on the physicochemical properties of the CNFs and CNF films. Furthermore, the effect of residual lignin and hemicellulose content on the thermal stability and UV resistance of CNF films was also investigated. This research provides the basic theory for CNFs in the application of thermal management and UV resistance fields.

2. Experiments

2.1. Materials

Sugarcane bagasse was provided by Guangxi Guitang (Group) Co., Ltd. (Guigang, China), spruce originated from European spruce (Alps, Bavaria, Germany). Sodium sulfide, sodium carbonate, sodium hydroxide, sodium chlorite (SC), glacial acetic acid, and sulfuric acid were purchased from Tianjin Zhiyuan Chemical Reagent Co., Ltd. (Tianjin, China).

2.2. Pretreatments

The scheme of lignocellulose pretreatments was shown in Figure 1. A six-pot digester (2201-6, Anderson Greenwood Instruments, South Plainfield, NJ, USA) was used for pretreatment experiments. For the hot water pretreatment process, 30 g of sugarcane bagasse (spruce) raw material was placed into a 1 L tank, 450 mL of water was added, and the pretreatment temperature and residence time were 170 °C and 90 min, respectively.

For the simulated green liquor pretreatment process, 30 g of sugarcane bagasse (spruce) raw material was placed into a 1 L tank, 180 mL of green liquor was added, and the pretreatment temperature and residence time were 120 °C and 60 min, respectively. The sulfidity and total alkali were 25% and 16% (based on Na_2O), respectively. The pretreated solids were washed with hot water for further use. Furthermore, the raw materials were pretreated with sodium chlorite [27,28]. Two grams of sugarcane bagasse (spruce) raw material was placed into a 250 mL conical flask, 200 mL distilled water was added into the conical flask, and a constant temperature was maintained in a water bath at 75 °C. One milliliter of glacial acetic acid and 1.2 g of sodium chlorite were added to the liquor every 1 h, and the reaction time was 3 h. After the reaction was completed, the solid was washed with hot water for further use.

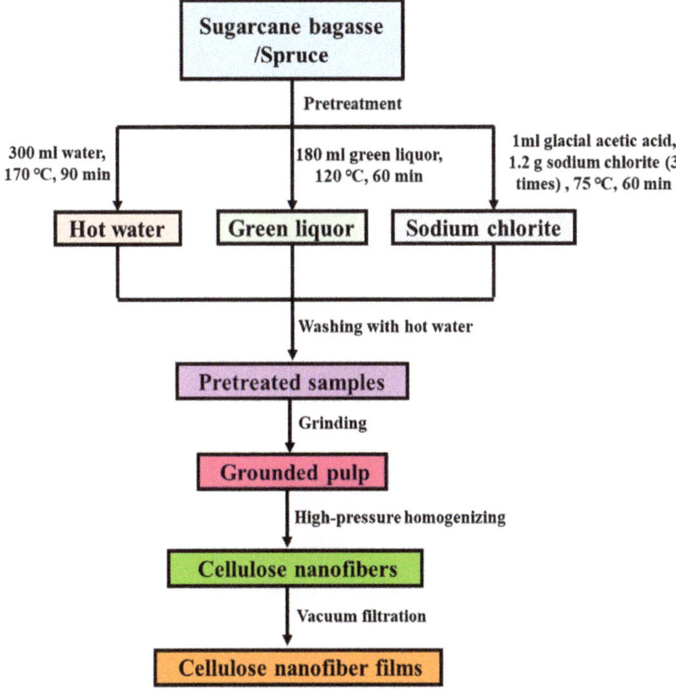

Figure 1. The scheme of lignocellulose pretreatments and cellulose nanofibers (CNFs) films' preparation.

2.3. Preparation of Cellulose Nanofibers

The pretreated sugarcane bagasse and spruce fibers were used to disc mills, and the grinding distances were 0.2, 0.1, and 0.1 μm, respectively, to prepare grounded pulp. Then, the ground pulp was pretreated by a high-pressure homogenizer (M-110EH-30, Microfluidizer, Westwood, CA, USA). The pulp concentration was 1%. First, the pulp was homogenized 20 times at a pressure of 350 bar and a hole diameter of 200 μm in the high-pressure homogenizer. Then, the pulp was homogenized 10 times at a pressure of 1500 bar and a pore size of 87 μm to obtain the CNFs. The CNF films were prepared by the vacuum filtration method [5]. The prepared CNFs were stirred with a magnetic stirrer to make them fully dispersed. The dispersed CNFs were converted into CNF films with a diameter of 90 mm and mass of 0.4 g and put into an automatic molding dryer (BBS-2, ESTANITHAAGE, Berlin, Germany) at 75 °C for 15 min. The prepared CNFs were nominated as hot water pretreated sugarcane bagasse CNFs (SCB-HW), green liquor pretreated sugarcane bagasse CNFs (SCB-GL), and sodium chlorite pretreated sugarcane bagasse CNFs (SCB-SC). The prepared spruce CNFs were named hot water pretreated spruce CNFs (SP-HW), green liquor pretreated spruce CNFs (SP-GL), and sodium chlorite pretreated spruce CNFs (SP-SC).

2.4. Chemical Composition Analysis of Cellulose Nanofibers

CNFs (0.25 g) were added into an autoclave, 1.5 mL of 72% concentrated sulfuric acid was added, and then carbonized at 30 °C for 1 h. After carbonization, 43 mL of distilled water was added, and an autoclave was used to react at 121 °C for 1 h. After the reaction was completed, the solids and filtrate were separated by filtration, and the solids were washed with distilled water to neutral. Then, the solid was dried at 105 °C for 4 h to obtain acid-insoluble lignin (Klason lignin). The acid-soluble lignin and carbohydrate (glucose and xylose) in the liquor were determined by a UV VIS spectrophotometer (Analytik Jena, Jena, Germany) and high-performance liquid chromatography (Agilent, 1260, Palo Alto, CA, USA), respectively, according to the previous report [29,30]. Each experiment was performed twice, and the statistical analysis was conducted with MS Excel 2016. The average values and confidence interval (level of significance = 0.05) were determined.

2.5. Characterization of Cellulose Nanofibers

Transmission electron microscope (TEM) analysis: The samples were prepared for TEM analysis. The CNFs were dispersed in water with a mass concentration of 0.05%, then a piece of filter paper was placed in a petri dish, a copper mesh was placed on the filter paper, and a certain amount of CNFs dispersion was dropped on the TEM copper grid and dried at room temperature for 12 h. Then, 1 mL of staining agent (2% phosphotungstic acid) was dropped on the copper mesh, the droplets completely covered the copper mesh, and it was put in a dark box for 30 min. Finally, the copper mesh was placed at room temperature to obtain CNFs dyed samples. A JEM-1200EX transmission electron microscope (JEOL, Tokyo, Japan) was used to analyze the CNF morphology. The test voltage and the current were 120 kV and 100 μA, respectively, and the magnification was 80,000 times for test analysis.

Fourier transform infrared (FTIR) spectroscopy: A German Bruker Tensor II was used for FT-IR measurement (Bruker, Karlsruhe, Germany) of CNFs. One milligram of sample and 100 mg of potassium bromide were mixed to prepare circular slices by pressing and then put into an infrared analyzer for testing. The scanning range was 4000–400 cm^{-1}, the number of scanning cycles was 32, and the scanning frequency was 4 cm^{-1} [31].

XPS analysis: The prepared CNF membrane was cut into a size of 0.5 × 0.5 cm^2, fixed on a sample device, and placed into an X-ray photoelectron spectrometer to analyze the sample. An X-ray spectrometer (Axis Ultra DLD, KRATOS, Manchester, UK) was used to analyze the surface element content of the CNFs. The working vacuum of the analysis was 5 × 10^{-9} torr, the monochromatic Al Kα source energy was 1486.6 eV (5 mA × 15 kV), and the beam spot size was 700 × 300 μm^2. A CAE scanning mode was adopted. Full-spectrum scan: The pass energy was 160 eV; the narrow-spectrum scan pass energy was 40 eV, and the number of scans was 3. The surface lignin content (SLC) of CNFs was calculated according to formula (1) [32].

$$\text{SLC } (\%) = \frac{O/C_{sample} - O/C_{carbohydrate}}{O/C_{lignin} - O/C_{carbohydrate}} \tag{1}$$

The values of $O/C_{carbohydrate}$ and O/C_{lignin} were 0.83 and 0.33, respectively.

X-ray diffraction (XRD) analysis: The prepared CNF film was cut into a size of 2 × 2 cm^2, placed into the groove of the glass sample plate, and analyzed with a high-resolution X-ray diffractometer (SMARTLAB 3KW, Tokyo, Japan). The scanning range was 5°–50° at a speed of 0.25 steps, each step was 10 s, the scanning speed was 0.18 min, and the step length was 0.018° for analysis. The crystallinity (CrI) of CNFs was calculated according to the Segal method [33] as follows.

$$\text{CrI } (\%) = \frac{(I_{002} - I_{am})}{I_{002}} \times 100\% \tag{2}$$

where I_{002} is the intensity of the diffraction peak at $2\theta = 22°$ and I_{am} is the intensity of the diffraction peak at $2\theta = 18°$.

2.6. Thermogravimetry (TGA) Analysis

The CNF samples were vacuum dried at 40 °C for 24 h, and then thermogravimetric analysis of the CNFs was performed with a synchronous thermal analyzer (STA 449F5, NETZSCH, Selbu, Germany) [27]. The mass of the sample was 10 mg, the nitrogen flow and heating rate were 40 mL·min^{-1} and 10 K·min^{-1}, respectively, and the heating temperature ranged from 25 to 700 °C.

2.7. UV Absorption Performance

The size of the prepared CNF films was 30×12 mm^2, and they were spectrally scanned using an ultraviolet-visible spectrophotometer (ANALYTIK Jena, Jena, Germany) with a scanning wavelength of 190–900 nm [34,35].

3. Results and Discussion

3.1. Chemical Composition of CNFs

The chemical composition analysis of different CNFs is shown in Table 1. The lignin content in hot water-pretreated SCB and spruce CNFs was higher than that in green liquor- and sodium chlorite-pretreated SCB and spruce CNFs. For example, the lignin contents of SCB-HW, SCB-GL, and SCB-SC were 22.5%, 8.4%, and 1.8%, respectively. These results indicate that the removal of lignin in sugarcane bagasse and spruce pretreated with green liquor was higher than that by hot water pretreatment. Furthermore, the lignin content in SCB-HW (SCB-GL) was lower than that in SP-HW (SP-GL) due to the compact structure of the high lignin content in the spruce cell wall, which resulted in the lignin difficultly removed under mild pretreatment conditions. After hot water pretreatment, the hemicellulose content in the SCB-HW and SP-HW seriously decreased, the lignin content in SCB-HW and SP-HW slightly decreased, indicating that hot water pretreatment mainly removed the hemicellulose from the CNFs.

Table 1. Chemical composition of different CNFs. SCB—sugarcane bagasse; SCB-HW—hot water pretreated sugarcane bagasse CNFs; SCB-GL—green liquor pretreated sugarcane bagasse CNFs; SCB-SC—sodium chlorite pretreated sugarcane bagasse CNFs; SP—spruce; SP-HW—hot water pretreated spruce CNFs; SP-GL—green liquor pretreated spruce; SP-SC—sodium chlorite pretreated spruce CNFs.

Samples	Cellulose (%)	Hemicellulose (%)	Lignin (%)
SCB	41.7 ± 0.6	26.6 ± 0.4	24.4 ± 0.7
SCB-HW	45.4 ± 0.6	5.5 ± 0.6	22.5 ± 0.6
SCB-GL	46.9 ± 0.5	7.6 ± 0.4	8.4 ± 0.3
SCB-SC	62.6 ± 0.3	3.5 ± 0.6	1.8 ± 0.3
SP	43.6 ± 0.2	23.9 ± 0.6	28.6 ± 0.7
SP-HW	47.4 ± 0.2	7.5 ± 0.6	26.0 ± 0.4
SP-GL	49.2 ± 0.2	9.1 ± 0.5	15.1 ± 0.7
SP-SC	55.6 ± 0.4	4.0 ± 0.5	2.8 ± 0.6

3.2. Photography and TEM Analysis

The photograph of different pretreated CNF films is shown in Figure 2, indicating the colors of different CNF films changed from brown to white. The hot water-pretreated SCB and SP CNF films were darker than the green liquor- and sodium chlorite-pretreated CNF films, due to the lignin content in the hot water-pretreated CNFs was higher than that in the green liquor- and sodium chlorite-pretreated CNFs. Compared to the SCB-HW (or SCB-GL) CNF films, the color of SP-HW (or SP-GL) CNF films was brighter, possibly due to the more unsaturated bonds of lignin was formed in SCB-HW (or SCB-GL) CNFs during hot water (or green liquor) pretreatment led to the color of SCB-HW (or SCB-GL) CNF films was darker. The SCB-SC CNF films was brighter than SP-SC CNF films due to the lignin of SCB-SC CNF was almost completely removed during sodium chlorite pretreatment.

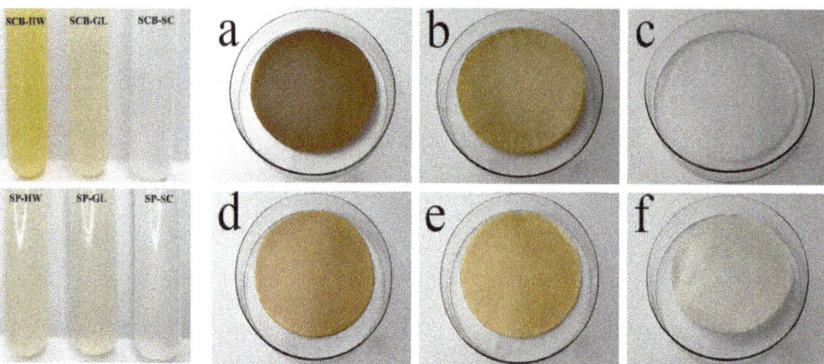

Figure 2. Photographs of different CNF films ((**a**) SCB-HW, (**b**) SCB-GL, (**c**) SCB-SC, (**d**) SP-HW, (**e**) SP-GL, (**f**) SP-SC).

TEM analysis of different pretreated CNFs is shown in Figure 3, indicating the size and morphology of CNFs were different. The diameters of SCB-HW, SCB-GL, and SCB-SC were 15.3, 13.7, and 12.8 nm, respectively. The diameters of SP-HW, SP-GL, and SP-SC were 19.5, 17.2, and 14.7 nm, respectively. Compared to hot water pretreatment, the green liquor- and sodium chlorite-pretreated CNFs were easily dispersed in water because more lignin was removed by the green liquor and sodium chlorite pretreatment, leading to more hydrophilic groups of CNFs, such as hydroxyl and carboxyl groups, exposed on the surface [36]. Furthermore, the diameter of sugarcane bagasse and spruce CNFs was different possibly due to the size of the original sugarcane bagasse raw material fiber was smaller than that of spruce raw material fiber.

3.3. FTIR Analysis

FTIR spectroscopy is widely used to investigate the change of functional groups in lignocellulose under different processing conditions [37]. The FTIR spectra of different CNFs are shown in Figure 4a, b. The peak intensity in FTIR was calibrated using the peak at $1110\ cm^{-1}$ as a reference peak [38]. It can be seen that the peak at $3356\ cm^{-1}$ was assigned to the intermolecular O–H stretching vibration on the cellulose carbon skeleton [38], the peak at $2860\ cm^{-1}$ was assigned to the C–H absorption peaks of methyl, methylene, and methine [39], and the peak at $1730\ cm^{-1}$ was assigned to the absorption peak of non-conjugated carbonyl with an aromatic ring and its ester and lactone. Compared to the SCB-HW and SP-HW, the peak intensities of the O–H stretching vibration and nonconjugated carbonyl in SCB-GL and SP-GL were decreased, possibly due to the oxidation and degradation of a large amount of lignin during the pretreatment process. The peak intensity of the O–H stretching vibration in SCB-SC and SP-SC was increased because the hydroxyl group was oxidized to a carboxyl group during the pretreatment process. The peak at $1700\ cm^{-1}$ was assigned to the ester bond of C=O, which is usually derived from the acetyl group in hemicellulose and the linkage between lignin and hemicellulose [40]. The peak at $1644\ cm^{-1}$ was assigned to the conjugated carbonyl of lignin [41], and the characteristic peaks at $1604\ cm^{-1}$, $1506\ cm^{-1}$, $1462\ cm^{-1}$, and $1422\ cm^{-1}$ were assigned to the aromatic ring of lignin [19,42]. Compared to the SCB-HW and SP-HW, these intensities of peaks in SCB-SC, SCB-GL, SP-SC, and SP-GL decreased or disappeared due to the removal of lignin after different pretreatments. The intensity of the peak at $809\ cm^{-1}$ decreased due to stretching of the homocyclic ring of mannose residues, which was related to hemicellulose removal.

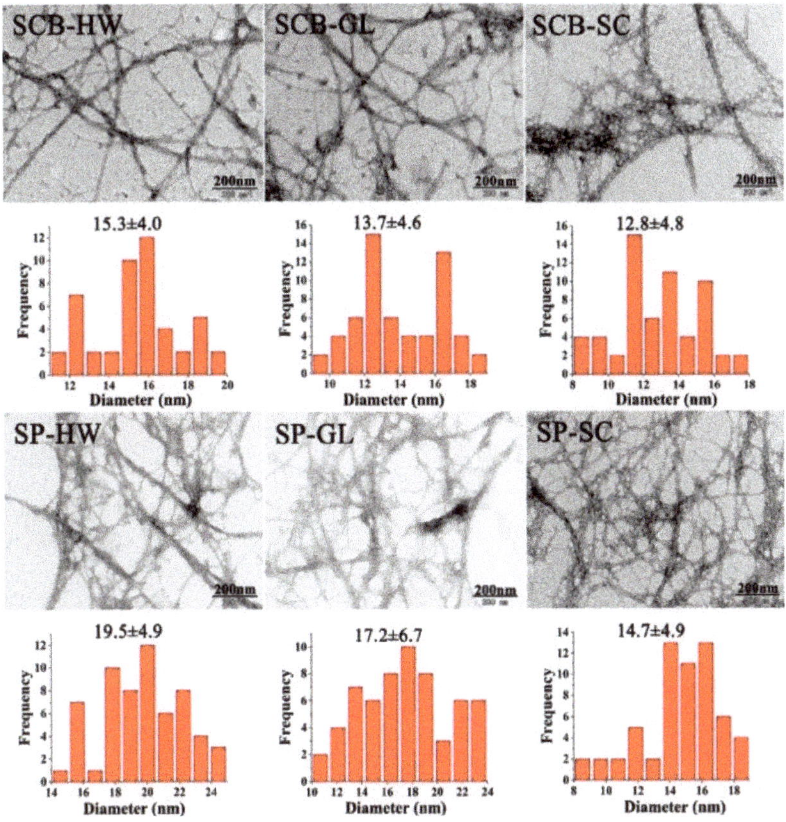

Figure 3. TEM and size analysis of CNFs.

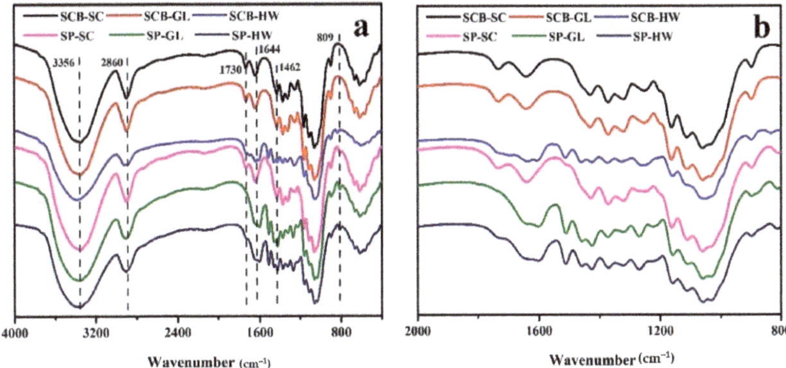

Figure 4. FTIR spectroscopy of CNFs (**a**) 4000–400 cm^{-1} (**b**) 2000–800 cm^{-1}.

3.4. XPS Analysis

To study the effect of pretreatment on the surface chemical properties of CNF films, XPS was used to analyze the C, O, and S elements on the surface of different CNF films [43,44], as shown in Table 2. Compared to hot water-pretreated CNF films, the content of C decreased, while the content of O increased in green liquor-pretreated CNF films. The small increase of sulfur due to S^{2-} or HS^- was attached on the surface of fiber during the green liquor pretreatment process. The increase in the oxygen-carbon ratio may be due to the lignin being removed during the green liquor pretreatment process. The content of C decreased, and the content of O increased in SCB-SC and SP-SC, leading to an increase in the ratio of O/C. The increase in the oxygen–carbon ratio due to the amount of lignin was removed by sodium chlorite pretreatment, resulting in more carbohydrates being exposed on the surface of the fiber. The change in the oxygen-carbon ratio impacted the surface lignin content (SLC) of the CNF films. Compared to hot water pretreated CNFs films, the SLC of green liquor and sodium chlorite pretreated CNFs films decreased, the SLC in green liquor pretreated CNFs films was higher than sodium chlorite pretreated CNFs films, due to a large amount of lignin being removed by sodium chlorite pretreatment. Furthermore, the SLC of SP-HW CNF films was higher than that of SCB-HW CNF films because the lignin content in SP-HW CNFs was higher than that in SCB-HW CNF films (Table 1).

Table 2. Surface element analysis of CNFs films.

Samples	C (%)	O (%)	S (%)	O/C	SLC (%)
SCB-HW	69.4	30.4	0.17	0.44	78
SCB-GL	66.5	33.2	0.23	0.50	66
SCB-SC	64.8	35.0	0.19	0.54	58
SP-HW	71.6	28.1	0.15	0.39	87
SP-GL	68.2	31.6	0.26	0.46	74
SP-SC	59.1	40.6	0.20	0.69	28

The change in lignin content on the surface of CNFs can also be characterized by the change in C1s. C1s can be divided into three combined types of C1 (C–C), C2 (C–O), C3 (O–C–O or C=O) [44–46], as shown in Figure 5. The binding energy of C1 at 285 eV was derived from lignin and extracts, the binding energy of C2 at 286.5 eV was derived from carbohydrates, and the binding energy of C3 at 288–288.5 eV was derived from lignin and fiber oxidation products [44]. The percentages of C1, C2, and C3 are shown in Table 3. The C1 content of SCB-HW was higher than SCB-GL and SCB-SC, and the C2 and C3 contents of SCB-HW were lower than SCB-GL and SCB-SC. The C1 content of SP-HW was higher than SP-GL and SP-SC, and the C2 and C3 contents of SP-HW were lower than SP-GL and SP-SC. These results indicated that the surface of hot water-pretreated CNF films had the high lignin content, the surface of sodium chlorite-pretreated CNF films had the low lignin content. The lignin content on the surface of the green liquor-pretreated CNF films was possibly correlated with small, soluble lignin fragments being reabsorbed on the surface of wood fibers during the green liquor pretreatment process [32]. The C2 and C3 contents of SCB-GL or SP-GL were higher than those of SCB-HW or SP-HW because more lignin was removed during the green liquor pretreatment process. The C2 and C3 contents of SCB-SC or SP-SC were higher than those of SCB-HW or SP-HW because lignin was removed by sodium chlorite, leading to more carbohydrates being exposed on the surface of CNFs. Furthermore, the increase in C3 content due to sodium chlorite increases oxygen-containing functional groups in fibers. These results were consistent with the XPS full spectrum analysis.

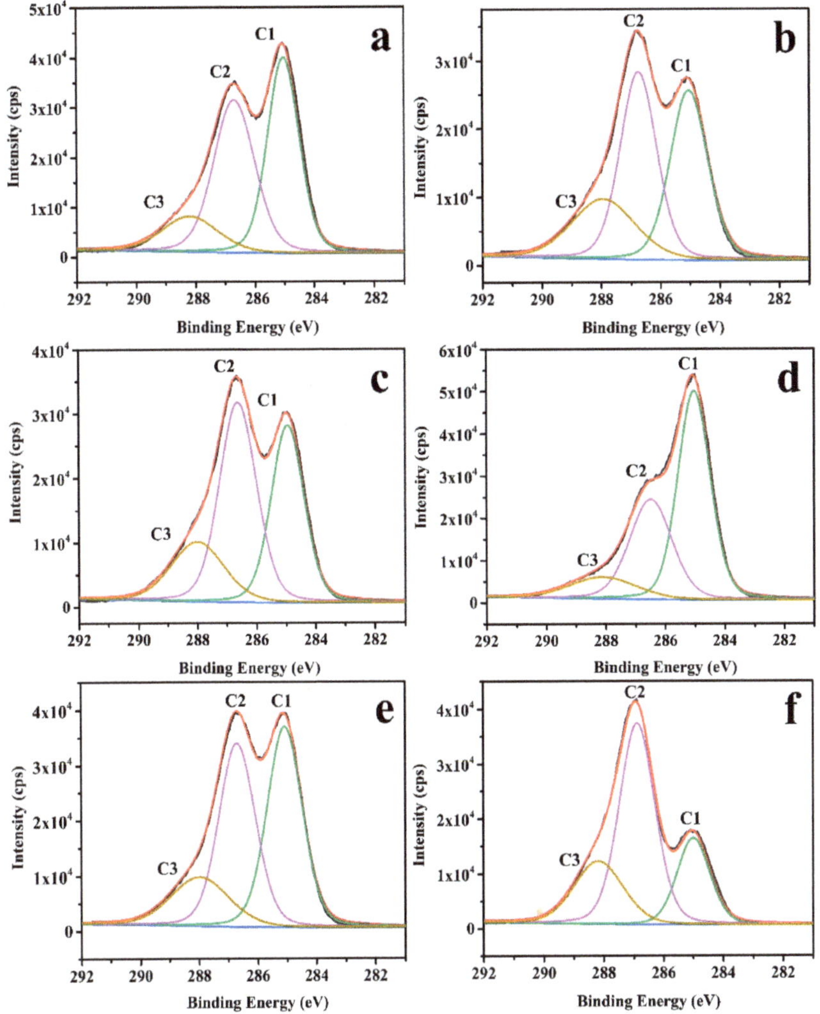

Figure 5. C1 s peaks of different CNFs films: (**a**) SCB-HW; (**b**) SCB-GL; (**c**) SCB-SC; (**d**) SP-HW; (**e**) SP-GL; (**f**) SP-SC.

Table 3. Elemental composition of C1s peak of CNFs films.

Samples	C1 (%)	C2 (%)	C3 (%)
SCB-HW	48.5	37.5	13.9
SCB-GL	39.5	42.4	18.2
SCB-SC	33.7	46.1	20.2
SP-HW	54.4	34.1	11.5
SP-GL	42.5	40.3	17.2
SP-SC	21.9	54.7	23.4

3.5. X-ray Diffraction (XRD) Analysis

The crystallinity of the pretreated sugarcane bagasse and spruce CNF films is shown in Figure 6. It can be seen that sugarcane bagasse and spruce before and after pretreatment have strong diffraction absorption peaks at $2\theta = 18°$ and $2\theta = 22°$, which indicate that the pretreated materials maintained the typical polymorphic form of cellulose I [47–49]. The relative crystallinity of cellulose was calculated according to the Segal formula, as shown in Table 4.

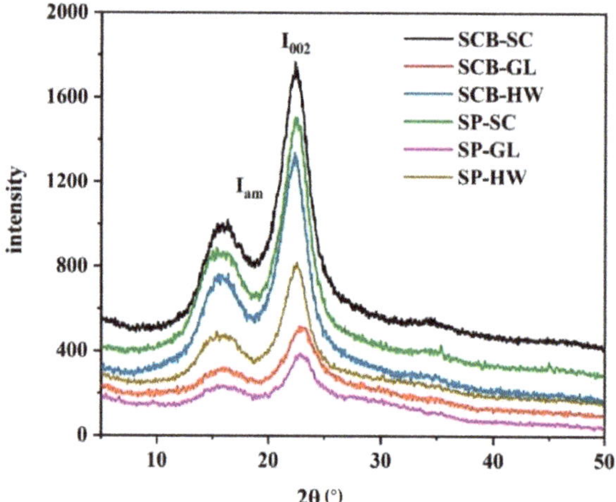

Figure 6. XRD analysis of CNFs films.

Table 4. Crystallinity of different CNFs films.

Samples	CrI (%)
SCB-HW	54.7 ± 0.5
SCB-GL	61.4 ± 0.7
SCB-SC	67.1 ± 0.4
SP-HW	51.2 ± 0.8
SP-GL	63.1 ± 0.4
SP-SC	65.5 ± 0.6

The crystallinities of SCB-HW and SP-HW were 54.72% and 51.23%, respectively, the crystallinities of SCB-GL and SP-GL were 61.35% and 63.09%, respectively, and the crystallinities of SCB-SC and SP-SC were 67.08% and 65.46%, respectively. It can be concluded that the crystallinity of the SC CNFs was higher than that of the hot water and green liquor pretreated CNFs because most of the lignin and hemicellulose were removed during the SC pretreatment process. The SCB-HW had higher crystallinity than SP-HW due to the removal of hemicellulose and lignin in sugarcane bagasse during the hot water pretreatment process was greater than that of spruce. Moreover, the aggregation state of cellulose and the supramolecular structure of cellulose formed a staggered combination of crystalline regions to amorphous regions. The cellulose molecular chains in the crystalline region have good orientation and high density and are not easily destroyed, while the cellulose chains in the amorphous region with poor orientation, disordered molecular arrangement, and irregular are easily destroyed [27]. After hot water pretreatment, the non-crystalline region of lignocellulose was destroyed, while the crystalline region of cellulose was relatively stable. Therefore, the proportion of the crystalline region of cellulose increased, finally leading to an increase in crystallinity. After green liquor pretreatment, the crystallinity of CNFs was higher than that of hot water-pretreated CNFs, possibly due to the lignin and hemicellulose in non-crystalline region were removed, and the molecular chain of the crystalline region of cellulose was broken to varying degrees under alkaline conditions.

3.6. Thermogravimetry (TGA) Analysis

The thermogravimetric analysis curves of different CNFs are shown in Figure 7, and the corresponding results are shown in Table 5. According to the change in mass loss shown in Figure 7a, the mass loss of CNFs at different temperatures can be divided into three regions. The temperature of region I ranged from 25 to 225 °C. The mass loss in this region was caused by the evaporation of water in the CNFs, and the average residual water was 5.64 wt.%. The temperature of region II ranged from 225 to 390 °C, which was the main stage of thermal degradation of different CNF samples, and the mass loss changed. The mass of region II suddenly decreased due to the depolymerization of cellulose [50]. The temperature of region III ranged from 390 to 700 °C. The cellulose in this zone was further pyrolyzed into coke and graphite. At this stage, the thermal degradation of each sample was relatively stable, and the mass loss was reduced. The residual mass of each sample at region III was consistently near 11%.

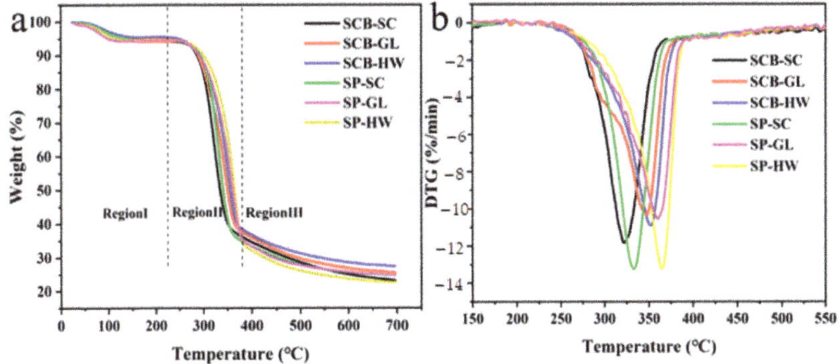

Figure 7. TG (**a**) and DTG (**b**) diagrams of different CNFs.

Table 5. Thermogravimetric analysis of different CNFs.

Sample	Region I		Region II		Region III	Tonset (°C)	Td (°C)
	Temperature (°C)	Residual Moisture (wt.%)	Temperature (°C)	Mass Residue (wt.%)	Mass Residue (wt.%)		
SCB-HW	251	5.11	387	38.11	10.63	297	352
SCB-GL	243	5.78	382	37.06	11.59	291	346
SCB-SC	255	6.51	369	37.23	14.11	284	323
SP-HW	252	5.50	392	32.86	10.20	321	364
SP-GL	240	5.45	385	33.97	9.13	320	360
SP-SC	239	5.49	372	35.68	10.87	301	333

The thermal stability of the SCB CNFs was as follows: SCB-HW > SCB-GL > SCB-SC. The thermal stability of the SP CNFs was as follows: SP-HW > SP-GL > SP-SC. Due to the high lignin content and the low hemicellulose content, there was a synergetic increase in the thermal stability of the CNFs. Furthermore, the thermal stability of SP-HW was higher than that of SCB-HW because the lignin content of SP-HW was higher than that of SCB-HW, and the hemicellulose contents of SP-HW and SCB-HW were basically equal. The thermogravimetric process of SP-GL as an example was analyzed in detail, as shown in Figure 7b. It can be concluded that when the heating temperature reached 240 °C, the free water of SP-GL CNFs was removed and began to undergo thermal degradation, resulting in a slight mass loss of SP-GL CNFs of 5.45%. When the temperature gradually increased to 320 °C, the mass loss of CNFs was obviously increased, and this temperature was defined as the initial degradation temperature (Tonset) of CNFs. When the temperature was continually raised to 360 °C, the mass of the CNFs rapidly degraded due to the formation of volatile substances such as methane, carbon monoxide, and carbon dioxide [19]. This temperature was defined as the maximum thermal degradation temperature (Td) of CNFs. When the temperature reached 393 °C, the mass loss of CNFs remained stable, the thermal degradation also tended to be gentle, and the residual mass was 32.68%. According to the previous analysis, the two important indicators of thermal stability were Tonset and Td, and both indicators showed an upward trend because the degradation temperature of cellulose and hemicellulose was 350 °C. While the degradation temperature of lignin was above 400 °C, the thermal stability of lignin was better than that of cellulose and hemicellulose [51]. Because lignin had aromatic groups, ether and carbon–carbon bonds, these structures could not decompose at 350 °C. Although cellulose and hemicellulose can decompose at low temperatures, increasing the lignin content can improve the thermal stability of CNFs. Conclusively, the thermal stability of the different CNFs was directly correlated to the synergy of lignin and hemicellulose content of CNFs.

3.7. Analysis of UV Absorption Performance

The results of ultraviolet spectrum analysis of the CNF films are shown in Figure 8. The analysis showed that the transmittance of SCB-SC and SP-SC in 400–760 nm (visible light region) was higher than that of other pretreatments, which was consistent with the change in lignin content in CNFs [52] and the transparency effect of the CNF films. The UV resistance of the SCB-HW, SCB-GL, SP-HW, and SP-GL (190–400 nm) CNFs was near 100%, and the SCB-SC and SP-SC CNFs had low absorption rates of UV light due to the low lignin content in the SCB-SC and SP-SC CNFs. Furthermore, the functional groups of lignin can absorb UV light, such as phenol, ketone, and other chromophore groups, and the ultraviolet absorption of CNFs was correlated to the lignin content and functional group of lignin [34,53]. Conclusively, the high content of lignin in different pretreatment CNFs led to the high UV resistance of the CNFs.

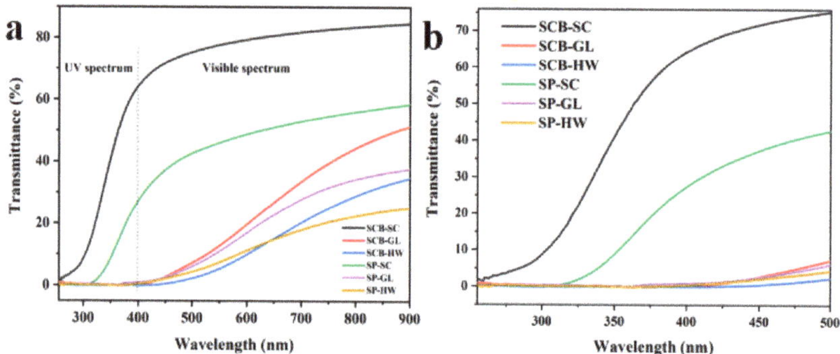

Figure 8. UV/Vis transmittance of different CNF films (**a**) 250–900 nm and (**b**) 200–500 nm.

4. Conclusions

Different pretreatment methods were used to pretreat sugarcane bagasse and spruce, and CNF films with different lignin and hemicellulose contents were prepared. Hot water pretreatment mainly removed the hemicellulose, and the green liquor mainly removed the lignin in the different CNFs. The thermal stability and UV resistance of all the spruce CNF films are higher than those of all the sugarcane bagasse CNF films. The thermal stability of the CNF films increased with increasing lignin content and decreasing hemicellulose content. The lignin content impacted the UV resistance of CNF films. Green liquor pretreatment can be used for the preparation of cellulose nanofibers. This study provides a theoretical basis for the application of lignin-containing cellulose nanofibers in thermal management and UV resistance fields.

Coatings **2021**, 11, 247

Author Contributions: Conceptualization M.L.; methodology, M.L., X.W., and S.Z.; validation, L.L. and M.L.; formal analysis, X.W., S.Z., and X.Y.; investigation, X.W.; resources, L.L., M.L., and S.W.; data curation, X.W. and M.L.; writing—original draft preparation, X.W.; writing—review and editing, L.L., M.L., and S.W.; visualization, X.W.; supervision, L.L. and M.L.; project administration, L.L. and S.W.; funding acquisition, L.L. and S.W. All authors have read and agreed to the published version of the manuscript.

Funding: This work was supported by the National Natural Science Foundation of China (31660182) and the Natural Science Foundation Project of Guangxi (2018GXNSFAA281336).

Institutional Review Board Statement: Not applicable.

Informed Consent Statement: Not applicable.

Data Availability Statement: Data available in a publicly accessible repository.

Acknowledgments: Thanks for the Guangxi Key Laboratory of Clean Pulp & Papermaking and Pollution Control provided the technology and financial support.

Conflicts of Interest: The authors declare no conflict of interest.

References

1. Thomas, B.; Raj, M.C.; Athira, K.B.; Rubiyah, M.H.; Joy, J.; Moores, A.; Drisko, G.L.; Sanchez, C. Nanocellulose, a versatile green platform: From biosources to materials and their applications. *Chem. Rev.* **2018**, *118*, 11575–11625. [CrossRef] [PubMed]
2. Wang, L.; Fu, H.; Liu, H.; Yu, K.; Wang, Y.; Ma, J. In-situ packaging ultra-uniform 3D hematite nanotubes by polyaniline and their improved gas sensing properties. *Mater. Res. Bull.* **2018**, *107*, 46–53. [CrossRef]
3. Farooq, A.; Jiang, S.; Farooq, A.; Naeem, M.A.; Ahmad, A.; Liu, L. Structure and properties of high quality natural cellulose nano fibrils from a novel material Ficus natalensis barkcloth. *J. Ind. Text.* **2019**, 1–17. [CrossRef]
4. Mariano, M.; Cercená, R.; Soldi, V. Thermal characterization of cellulose nanocrystals isolated from sisal fibers using acid hydrolysis. *Ind. Crop. Prod.* **2016**, *94*, 454–462. [CrossRef]
5. Zhang, N.; Tao, P.; Lu, Y.; Nie, S. Effect of lignin on the thermal stability of cellulose nanofibrils produced from bagasse pulp. *Cellulose* **2019**, *26*, 7823–7835. [CrossRef]
6. Li, S.; Lee, P.S. Development and applications of transparent conductive nanocellulose paper. *Sci. Technol. Adv. Mater.* **2017**, *18*, 620–633. [CrossRef] [PubMed]
7. Liao, Q.; He, M.; Zhou, Y.; Nie, S.; Wang, Y.; Hu, S.; Yang, H.; Li, H.; Tong, Y. Highly cuboid-shaped heterobimetallic metal-organic frameworks derived from porous Co/ZnO/C microrods with improved electromagnetic wave absorption capabilities. *ACS Appl. Mater. Interfaces* **2018**, *10*, 29136–29144. [CrossRef]
8. Oikonomou, E.K.; Christov, N.; Cristobal, G.; Bourgaux, C.; Heux, L.; Boucenna, I.; Berret, J.F. Design of eco-friendly fabric softeners: Structure, rheology and interaction with cellulose nanocrystals. *J. Colloid Interface Sci.* **2018**, *525*, 206–215. [CrossRef]
9. Wu, X.; Li, S.; Coumes, F.; Darcos, V.; Him, J.L.K.; Bron, P. Modeling and self-assembly behavior of PEG-PLA-PEG triblock copolymers in aqueous solution. *Nanoscale* **2013**, *5*, 9010–9017. [CrossRef] [PubMed]
10. Tao, P.; Wu, Z.; Xing, C.; Zhang, Q.; Wei, Z.; Nie, S. Effect of enzymatic treatment on the thermal stability of cellulose nanofibrils. *Cellulose* **2019**, *26*, 7717–7725. [CrossRef]
11. Lin, X.; Wu, Z.; Zhang, C.; Liu, S.; Nie, S. Enzymatic pulping of lignocellulosic biomass. *Ind. Crop. Prod.* **2018**, *120*, 16–24. [CrossRef]
12. Huang, C.; Dong, H.; Su, Y.; Wu, Y.; Narron, R.; Yong, Q. Synthesis of carbon quantum dot nanoparticles derived from byproducts in bio-refinery process for cell imaging and in vivo bioimaging. *Nanomaterials* **2019**, *9*, 387. [CrossRef] [PubMed]
13. Atifi, S.; Miao, C.; Hamad, W.Y. Surface modification of lignin for applications in polypropylene blends. *J. Appl. Polym. Sci.* **2017**, *134*, 45103–45112. [CrossRef]
14. Cusola, O.; Rojas, O.J.; Roncero, M.B. Lignin particles for multifunctional membranes, antioxidative microfiltration, patterning, and 3D structuring. *ACS Appl. Mater. Interfaces* **2019**, *11*, 45226–45236. [CrossRef]
15. Rukmanikrishnan, B.; Rajasekharan, S.K.; Lee, J. K-carrageenan/lignin composite films: Biofilm inhibition, antioxidant activity, cytocompatibility, UV and water barrier properties. *Mater. Today Commun.* **2020**, *24*, 101346–101355. [CrossRef]
16. Nair, S.S.; Yan, N. Effect of high residual lignin on the thermal stability of nanofibrils and its enhanced mechanical performance in aqueous environments. *Cellulose* **2015**, *22*, 3137–3150. [CrossRef]
17. Vanholme, R.; Storme, V.; Vanholme, B. A systems biology view of responses to lignin biosynthesis perturbations in Arabidopsis. *Plant Cell* **2012**, *24*, 3506–3529. [CrossRef] [PubMed]
18. DeMartini, J.D.; Pattathil, S.; Miller, J.S. Investigating plant cell wall components that affect biomass recalcitrance in poplar and switchgrass. *Energy Environ. Sci.* **2013**, *6*, 898–909. [CrossRef]
19. Isaac, A.; de Paula, J.; Viana, C.M. From nano- to micrometer scale: The role of microwave-assisted acid and alkali pretreatments in the sugarcane biomass structure. *Biotechnol. Biofuels* **2018**, *11*, 73. [CrossRef]

20. Chirayil, C.J.; Mathew, L.; Thomas, S. Review of recent research in nano cellulose preparation from different lignocellulosic fibers. *Rev. Adv. Mater. Sci.* **2014**, *37*, 20–28.
21. Qing, Y.; Yi, J.; Wu, Y.; Cai, Z.; Wu, Q. Research progress in cellulose nanofiber films. *Trans. China Pulp. Pap.* **2016**, *31*, 55–62.
22. Lê, H.Q.; Dimic-Misic, K.; Johansson, L.; Maloney, T.; Sixta, H. Effect of lignin on the morphology and rheological properties of nanofibrillated cellulose produced from γ-valerolactone/water fractionation process. *Cellulose* **2018**, *25*, 179–194. [CrossRef]
23. Herrera, M.; Thitiwutthisakul, K.; Yang, X.; Rujitanaroj, P.; Rojas, R.; Berglund, L. Preparation and evaluation of high-lignin content cellulose nanofibrils from eucalyptus pulp. *Cellulose* **2018**, *25*, 3121–3133. [CrossRef]
24. Jiang, Y.; Liu, X.; Yang, Q.; Song, X.; Qin, C.; Wang, S.; Li, K. Effects of residual lignin on composition, structure and properties of mechanically defibrillated cellulose fibrils and films. *Cellulose* **2019**, *26*, 1577–1593. [CrossRef]
25. Wongjaiyen, T.; Brostow, W.; Chonkaew, W. Tensile properties and wear resistance of epoxy nanocomposites reinforced with cellulose nanofibers. *Polym. Bull.* **2017**, *75*, 2039–2051. [CrossRef]
26. Zainuddin, S.Y.; Ahmad, I.; Kargarzadeh, H. Potential of using multiscale kenaf fibers as reinforcing filler in cassava starch-kenaf biocomposites. *Carbohydr. Polym.* **2013**, *92*, 2299–2305. [CrossRef] [PubMed]
27. Li, M.; Guo, C.; Luo, B. Comparing impacts of physicochemical properties and hydrolytic inhibitors on enzymatic hydrolysis of sugarcane bagasse. *Bioprocess Biosyst. Eng.* **2020**, *43*, 111–122. [CrossRef]
28. Jiang, B.; Wang, W.; Gu, F.; Cao, T.; Jin, Y. Comparison of the substrate enzymatic digestibility and lignin structure of wheat straw stems and leaves pretreated by green liquor. *Bioresour. Technol.* **2016**, *199*, 181–187. [CrossRef] [PubMed]
29. BH Sluiter, A.; Hames, B.; Ruiz-Peinado, R.; Scarlata, C.; Sluiter, W.J.; Templaton, D.; Crocker, A.; Sluiter, M.D. Determination of structural carbohydrates and lignin in biomass. *Natl. Renew. Energy Lab.* **2008**, *510*, 1–16.
30. Herrera-Morales, J.; Turley, T.A.; Betancourt-Ponce, M.; Nicolau, E. Nanocellulose-block copolymer films for the removal of emerging organic contaminants from aqueous solutions. *Materials* **2019**, *12*, 230. [CrossRef]
31. Chen, Y.W.; Hasanulbasori, M.A.; Chiat, P.F.; Lee, H.V. Pyrus pyrifolia fruit peel as sustainable source for spherical and porous network based nanocellulose synthesis via one-pot hydrolysis system. *Int. J. Biol. Macromol.* **2019**, *123*, 1305–1319. [CrossRef]
32. Yue, Z.; Hou, Q.; Liu, W.; Yu, S. Autohydrolysis prior to poplar chemi-mechanical pulping: Impact of surface lignin on subsequent alkali impregnation. *Bioresour. Technol.* **2019**, *282*, 318–324. [CrossRef]
33. Segal, L.; Creely, J.J.; Martin, A.E.; Conrad, C.M. An empirical method for estimating the degree of crystallinity of native cellulose using the X-ray diffractometer. *Text. Res. J.* **1959**, *29*, 786–794. [CrossRef]
34. Bian, H.; Gao, Y.; Wang, R.; Liu, Z.; Wu, W.; Dai, H. Contribution of lignin to the surface structure and physical performance of cellulose nanofibrils film. *Cellulose* **2018**, *25*, 1309–1318. [CrossRef]
35. Errokh, A.; Magnin, A.; Putaux, J.L.; Boufi, S. Hybrid nanocellulose decorated with silver nanoparticles as reinforcing filler with antibacterial properties. *Mater. Sci. Eng. C Mater. Biol. Appl.* **2019**, *105*, 110044. [CrossRef]
36. Hong, S.; Song, Y.; Yuan, Y. Production and characterization of lignin containing nanocellulose from luffa through an acidic deep eutectic solvent treatment and systematic fractionation. *Ind. Crop. Prod.* **2020**, *143*, 111913. [CrossRef]
37. Sim, S.F.; Mohamed, M.; Lu, N.; Sarman, N.S.P.; Samsudin, S.N.S. Computer-assisted analysis of fourier transform infrared (ftir) spectra for characterization of various treated and untreated agriculture biomass. *Bioresources* **2012**, *7*, 5367–5380. [CrossRef]
38. Sun, B.; Zhang, Y.; Li, W.; Xu, X.; Zhang, H.; Zhao, Y.; Lin, J.; Sun, D. Facile synthesis and light-induced antibacterial activity of ketoprofen functionalized bacterial cellulose membranes. *Colloids Surf. A Physicochem. Eng. Asp.* **2019**, *568*, 231–238. [CrossRef]
39. Zhang, X.; Sun, H.; Tan, S.; Gao, J.; Fu, Y.; Liu, Z. Hydrothermal synthesis of Ag nanoparticles on the nanocellulose and their antibacterial study. *Inorg. Chem. Commun.* **2019**, *100*, 44–50. [CrossRef]
40. Hsu, T.C.; Guo, G.L.; Chen, W.H.; Hwang, W.S. Effect of dilute acid pretreatment of rice straw on structural properties and enzymatic hydrolysis. *Bioresour. Technol.* **2010**, *101*, 4907–4913. [CrossRef] [PubMed]
41. Zhang, S.; Zhang, F.; Jin, L.; Liu, B.; Mao, Y.; Liu, Y.; Huang, J. Preparation of spherical nanocellulose from waste paper by aqueous NaOH/thiourea. *Cellulose* **2019**, *26*, 5177–5185. [CrossRef]
42. Hamed, S.A.A.K.M.; Hassan, M.L. A new mixture of hydroxypropyl cellulose and nanocellulose for wood consolidation. *J. Cult. Herit.* **2019**, *35*, 140–144. [CrossRef]
43. Gan, I.; Chow, W.S. Synthesis of phosphoric acid-treated sugarcane bagasse cellulose nanocrystal and its thermal properties enhancement for poly(lactic acid) nanocomposites. *J. Thermoplast. Compos. Mater.* **2018**, *32*, 619–634. [CrossRef]
44. Li, M.; Yin, J.; Hu, L.; Chen, S.; Min, D.; Wang, S.; Luo, L. Effect of hydrogen peroxide bleaching on anionic groups and structures of sulfonated chemo-mechanical pulp fibers. *Colloids Surf. A Physicochem. Eng. Asp.* **2020**, *585*, 124060–124068. [CrossRef]
45. Koljonen, K.; Österberg, M.; Johansson, L.S.; Stenius, P. Surface chemistry and morphology of different mechanical pulps determined by ESCA and AFM. *Colloids Surf. A Physicochem. Eng. Asp.* **2003**, *228*, 143–158. [CrossRef]
46. Hon, D.N.-S. ESCA study of oxidized wood surfaces. *J. Appl. Polym. Sci.* **1984**, 2777–2784. [CrossRef]
47. Supian, M.A.F.; Amin, K.N.M.; Jamari, S.S.; Mohamad, S. Production of cellulose nanofiber (CNF) from empty fruit bunch (EFB) via mechanical method. *J. Environ. Chem. Eng.* **2020**, *8*, 103020–103024. [CrossRef]
48. French, A.D. Idealized powder diffraction patterns for cellulose polymorphs. *Cellulose* **2014**, *21*, 885–896. [CrossRef]
49. Gwon, J.G.; Lee, S.Y.; Doh, G.H.; Kim, J.H. Characterization of chemically modified wood fibers using FTIR spectroscopy for biocomposites. *J. Appl. Polym. Sci.* **2010**, *116*, 3212–3219. [CrossRef]
50. Trilokesh, C.; Uppuluri, K.B. Isolation and characterization of cellulose nanocrystals from jackfruit peel. *Sci. Rep.* **2019**, *9*, 16702–16709. [CrossRef] [PubMed]

51. Peng, Y.; Nair, S.S.; Chen, H.; Yan, N.; Cao, J. Effects of lignin content on mechanical and thermal properties of polypropylene composites reinforced with micro particles of spray dried cellulose nanofibrils. *ACS Sustain. Chem. Eng.* **2018**, *6*, 11078–11086. [CrossRef]
52. Hambardzumyan, A.; Foulon, L.; Chabbert, B.; Aguie-Beghin, V. Natural organic UV-absorbent coatings based on cellulose and lignin: Designed effects on spectroscopic properties. *Biomacromolecules* **2012**, *13*, 4081–4088. [CrossRef] [PubMed]
53. Habibi, Y.; Lucia, L.A.; Rojas, O.J. Cellulose nanocrystals: Chemistry, self-assembly, and applications. *Chem. Rev.* **2010**, *110*, 3479–3500. [CrossRef] [PubMed]

Article

An Experimental Study for the Improvement of the Stain Resistance for Exterior Wall Paints in a Western City in China

Xiao Huang [1], Caixia Wang [2,*] and Dengling Zhu [3]

[1] State Key Laboratory of Geomechanics and Geotechnical Engineering, Institute of Rock and Soil Mechanics, Chinese Academy of Sciences, Wuhan 430071, China; huangxiao@mail.whrsm.ac.cn
[2] Institute for Disaster Management and Reconstruction, Sichuan University, Chengdu 610041, China
[3] Henan Provincial Academy of Building Research, Zhengzhou 450000, China; ling123abc666@126.com
* Correspondence: caixia.wang@connect.polyu.hk

Abstract: In recent years, with the rapid development of the building industry in western cities during the period of "Western Development" in China, the paints industry has developed rapidly and become more mature. In addition, waterborne inorganic exterior wall paints have been a great choice in the building industry because of their reduced volatile organic compounds (VOCs) and less toxicity and odor. However, the problem of stain resistance for exterior wall paints in western cities has not been solved, which has become a major obstacle to the application and promotion of exterior wall paints in western cities in China. Therefore, effective measures should eventually be carried out for improving the stain resistance of exterior wall paints in western cities in China. In this paper, an experimental study on improving stain resistance for exterior wall paints in a typical western city in China, Chongqing, is reported. In the three defined designs, the effects of the paint structure type, the pigment volume concentration (PVC), thickeners, cosolvents and wetting and dispersing agents on the stain resistance of exterior wall paints in a typical western city in China, Chongqing, were examined. The experimental results suggest that the stain resistance of silicone–acrylic paint was the most suitable among the three kinds of tested paints (silicone–acrylic paint, styrene–acrylic paint and pure acrylic paint). In addition, the PVC had a great influence on the stain resistance of the exterior wall paints. The compactness, water absorption and stain resistance of the paint's film were the most suitable when the PVCs of the paints reached 45%. Furthermore, the tested wetting and dispersing agents made the same contributions to the paints' stain resistance, as their decline rates for reflectivity were similar. The reflectivity of the film significantly decreased when the ratio of associating thickener to non-associating thickener reached 4:1, and also significantly declined when the content of propylene glycol reached 5%.

Keywords: exterior wall paints; stain resistance; western city; volatile organic compounds (VOCs)

Citation: Huang, X.; Wang, C.; Zhu, D. An Experimental Study for the Improvement of the Stain Resistance for Exterior Wall Paints in a Western City in China. *Coatings* **2021**, *11*, 220. https://doi.org/10.3390/coatings 11020220

Academic Editor: Philippe Evon

Received: 21 January 2021
Accepted: 9 February 2021
Published: 12 February 2021

Publisher's Note: MDPI stays neutral with regard to jurisdictional claims in published maps and institutional affiliations.

1. Introduction

Architectural paints account for the largest proportion of the paints industry [1]. With the vigorous development of China's construction industry in recent years, the architectural paints industry has also entered a stage of rapid development [2–4]. Additionally, national and local policies and regulations actively promote the development of architectural paints in terms of the quality of products, the technical level of the paints and application methods [5].

Compared with solvent-based paints, waterborne paints show the advantages of less toxicity and odor and reduced volatile organic compounds (VOCs). Taking into account the harsh conditions due to increasing environmental pollution, exterior wall paints should also play important roles in protecting buildings against air pollution [6,7]. However, the stain resistance of waterborne paints is lower than that of other exterior wall decoration materials such as glass and tiles [8,9]. Some oily substances emitted from factories, smoke from automobile exhaust and dust in the air accelerate the deterioration of the exterior wall

259

surfaces of buildings [10,11]. As shown in Figure 1, the exterior wall paints of a building are seriously contaminated. The problem of stain resistance for exterior wall paints has become a serious constraint to the application and popularization of exterior wall paints in western cities in China [12–14].

Figure 1. Contaminated exterior wall paints in a typical western city in China, Chongqing.

In a field investigation of architectural exterior wall paints in a typical western city in China, Chongqing, 75 projects were randomly selected. Factors such as the building years, types of building exterior wall paints, contaminated parts of exterior wall paints, smoothness of exterior walls, color matching of exterior wall paints, orientation of building exterior walls, exterior wall materials and the structural design of exterior walls were considered. From the results of the field investigation, flow-hanging contamination, dust accumulation contamination [15] and paint cracking were regular contamination patterns for exterior wall paints in the typical western city, Chongqing. In terms of exterior wall paint flow-hanging contamination, mild-degree contamination accounted for about 31.2%, and moderate and serious flow-hanging contamination were about 69%. For the dust-accumulation contamination of exterior wall paint, the proportion of slight contamination was as low as 20.7%, and the degree of moderate and serious dust-accumulation contamination was as high as 79.3%. Additionally, the exterior wall paint surface cracked soon after completion, which seriously affected the surface of the building. These observations indicate that the overall contamination situation for exterior wall paint in this typical western city is relatively severe.

Finance and labor support have been investigated for improving the stain resistance of paints, therefore increasing the cost of buildings' construction [16,17]. For example, studies showed that the stain resistance of the paints could be improved by changing the parameters of the nanometer zinc oxide and nanometer titanium dioxide in paints [18,19]. However, it is hard to solve various types of contamination of exterior wall paints by using a single and identical method. The problem of stain resistance is becoming the main challenge during the process of the popularization and application of exterior wall paints [20].

The aim of the present work was the improvement of the stain resistance for exterior wall paints in a typical western city in China, Chongqing. In this paper, firstly, the experimental results for the effect of the type of paint on stain resistance are presented. This is followed by reporting the experimental results for the effect of the pigment volume concentration (PVC) on the stain resistance of exterior wall paints. Thirdly, the effects of thickeners, cosolvents and wetting and dispersing agents on the stain resistance of exterior wall paints are presented. Finally, a conclusion is given.

2. Materials and Methods

2.1. Paint Formulation

In the experimental study presented in this paper, three different designs, named Design 1, Design 2 and Design 3, were defined. In Design 1, three different types of paints (pure acrylic paint, styrene–acrylic paint and silicone–acrylic paint) were used; in Design 2, paints with 7 different PVCs were examined, and in Design 3, the effects of thickeners, cosolvents and wetting and dispersing agents were compared. The formulations of the paints for the three designs are detailed in this Section.

Firstly, the sources of the raw materials of the experimental paints are shown in Table 1.

Table 1. Sources of raw materials of the experimental paints.

Name of Raw Material	Model	Manufacturer
Propylene glycol	-	Dow Chemical Co., Ltd., Midland, MI, USA
Dispersing agent	2500	Japan San Nuopco Co., Ltd., Tokyo, Japan
Fungicide	LXE	Dow Chemical Co., Ltd., Midland, MI, USA
Wetting agent	X450	Rohm and Haas Co., Ltd., Guangdong, China
Defoamer	F111	Haichuan Co., Ltd., Shenzhen, China
Defoamer	1340	Haichuan Co., Ltd., Shenzhen, China
Hydroxyethyl cellulose	3w	Shanghai Huguang Co., Ltd., Shanghai, China
Multifunctional auxiliaries	APM-95	Japan San Nuopco Co., Ltd., Tokyo, Japan
Film-forming agent	Texanol	Eisman Co., Ltd., Japan
Thickening agent	2020	Rohm and Haas Co., Ltd., Foshan, China
Thickening agent	ASE60	Rohm and Haas Co., Ltd., Foshan, China
Titanium dioxide	R-998	Yugang Titanium dioxide Co., Ltd., Chongqing, China
Calcined kaolin	-	Market sale, Chongqing, China
Water	Ultra-pure water	Market sale, Chongqing, China
Silicone–acrylic paint	TD-686	Jiangsu Sunrise Co., Ltd., Suzhou, China
Styrene–acrylic paint	DC-420	Jiangsu Sunrise Co., Ltd., Suzhou, China
Pure acrylic paint	TRC-4369	Jiangsu Sunrise Co., Ltd., Suzhou, China
Pure acrylic paint	2468	Dow Chemical Co., Ltd., Midland, MI, USA

Secondly, for Design 1, three water-based paints, pure acrylic paint, styrene–acrylic paint and silicone–acrylic paint, were prepared. The compositions of the paints for Design 1 are given in Table 2.

Table 2. Compositions of the paints for Design 1.

No.		Ingredients (wt.‰)	Content
	1	Deionized water	281.50
	2	Propylene glycol	15.00
	3	Dispersing agent SN5040	4.00
	4	Dispersing agent 2500	2.00
	5	Fungicide LXE	1.00
	6	Wetting agent X405	1.00
Pulping stage	7	Defoamer F111	2.00
	8	Cellulose 3W	1.50
	9	AMP-95	2.00
	10	Titanium dioxide R998	50.00
	11	Calcined kaolin	120.00
	12	Wollastonite	150.00
	13	Texanol	8.00
	14	Paint A-I	350.00
Painting stage	15	Defoamer 1340	2.00
	16	Thickener 2020	2.00
	17	Thickener ASE60	8.00
Total			1000.00

In Design 1, three types of paints with different structures were selected as variables. Paints A, B and C are silicone–acrylic paints; paints D, E and F are styrene–acrylic paints, and paints G, H and I are pure acrylic paints.

Thirdly, in Design 2, paints with 7 different PVCs were examined. The compositions of the paints for Design 2 are given in Tables 3 and 4.

Table 3. Compositions of the paints for Design 2.

Materials	Mixing Amount (g)
Deionized water	145.00
Propanediol	7.50
Dispersing agent SN5040	2.00
Dispersing agent 2500	1.00
Fungicide LXE	0.50
Wetting agent X405	0.50
Defoamer NXZ	1.00
Cellulose	0.75
AMP-95	1.00
Pigment and filler	Table 4
Texanol	4.00
Silicone–acrylic paint	175.00
Defoamer1340	1.00
Thickener 2020	1.00
Thickener ASE60	4.00

Table 4. The pigment volume concentrations (PVCs) in formulations for Design 2.

PVC (%)	25%	35%	40%	45%	50%	55%	60%
Titanium dioxide	6.50	9.20	15.10	22.50	29.20	36.10	42.30
Calcined kaolin	32.90	43.10	50.20	57.70	65.40	72.20	75.90
Wollastonite	51.10	63.20	70.10	77.90	85.80	91.40	96.20

Fourthly, in Design 3, paints with different thickeners, wetting and dispersing agents and cosolvents were used. The compositions of the paints for Design 3 are given in Tables 5 and 6.

Table 5. Compositions of the paints for Design 3 (thickeners).

Thickener System	Content of Hydroxyethyl Cellulose Ether			Collocation of Associating Thickeners and Non-Associating Thickeners (Associating Thickeners:Non-Associating Thickeners)		
	1.2‰	1.5‰	1.8‰	1:4	1:1	4:1
Deionized water	277.50	277.50	277.50	277.50	277.50	277.50
Propanediol	15.00	15.00	15.00	15.00	15.00	15.00
Dispersing agent APC	6.00	6.00	6.00	6.00	6.00	6.00
Fungicide LXE	1.00	1.00	1.00	1.00	1.00	1.00
Wetting agent 070	1.00	1.00	1.00	1.00	1.00	1.00
Defoamer 101	4.00	4.00	4.00	4.00	4.00	4.00
Cellulose	1.20	1.50	1.80	1.50	1.50	1.50
AMP-95	2.00	2.00	2.00	2.00	2.00	2.00
Titanium dioxide R258	80.00	80.00	80.00	80.00	80.00	80.00
Calcined kaolin	100.00	100.00	100.00	100.00	100.00	100.00
Wollastonite	140.00	140.00	140.00	140.00	140.00	140.00
Texanol	8.00	8.00	8.00	8.00	8.00	8.00
TRC-4369	250.00	250.00	250.00	250.00	250.00	250.00
Defoamer 732	4.00	4.00	4.00	4.00	4.00	4.00
Thickener 8W	2.00	2.00	2.00	2.00	5.00	8.00
ASE60	8.00	8.00	8.00	8.00	5.00	2.00
Total			1000.00			

Table 6. Compositions of the paints for Design 3 (cosolvents).

Content of Propanediol	5%	10%	15%
Deionized water	291.50	286.50	281.50
Propanediol	5.00	10.00	15.00
Dispersant APC	6.00	6.00	6.00
Fungicide LXE	1.00	1.00	1.00
Wetting agent 070	1.00	1.00	1.00
Defoamer 101	2.00	2.00	2.00
Cellulose	1.50	1.50	1.50
AMP-95	2.00	2.00	2.00
Titanium dioxide R258	80.00	80.00	80.00
Calcined kaolin	100.00	100.00	100.00
Wollastonite	140.00	140.00	140.00
Texanol	8.00	8.00	8.00
TRC-4369	250.00	250.00	250.00
Defoamer 732	2.00	2.00	2.00
Thickener 8W	2.00	2.00	2.00
ASE60	8.00	8.00	8.00
Total		1000.00	

2.2. The Test Equipment and Instruments

The main test equipment and analytical instruments used in the experiments are shown in Table 7.

Table 7. Main test equipment and analytical instruments.

Name of Instrument and Equipment	Manufacturer
QBY-II swing rod type paint film hardness tester	Tianjin Material Testing Machine Factory, Tianjin, China
QHQ-A film-coated pencil scratch hardness tester	Tianjin Material Testing Machine Factory, Tianjin, China
QRS paint film reflectance tester	Tianjin Material Testing Machine Factory, Tianjin, China
LXY-IV paint film washing resistance tester	Tianjin Material Testing Machine Factory, Tianjin, China
SBDY-1 whiteness meter	Hebei Rongda Luqiao Science Equipment Factory, Cangzhou, China
QFD electric paint film adhesion tester	Tianjin Material Testing Machine Factory, Tianjin, China
JC2000A static drop contact angle/interfacial tension-measuring instrument	Shanghai Zhongchen Digital Technology equipment Co., Ltd., Shanghai, China
JL-1155 laser particle size distribution tester	Chengdu Jingxin Powder testing equipment Co., Ltd., Chengdu, China
Nano ZS90 laser particle size distribution measurement and potential analyzer	British Malvern Co., Ltd., Malvern, UK
Optical microscope	Shanghai Optical instrument Factory, Shanghai, China
Ultraviolet spectrophotometer	Japan Hitachi Co., Ltd., Tokyo, Japan

2.3. The Experimental Methods

As previously reported in Section 2.1, three different designs, named Design 1, Design 2 and Design 3, were examined, as shown in Table 8. The experimental methods for the three designs are described in detail in this Section.

All the related tests were carried out under the standard test conditions of $23 \pm 2\,°C$ and $50\% \pm 5\%$ relative humidity.

In Design 1, three kinds of paints (pure acrylic paint, styrene–acrylic paint and silicone–acrylic paint) were compared in three aspects (the water resistance, alkali resistance and stain resistance). Based on Design 1, a paint with a high stain-resistance property was selected. The experimental methods of the water-resistance test, the alkali-resistance test and the stain-resistance test for the paints were conducted as follows.

The water resistance of the paint was tested according to GB/T1733-1993 [21]. For the experimental methods of the water-resistance test, asbestos cement boards were used as the test boards, and they had identical dimensions of $70 \times 150 \times 5\ mm^3$. The test boards used in the experiments adopted the high-density grade IV chrysotile fiber cement plate specified in JC/T412.2-2006 [22]. The treatment of the test plate was carried out in accordance with

GB/T9271-1988 [23]. In addition, the average pH of the rainfall in this typical western city in China, Chongqing, was 4.87; the main acidifying substance of the rainfall was sulfate, and the main anions in the precipitation were sulfate and nitrate, accounting for 45.6 and 6.9% of the total ions, respectively. Therefore, in the water-resistance test, analytically pure reagents, such as nitric acid, sulfuric acid and deionized water, were used to make an acid solution that was similar to the condition of acid rainfall in Chongqing, with a pH value of 4.87 (2% sulfuric acid solution and 0.5% nitric acid solution). To follow up, three standard test boards, with the edges and backs sealed by mixtures of rosin and paraffin wax according to the mass ratio of 1:1, were used. Two of the three test boards were soaked in the acidic solution. The observation continued until the surfaces of the test boards powdered, peeled off, blistered or became exposed.

Table 8. Detailed description of Design 1, Design 2 and Design 3.

Designs	Experimental Content	Comparison
Design 1	Pure acrylic paint Styrene–acrylic paint Silicone–acrylic paint	Water resistance Alkali resistance Stain resistance
Design 2	PVC concentrations	Compactness Water absorption Stain resistance
Design 3	Thickeners Cosolvents Wetting and dispersing agents	Stain resistance

The alkali resistance of the paint was tested according to GB/T 9265-2009 [24]. Two of the three test boards that had been treated in the same way as in the previous water-resistance test were soaked in a calcium hydroxide-saturated solution. The observation continued until the surfaces of the test boards blistered, peeled off, powdered, became soft or became exposed.

The stain resistance of exterior wall paints was tested according to GB/T9780-2005 [25]. The test boards were treated in the same way as in the previous water-resistance test. In terms of the configuration of fly-ash water, the fly ash was mixed with water in the ratio of 1:1 (mass). Following up were tests of the original reflection coefficients of the cured test boards and the reflection coefficient of the paint template after brushing and rinsing according to GB/T9780-2005The decline rate of the reflection coefficient of the paint is expressed by Equation (1):

$$X = \frac{(A - B)}{A} \times 100\% \tag{1}$$

where X is the reflection coefficient of the paint, %; A, the original reflection coefficient of the cured test boards, %; and B, the reflection coefficient of the paint template after brushing and rinsing, %.

In Design 2, paints with 7 different PVCs were examined in three aspects (compactness, water absorption and stain resistance). The PVC refers to the ratio of the volume of the pigments and fillers in the paint to the total volume of all the nonvolatile components in the formula (including resins, solid components of the paints, pigments, fillers, etc.). The critical pigment volume concentration (CPVC) means that the volume of the base material just covers the surface of the pigment particles and fills the stacking space of the pigment particles. At this point, the pigment volume concentration is generally expressed by the CPVC. At low PVC values, the surface contact of the pigment particles is considerably low, but with an increase in pigments, when the PVC exceeds a certain extreme value, the base material is not able to completely fill the gaps between the pigment particles. These unfilled voids are left in the film, so the physical properties of the film begin to decrease sharply when the PVC exceeds the extreme value, and the PVC value at this point is called the CPVC. Therefore, the performance of the paints changes when they are near the CPVC values. In the experiments, by measuring the properties of the paints at different PVC values, such as the density, tensile strength, adhesion, coloring power, permeability, gloss and coverage, the sudden-change points could be found, allowing the CPVC values to be determined.

In terms of Design 3, the effects of thickeners, cosolvents and wetting and dispersing agents were studied on exterior wall paints' stain resistance. Firstly, to study the performance of thickeners against contamination in paint, it was mainly compared according to the content of hydroxyethyl cellulose ether, and the ratio of associative and non-associative thickeners. The contents of hydroxyethyl cellulose ether in the three groups were 1.2, 1.5 and 1.8‰, respectively. Furthermore, the ratios of associating and non-associating thickeners in the three groups were 1:4, 1:1 and 4:1, respectively. Secondly, in order to study the effect of wetting and dispersing agents on the stain resistance of exterior wall paints, the Clariant LCN070 and ED3060 wetting agents were used in combination with a Clariant dispersant, APC, in a group of experiments. The content of the wetting agent was 0.1%, and that of the dispersant was 0.6%. Thirdly, in the study of the effects of cosolvents on paints' stain resistance, propylene glycol was selected as the cosolvent, and the contents of the cosolvent propylene glycol were 0, 5, 10 and 15%, respectively.

3. Results and Discussion

As previously reported in Section 2, three different designs, named Design 1, Design 2 and Design 3, were examined in this paper. To follow up, the results and discussion for the three designs are detailed in this Section.

3.1. Design 1

In Design 1, three water-based paints, pure acrylic paint, styrene–acrylic paint and silicone–acrylic paint, were prepared by varying three aspects (the water resistance, alkali resistance and stain resistance) to study the effects on exterior wall paints' stain resistance. The formulation of the paints in Design 1 is described in detail in Section 2.1.

Firstly, the experimental results for the water resistance of the paints are given in Tables 9 and 10. On the one hand, Table 9 shows that the water resistance of the nine experimental exterior wall paints showed no changes after being soaked in water for 30 days. The reason was that the resin molecules in the paints had a dense internal structure and good surface properties such as water repellent and hydrophobic properties. On the other hand, in the water-resistance tests, the acid-rainwater condition of a typical western city in China, Chongqing, was simulated. As previously reported in Section 2.3, analytically pure reagents, such as nitric acid, sulfuric acid and deionized water, were used to make an acid solution that was similar to the condition of acid rainfall in Chongqing, with a pH value of 4.87 (2% sulfuric acid solution and 0.5% nitric acid solution). Table 10 shows that, in the experiment of simulating the composition of the rainfall in Chongqing, the rainfall resistance of the A, B and C exterior wall paints was the most suitable. Paints A, B and C were all silicone–acrylic exterior wall paints, and the excellent properties of the silicone–acrylic paints benefited from the silicone resin in them. The main chain of the silicone resin was Si–O–Si, in which the bond energy of Si–O was 422.5 kJ/mol, and the bond energy of the side chain C–Si was 318 kJ/mol, so the silicone–acrylic paint had good acid-rainfall resistance performance. The results also show that the acid-rainfall resistance of the silicone–acrylic paint was better than that of the styrene–acrylic paint, and that of the styrene–acrylic paint was better than that of the pure acrylic paint.

Table 9. Results for Design 1 (water resistance I).

No.	A	B	C	D	E	F	G	H	I
Time/d	30	30	35	30	30	35	30	30	35
Performance	No change	No change	No change	No change	No change	No change	No change	No change	No change

Table 10. Results for Design 1 (water resistance II).

No.	A	B	C	D	E
Time/h	105	93	120	70	82
Performance	Blistering	Small holes	Underlying paint exposure	Blistering	Underlying paint exposure
No.	F	G	H	I	-
Time/h	67	39	62	45	-
Performance	Small holes	Blistering	Underlying paint exposure	Blistering	-

Secondly, the experimental results for the alkali resistance of the nine exterior wall paints A–I are shown in Table 11. Table 11 shows that the alkali resistance of the nine exterior wall paints in the experiment met the criterion of the national standard for alkali resistance GB/T 9265-2009. The silicone–acrylic paint, styrene–acrylic paint and pure acrylic paint all had good alkali-resistance properties and were suitable for use as exterior wall materials.

Table 11. Results for Design 1 (alkali resistance).

No.	A	B	C	D	E	F	G	H	I
Time/d	35	35	40	35	35	40	35	35	40
Performance	No change	No change	No change	No change	No change	No change	No change	No change	No change

Thirdly, for the experiment for stain resistance, after brushing two times, four times and five times, the changes in the reflection rates and the decline rates for the paints' films for the nine exterior wall paints A–I are shown in Figures 2–4, respectively.

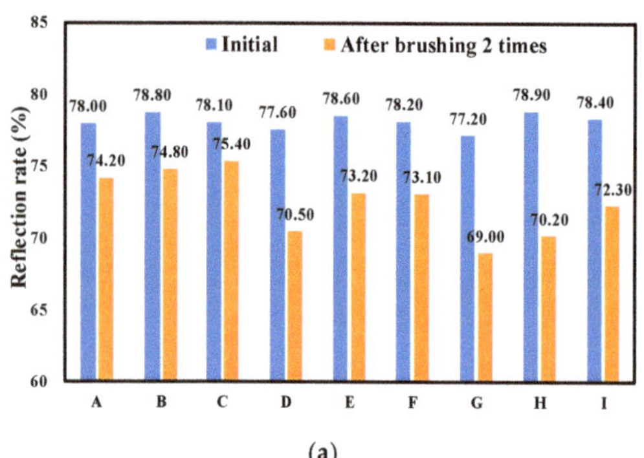
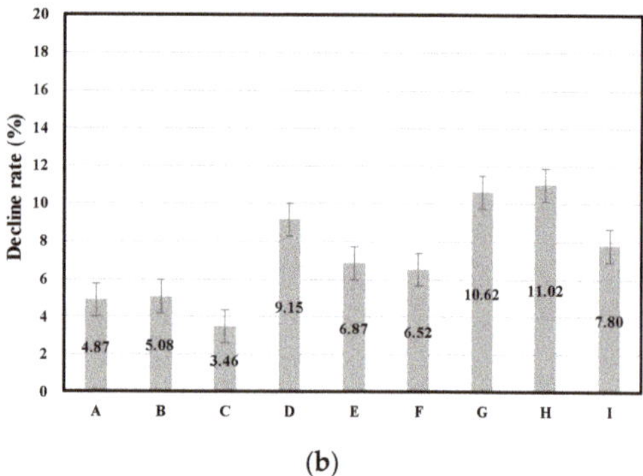

Figure 2. Design 1: (**a**) the change in the reflection rate and (**b**) the decline rate (after brushing 2 times).

As shown in Figure 2, after brushing two times, the highest decline rate for reflection occurred in group H (11.02%), from 78.90% to 70.20%, and the lowest decline rate for reflection occurred in group C (3.46%), from 78.10% to 75.40%. In addition, the average decline rate for reflection for groups A–C (silicone–acrylic paint) was 4.47%, and those for groups D–F (styrene–acrylic paint) and groups G–I (pure acrylic paint) were 7.51% and 9.81%, respectively.

Figure 3 demonstrates that after brushing four times, the highest decline rate for reflection occurred in group H (13.68%), from 78.9% to 68.10%, and the lowest decline rate for reflection occurred in group C (4.10%), from 78.10% to 74.90%. In addition, the average decline rate for reflection for groups A–C (silicone–acrylic paint) was 5.57%, and those for groups D–F (styrene–acrylic paint) and groups G–I (pure acrylic paint) were 9.69% and 12.66%, respectively.

Figure 4 shows that after brushing five times, the highest decline rate for reflection occurred in group H (18.25%), from 78.90% to 64.50%, and the lowest decline rate for reflection occurred in group C (4.61%), from 78.10% to 74.50%. In addition, the average decline rate for reflection for groups A–C (silicone–acrylic paint) was 6.42%, and those for groups D–F (styrene–acrylic paint) and groups G–I (pure acrylic paint) were 12.21% and 17.23%, respectively.

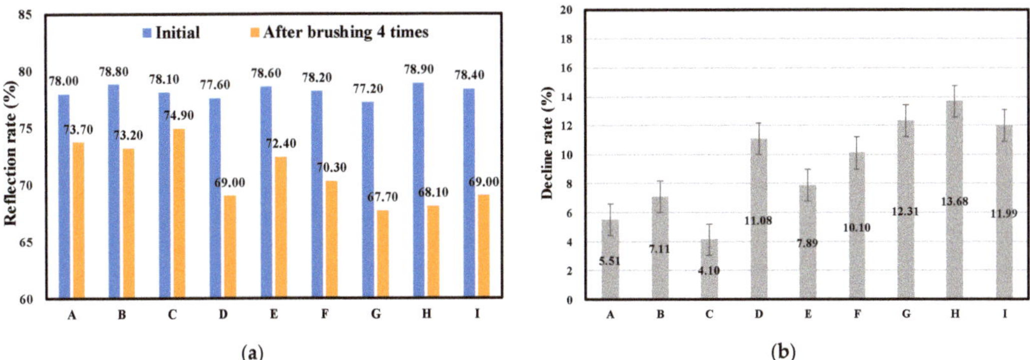

Figure 3. Design 1: (**a**) the change in the reflection rate and (**b**) the decline rate (after brushing 4 times).

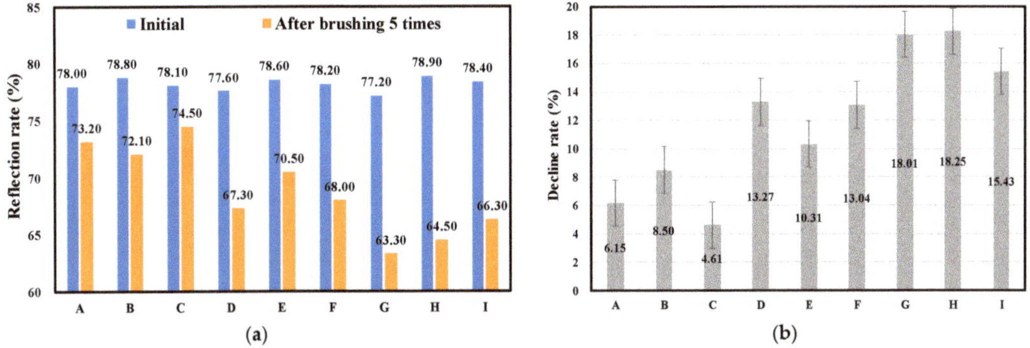

Figure 4. Design 1: (**a**) the change in the reflection rate and (**b**) the decline rate (after brushing 5 times).

The experimental results shown in Figures 2–4 demonstrate that the average decline rate for groups G–I (pure acrylic paint) was the highest, followed by that for groups D–F (styrene–acrylic paint), and that for groups A–C (silicone–acrylic paint) was the lowest. The experimental results demonstrate that the stain resistance of the silicone–acrylic paint was the most suitable, followed by that of the styrene–acrylic paint, and the stain-resistance ability of the pure acrylic paint was the lowest. The reason was that the silicone–acrylic paint was modified by introducing a certain number of organosilicon groups into the acrylic resin structure, and the surface tension was low. The film had non-adhesive properties. The silicone-modified acrylic resin had the mentioned advantages and thus can well comply with the requirements for exterior wall paints. The main chain of the acrylic resin in the G, H and I pure acrylic paints was C–C bonds; the bond energy was lower than that of Si–O–Si, though. Its continuous film had a lower surface energy, so the stain resistance of the pure acrylic paint was lower than that of the silicone–acrylic paint.

3.2. Design 2

For Design 2, paints with seven different PVCs were examined. The results for Design 2 are reported in terms of three aspects: the effects of the PVC on the compactness, water absorption and stain resistance of a paint's film. The formulations of the paints in Design 2 are detailed in Section 2.1.

Figure 5 shows the effect of the PVC on the compactness of the paint's film. The compactness of the paint's film was closely related to its porosity. The higher the porosity is, the lower the compactness of the paint's film is. With a continuous increase in PVC, the porosity of the film increased gradually. As shown in Figure 5, when the PVC reached 45%, the porosity changed suddenly, and then, the porosity of the film increased sharply. This is explained by the fact that when the PVC was less than the CPVC, the pigments and fillers in the paint system were wrapped by enough paint particles. After the moisture volatilization in the film-forming process for the paint, the paint particles, pigments and fillers were squeezed by capillary forces and other forces to form a complete film. The evaporation of water was bound to leave a certain gap. When the PVC was less than the CPVC, there were more paint particles around the pigments and fillers. However, once the PVC exceeded the paint particles, a large number of air and insufficient paint particles occupied the gap between the pigments and fillers. Figure 5 shows that the CPVC of the experimental paint in the test was about 45%.

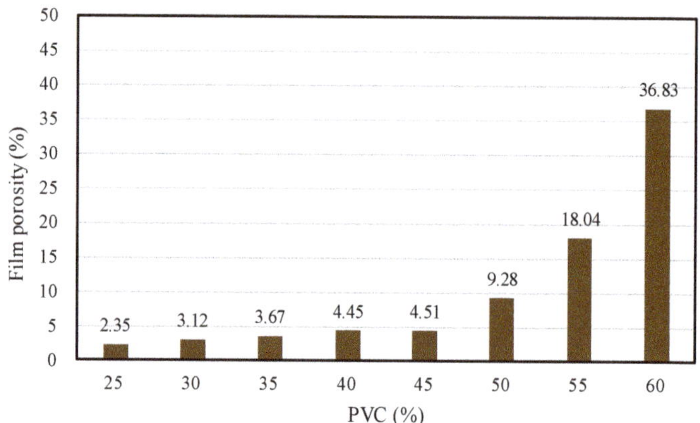

Figure 5. Effect of PVC on porosity of paint's film.

Figure 6 shows the effects of different PVCs on the water absorption of the paint's film. Seven groups of samples were tested. The pore was the main means of water absorption by the film when the surface energy of the film was certain. The porosity was the main factor affecting the water absorption of the film, and an increase in the porosity led to an increase in the water absorption of the film. Figure 6 shows that the water absorption of the film increased with an increase in PVC. When the PVC reached 45%, the curve had an inflection point, and then, the water absorption increased sharply with an increase in PVC.

Figure 7 shows the effects of different PVCs on the stain resistance of the paint's film. The decreases in the values of reflectivity of seven groups of different PVC samples were tested to characterize the stain resistance of the film. This is because the reflectivity value of the paint's film had been shown to be closely related to its stain resistance. The higher the reflectivity value of the film was, the worse the stain resistance rate of the film was, and vice versa. As can be observed in Figure 7, the decline rate for the reflection value of the film decreased gradually with an increase in PVC. When the PVC increased to about 45%, the decline rate for the reflection value decreased to the lowest point and then increased sharply with an increase in PVC. At the beginning, the stain resistance of the film gradually improved with an increase in PVC; it began to decrease sharply when the PVC exceeded a certain point.

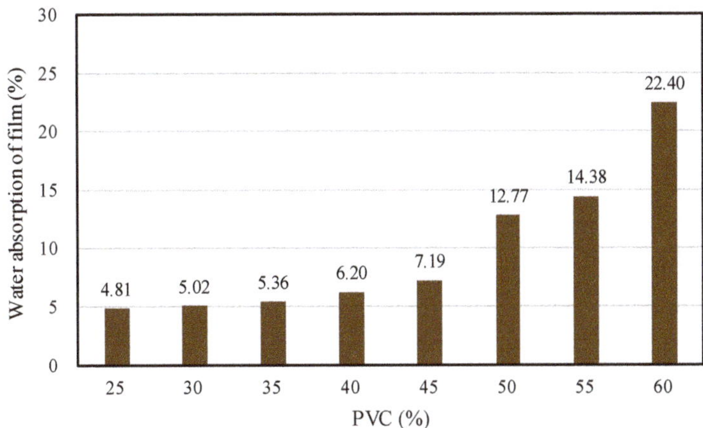

Figure 6. Effect of PVC on water absorption of paint's film.

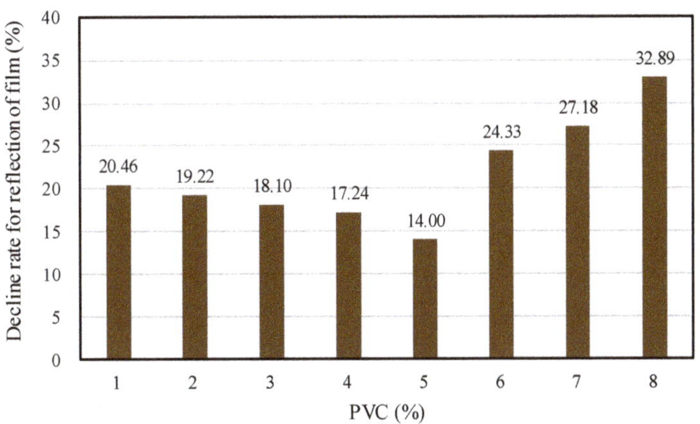

Figure 7. Effect of PVC on decrease rate for reflection of paint's film.

The results for Design 2 suggest that the PVC was an important factor affecting the stain resistance of the paints. When the PVC reached 45%, the compactness, water absorption and stain resistance of the paint's film were the most suitable.

3.3. Design 3

In terms of Design 3, the effects of thickeners, cosolvents and wetting and dispersing agents were compared. The results for Design 3 are reported in terms of three aspects: the effects of thickeners, cosolvents and wetting and dispersing agents on the stain resistance of the paint's film. The formulations of the paints in Design 3 are detailed in Section 2.1.

Figures 8 and 9 show the effects of thickeners on the stain resistance of the paints. Paints with seven different contents of thickeners were examined. Figure 8 shows that after brushing one time, the average decline rates for the reflection value decreased, from 52.23% to 50.98%, with an increase in the content of hydroxyethyl cellulose ether from 1.2 to 1.8‰. In addition, as illustrated in Figure 9, the average decline rates for the reflection value varied from 51.25% and 51.44% to 50.01%, with changes in the ratio of associating thickener to non-associating thickener from 4:1 and 1:1 to 1:4, respectively. Figures 8 and 9 suggest that the reflection decline rate decreased with an increase in the content of hydroxyethyl cellulose ether, and the most suitable ratio of associating thickener to non-associating thickener for the experimental paint was 4:1.

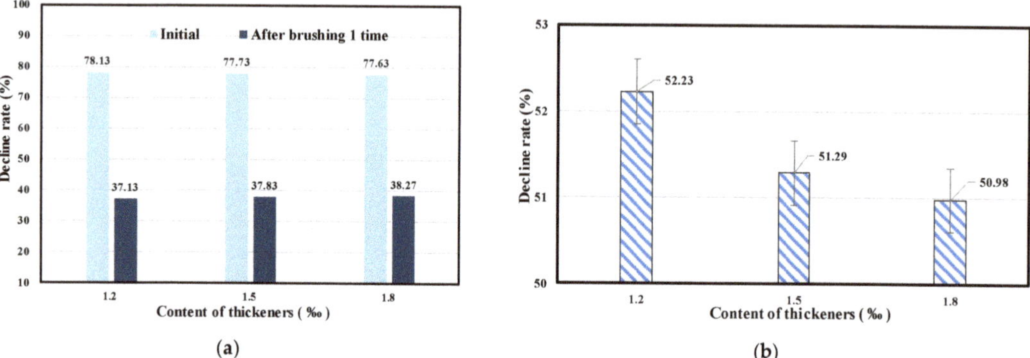

Figure 8. Design 3: (**a**) the change in the reflection rate and (**b**) the decline rate (content of thickeners).

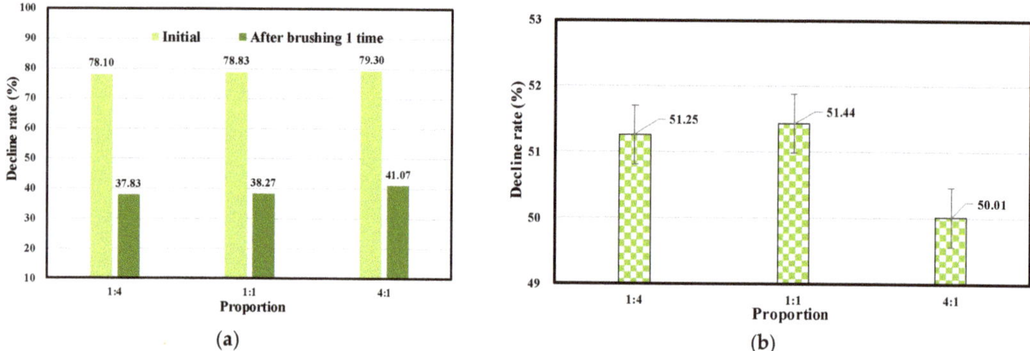

Figure 9. Design 3: (**a**) the change in the reflection rate and (**b**) the decline rate (different ratios of associating thickener and non-associating thickener).

Table 11 shows the effects of wetting and dispersing agents on the stain resistance of the paints. In order to study the effect of wetting and dispersing agents on the stain resistance of exterior wall paints, the Clariant LCN070 and ED3060 wetting agents were used in combination with the Clariant dispersing agent APC in experiments. The content of the wetting agent was 0.1%, and that of the dispersant was 0.6%. The experimental results for the wetting agents using LCN070 and ED3060 and the Clariant dispersing agent APC are shown in Table 12.

Table 11 demonstrates that the decline rates for the two paints with different wetting and dispersing agents were basically the same, indicating that the contributions of the two wetting and dispersing agents in the stain resistance of the paints in this group of experiments were basically the same.

The next part was the study of the effect of the amount of cosolvent on the stain resistance of exterior wall paints. The function of propanediol as a cosolvent was to fully dissolve the filler in the liquid phase. In this group of experiments, the amounts of the cosolvent propylene glycol were 5, 10 and 15%, respectively. Figure 10 shows the effect of the cosolvent on the stain resistance of the paints. Propylene glycol was the cosolvent in the experiments. It can be seen from Figure 10 that when the content of propylene glycol was 5%, the decline rate for the reflectivity, the stain resistance, of the paints was lower, indicating that the stain resistance of the paints was the highest in this group when the content of propylene glycol was 5%.

Table 12. The effects of wetting and dispersing agents on stain resistance of paints.

Type of Wetting Agent	Type of Dispersing Agent	A (%)	B (%)	Decline Rate for Reflection Coefficient (%)	Average Decline Rate
LCN070	APC	80.00	42.43	46.96	48.76%
		79.63	40.17	49.56	
		79.30	39.83	49.77	
ED3060	APC	80.50	40.77	49.36	48.70%
		80.13	40.47	49.50	
		79.97	42.18	47.25	

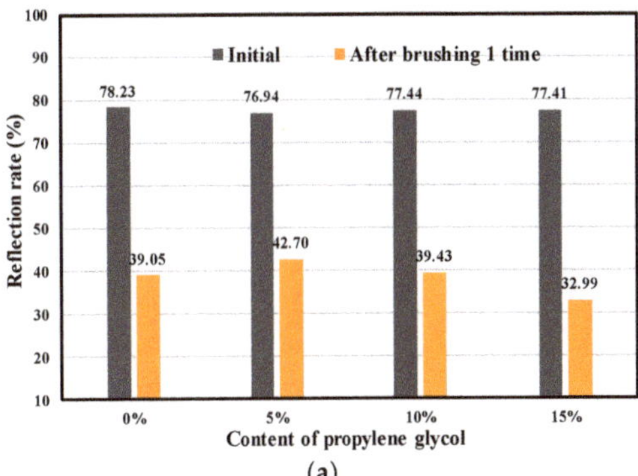

(a)

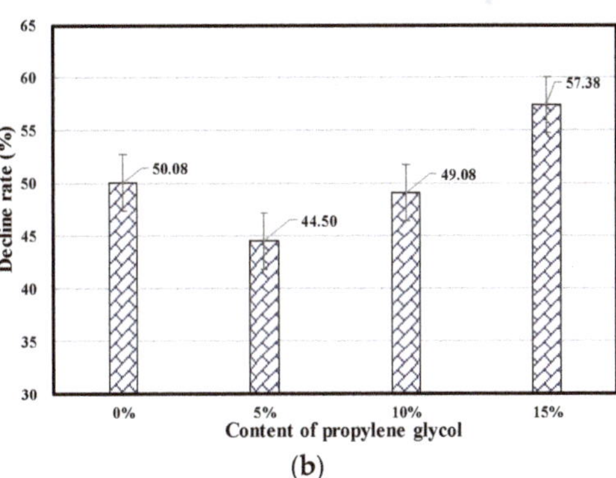

(b)

Figure 10. Design 3: (**a**) the change in the reflection rate and (**b**) the decline rate (according to content of propylene glycol).

4. Conclusions

This paper experimentally studied the improvement of stain resistance for exterior wall paints in a typical western city in China, Chongqing. Three different designs, named Design 1, Design 2 and Design 3, were studied in this paper. The results for Design 1 suggest that the stain resistance of architectural paints with different paint structures was different. After comparing the water resistance, alkali resistance and stain resistance of silicone–acrylic paint, styrene–acrylic paint and pure acrylic paint, the results show that the stain resistance of the silicone–acrylic-paint-based exterior wall paints was the most suitable among the three. The results for Design 2 show that the exterior wall paints' PVC had a great influence on the stain resistance, and the film reflectivity decreased gradually with an increase in PVC. When the PVC increased to about 45%, the film reflectivity significantly decreased and then increased sharply with an increase in PVC, indicating that when PVC $\leq$ CPVC, the stain resistance of the paint increased with an increase in PVC. In Design 3, the effects of thickeners, cosolvents and wetting and dispersing agents were compared. When the content of a thickener (hydroxyethyl cellulose ether) increased, the reflectivity decline rate also decreased, and hence, the stain resistance of the paint was enhanced. When the ratio of the associated thickener to the non-associated thickener was 4:1, the reflectivity decline rate was the smallest, and the stain resistance of the paint was the most suitable. When the content of a cosolvent (propylene glycol) was 5%, the reflectivity decline rate, the contamination resistance, decreased significantly, suggesting that the stain resistance of the paints with 5% propylene glycol was the most suitable in the comparison group. The reflectivity decline rates for the two paints with different wetting and dispersing agents were basically the same, indicating that the contributions of the two wetting and dispersing agents in the stain resistance of the paints in this group of experiments were basically the same.

Author Contributions: Data curation, D.Z.; formal analysis, X.H.; investigation, D.Z.; writing – original draft, C.W.; writing – review & editing, X.H. All authors have read and agreed to the published version of the manuscript.

Funding: This research received no external funding.

Institutional Review Board Statement: Not applicable.

Informed Consent Statement: Not applicable.

Acknowledgments: The authors gratefully acknowledge financial support from the Hong Kong Polytechnic University and Sichuan University.

Conflicts of Interest: The authors declare no conflict of interest.

References

1. Challener, C. The Paint and Coatings Industry in the Age of Digitalization. *Coatings* **2018**, *15*, 54–60.
2. Hsu, Y.-T.; Wang, W.-H.; Hung, W.-H. Architectural Sustainability and Efficiency of Enhanced Waterproof Coating from Utilization of Waterborne Poly (Siloxane-Imide-Urethane) Copolymers on Roof Surfaces. *Sustainability* **2020**, *12*, 4411. [CrossRef]
3. Kellogg, S. Developing and Evaluating Early Rain Resistance in Exterior Architectural Coatings. *Coatings* **2018**, *15*, 46–49.
4. Wieroniey, T. Life-Cycle Assessment of Architectural Coatings Considering Different Preservative Scenarios. *Coatings* **2018**, *15*, 41–44.
5. Dalawai, S.P.; Aly, M.A.S.; Latthe, S.S.; Xing, R.; Sutar, R.S.; Nagappan, S.; Ha, C.-S.; Sadasivuni, K.K.; Liu, S. Recent Advances in durability of superhydrophobic self-cleaning technology: A critical review. *Prog. Org. Coat.* **2020**, *138*, 105381. [CrossRef]
6. Faraldos, M.; Kropp, R.; Anderson, M.A.; Sobolev, K. Photocatalytic hydrophobic concrete coatings to combat air pollution. *Catal. Today* **2016**, *259*, 228–236. [CrossRef]
7. Striegel, M.F.; Guin, E.B.; Hallett, K.; Sandoval, D.; Swingle, R.; Knox, K.; Best, F.; Fornea, S. Air pollution, coatings, and cultural resources. *Prog. Org. Coat.* **2003**, *48*, 281–288. [CrossRef]
8. Assaad, J.J. Value-added waste latex paints in masonry cement for plastering applications. *J. Adhes. Sci. Technol.* **2020**, *34*, 2703–2724. [CrossRef]
9. Qureshi, S.A.; Shafeeq, A.; Ijaz, A.; Butt, M.M. Development and Regression Modeling of Dirt Resistive Latex Façade Paint. *Coatings* **2019**, *9*, 150. [CrossRef]
10. Rosso, F.; Pisello, A.L.; Castaldo, V.L.; Ferrero, M.; Cotana, F. On Innovative Cool-Colored Materials for Building Envelopes: Balancing the Architectural Appearance and the Thermal-Energy Performance in Historical Districts. *Sustainability* **2017**, *9*, 2319. [CrossRef]
11. De Masi, R.F.; Ruggiero, S.; Vanoli, G.P. Acrylic white paint of industrial sector for cool roofing application: Experimental investigation of summer behavior and aging problem under Mediterranean climate. *Sol. Energy* **2018**, *169*, 468–487. [CrossRef]
12. Guo, M.Z.; Maury-Ramirez, A.; Poon, C.S. Self-cleaning ability of titanium dioxide clear paint coated architec-tural mortar and its potential in field application. *J. Clean. Prod.* **2016**, *112*, 3583–3588. [CrossRef]
13. Ibrahim, M.; Bianco, L.; Ibrahim, O.; Wurtz, E. Low-emissivity coating coupled with aerogel-based plaster for walls' internal surface application in buildings: Energy saving potential based on thermal comfort assessment. *J. Build. Eng.* **2018**, *18*, 454–466. [CrossRef]
14. Song, H.; Lei, X.; Xue, F.; Wu, H.; Cheng, F. Composite phase-change coating with coal-fly-ash-based zeolite as carrier. *Prog. Org. Coatings* **2019**, *136*, 136. [CrossRef]
15. Jiang, Y. Study on the Influential Factors and Improvement Measures of Rain Prints of the Exterior Latex Paint. Master's Thesis, Chongqing University, Chongqing, China, 2015.
16. Jankolovits, J.; Kusoglu, A.; Weber, A.Z.; Van Dyk, A.; Bohling, J.; Roper, I.J.A.; Radke, C.J.; Katz, A. Stable Aqueous Dispersions of Hydrophobically Modified Titanium Dioxide Pigments through Polyanion Adsorption: Synthesis, Characterization, and Application in Coatings. *Langmuir* **2016**, *32*, 1929–1938. [CrossRef] [PubMed]
17. Parvate, S.; Mahanwar, P. Insights into the preparation of water-based acrylic interior decorative paint: Tuning binder's properties by self-crosslinking of ally! Acetoacetate-hexamethylenediamine. *Prog. Org. Coat.* **2019**, *126*, 142–149. [CrossRef]
18. Noman, M.T.; Petrů, M. Functional Properties of Sonochemically Synthesized Zinc Oxide Nanoparticles and Cotton Composites. *Nanomaterials* **2020**, *10*, 1661. [CrossRef] [PubMed]
19. Noman, M.T.; Petru, M.; Militky, J.; Azeem, M.; Ashraf, M.A. One-Pot Sonochemical Synthesis of ZnO Nano-particles for Photocatalytic Applications, Modelling and Optimization. *Materials* **2020**, *13*, 14. [CrossRef] [PubMed]
20. Pandit, S.K.; Tudu, B.K.; Mishra, I.M.; Kumar, A. Development of stain resistant, superhydrophobic and self-cleaning coating on wood surface. *Prog. Org. Coat.* **2020**, *139*, 105453. [CrossRef]

21. GB/T1733-1993, Determination of Resistance to Water of Films. National standard of the People's Republic of China: Beijing, China, 1993.
22. JC/T412.2-2006, Fiber Cement Flat Sheets Part 2: Asbestos Fiber Cement Flat Sheets. Building materials industry standard of the people's Republic of China: Beijing, China, 2006.
23. GB/T9271-1988, Paints and Varnishes–Standard Panels for Testing. National standard of the People's Republic of China: Beijing, China, 1988.
24. GB/T9265-2009, Determination for Alkali Resistance of Film of Architectural Paints and Coatings. National standard of the People's Republic of China: Beijing, China, 2009.
25. GB/T9780-2005, Test Method for Dirt Pickup Resistance of Film of Architectural Coatings and Paints. National standard of the People's Republic of China: Beijing, China, 2005.

 *coatings*

Article

Enhancing Flame Resistance of Cellulosic Fibers Using an Ecofriendly Coating

Riadh Zouari and Sondes Gargoubi *

Textile Engineering Laboratory-LGTex, Textile Department, Iset Ksar Hellal, University of Monastir, Monastir 5000, Tunisia; riadh.zouari@ksarhelal.r-iset.tn
* Correspondence: gargoubisondes@yahoo.fr

Abstract: Among the various advanced materials, flame-retardant cellulosic textiles are important as they directly relate to human health and hazards. The use of environmentally friendly flame-retardant coatings is currently one of the major concerns in the textile coating industry. In this work, acrylic acid was grafted onto the surface of cotton using plasma technology to enhance the attachment of acrylate phosphate monomer. Surface analyses, such as scanning electron microscopy (SEM), energy dispersive x-ray (EDX) and attenuated total reflectance Fourier-transform infrared (ATR-FTIR), were carried out to characterize the coating. Textile properties such as wettability and mechanical properties of untreated and treated cotton samples were investigated. A laundering test was also performed to predict the durability of the finishing. The outcomes revealed that acrylic acid-grafted samples treated with acrylate phosphate monomer have good flame-retardant properties.

Keywords: cellulosic; fiber; flame retardant; ecofriendly; cotton; coating

Citation: Zouari, R.; Gargoubi, S. Enhancing Flame Resistance of Cellulosic Fibers Using an Ecofriendly Coating. *Coatings* **2021**, *11*, 179. https://doi.org/10.3390/coatings11020179

Academic Editor: Philippe Evon
Received: 31 December 2020
Accepted: 28 January 2021
Published: 3 February 2021

Publisher's Note: MDPI stays neutral with regard to jurisdictional claims in published maps and institutional affiliations.

1. Introduction

Cotton is a natural fiber in which cellulose represents the major component of its chemical structure.

Nowadays, cotton fabrics are widely used in our daily lives due to their remarkable combination of properties: soft to the touch, good moisture vapor transmission and good moisture absorbency [1]. However, being organic in nature, cotton fabrics and other cellulose-based fabrics pose considerable fire risks [2]. Therefore, the flame-retardant finishing of cotton fabrics is important as it is strongly related to the protection of human beings and textile substrates from fire hazards.

In this context, the assessment of the flammability of cotton substrates has been approached by applying suitable flame-retardant finishes that are able to suppress or delay the appearance of a flame or reduce the flame-spread rate [3]. These finishes are applied to textiles by using a variety of agents, including borax and boric acid mixture, nitrogen- and phosphorus-based chemicals, antimony and halogen based chemicals [4].

Coating technology is regarded as the most practical method to impart functional agents to textile substrates [5]. Based on the chemicals used in the coating formulations and also on the desired performance of the coated surfaces, various types of coating techniques are used in textile functionalization, the main types including pad coating, direct or knife coating, sol gel coating, digital coating, electrospining coating and combined coating [6].

The current textile coating market is dominated by the use of binding agents or binders to fix the chemicals forming the coating layer, which have no affinity for fibers [7].

Numerous chemicals are used as binders. However, in most of them, formaldehyde is used as a crosslinker to apply the resistant coatings that are required by many advanced applications [8]. Formaldehyde is a toxic and carcinogenic chemical that has come under survey [9,10]. In addition, binders can contain solvents that are used as carriers during the manufacturing process. These solvents end up in the environment through waste water and air [11–13]. The major challenge facing textile manufacturers developing coated

275

textiles is how to rapidly switch to safer chemicals and processes. Recent research on functional coatings on textiles has focused on the use of environmentally friendly technologies and chemicals [6]. In this context, plasma technology was reported as an ecofriendly coating technique.

Plasma is a partially ionized gas composed of various species, such as electrons, negative and positive ions, radicals, excited molecules, neutrals, and UV photons. These different species are useful for textile surface modification. The plasma gas reacts with a small non-polymerizing molecule to achieve surface activation, cleaning, oxidation, modification in surface energy, an increase in surface roughness and etching. Similarly, plasma reacts with a bigger molecule to achieve plasma polymerization, coating, deposition, and creation of nanostructures [14]. These altogether improve the functional value of the textile materials. For fire-retardant coatings, plasma is applied either as a pretreatment, to increase the uptake of flame-retardant chemicals; for graft polymerization of these chemicals; or as a post-treatment to form strong bindings [4]. This technology also has other advantages, such as leaving the characteristics of the textile bulk unaffected, working in the gas phase without the need for water use, reducing processing time and achieving energy saving [15].

The aim of this study was to combine plasma pretreatment and acrylic acid grafting on cotton fabric to enhance the attachment of acrylate phosphate in order to obtain fire-retardant cotton fabric. The acrylic acid finish on cotton highlights a formaldehyde-free route for achieving surface modification of cotton, with high scope for incorporation of much improved physical and mechanical properties [16]. The combination of physical and chemical surface treatments was expected to enhance cellulosic fibers functionality and to contribute to a green and safe production process and end product.

2. Materials and Methods

2.1. Materials

Twill weave cotton fabric with the following specifications was procured commercially: the weight was 280 g m^{-2} and the density was 20 ends cm^{-1} in the warp direction and 14 picks cm^{-1} in the weft direction. The fabric was additionally treated with sodium hydroxide to remove hydrophobic compounds from the fiber surfaces. Analytical grade chemicals purchased from Sigma Aldrich were used for all experiments without further purification: Bis [2-(methacryloyloxy) ethyl] phosphate, molecular weight of 322.25 g/mol, and acrylic acid, molecular weight of 102.13 g/mol. Non-ionic low-sudsing detergent was kindly provided by S2C Company.

2.2. Methods

2.2.1. Plasma Treatment

The cotton samples were treated using an atmospheric pressure plasma jet technology system from Plasmatreat Company (Plasmatreater AS400, Steinhagen, Germany) for surface activation at atmospheric pressure. A high-frequency (23 kHz) pulsed voltage was applied between two tubular electrodes separated by a dielectric material. The PT400 generator delivered a pulse-pause modulated current. The current modulation was controlled by adjusting the plasma cycle time (PCT). With a PCT of 100%, the pulse duration was equal to the pause duration. The main gas flow consisted of dry air, which was introduced through the torch at a pressure of 5 Bar. The torch could be moved in the x- and y-directions to activate surfaces or to deposit films over large surfaces, with a typical velocity and substrate-to-nozzle distance of 5 m/min and 40 mm, respectively. These parameters were fixed after many treatments with various experimental conditions. PCT, frequency and dry air pressure were kept fixed at the standard conditions recommended by the machine constructor to guarantee the long life of the plasma generator. We focused on the variation of the torch velocity as well as the distance between the nozzle and the substrate. For cotton material, when this distance was less than 40 mm, the plasma air blown from the nozzle at 5 bar touched the substrate and induced some burned areas with a brown color.

The optimal torch velocity at a "nozzle/substrate" distance of 40 mm was 5 m/min. For higher velocities, the wettability was not significantly improved and for lower velocities, we encountered the problem of burned areas due to high exposure time to the plasma.

When applied to cellulosic fibers, plasma treatment induces the formation of free radicals on the fibers (Figure 1). These active entities act as initiator agents for chemical reactions that otherwise would not occur on a non-treated fiber.

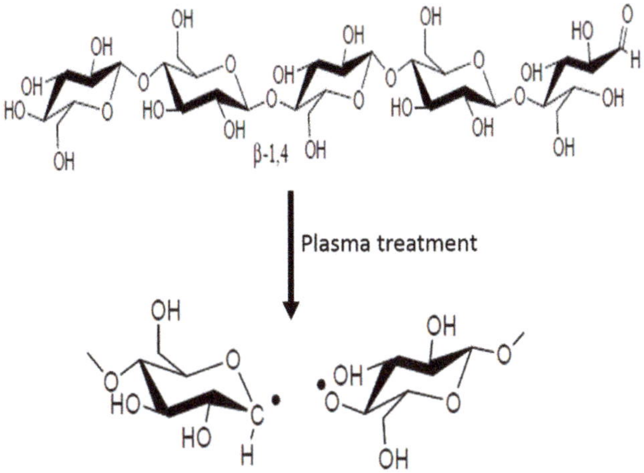

Figure 1. Formation of free carbon and oxygen radicals after cotton plasma treatment.

The literature indicates that the formation of free radicals occurs through ionization or excitation of the cellulosic polymer through electrostatic interaction between the orbital electrons in the polymer and fast-moving electrons induced by plasma. The ionization leads to molecular fragmentation and the formation of free radicals, while excitation leads to dissociation of the polymer to form free radicals [17].

2.2.2. Grafted Polymerization

The samples exposed to air plasma treatment were dipped in acrylic acid aqueous solution (0.5 M) for 4 h at 80 °C. Next, the grafted cotton fabrics were taken out of the solution and washed with 95% ethanol to remove polyacrylic acid homopolymer.

The monomer solution was applied at pH = 7. Under the neutral condition, we achieved optimum monomer grafting and esterification and retained 80% of the initial fabric tenacity [16]. Cellulosic materials degrade upon hydrolysis of glucoside bonds in highly acidic conditions, since the cellulosic chain length (DP) decreases.

2.2.3. Phosphorus Monomer Immobilization Assay

Phosphorus monomer immobilization was performed in an aqueous, heterogeneous mixture with cotton fabrics grafted with polyacrylic acid and monomer (50%) (w/w) for 5 min at ambient temperature. Then, a plasma treatment was performed to enable the activation mechanism of the radicals and thus the grafting of the monomer. In addition, the crosslinking and polymerization of the monomer were enabled.

2.3. Characterization and Measurements

2.3.1. Scanning Electron Microscopy

A scanning electron microscope (Model HITACHI TM 3000) was used for morphological characterization of the cotton fiber surface at high magnification. The metallized specimens were analyzed in partial vacuum conditions (0.1–0.15 torr) and under an accelerating voltage of 10 or 15 KV. Scanning electron microscopy with energy dispersive x-ray

(SEM-EDX) analyses were performed to determine the elementary chemical composition or to present the cartography of the distribution of the elements in the form of an image. To increase the SEM image quality, the conductivity of the samples was improved with a nanometric gold film coating. For each treatment, two different samples were used. For each sample, SEM graphs were produced for different areas.

2.3.2. Attenuated Total Reflectance Fourier-Transform Infrared Spectroscopy

Attenuated total reflectance Fourier-transform infrared (ATR-FTIR) spectroscopy was used to monitor the chemical composition of the surfaces. FTIR spectra were recorded using a commercial ATR-FTIR attachment (Spectrum Two™ FTIR, Perkin Elmer, Waltham, MA, USA). Spectra were recorded between 4000 and 400 cm^{-1}.

2.3.3. Water Contact Angle

The wettability of untreated and treated surfaces was measured using a DSA25 Drop Shape Analyzer. A droplet of 2 µL of ultrapure water was placed on the fabric. The results shown for each sample are the averages of three measurements.

2.3.4. Mechanical Properties

Tensile tests were carried out according to NF EN ISO 13934-2 standards using a Lloyd tensile testing machine (Lloyd LR 5k, Lloyd Instruments Ltd., Largo, FL, USA) at an extension speed of 50 mm/min with 5 kN load cell. Test specimens had dimensions of 75 mm in gauge length with a thickness of 1 mm. Tests were conducted at the warp direction of the fabric and run in triplicate to reduce experimental error. The samples were conditioned before testing.

2.3.5. Flame Retardant Test

The flammability test in vertical configuration was carried out by applying a methane flame for 10 s at the bottom of a fabric specimen (10 cm × 9 cm). The test was repeated twice for each sample to evaluate burning behavior: time, rate and the final residue. Vertical flammability tests were repeated twice. Photographs showing fabric behaviors were taken and subsequently analyzed with ImageJ software to quantitatively assess the fabric flammability. ImageJ provides a filtered binary image of the targeted area, adjusted near the edges of the sample, and calculates its total area in terms of the number of pixels. The threshold levels were adjusted so that the burned area appeared white in the processed image. ImageJ measures the total surface in all black colored areas in pixels. Therefore, the percentage of the burned area was calculated as the ratio of the total targeted area and the white colored areas.

2.3.6. Evaluation of Laundering Durability

The permanence of flame-retardant treatment after washing was investigated by subjecting flame-retardant fabric to a washing cycle. The treated cotton fabric was washed in Autowash, using the ISO 105-C06 standard method. Laundering involved a washing temperature of 40 °C for 30 min with 150 mL of water containing 0.6 g of non-ionic detergent. Samples were then dried for 10 min at 70 °C [18].

3. Results and Discussion

3.1. Morphological Structure

Figure 2 shows the scanning electron microscopy images for crude cotton as well as samples treated with acrylic acid before and after plasma activation. At 1000× *g* magnification, the crude cotton showed a smooth surface with some visible macrofibrils oriented predominantly in the direction of the fiber axis. The surface of the plasma-treated cotton showed a striped and more distinct macrofibrilar structure.

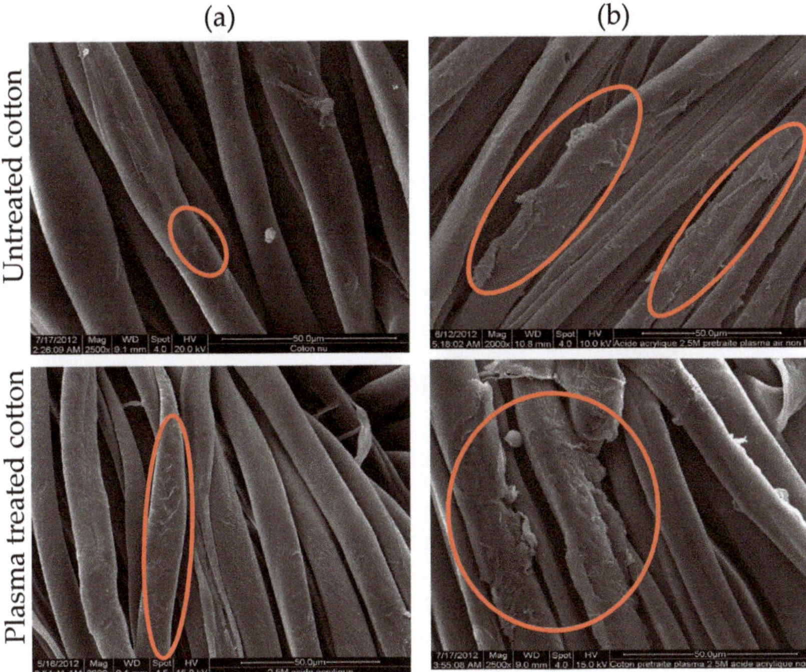

Figure 2. Surface morphology of crude and plasma-treated cotton fabrics (**a**) before acrylic acid deposition and (**b**) after acrylic acid deposition.

After treatment with acrylic acid, the plasma-treated fibers clearly showed significant presences of entities on their surfaces which totally covered the fibrillary structure. These results corroborate with those previously presented by Garcia-Torres et al. [19] and prove that the plasma activation was able to graft more acrylic acid onto cotton fibers.

Following phosphorus monomer immobilization (Figure 3a), the fibers clearly showed a grafted coating on their surfaces. A uniform layer appeared on the fibers and in the gaps between individual fibers linking them together. Plasma graft polymerization is an ecological technique capable of simultaneously grafting and polymerizing the monomer onto cotton fabric, forming a polymer layer covalently bonded to the fibers [20].

EDX characterization results show the composition of cellulosic fibers. As shown in Figure 3a, the main constituent elements of flame-retardant fibers were C, O and P elements, indicating that the phosphorus monomer existed on the surface of the treated cotton fabric. The content of the P element was about 4%.

3.2. ATR-FTIR Results

Characterization ATR-FTIR spectra of the cotton surface before and after plasma graft polymerization and monomer immobilization are presented in Figure 4. Spectra of all samples showed characteristic peaks of cellulose. The broad peak at 3291 cm^{-1} was attributed to hydroxyl groups (OH) stretching absorption. The broad stretching vibration of C-H was centered at 2906 cm^{-1}, while C-H bending vibrations were detected at 1434, 1373 and 1319 cm^{-1}, asymmetric stretching of C-O-C appeared at 1162 cm^{-1} and the vibration involving the C-O stretching appeared at 1058 cm^{-1} and 1027 cm^{-1} [21]. The spectra of surfaces before monomer immobilization did not show a remarkable difference when compared to the untreated sample except for acrylic acid grafting where a new peak appeared at 1715 cm^{-1} (Figure 4b). This peak was attributed to the C=O adsorption band. It indicates that a successfully induced grafting polymerization of acrylic acid was achieved by plasma activation [22]. The ATR-FTIR was also used to observe the changes of the chemical structure of the cotton fabric after monomer immobilization according to methacrylate functionality and phosphorus content. The FTIR spectrum is shown in Figure 4c. The phosphate structure of the monomer was observed at 1250 cm^{-1} and the bands of both 1029 cm^{-1} and 987 cm^{-1}, which were assigned to stretching vibrations of P=O and P-O-C, respectively, were also observed. The photoreactive methacrylate groups at 1716, 1166 and 1052 cm^{-1}, corresponding to the vibrations of C=O, C-C-O, and O-C-C, respectively [23], were also detected.

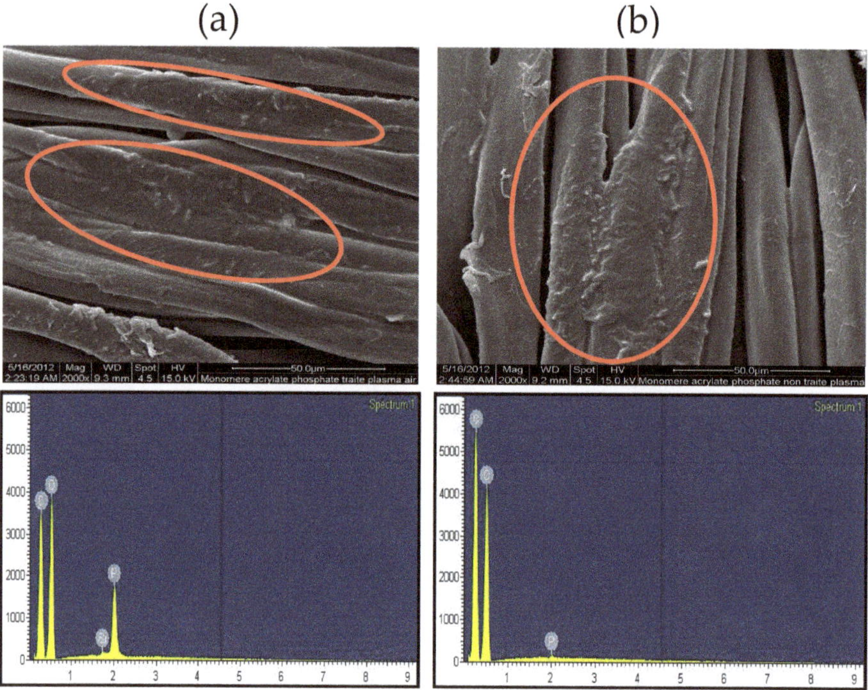

Figure 3. Scanning electron microscopy (SEM) and energy dispersive x-ray (EDX) images of flame-retardant cellulosic fibers: (**a**) before washing and (**b**) after washing.

3.3. Assessment of the Surface Wettability

The wettability of the cellulose textiles measured as the water contact angle (Figure 5) indicated the hydrophilic behavior of the untreated cotton (UC) and the cotton treated with plasma and acrylic acid (PCA). A contact angle value of 132° indicated the hydrophobic nature of the flame-retardant treatment (flame-retardant coating (FRC)).

The hydrophilicity of the cotton treated with acrylic acid (PCA) was improved. Similar behavior of cellulose surface wettability was reported by Garcia-Torres et al. [19]. Acrylic acid improves the hydrophilicity of cellulosic fibers owing to its high volatility, solubility in water and its bearing of a high ratio of a carboxyl polar group [24].

Side effects associated with the flame-retardant treatment include reduced fabric strength, poorer fabric handle and declined water absorbency. Some of these side effects may be overcome by the use of softeners and wetting agents during the post-treatment process of fabrics [25].

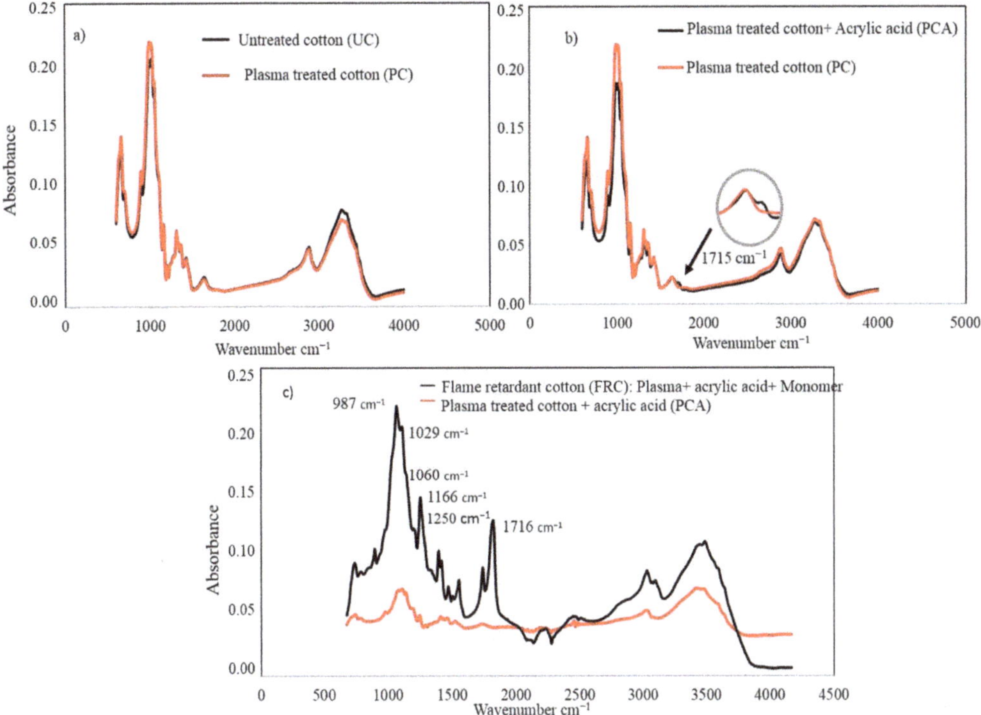

Figure 4. Attenuated total reflectance Fourier-transform infrared (ATR-FTIR) spectra of untreated and plasma-treated cotton (**a**) before grafting with acrylic acid, (**b**) after grafting with acrylic acid and (**c**) after monomer immobilization.

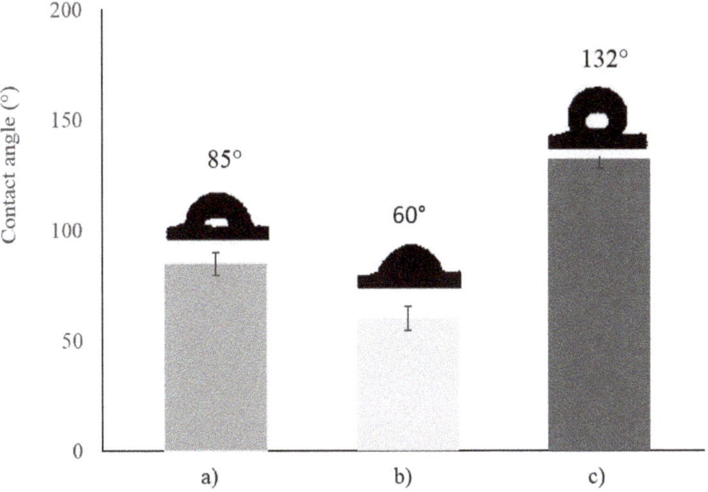

Figure 5. Water contact angles on the cotton surfaces: (**a**) untreated cotton, (**b**) plasma-treated cotton after grafting with acrylic acid and (**c**) treated cotton after immobilization of phosphorus monomer. The results shown are averages of three measurements and the error bar represents the standard deviation.

3.4. Mechanical Properties

It is important to examine the effect of flame-retardant treatment on the mechanical properties of the textiles since they are essential for the material overall function. Tensile strength is one of the most relevant textile mechanical properties. For this reason, tensile strength performance was evaluated after the finishing processes. Figure 6 shows the load–extension curves of untreated and flame-retardant cotton. The tested fabrics failed at the middle of the sample, away from the grips. It can be seen from the curves that the resistance of treated cotton did not change remarkably after flame-retardant finishing. Treated cotton retained about 94% of its initial mechanical properties and the difference was in the error limits. It can be concluded that the coating only occurred on the surface and did not tackle the bulk of cellulosic fibers.

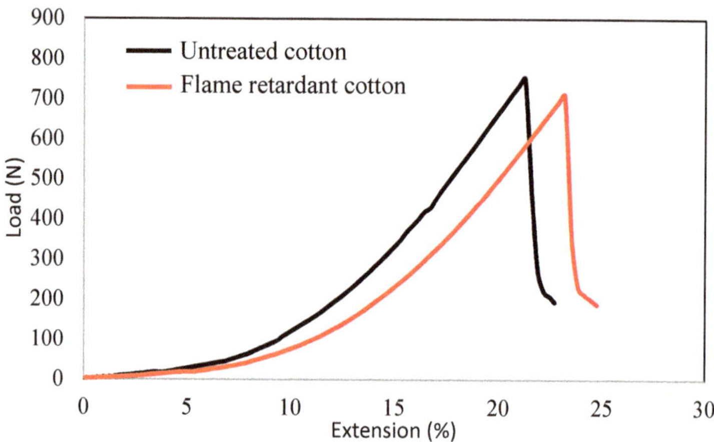

Figure 6. Load–extension curves of cotton fabrics.

For quantitative analyses, we defined two characteristic parameters extracted from the force vs. extension curves: elongation at break and maximum breaking strength (Table 1). The higher the maximum breaking strength value is, the higher the mechanical properties of the cotton fabric are. As shown in Table 1, the maximum breaking strength of untreated cotton was higher than that with the flame-retardant treatment. The difference between these two values was not significant, being lower than 1. The fabric elongation at break did not change remarkably after flame-retardant treatment and the difference between average values was within the error limits.

Table 1. Tensile characteristic parameters.

	Untreated Cotton	Flame-Retardant Cotton
Maximum breaking strength (N)	750 ± 6	710 ± 9
Elongation at break (%)	20 ± 4	23 ± 5

$\pm$ Standard deviation.

3.5. Flammability

Figure 7 represents the vertical burning behavior of the control and the flame-retardant cotton samples at different intervals of time. It was found that, for the untreated cotton, after removing the ignition source, the flame quickly spread and ignited the whole fabric with no residue remaining. However, for the treated cotton, after removing the ignition source, the flame spreading was slowed down and the char length was less important.

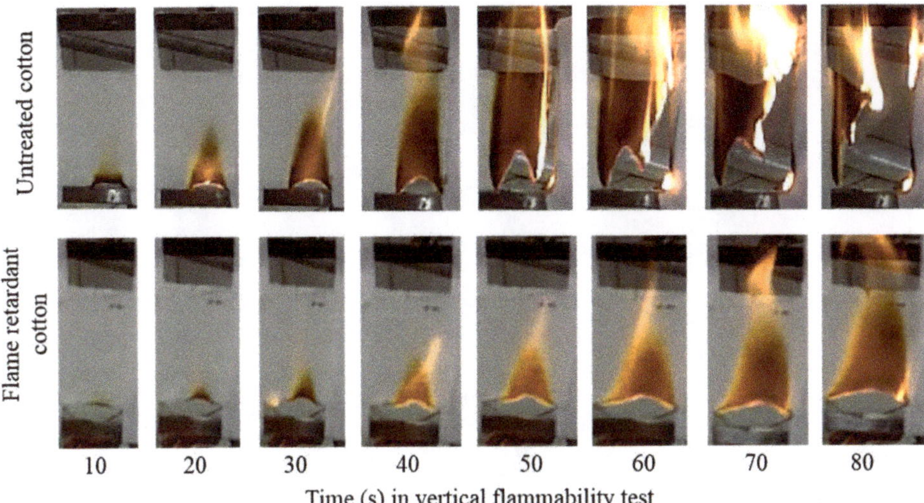

Figure 7. Flammability of untreated and flame-retardant cotton at different intervals of time.

Phosphorus-based chemicals exhibit both condensed phase and gas phase activities. They work in the condensed phase by raising char formation and depriving the gas phase of further volatile decomposition compounds. They also work in the gas phase as flame inhibitors, breaking the cycle of free radical generation [26].

Phosphorus-based chemicals are particularly effective in oxygen-containing polymers, such as cellulose. While heating in the presence of oxygen, the phosphorus decomposes into phosphonic and phosphoric acids which accelerate chain stripping processes and the removal of side groups that react with the phosphorus to become less flammable entities than in a phosphorus-free substrate [27].

It is important to note that, in a real-life situation, the user of the flame-retardant cellulosic material will get a longer time either to extinguish the fire or to escape from the fire zone. In addition, the user will find it easier to escape from fire hazards with slow flame spreading.

Figure 8, showing the output of the image analyzed by ImageJ software, clearly provides evidence of a significant flame-retardant effect on treated cotton fabrics compared to untreated ones. Indeed, the treated samples presented a burned surface of 2.4% ± 1.12% after 10 s of exposing the textile to the fire and 44% ± 1.8% after 80 s, whereas untreated samples presented a burned surface of 7% ± 0.8% after 10 s and 70% ± 1.89% after 80 s.

3.6. Laundering Durability

Figure 2b shows the flame-retardant cellulosic fibers after wash and it is obvious that the flame-retardant agent agglomerates are less homogeneous on the surface of the fibers. The content of the P element on the flame-retardant treated cotton fabric surface decreased (0.13%). Most of the product was washed off from the cotton surface.

The significant loss of the flame-retardant coating, resulting from one washing cycle, affected the flame-retardant properties of the washed samples. However, the flame resistance remained slightly better than that of untreated cotton. This is due to the protective non-homogeneous coating that resisted washing.

The obtained results clearly indicate that the coating exhibited low washing fastness.

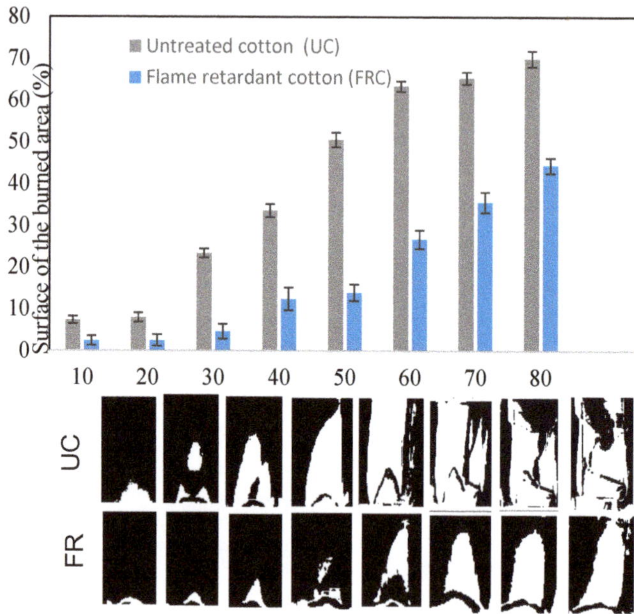

Time (s) in vertical flammability test

Figure 8. Vertical flammability of untreated cotton (UC) and flame-retardant cotton (FRC). Vertical flammability was quantified by calculating the average percentage ± (standard deviation) of the surface burned area. ImageJ software was used to calculate the burned area (white area) as well as the total surface area.

4. Conclusions

The goal of the present work was to investigate an ecofriendly flame-retardant treatment for cellulosic fibers. An organic phosphorus monomer was immobilized onto cotton fabrics grafted with polyacrylic acid to improve tethering. Polyacrylic acid was first introduced onto the cellulosic fibers by plasma-induced grafting polymerization. FTIR, SEM, EDX, wettability and mechanical property analysis were conducted to characterize the coated fabric samples. The washing durability was also evaluated.

The important results of the study are summarized as follows:

- The coating showed significant flame-retardant properties compared to untreated fibers.
- The plasma treatment enhanced the graft polymerization of acrylic acid onto cellulosic fibers.
- The coating of cellulosic fabric appeared not to affect the mechanical behavior of the cotton.

Thus, plasma-treated cotton grafted with acrylic acid and linked with phosphorus monomer holds promising potential for firefighting applications. Further experimental investigations are, however, needed to improve the low washing fastness of the coated surfaces.

Author Contributions: Conceptualization, R.Z. and S.G.; methodology, R.Z.; software, S.G.; validation, R.Z. and S.G.; formal analysis, R.Z.; investigation, S.G.; resources, R.Z.; data curation, S.G.; writing—original draft preparation, R.Z and S.G.; writing—review and editing, R.Z and S.G.; visualization, S.G.; supervision, R.Z.; project administration, R.Z. All authors have read and agreed to the published version of the manuscript.

Funding: This work was funded by the Tunisian Ministry of Higher Education and Scientific Research in the framework of the PROJECT 18PJEC12-22.

Institutional Review Board Statement: Not applicable.

Informed Consent Statement: Not applicable.

Data Availability Statement: Data sharing is not applicable to this article.

Conflicts of Interest: The authors declare no conflict of interest.

References

1. Khanzada, H.; Khan, M.Q.; Kayani, S. Cotton Based Clothing. In *Cotton Science and Processing Technology*; Springer: Cham, Switzerland, 2020; pp. 377–391.
2. Salmeia, K.A.; Jovic, M.; Ragaisiene, A.; Rukuiziene, Z.; Milasius, R.; Mikucioniene, D.; Gaan, S. Flammability of cellulose-based fibers and the effect of structure of phosphorus compounds on their flame retardancy. *Polymers* **2016**, *8*, 293. [CrossRef] [PubMed]
3. Salmeia, K.A.; Gaan, S.; Malucelli, G. Recent advances for flame retardancy of textiles based on phosphorus chemistry. *Polymers* **2016**, *8*, 319. [CrossRef] [PubMed]
4. Samanta, K.K.; Basak, S.; Chattopadhyay, S. Sustainable Flame-Retardant Finishing of Textiles: Advancement in Technology. In *Handbook of Sustainable Apparel Production*; CRC Press: Boca Raton, FL, USA, 2015; pp. 64–89.
5. Moiz, A.; Padhye, R.; Wang, X. Durable superomniphobic surface on cotton fabrics via coating of silicone rubber and fluoropolymers. *Coatings* **2018**, *8*, 104. [CrossRef]
6. Billah, S.M.R. Textile Coatings. In *Functional Polymers*; Jafar Mazumder, M.A., Sheardown, H., Al-Ahmed, A., Eds.; Springer International Publishing: Cham, Switzerland, 2019; pp. 825–882. [CrossRef]
7. Mazumder, M.A.J.; Sheardown, H.; Al-Ahmed, A. *Functional Polymers*; Springer: Cham, Switzerland, 2019.
8. Munna, M.K.H. Eco-friendly modified silicone poly-acrylate binder synthesis and application of textiles pigment printing. *Am. J. Appl. Chem.* **2016**, *4*, 201–206.
9. David, D.; Arkerman, H. Evaluation of the oral toxicity of formaldehyde in rats. *Universa Med.* **2008**, *27*, 106–112.
10. El Hage, R.; Khalaf, Y.; Lacoste, C.; Nakhl, M.; Lacroix, P.; Bergeret, A. A flame retarded chitosan binder for insulating miscanthus/recycled textile fibers reinforced biocomposites. *J. Appl. Polym. Sci.* **2019**, *136*, 47306. [CrossRef]
11. Freedman, D.; Payauys, A.; Karanfil, T. The effect of nutrient deficiency on removal of organic solvents from textile manufacturing wastewater during activated sludge treatment. *Environ. Technol.* **2005**, *26*, 179–188. [CrossRef]
12. Brik, M.; Schoeberl, P.; Chamam, B.; Braun, R.; Fuchs, W. Advanced treatment of textile wastewater towards reuse using a membrane bioreactor. *Process Biochem.* **2006**, *41*, 1751–1757. [CrossRef]
13. El-Molla, M.; Schneider, R. Development of ecofriendly binders for pigment printing of all types of textile fabrics. *Dyes Pigment.* **2006**, *71*, 130–137. [CrossRef]
14. Abd Jelil, R. A review of low-temperature plasma treatment of textile materials. *J. Mater. Sci.* **2015**, *50*, 5913–5943. [CrossRef]
15. Parvinzadeh, M.; Ebrahimi, I. Atmospheric air-plasma treatment of polyester fiber to improve the performance of nanoemulsion silicone. *Appl. Surf. Sci.* **2011**, *257*, 4062–4068. [CrossRef]
16. Ghosh, P.; Das, D. Modification of cotton by acrylic acid (AA) in the presence of NaH_2PO_4 and $K_2S_2O_8$ as catalysts under thermal treatment. *Eur. Polym. J.* **2000**, *36*, 2505–2511. [CrossRef]
17. Mohamed, H.; El-Halwagy, A.; Garamoon, A. Plasma Application in Textiles. *J. Text. Color. Polym. Sci.* **2019**, *16*, 15–32. [CrossRef]
18. Gargoubi, S.; Chaouch, W.; Stambouli, M.; Bhouri, N.; Boudokhane, C.; Zouari, R. Getting rid of the unpleasant odor in new artificial leather using natural and synthetic fragrances. *Chem. Ind. Chem. Eng. Q.* **2019**, *25*, 141–151. [CrossRef]
19. Garcia-Torres, J.; Sylla, D.; Molina, L.; Crespo, E.; Mota, J.; Bautista, L. Surface modification of cellulosic substrates via atmospheric pressure plasma polymerization of acrylic acid: Structure and properties. *Appl. Surf. Sci.* **2014**, *305*, 292–300. [CrossRef]
20. Edwards, B.; El-Shafei, A.; Hauser, P.; Malshe, P. Towards flame retardant cotton fabrics by atmospheric pressure plasma-induced graft polymerization: Synthesis and application of novel phosphoramidate monomers. *Surf. Coat. Technol.* **2012**, *209*, 73–79. [CrossRef]
21. Liu, P.-S.; Chen, Q.; Wu, S.-S.; Shen, J.; Lin, S.-C. Surface modification of cellulose membranes with zwitterionic polymers for resistance to protein adsorption and platelet adhesion. *J. Membr. Sci.* **2010**, *350*, 387–394. [CrossRef]
22. Lee, S.-D.; Hsiue, G.-H.; Chang, P.C.-T.; Kao, C.-Y. Plasma-induced grafted polymerization of acrylic acid and subsequent grafting of collagen onto polymer film as biomaterials. *Biomaterials* **1996**, *17*, 1599–1608. [CrossRef]
23. Jang, J.; Jeong, Y.-K. Synthesis and flame-retardancy of UV-curable methacryloyloxy ethyl phosphates. *Fibers Polym.* **2008**, *9*, 667. [CrossRef]
24. Wang, J.; Chen, X.; Reis, R.; Chen, Z.; Milne, N.; Winther-Jensen, B.; Kong, L.; Dumée, L.F. Plasma modification and synthesis of membrane materials—A mechanistic review. *Membranes* **2018**, *8*, 56. [CrossRef]
25. Tang, K.-P.M.; Kan, C.-W.; Fan, J.-T.; Sarkar, M.K.; Tso, S.-L. Flammability, comfort and mechanical properties of a novel fabric structure: Plant-structured fabric. *Cellulose* **2017**, *24*, 4017–4031. [CrossRef]
26. Gallo, E. Progress in Polyesters Flame Retardancy: New Halogen-Free Formulations. Ph.D. Thesis, Universita Degli Studi Di Napoli Federico II, Napoli, Italy, 2009.
27. Kind, D.J. Formulation and Burning Behaviour of Fire Retardant Polyisoprene Rubbers. Ph.D. Thesis, University of Central Lancashire, Preston, UK, 2011.

Article

Mucilages from Different Plant Species Affect the Characteristics of Bio-Mortars for Restoration

Chiara Alisi [1], Loretta Bacchetta [1,*], Emanuel Bojorquez [2], Mauro Falconieri [1], Serena Gagliardi [1], Mirta Insaurralde [3], Maria Fernanda Falcon Martinez [4], Alejandro Meza Orozco [3], Franca Persia [1], Anna Rosa Sprocati [1], Silvia Procacci [1] and Angelo Tatì [1]

[1] ENEA, Via Anguillarese 301, S. Maria di Galeria, 00123 Rome, Italy; chiara.alisi@enea.it (C.A.); mauro.falconieri@enea.it (M.F.); serena.gagliardi@enea.it (S.G.); franca.persia@enea.it (F.P.); annarosa.sprocati@enea.it (A.R.S.); silvia.procacci@enea.it (S.P.); angelo.tati@enea.it (A.T.)
[2] CONACYT-El Colegio De Michoacán, La PiedadCP 59379, Mexico; tochis_83@hotmail.com
[3] El Colegio De Michoacán, La PiedadCP 59379, Mexico; m.insaurralde@gmail.com (M.I.); alejandro.meza@colmich.edu.mx (A.M.O.)
[4] Sovrintendenza Archeologia Belle Arti e Paesaggio di L'Aquila, 67100 L'Aquila, Italy; falconfernanda@libero.it
* Correspondence: loretta.bacchetta@enea.it

Abstract: The need for compatible materials for the preservation of cultural heritage has resulted in the revival of lime-based mortar technology and other applications. This work investigates the cohesion and integrity of lime mortars added with fresh mucilage extracted from five plants and evaluates their bioreceptivity for long-term durability. Specimens of lime mortars added with 2.5% of fresh mucilage extracted from *Aloe vera*, *Cylindropuntia californica*, *Opuntia engelmannii*, *Opuntia ficus-indica* and *Salvia hispanica* mucilages were analyzed for color change (colorimetry), cohesion (ultrasound measurements), integrity (X-rays) and bioreceptivity (microbiological tests).The internal structure of the specimens added with Cacti mucilages shows better compactness, and no color change was noticed in the bio-mortars also after aging. The bioreceptivity response of mortars inoculated with bacteria, fungi and a photosynthetic biofilm was quite different. Specimens added with *Aloe* and *Cylindropuntia* mucilages showed a higher extent of bioreceptivity than the control; the specimens of bio-mortars added with *Opuntia engelmannii*, *Opuntia ficus-indica* and *Salvia hispanica* mucilages did not appear, up to threemonths after the contamination, any microbial growth. These results indicate that the addition of mucilage improves the mortar qualities, but the choice of the plant mucilage must be carefully evaluated since it can be responsible for changes in the bioreceptivity of the mortar.

Keywords: lime mortar; mucilaginous plants; bio-products; Fourier-transform infrared (FTIR) characterization

Citation: Alisi, C.; Bacchetta, L.; Bojorquez, E.; Falconieri, M.; Gagliardi, S.; Insaurralde, M.; Martinez, M.F.F.; Orozco, A.M.; Persia, F.; Sprocati, A.R.; et al. Mucilages from Different Plant Species Affect the Characteristics of Bio-Mortars for Restoration. *Coatings* **2021**, *11*, 75. https://doi.org/10.3390/coatings11010075

Received: 24 December 2020
Accepted: 6 January 2021
Published: 11 January 2021

Publisher's Note: MDPI stays neutral with regard to jurisdictional claims in published maps and institutional affiliations.

1. Introduction

The need for compatible materials for the preservation of cultural heritagehas resulted in the revival of lime-based mortar technology and other applications. Quick lime is the traditional ingredient used in ancient mortars and plasters; early examples have been found in Palestine and Turkey dating back to 12,000 before present (BP) as well as in Asia, Egypt, Greece and throughout the Roman Empire, in North and South America [1,2]. Currently, lime is the binder of choice for the preparation of compatible mortars used to construct masonry and monuments and also in the conservation of architectural heritage. Despite the apparent simplicity of the technology involved in mortar preparation, the different chemical reactions occurring during the various steps are quite complex. In particular, the hardening of slaked lime (reaction with air and progressive drying) and the choice of the filler are two important steps that influence both durability and workability. Ancient recipes describe the use of animal glues, such as casein and other dairy products, fat, albumen, and blood, but also vegetable glues, such as linseed oil, beer, and resins [3,4] as an additive to improve quality mortars.

Modern diagnostic techniques are now able to identify the large variety of natural products used in historical mortars; for example, animal glue, lipid and proteinaceous materials were identified in medieval mortars in Southern Italy [5]; egg white, juice of plants, drying oil, molasses were found in mortars from Spanish Colonial Period in the Philippines [6]; blood, crop flour, sticky-rice have been used as binders in the mortars of Chinese wooden buildings [7]. The Mesoamerican historical buildings tradition [8], a pre-Hispanic practice, includes the addition of mucilage from different plant species. The use of cacti mucilage as an additive to the plasters and as a binder for the wall paintings was discovered in the mural paintings of Cacaxtla, showing a good state of conservation [9]. The use of Opuntia mucilage has continued over Mexican history to modern times through traditional construction techniques, being transmitted from generation-to-generation. Apart fromempirical knowledge, very few data are available on mucilage-based mortar for restoration and their physical and mechanical properties. From the chemical point of view, Opuntia mucilage is a complex carbohydrate with a molecular weight between 1.5 and 4×10^6 g/mol, composed of a variable amount of arabinose (24.6–42%), galactose (21–40.1%), galacturonic acid (8–12.7%), rhamnose (7–13.1%), and xylose (22–22.2%) [10]. In general, the composition and percentage of carbohydrates varyin the mucilage ofdifferent species. However, their presence is important because the carboxylic groups react with divalent cations to form an egg-box structure, which allows water retention in plants [11,12]. As an additional component, it contains mono and divalent cations, DNA, proteins in small amounts and phytomolecules, such as polyphenols. This chemical composition makes the mucilage an excellent additive to lime, which may prevent surface breaks of mortars guaranteeing the desirable moisture [13]. Other functions of the vegetable mucilage used in the preparation of lime mortar areto promote the plasticity of the mortar allowing its use in thin layers [14] and to act as a consolidator thanks to the high viscosity.

On these premises, the aim of our work was the study of the cohesion and integrity of lime mortars added with fresh mucilage from five plants and the evaluation of their bioreceptivity to substantiate the improvement in durability.

Therefore, in order to fulfill our aim, we: (i) extracted and characterized by FTIR the mucilage from fiveplant species; (ii) prepared samples of bio-mortars with the different mucilages; (iii) evaluated physical properties of the bio-mortars after threemonths of drying and after threeyears aging using non-destructive tests (ultrasound and X-rays measurements, colorimetry); (iv) assessed the bioreceptivity of the bio-mortars inoculated with fungi, bacteria and photosynthetic biofilm isolated from the walls of archeological sites in order to employ the mucilages in restoration and conservation intervention.

2. Materials and Methods

2.1. Plant Materials and Mucilage Extraction

Cladodes (1–3 years old) of *Opuntia ficus-indica* and *Opuntia engelmannii*, stems of *Cylindropuntia californica* and leaves of *Aloe vera* were collected from a wild bush at early hours in the morning in Maccarese (Rome) and immediately used for the experimental trials. The plant materials were washed with tap water in order to remove impurities and spines, peeled from epidermis on both sides and chopped into pieces of 1–2 cm. Commercial seeds from Chia (*Salvia hyspanica*) were bought from a local retailer (NaturaSi shop, Rome, Italy). Fresh chopped plant material of *Opuntia ficus-indica*, *Opuntia engelmannii*, *Cylindropuntia californica*, and *Aloe vera* (200 g each) was mixed with distilled water (200 mL each) in different flasks (biomass to water ratios: 1:1, *w/v*). The samples were left to soak for 24 h at 25 °C, in dark conditions without stirring. The mucilage was removed by percolation through a 0.5 mm mesh. Seeds from *Salvia hyspanica* were mixed with distilled water in biomass to water ratioof 1:5 (*w/v*). Seeds were removed from mucilage by centrifugation at 5000 rpm for 10 min.

2.2. Mucilages Characterization

The mucilages extracted from the different plants were immediately analyzed with a digital refractometer DBR35/45/SALT (Metricon Corporation, Pennington, NJ, USA) in order to measure the Brix degrees (°Bx), i.e., the percentage of sugar content by weight (grams per 100 mL of water). Furthermore, the pH of each mucilage was measured with a pH-meter (Mettler Toledo, Columbus, OH, USA). For further analyses, the extracts were kept at −20 °C.

For the infrared analysis, the mucilages were freeze-dried in order to obtain anhydrous powders. Infrared absorption spectra were measured with a Spectrum 100 Fourier-transform infrared (FTIR) spectrometer (PerkinElmer, Waltham, MA, USA) in the wavenumber range between 400 and 4000 cm^{-1} with a spectral resolution of 4 cm^{-1}. Absorbance spectra of the dried powders were calculated as:

$$A = -\log \frac{T}{100} \tag{1}$$

from the transmissivity measured with attenuated total reflectance (ATR) accessory equipped with a multipass KRS5 crystal. After the blank subtraction, absorbance spectra were normalized in order to facilitate the comparison.

Since with the ATR technique it is impossible to know the exact quantity of the sample which is actually probed, the resulting spectra cannot be used for the quantitative analysis of the chemical content, i.e., the absolute value of the intensity of the absorption peaks cannot be related to the number of the corresponding chemical bonds in the sample; nevertheless, the positions and the relative intensities of the bands give useful information about the composition of the specimen. Hence, after the blank subtraction, each absorbance spectrum was normalized to one by dividing by the maximum value of the intensity of a selected band (in this work, we selected the intense band centered at 1040 cm^{-1}) in order to facilitate the comparison to the other spectra.

2.3. Bio-Mortar Samples Preparation

Samples were prepared with a ratio of aged lime putty and marble powder: 1:2 (*v/v*) and added with mucilage from the different plant species. Optimal additive concentration for the improvement of mechanical resistance was assessed in previous work [7]; for this reason, in this study, a 2.5% (*v/w*) of mucilage was used. The bio-mortar samples were then put in a mold (size 5 cm × 5 cm × 5 cm) according to normal UNE-EN-1015-11 [10] and dried at room temperature for 30 days.

Six replicas were prepared with each kind of mucilage to be used for colorimetric, radiographic, ultrasound and bioreceptivity tests. The tests were performed on the fresh samples and after 3 years of natural aging.

In order to compare the cohesive performance of the bio-mortars prepared with *Opuntia ficus-indica*, 10% *v/w* (OFI) with respect to mortars prepared with synthetic additive, a preliminary trial was carried out on replicated mortar samples added with 10% of Primal AC 33, an acrylic polymer widely used for restoration and conservation purposes.

2.4. Evaluation of Physical Properties of Bio-Mortars by Non-Destructive Tests

Colorimetric measurements were carried out to evaluate color changes induced by the presence of the plant mucilage with respect to the mortars prepared without vegetable additives. Color values, reported in the *CIEL*a*b** space, were measured with a Techkon SP 820 λ (TECHKON GmbH, Königstein, Germany), with a 3 mm spot, using a standard D65 illumination. In the color representation space, $L^*\ a^*\ b^*$, L^* is the lightness, a^* is the red/green position, and b^* is the blue/yellow position. In the *CIEL*a*b** system, the distance between two color points represents their color difference ΔE^* and it is calculated from the differences of its components ΔL^*, Δa^*, Δb^*, according to the Equation (2):

$$\Delta E = \sqrt{(\Delta L^*)^2 + (\Delta a^*)^2 + (\Delta b^*)^2} \tag{2}$$

The measurements were performed on the same face of each cubic sample in three points (a mask was used to repeat the measurement). Reported data are the mean of 10 replicas. These measurements allowed calculating the total color difference ΔE, relative to the same area of the sample.

Ultrasonic velocity (US) measurements were carried out with the aim of verifying the internal cohesion in each sample. The US in a mortar sample is related to its compactness due to density, porosity and composition. A portable instrument, a Krautkramer USM 23 (GE Inspection Technologies GmbH, Wunstorf, Germany) equipped with a low-frequency probe (50 kHz), was used in the transmission method, and no sample preparation was necessary. Ultrasonic wave velocity was measured in all specimens in the three directions X, Y, and Z. For each direction, three values were recorded, and the average was calculated. A thin layer of water was used as an acoustic coupling medium between the stone and the transducer.

The compressive strength (R_{CK}) obtained from ultrasonic tests was estimated as for the Equation (3):

$$R_{ck} = a \times e^{1,1 \times VelUS}[MPa] \tag{3}$$

In which *Vel US* is the measured US expressed in m/s, and *a* is a suitable constant that must be calibrated on a reference sample [15].

Digital radiography was carried out with Seifert ISOVOLT Titan 160 M2 0.4–1.5 (GE Inspection Technologies GmbH, Wunstorf, Germany) to study the density of the mortar. This method allows the internal vision of the object in a non-invasive way [16]. X-rays, electromagnetic radiation more energetic than visible light, cross the objects highlighting the macro structural nonuniformity and the defects present inside the object's volume. Therefore, if defects exist in the object under investigation such as cavities, cracks, large inclusions less absorbent than the matrix, or discontinuities of denser and more absorbent material, spots of different tones and proportional to the thickness of the defect will appear on the plate, and the defect will appear delimited by its perspective projection. The shades of gray represent the absorption of the rays proportional to the density of the sample passed through. In black, the denser objects and inclusions;in white, the pores. The elaboration of the digital radiography images isnondimensional values.

2.5. Bioreceptivity Tests

Bioreceptivity is the aptitude of a material to be colonized by living organisms, as conceived by Guillitte [17]. For bio-receptivity tests, the bio-mortar samples were prepared following the procedure described in Section 2.3, but adding 10% of mucilage (*v/w*) instead of 2.5% to enhance the organic content in the mortar and increase the possibility of biodeteriogens growth. The mortar was distributed into sterile Petri dishes and set to dry for 30 days. To perform the contamination, three fungi and three bacteria strains were chosen among those commonly found in bio-deterioration studies, isolated from Domus Aurea (Rome, Italy) and an Etruscan tomb in Cerveteri (Cerveteri, Italy), as well as a photosynthetic biofilm collected in the Catacomb of San Callisto in Rome (courtesy of prof. Laura Bruno). *Penicillium commune* Fcont, *Cladosporium* sp. DAF6 and *Fusarium* sp. CERV14F1 spores were inoculated onto the samples (10^6 cells/mL, in triplicate) and incubated at 28 °C and 96% RH. *Rhodococcus opacus* CER14.3, *Bacillus amyloliquefacens* CER14.4 and *Virgibacillus* sp. NOT1 bacterial cells were inoculated onto the samples (10^6 cells/mL, in triplicate) and incubated at about 26 °C (room temperature). Phototrophic biofilm CSC3W (composed ofalgae, cyanobacteria, moss protonemata) was inoculated onto the samples (in triplicate) and incubated at room temperature in the daylight.

The growth of the biodeteriogens was daily monitored by stereomicroscope and digital microscope, and recorded by photograph. Positive and negative controls were included in the experimental trial: negative controls were mortars and bio-mortars without artificial contamination to check if any spontaneous biological growth could be detected; positive controls were solid agar media inoculated with the same microorganisms used in the artificial contamination to assess the viability of the contaminants in optimal conditions.

2.6. Statistical Analyses

Statistical analyses were used to compare results obtained in replicates (9 values for each variable), and data were expressed as mean $\pm$ standard deviation (SD). All analyses were performed by SPSS statistics software v 25 (IBM, Armonk, New York, USA). Using the analysis of variance (ANOVA), differences among means were determined for significance at $p < 0.05$. A comparison of means was made by Duncan and Tukey's tests ($n = 5$ replicates).

3. Results and Discussion

The extraction method of mucilage from succulent plants was optimized and standardized in a previous study [10,18]. The water extraction for 24 h at room temperature yielded a mucilage that showed visible viscosity and was colorless, indicating the absence of chlorophyll in the samples. Usually, the °Bx is a measure of the sugar content of an aqueous solution with one Brix degree equivalent to 1 g of sucrose in 100 g of the solution; it represents the strength of the solution as a percentage by mass. Since mucilage in water is a complex solution containing different solids other than pure sucrose, the °Bx is an indicative measure thatonly approximates the dissolved solid content.

Values of pH and°Bx of the mucilage extract from *Opuntia ficus-indica*, *Opuntia engelmannii*, *Cylindropuntia californica*, *Aloe vera* and *Salvia hispanica* were compared, and the results are shown in Table 1. Mucilage extract of *Salvia hispanica* showed alower value in soluble sugar content and the highest in pH. The cacti mucilages (Opuntias and *Cylindropuntia*) showed similar values, 0.9 to 1.0 °Bx, and *Aloe vera* exhibited the highest soluble sugar content. Among the succulent species, whose mucilage is characterized by a pH around 4, only *Cylindropuntia* showed a sub-acidic value.

Table 1. °Bx and pH values of mucilage's extract from different plant species. The data represent the mean value $\pm$ SE. (n = five biological replicates). Different letters indicate significant differences ($p < 0.001$; Tukey's test).

Plant Materials	°Bx $\pm$ SE	pH $\pm$ SE
Opuntia ficus-indica	1.02 $\pm$ 0.03 b	4.60 $\pm$ 0.04 a
Aloe vera	1.52 $\pm$ 0.03 a	4.55 $\pm$ 0.02 a
Cylindropuntia californica	0.9 $\pm$ 0.03 b	5.18 $\pm$ 0.003 b
Opuntia engelmannii	1.0 $\pm$ 0.07 b	4.30 $\pm$ 0.03 c
Salvia hispanica (seeds)	0.3 $\pm$ 0.03 c	6.1 $\pm$ 0.01 d

Infrared spectroscopy can help in the identification of the chemical composition of a sample. The normalized absorbance spectra are reported in Figure 1 in the most characteristic spectral interval between 800 and 1800 cm^{-1}, the so-called sugar fingerprint region. Comparison of the experimental data with spectra reported in literature enables to attribute some diagnostic peaks to specific sugars, and hence identify the composition of the samples [19–22]. In particular, the peak around 1051 cm^{-1} is due to the presence of arabinose and rhamnose, the one around 940 cm^{-1} is due to the presence of galactose, while the peaks at 1104 cm^{-1} and 1179 cm^{-1} to the xylose. Furthermore, in the spectra reported in Figure 1, the peaks at 1104 cm^{-1} and 1118 cm^{-1}are due to the vibrations of C-O bonds in uronic acids, and the band centered at 1429 cm^{-1} can be ascribed to the carboxylic group of the galacturonic acid. The infrared spectra of the different *Opuntia* and *Cylindropuntia* mucilages are quite similar, demonstrating similarity in the sugar content, as also measured with the refractometric technique. On the other hand, the peaks centered at 1400 cm^{-1}and 1573 cm^{-1} in the spectrum of the mucilage extracted from *Aloe vera* are much more intense than those in the spectra of *Opuntia* and *Cylindropuntia* species, indicating higher content of sugars, as also confirmed by the presence of two broad, intense peaks at around 1180 cm^{-1}and 1240 cm^{-1} and by the measured Brix degrees. The spectrum of the *Salvia hispanica* mucilage, on the contrary, shows poorly intense bands in the sugar fingerprints region, demonstrating a low content of carbohydrates with respect to the other samples reported in this work. All in all, the results of the FTIR analysis are in very good agreement with those obtained by the refractometric analysis.

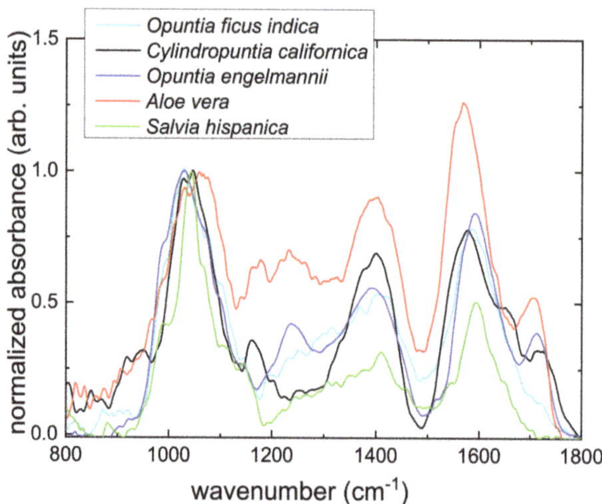

Figure 1. Normalized infrared absorption spectra of mucilages extracted from the different species reported in the label.

Since it is recognized that the gelling capacity of mucilage and its functional value as an additive in mortars is principally due to the presence of uronic acids [11,19], the analysis of the carboxylic acids is of the utmost importance. In particular, in all the spectra of mucilages extracted from *Opuntia* and *Cylindropuntia* species, the diagnostic peaks at 1118 cm^{-1} and 1429 cm^{-1} have the same intensity, indicating that the use of the extracts as an additive in bio-mortars could have almost equivalent positive effects, as also demonstrated by non-destructive tests (see below). Furthermore, the very high-intensity of the same peaks in the spectrum of *Aloe vera* mucilage suggests that its use could result in even higher mortar mechanical performances. On the other hand, the spectrum of the *Salvia hispanica* extract, with less pronounced peaks indicates a poor gelling capacity.

Colorimetric analyses showed very small variations of chromatic coordinates values induced by the presence of the mucilages in the mortars, which did not significantly change the colors; indeed,ΔE^* remains always <4. Hence, since, according to [23], only a $\Delta E^* > 5$ is perceived by the human eye, no color alteration occurs. Furthermore, no color changes result after samples of aging.

The results of non-destructive tests on bio-mortars supplemented with 2.5% of mucilage from *Opuntia ficus-indica* (OFI) *Opuntia engelmannii* (OE), *Cylindropuntia californica* (CC), *Aloe vera* (Aloe) and *Salvia hispanica* (SH) are reported in Figures 2–4. The analysis wasperformed after three months of drying and after three years of natural aging of the bio-mortars supplemented by the different mucilages.

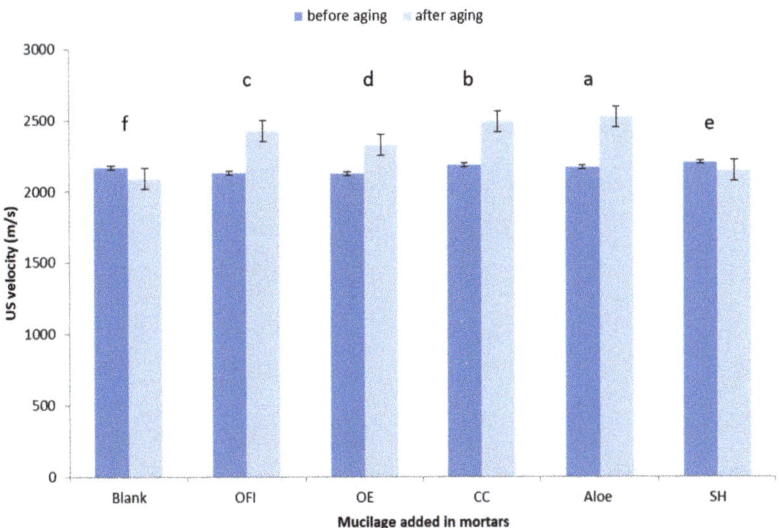

Figure 2. Ultrasound speeds into the mortars before (dark blue) and after (light blue) 3 years of natural aging. Values with different letters are significantly different at $p < 0.05$. The bars represent the standard error. *Opuntia ficus-indica* (OFI), *Opuntia engelmannii* (OE), *Cylindropuntia californica* (CC), *Aloe vera* (Aloe) and *Salvia hispanica* (SH).

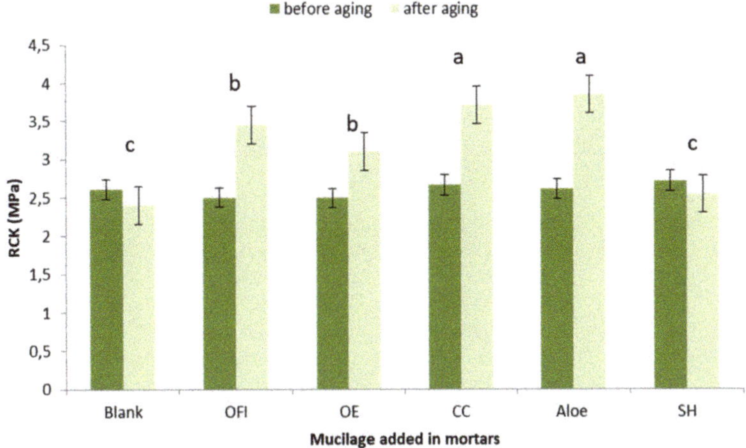

Figure 3. Elastic modulus estimated with the ultrasonic method before (dark green) and after (light green) 3 years of natural aging. Values with different letters are significantly different at $p < 0.05$. The bars represent the standard error. *Opuntia ficus-indica* (OFI), *Opuntia engelmannii* (OE), *Cylindropuntia californica* (CC), *Aloe vera* (Aloe) and *Salvia hispanica* (SH).

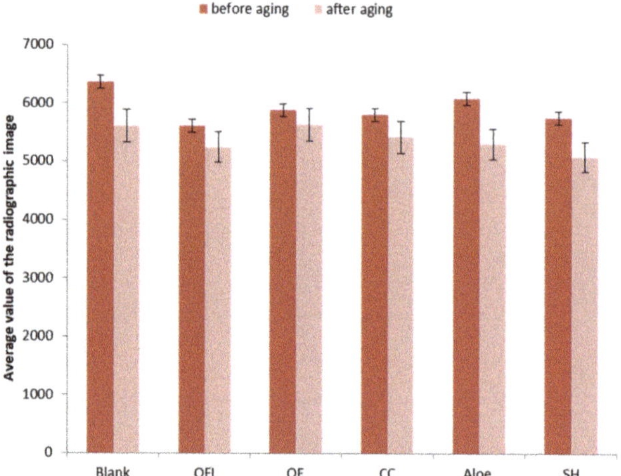

Figure 4. Radiographic density measurements before (dark red) and after (light red) 3 years of natural aging. The bars represent the standard error. *Opuntia ficus-indica* (OFI), *Opuntia engelmannii* (OE), *Cylindropuntia californica* (CC), *Aloe vera* (Aloe) and *Salvia hispanica* (SH).

The ultrasonic pulse velocity is an indirect method for quantifying homogeneity in the samples [24]. The results of the US measurements performed three months after hardening showed that the addition of mucilage to mortars promote a small variation in the propagation velocity of the ultrasonic waves into the samples (Figure 2); after three years of natural aging, a slight increase was observed in those bio-mortars containing succulent plant extracts: Aloe, CC and OFI mucilages seem to induce the more significant change in the ultrasound propagation. Similar results were reported by [25] showing the addition of *Opuntia ficus-indica* mucilage at water replacement concentration between 4–8% as suitable for durability enhancing applications in cement-based mortar.

The measure of the elastic module in all the samples just after hardening showed a similar behavior (Figure 3); a significant increase in the samples of bio-mortars added with Cacti mucilages (Aloe, CC, OFI, OE) was observed after aging. A possible explanation of this behavior can be due to the delay of the hardening (carbonating) of the bio-mortars, thus favoring the formation of a more regular crystalline structure.

Radiographic density showed few differences in the bio-mortars in comparison with the blank three months after drying. Nevertheless, differences were observed in the samples supplemented by mucilage after long term aging. These materials seem to be more homogenous with high-density. These results are also in agreement with those of the ultrasound pulse velocity test, which provides an indirect evaluation of the porous material void content because sound travels faster in a homogeneous solid material than through air. Highly porous material shows a low velocity of the ultrasound wave, as shown in the blank (Figure 2).

Furthermore, a preliminary radiographic test performed to compare the use of Primal AC 33 with OFI as an additive (Figure 5) produced a promising result. The analysis of mortar added with primal acrylic resin showed comparable compactness to the mortar added with the same percentage of *Opuntia ficus-indica* mucilage.

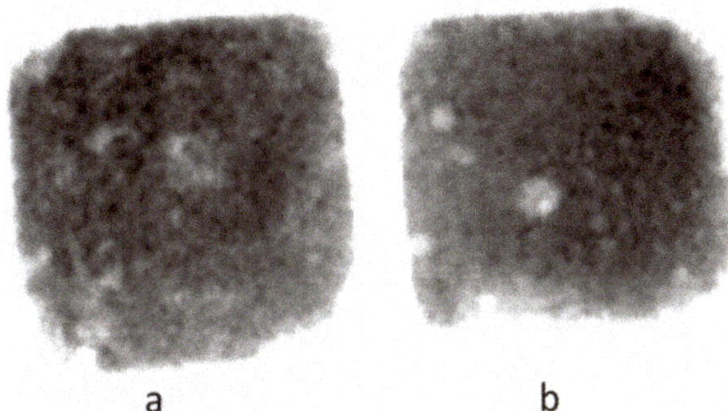

Figure 5. Radiographic images of bio-mortars realized with 10% of primal (**a**) and 10% of OFI (**b**) showing in white the porosity.

These encouraging results contribute to the idea of the potential utilization of *Opuntia ficus indica* cladodes as raw material for colloid production. In Italy, for instance, where Sicily is the first cactus pear supplier, the cladodes are not utilized byproducts. Pruning in the specialized orchards produces annually a range of 6–8 tons/hectare of cladodes and immature fruits, which are one of the main costs for farmers, not easy to dispose of. The possibility of recycling this biomass from waste into high added value byproducts implies non only a reduction of crop management costs, but it could generate income diversification for farmers.

The bioreceptivity trials were performed with a variety of biodeteriogens (bacteria, fungi and mixed photosynthetic biofilm), but only a few of them were able to grow on the samples of bio-mortars, as shown in Table 2. A little growth of fungi was detected on the Control (mortar without additives), but no evidence of biodeterioration was observed in the bio-mortar samples inoculated with bacteria and fungi.

Table 2. Response of inoculation on mortar and bio-mortars after 3 months of testing. (-) no growth; (+/-) little growth; (+) full growth.

Samples	Bacteria	Fungi	Biofilm CSC3W
Control	-	+/-	+
OFI	-	-	-
OE	-	-	-
CC	-	-	+
Aloe	-	-	+
SH	-	-	-

The bio-mortars artificially contaminated with the photosynthetic biofilm showed a different bioreceptivity, as shown in Table 2 and Figure 6. A major biofilm development was observed in the control sample, as well as in the bio-mortars containing CC and Aloe. The result indicates the natural bioreceptivity of the mortar, which can be colonized by microorganisms if kept in suitable environmental conditions. The bio-mortars added with OFI, OE and SH were not susceptible to the microbiological attack nor of bacteria, or fungi or photosynthetic biofilm, during all the time of investigation (3 months), while the CC and Aloe mucilage addition increased the bioreceptivity of the mortar since the biofilm development started at 15 days of incubation and went on up to 120 days when started to die as indicated by the change of color for green to brown.

Figure 6. Mortar (blank) and bio-mortar samples (added with Aloe and OFI) 15 days, 60 days and 120 days after inoculation with a photosynthetic biofilm.

4. Conclusions

Lime mortars prepared with the addition of organic additives have been used for centuries in building constructions and, more recently, for restoring and protecting historical buildings. In Mexico, the use of *Opuntia ficus-indica* mucilage in mortars is a traditional procedure, but despite the large use, only in recent years, a scientific approach was used to evaluate the effectiveness of this practice and to elucidate the physicochemical mechanisms involved. In our work, we demonstrated the formation of a more regular crystalline structure in bio-mortars containing succulent plant extracts in respect to mortars prepared without vegetable extracts or with *Salvia hispanica* mucilage. The biochemical analysis showed the different composition of mucilage of Cacti species versus *Salvia hispanica*, confirming the role of the gelling species in the mortar structure formation. Furthermore, the radiographic tests confirmed a similar performance between *Opuntia ficus-indica* mucilage and Primal acrylic resin when added at the same percentage to the lime mortars, indicating that a green product could efficiently substitute a synthetic component. Moreover, the bioreceptivity studies indicated that, although the role of the different mucilages can be similar in enhancing the durability of the bio-mortars, they are not suitable as additives in the same way: while the *Aloe vera* mucilage gives the bio-mortar a high susceptibility to biological attacks, the bio-mortars added with *Opuntia ficus-indica* and *Opuntia engelmannii* mucilage show a very low bioreceptivity, probably due to the presence of a higher content of hydro-soluble phytomolecules such as polyphenols, thus making the *Opuntia* mucilage a good candidate as a sustainable alternative to chemical additives. More detailed biochemical composition studies will help to elucidate the role of the different components of the vegetable mucilage additives, paving the way for their widespread uses in restoration as a modifier and/or consolidant.

Author Contributions: Conceptualization, C.A., L.B. and F.P.; Data curation, E.B., S.P. and A.T.; Formal analysis, L.B. and E.B.; Funding acquisition, C.A. and F.P.; Investigation, A.R.S. and S.P.; Methodology, M.F., M.I. and M.F.F.M.; Project administration, C.A., L.B. and F.P.; Resources, C.A., L.B. and A.M.O.; Software, A.T.; Supervision, C.A., L.B., M.I. and F.P.; Validation, C.A. and A.M.O.;

Visualization, C.A. and L.B.; Writing—original draft, C.A. and M.I.; Writing—review & editing, C.A., L.B. and S.G. All authors have read and agreed to the published version of the manuscript.

Funding: The research was funded by a grant fromthe Italian Ministry of Foreign Affairs and International Cooperation (Progetto Grande Rilevanza PGR01082).

Institutional Review Board Statement: Not applicable

Informed Consent Statement: Not applicable.

Data Availability Statement: Data sharing not applicable.

Conflicts of Interest: The authors declare no conflict of interest. The funders had no role in the design of the study; in the collection, analyses, or interpretation of data; in the writing of the manuscript, or in the decision to publish the results.

References

1. Ventolà, L.; Vendrell, M.; Giraldez, P.; Merino, L. Traditional organic additives improve lime mortars: New old materials for restoration and building natural stone fabrics. *Constr. Build. Mater.* **2011**, *25*, 3313–3318. [CrossRef]
2. Moropoulou, A.; Bakolas, A.; Bisbikou, K. Investigation of the technology of historic mortars. *J. Cult. Herit.* **2000**, *1*, 45–58. [CrossRef]
3. Sickels, L.B. Organics vs. synthetics: Their use as additives in mortars. In Proceedings of the Mortars, Cements and Grouts Used in the Conservation of Historic Buildings, Symposium (ICCROM), Rome, Italy, 3–6 November 1981; pp. 25–52.
4. Arcolao, C. *Le Ricette del Restauro. Malte, Intonaci, Stucchi dal XV al XIX Secolo*; Marsilio: Venezia, Italy, 1998.
5. Rampazzi, L.; Colombini, M.P.; Conti, C.; Corti, C.; Lluveras-Tenorio, A.; Sansonetti, A.; Zanaboni, M. Technology of Medieval Mortars: An Investigation into the Use of Organic Additives. *Archaeometry* **2016**, *58*, 115–130. [CrossRef]
6. Cayme, J.; Asor, A. Characterization of Historical Lime Mortar from the Spanish Colonial Period in the Philippines. *Conserv. Sci. Cult. Herit.* **2017**, *16*, 33–57. [CrossRef]
7. Fang, S.Q.; Zhang, H.; Zhang, B.J.; Zheng, Y. The identification of organic additives in traditional lime mortar. *J. Cult. Herit.* **2014**, *15*, 144–150. [CrossRef]
8. Kita, Y. The functions of vegetable mucilage in lime and earth mortars—A review. In Proceedings of the 3rd Historic Mortars Conference, Glasgow, UK, 9–14 September 2013.
9. Magaloni, D.; Brittenham, C.; Baglioni, P.; Giorgi, R.; Bernini, L. *Cacaxtla: La Elocuencia de los Colores en La Pintura Mural Prehispánica en México*; Universidad Nacional Autónoma de México: Mexico City, Mexico, 2013.
10. Bacchetta, L.; Maccioni, O.; Martina, V.; Bojorquez-Quintal, E.; Persia, F.; Procacci, S.; Zaza, F. Quality By Design Approach To Optimize Cladodes Soluble Fiber Processing Extraction In Opuntia ficus indica (L.) Miller. *J. Food Sci. Technol.* **2019**, *56*, 3627–3634.
11. Grant, G.T.; Morris, E.R.; Rees, D.A.; Smith, P.J.C.; Thom, D. Biological interactions between polysaccharides and divalent cations: The egg-box model. *FEBS Lett.* **1973**, *32*, 195–198. [CrossRef]
12. Borgogna, M.; Skjåk-Bræk, G.; Paoletti, S.; Donati, I. On the Initial Binding of Alginate by Calcium Ions. The Tilted Egg-Box Hypothesis. *J. Phys. Chem.* **2013**, *117*, 7277–7282. [CrossRef] [PubMed]
13. Cárdenas, A.; Arguelles, W.M.; Goycoolea, F.M. On the Possible Role of *Opuntia ficus-indica* Mucilage in Lime Mortar Performance in the Protection of Historical Buildings. *J. Prof. Assoc. Cactus Dev. Évora* **1998**, *3*, 1–8.
14. Sandoval, D.C.G.; Sosa, B.L.; Martinez-Avila, G.C.G.; Rodriguez Fuentes, H.; Victor, H.; Avendaño Abarca, V.H.; Romeo Rojas, R. Formulation and Characterization of Edible Films Based on Organic Mucilage from Mexican *Opuntia ficus-indica*. *Coatings* **2019**, *9*, 506. [CrossRef]
15. Watanabe, S.; Hishikawa, K.; Kamae, K.; Namiki, S. Study on estimation of compressive strength of concrete in structure using ultrasonic method. *Jpn. Architect Rev.* **2018**, *1*, 1–9. [CrossRef]
16. Bufarini, S.; D'aria, V.; Giacchetti, R. *Il Controllo Strutturale Degli Edifici in Cemento Armato e Muratura. Circolare Esplicativa n. 617/2009*; EPC srl: Milano, Italy, 2010; ISBN1 8863102422. ISBN2 9788863102420.
17. Guillitte, O. Bioreceptivity: A new concept for building ecology studies. *Sci. Total Environ.* **1995**, *167*, 215–220. [CrossRef]
18. Persia, F.; Alisi, C.; Bacchetta, L.; Bojorquez, E.; Colantonio, C.; Falconieri, M.; Insaurralde, M.; Meza Orozco, A.; Sprocati, A.R.; Tatì, A. Nopalasorganic additive for bio-compatible and eco-sustainable lime mortars. In Proceedings of the VIIth International Conference "Diagnosis, Conservation and Valorization of Cultural Heritage, Naples, Italy, 15–16 December 2016; pp. 245–251, ISBN 978-88-942118-0-1.
19. Rodriguez-Navarro, C.; Ruiz-Agudo, E.; Burgos-Cara, A.; Elert, K.; Hansen, E.F. Crystallization and Colloidal Stabilization of $Ca(OH)_2$ in the presence of Nopal Juice (*Opuntia ficus indica*): Implications in Architectural Heritage Conservation. *Langmuir* **2017**, *33*, 10936–10950. [CrossRef] [PubMed]
20. Lefsih, K.; Delattre, C.; Pierre, G.; Michaud, P.; Aminabhavi, T.M.; Dahmoune, F.; Madani, K. Extraction, characterization, and Gelling Behavior Enhancement of Pectins from the Cladodes of *Opuntia ficus indica*. *Int. J. Biol. Macromol* **2016**, *82*, 645–652. [CrossRef] [PubMed]

21. Coimbra, M.A.; Barros, A.; Barros, M.; Rutledge, D.N.; Delgadillo, I. Multivariate analysis of uronic acid and neutral sugars in whole pectic samples by FT-IR spectroscopy. *Carbohydr. Polym.* **1998**, *37*, 241–248. [CrossRef]
22. Coimbra, M.A.; Barros, A.; Rutledge, D.N.; Delgadillo, I. FTIR spectroscopy as a tool for the analysis of olive pulp cell-wall polysaccharide extracts. *Carbohydr. Res.* **1981**, *317*, 145–154. [CrossRef]
23. Mokrzycki, W.; Tatol, M. Color difference Delta E—A survey. *Mach. Graph. Vision* **2011**, *20*, 383–411.
24. Trocónis-Rincón, O.; Romero-Carruyo, A.; Andrade, C.; Helene, P.; Díaz, I. *Manual for Inspecting, Evaluating and Diagnosing, Corrosion in Reinforced Concrete Structures*; CYTED: Maracaibo, Venezuela, 2000; pp. 82–117.
25. Molina, W.M.; Alonso-Guzman, E.M.; Torres-Acosta, A.A.; Mendoza Pérez, I.N. Cement-Based Materials Enhanced Durability from Opuntia Ficus Indica Mucilage Additions. *ACI Mater. J.* **2015**, *112*, 165–172.

 *coatings*

Article

A New Composite Biomaterial Made from Sunflower Proteins, Urea, and Soluble Polymers Obtained from Industrial and Municipal Biowastes to Perform as Slow Release Fertiliser

Philippe Evon [1,*], Laurent Labonne [1], Elio Padoan [2], Carlos Vaca-Garcia [1], Enzo Montoneri [2,*], Valter Boero [2] and Michéle Negre [2]

[1] INP, Laboratoire de Chimie Agro-Industrielle, Université de Toulouse, ENSIACET, 4 Allée Emile Monso, BP 44362, 31030 Toulouse, France; Laurent.Labonne@ensiacet.fr (L.L.); carlos.vacagarcia@ensiacet.fr (C.V.-G.)
[2] Dipartimento di Scienze Agrarie, Università di Torino, Forestali e Alimentari, Largo P. Braccini 2, 10095 Grugliasco, Italy; elio.padoan@unito.it (E.P.); valter.boero@unito.it (V.B.); michele.negre@unito.it (M.N.)
* Correspondence: philippe.evon@ensiacet.fr (P.E.); enzo.montoneri@gmail.com (E.M.)

Abstract: Controlled-release fertilizers (CRF) are needed under current agriculture practice to decrease the environmental impact caused by fertilizer doses applied in excess of plants' uptake rate. Commercial CRF are available. They are manufactured from mineral fertilizers encapsulated into a synthetic polymer matrix or coated by a polymer layer. However, substitution of fossil sourced organic polymers with biopolymers is a major environmental concern. In the present paper, we describe the manufacture by a continuous twin-screw extrusion process, and the mechanical and chemical properties of injection-molded composite pellets containing 90% sunflower protein concentrate (SPC) matrix, and 5–10% of a biopolymer (BP) obtained from municipal biowastes (MBW), and/or urea (U). The reported results show that SPC-BP-U behaves as an efficient eco-friendly CRF. BP contributes to several benefits to the performance of the composite pellets, upon increasing surface hardness, and controlling the formation of ammonia from urea hydrolysis and the release of organic nitrogen. The SPC-BP-U appears a powerful eco-friendly CRF to supply organic C and the three major N, P, and K nutrients to soil and plants. It offers worthwhile scope for being tested in the cultivation of specific plants under the real operational conditions of agriculture practices.

Keywords: biopolymers; sunflower protein concentrate; municipal bio-waste; urea; slow-release fertilizers

Citation: Evon, P.; Labonne, L.; Padoan, E.; Vaca-Garcia, C.; Montoneri, E.; Boero, V.; Negre, M. A New Composite Biomaterial Made from Sunflower Proteins, Urea, and Soluble Polymers Obtained from Industrial and Municipal Biowastes to Perform as Slow Release Fertiliser. *Coatings* **2021**, *11*, 43. https://doi.org/10.3390/coatings11010043

Received: 17 December 2020
Accepted: 30 December 2020
Published: 2 January 2021

Publisher's Note: MDPI stays neutral with regard to jurisdictional claims in published maps and institutional affiliations.

1. Introduction

The fertilizer industry faces a continuing challenge to improve its products and increase the efficiency of their use, particularly of nitrogenous fertilizers, and to minimize any possible adverse environmental impact. This is done either through improvement of fertilizers already in use, or through development of new specific fertilizer types [1]. One specific objective in Europe is the development of new sustainable fertilizers with the aim of reducing the consumption and/or import of conventional mineral fertilizers from third-world countries, and of N fertilizers based on energy-intensive production process. The development of innovative controlled-released fertilizers (CRFs) could limit the consumption of the traditional ones. Indeed, CRFs present the interest to better match the plants' need for nutrients over time.

Several CRFs products are already available in the market. They are manufactured from blends of fertilizers as active principle and organic polymers. There are two major CRFs types [2]. In the first, the fertilizer is encapsulated into the polymer matrix. In the second, the fertilizer is coated by a polymer layer. The fertilizer bioavailability for the plant uptake is controlled by the diffusion rate through the pores of the polymer matrix or through the surrounding polymer shell. Although several commercial CRFs can achieve satisfactory fertilizers release rate, the plastic polymer in the blend raises concern for its low biodegradability and the accumulation of plastic impurities in soil. The new EU regulation on fertilizing products addresses the degradability of the organic matter of CRFs by stipulating 90% conversion of the organic carbon into CO_2 in a maximum of 24 months. The specific challenge focuses on finding organic polymers that comply with the biodegradability EU requirements, while achieving the controlled release of nutrients in the best possible manner.

Much research work has been carried out to develop new CRFs containing natural polymers, instead of synthetic polymers. The substitution of synthetic polymers derived from fossil sources with natural polymers is a current major environmental issue [3]. Although biodegradable polymers can be also synthesized from monomers obtained from the petrochemical industry, the consumption of chemicals from fossil sources is a major source of GHG emission. In addition, it contributes to the depletion of fossil sources. Natural polymers such as lignin, starch, chitin, cellulose, and other polysaccharides (e.g., *k*-carrageenan) have been investigated to make CRFs. Generally, these cannot be used in natural pure form as effective coating material due to their inherent properties, such as poor mechanical properties, hydrophobicity, high industrial demand, poor solubility, and processability. These imply limitations for their use.

To overcome the above criticalities, researchers have focused their work on chemical and physical modifications of natural polymers. Biopolymers from monomers obtained from renewable source have also been obtained, e.g., polylactic acid (PLA) or poly hydroxyl alkanoates (PHA). Aims of current research on CRFs are to reduce production cost, improve nutrient release rate, optimize organic matter biodegradability rate, and guarantee crushing strength to bear crack resistance under environmental stresses. These three properties need to be compatible with each other. Indeed, a poor mechanical strength or high biodegradability may result in faster nutrient release rate, excess nutrient availability over the plant uptake rate, high nutrient leaching from soil into ground, and low CRFs efficiency. High crushing strength also facilitates the handling and transportation and reduces the water absorption [4].

In the present work, the current CRFs issues have been addressed by fabricating a new material out of three components. These are a sunflower protein concentrate (SPC) obtained from sunflower oil cake (SOC), a biopolymer (BP) obtained from municipal biowaste (MBW), and commercial urea (U). All three components contain N. Three different blends were manufactured, i.e., SPC-10% U, SPC-10% BP, SPC-5% U-5% BP. The underlying rationale of the work was to use SPC as polymer matrix, and BP and U as fillers/active principles. The reason for this approach is based on the following literature data reported for SPC, BP and U.

The SPC is well known biopolymer used in human nutrition for its nutritional value [5], and potential antioxidant properties and benefits for disease prevention and aging retardation [6]. SPC are reported also to have also good processability to manufacture films by solvent casting [7] and extrusion at 160 °C [8]. This material therefore seemed a potential eco-friendly matrix to be used for manufacturing new CRFs.

The BP is obtained by chemical hydrolysis of fermented unsorted urban food wastes. The BPs sourcing food wastes contain all major animal and vegetable natural polymers. Upon anaerobic fermentation, the most readily degradable fats, polysaccharides, and proteins are converted to methane and carbon dioxide. The solid anaerobic digestate (AD) contains the recalcitrant lignocellulosic fraction. The BP, obtained from AD chemical hydrolysis, contains a mix of water-soluble molecules with molecular weight from 5 to above 750 kDa [9]. These molecules are water soluble lignocellulosic fragments. They keep the memory of the macromolecularity and functional groups, and of the mechanical properties of the proximates in the parent food wastes. However, BP is water soluble and thermally stable up to 200 °C. This allows for fabricating, by solvent casting and melt extrusion, composite articles containing BP in blends with synthetic polymers such as polyethylene-co-vinyl alcohol [10] and polyethylene-co-acrylic acid [11]. The blends have been proven to have higher mechanical resistance, compared to the article manufactured from the neat synthetic polymer only. The BP is not biodegradable. However, based on its sourcing material, no adverse effect is expected from its accumulation in soil. Indeed, MBW anaerobic digestates and/or compost are used as soil amending agents and/or fertilizers. The BP has also been proven [9] more efficient soil fertilizer and plant growth biostimulant in the cultivation of several food and ornamental plants, in comparison with its sourcing fermented MBW materials and with commercial organo-inorganic peat and leonardite derived products claimed by the vendor as plant biostimulants. This material seemed therefore a potential processable filler, capable of increasing the mechanical resistance of the SPC matrix and to perform at the same time as active fertilizer principle in CRFs.

Urea is a main commercial fertilizer. Its world consumption amounts to approximately 51 Mt/yr [12]. In the present work, urea has been chosen as reference commercial N fertilizer. Urea and lignin materials have also been investigated for use in the CRF field. Urea coated with pine lignin [13] mixed with various types of additives exhibited 59% higher crushing strength than the uncoated fertilizer. Urea coated with four types of lignin, recovered from the effluent liquor of the paper and paper industry [14], gave products with film forming properties, but unsatisfactory release rate. The same occurred for urea granules fabricated by mixing urea with kraft lignin under melting conditions [15]. Kraft lignin was added to a tapioca starch-urea-borate matrix to modulate the starch matrix hydrophilicity properties and reduce the urea release in water [16]. The final film product remained intact after one month of contact time with water. The retardation of the urea release rate by Kraft lignin was confirmed also for other formulations [15]. Based on these results, the BP keeping the memory of the pristine lignin matter present in the sourcing food waste seemed to the authors of the present work to offer intriguing perspectives for use together with urea and the SPC matrix in the manufacture of new CRFs.

The main aims of the present work were to develop a twin-screw extrusion process tailored to the fabrication of the new SPC-BP-U pellets and to test the pellets mechanical behavior and release rate in solution of organic and inorganic nitrogen. The principal conclusions of the work are that the developed twin-screw extrusion process successfully produces the SPC-BP-U pellets and that the pellets behave as CRFs. The results also suggest that BP has positive effects on the mechanical properties of the pellets and on the chemical behavior of the organic and inorganic N species released in solution.

2. Materials and Methods

2.1. SPC, BP, U, and Other Reagents

The BP was available from previous work [9]. It was obtained from the anaerobic digestate of the MBW processed in the ACEA waste treatment plant in Pinerolo, Italy. The sampled digestate was hydrolyzed at 60 °C in pH 13 water. The liquid hydrolysate was separated from the insoluble residue by centrifugation and then filtered through polysulphone membrane with 5 kDa cut off. The membrane retentate was dried at 60 °C to yield the solid BP.

The SPC was obtained from a sunflower oil cake (SOC). To obtain it, SOC was sieved using a Ritec (France) 600 vibrating sieve shaker fitted with a 1 mm grid. The oversize was mainly composed of solid particles from the seed hull, and therefore rich in fibers. Conversely, the undersize (SPC) was enriched with the smaller particles coming from the kernel of the seed, thus having a high protein content. This resulted in a higher content of SPC in proteins, the latter having been evaluated at 50.7% ± 0.1% (in proportion to the SPC dry weight) using the Kjeldahl method [17]. The other chemicals inside SPC were minerals, lipids, cellulose, hemicelluloses, lignins, and water-solubles, with contents of 8.6% ± 0.1%, 1.4% ± 0.1%, 11.0% ± 0.9%, 11.2% ± 1.7%, 0.8% ± 0.2%, and 26.3% ± 0.1%, respectively. The SPC components were determined according to the following literature methods: ISO 749 standard [18] for minerals, ISO 659 standard [19] for lipids, and the ADF-NDF method [20,21] for cellulose, hemicelluloses, and lignin. The content in water-soluble components was estimated by measuring the mass loss of the test sample after 1 h in boiling water. All analyses were carried out in duplicate.

2.2. Fabrication of Extruded Pellets and Injection-Molded Pieces

2.2.1. Extruded Pellets

The extruded pellets were obtained by destructuring and then plasticizing the proteins present in SPC. Thermo-mechano-chemical destructuring was carried out [22,23]. Protein plasticization was conducted in the presence of an aqueous solution of sodium sulfite (1 kg sodium sulfite for 10 kg water) or even urea (U) using a Clextral (France) Evolum HT 53 co-rotating and co-penetrating twin-screw extruder (Figure 1). The latter was composed of nine modules, each 4D in length, D corresponding to the screw diameter (i.e., 53 mm). The total barrel length was thus 36D (i.e., 1.908 m).

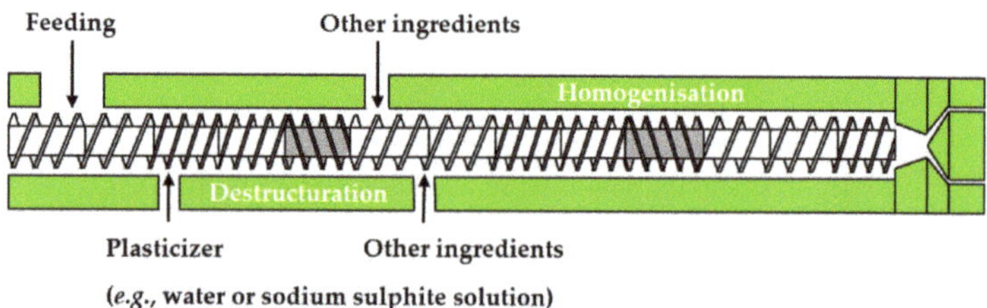

Figure 1. Twin-screw extrusion scheme.

The screw rotation speed was 300 rpm for all the formulations produced. The temperature profile along the barrel was 20 °C for the feeding module (module 1), 80 °C for module 2, and 100 °C for the seven other modules (i.e., modules 3 to 9).

SPC was introduced at the level of module 1 using a Coperion K-Tron (Coperion K-Tron (Schweiz) GmbH, Niederlenz, Switzerland) SWB 300-N weight feeder at a 64.5 kg/h inlet flow rate. The sodium sulphite solution was injected at the end of the second module at a 23.3 kg/h inlet flow rate using a DKM (Clextral, Firminy, France) Super K CAMP 112 piston pump. The resulting formulation was referred to as SPC. When incorporated in the formulation, BP and U were also introduced at the level of module 1 using a Coperion K-Tron K-ML-KT20 weight feeder (Coperion K-Tron (Schweiz) GmbH, Niederlenz, Switzerland) at a 6.5 kg/h inlet flow rate (i.e., 10% (w/w) in proportion to SPC). The corresponding formulations have been designated as follows: SPC-BP and SPC-U, respectively. For its part, the SPC-BP-U formulation was obtained by adding simultaneously BP and U in module 1, both introduced at a 3.25 kg/h inlet flow rate.

The intimate mixing of the solids each other, and the impregnation of the liquid into the solid(s) was made possible by the use of two consecutive pairs of bilobe paddles (BL22 type), each 1D in length, positioned at the beginning of module 4. The plasticization of sunflower proteins was obtained through intense mechanical shear using four consecutive pairs of reversed elements (CF2C type), each 0.5D in length, positioned at the end of module 7 and beginning of module 8. A die equipped with six holes, each 3 mm in diameter, was positioned at the end of the barrel, and the extruded pellets were obtained using a Clextral HC 45 granulating system. They were then dried using a Clextral Evolum 600 continuous belt dryer up to a 10% moisture content before their packaging inside sealed plastic bags.

2.2.2. Injection-Molded Pieces

The extruded pellets were molded into standard bending and tensile specimens by thermoplastic injection using a Negri Bossi (Cologno Monzese, Italy) VE 160–720 machine with a clamping force of 160 ton, and a mold with two cavities. All formulations were rehydrated by adding water to the extruded pellets prior to thermoplastic injection. The moisture content of the extruded pellets was 20% at the time of molding.

The conditions used to produce the test specimens were as follows: 70 °C/90 °C/110 °C for the temperature profile along the plasticizing screw, 130 °C for the temperature of the nozzle, 150 rpm for the rotation speed of the plasticizing screw up to a 21 mm length for the shot building, 30 bar for the counter pressure, 150 mm/s for the injection speed, 800 bar for the follow-up pressure applied during a 2.5 s duration, 1600 kN for the clamping force, 50 °C for the mold temperature, and 20 s for the cooling time before opening the mold and ejecting the test specimens from the cavities.

Once obtained, the test specimens were placed in a climatic chamber at 60% relative humidity (RH) and 25 °C for three weeks for equilibration. Once equilibrated, they were then used for characterization.

2.3. Characterization of Mechanical Properties and Water Sensitivity Test for SPC Composites

2.3.1. Density of Extruded Pellets and Injection-Molded Pieces

The density of the extruded pellets was measured using a 50 mL pycnometer, and cyclohexane as immersion liquid. Cyclohexane was chosen because of its marked hydrophobic character. Indeed, cyclohexane contains only carbon and hydrogen atoms whose bonds are not polarized, thus classifying this solvent in the category of the apolar solvents. On the contrary, the granules are hydrophilic as they are composed of biomolecules and biopolymers with many polarized bonds (e.g., CO and NH bonds for proteins, OH bonds for cellulose, etc.). In consequence, there is no absorption of cyclohexane by the granules and no change in their volume by swelling. All determinations were carried out in duplicate.

The density of the injection-molded pieces was measured from three test specimens with an 80 mm length, a 10 mm width, and a 4 mm thickness. Their thickness and their width were measured at three points, and their length at two points, with a 0.01 mm resolution electronic digital sliding caliper. Thickness (t), width (b), and length (l) mean values were recorded to calculate the specimen volume, and test specimens were all weighed to calculate their density (d). Mean apparent density of the injection-molded pieces from the same formulation was the mean value of measurements made on the three test specimens.

2.3.2. Resistance to Mechanical Abrasion

The resistance to mechanical abrasion of both pellets and injection-molded pieces was estimated from an unstandardized test that was specifically developed during this study. For this test, a cylindrical plastic container having a 115 mm diameter and a 130 mm height was used. The bottom of the container was fixed to an axis, inclined at 15° to the top with respect to the horizontal direction, and rotating by means of a motor at an 80 rpm rotation speed. About 30 g of pellets or injection-molded pieces having a $10 \times 10 \times 4 \text{ mm}^3$ volume were positioned inside the container before the beginning of the test. During the test, 10 metal parts for a total mass of 125 g were also placed inside the container, their addition aiming to simulate the mechanical abrasion to which the pellets or the injection-molded pieces will be subjected during their spreading to the field. Two wooden blades having the same length as the height of the container, both 1 cm high and 1 cm wide, were fixed inside the cylinder, 180° to each other. Here, their objective was to amplify the effect of mechanical abrasion on the tested sample by the metal parts. The test duration was 1 h. At the end of the test, the plant objects were recovered and then sieved on a 2 mm grid to quantify the fines generated during the mechanical abrasion test. The result was expressed as the ratio of fines to the initial mass of the sample analyzed (%). The higher this ratio, the more sensitive the sample was to mechanical abrasion. All determinations were carried out in duplicate.

2.3.3. Tensile and Bending Properties

The tensile properties of the injection-molded pieces were determined according to ISO 527-4 standard [24]. In particular, the two points chosen at the beginning of the stress-strain curve for the Young's modulus calculation were associated with the following elongations: 0.0005 (i.e., 0.05%) and 0.0025 (i.e., 0.25%).

The flexural properties were determined according to the ISO 178 standard [25], i.e., using the three points bending method. For both tensile and bending properties, an Instron (Instron, Norwood, MA, USA) 33R 4204 universal testing system fitted with a 5 kN load cell was used.

The distance between jaws for the tensile tests was 105 mm, and a testing speed of 3 mm/min was applied. The grip separation for the bending tests was 40 mm, and a testing speed of 2 mm/min was applied. All determinations were carried out through four repetitions.

2.3.4. Shore D Surface Hardness

The Shore D surface hardness of the injection-molded pieces was determined according to the ISO 868 standard [26] with a Bareiss (Bareiss Prüfgerätebau GmbH, Oberdischingen, Germany) durometer. The indentation direction was perpendicular to the upper face of the injection-molded piece. For each formulation, determinations were carried out from four different bending test specimens, and measurements were made 10 times per each test specimen (five times for each of its sides).

2.3.5. Water Sensitivity

The water sensitivity of the injected pieces was determined according to ISO 16983 standard [27]. Measurements were conducted after soaking of the analysed samples in water during 1, 3, 6, and 24 h, respectively. Thickness swelling (TS, %) and water absorption (WA, %) were calculated. All determinations were carried out in triplicate.

2.4. Measurements of N Release from SPC Composites

The study of the release of urea and other N-containing compounds was performed according to a previously reported procedure [1]. The samples (4 g each) were immersed in 80 mL demineralized water in a 100 mL glass flask. Three replicates were carried out for each material. The flasks were stored at 25 °C in the dark. At given sampling times (1, 2, 4, 10, 15, 25 days), the liquid and solid phases were separated by centrifugation (15 min at 3000 rpm). The liquid phase was used for the determination of the concentrations of urea, ammonia, and total nitrogen with the methods described below. At each sampling, the same amount of demineralized water as that of the discarded supernatant was added to the solid phase.

2.4.1. Determination of the Urea Concentration

The concentration of urea in the supernatant was measured following the spectrophotometric method of Chen et al. [28]. One ml aliquot solution was diluted with water (1:1000 (v/v) after 1 day extraction, 1:500 (v/v) after 2, 4, and 7 days of extraction, 1:100 (v/v) after 10 days of extraction, and 1:50 (v/v) after 15 and 25 days of extraction).

To the diluted samples (18 mL), transferred in a 100 mL amber glass bottles, 0.9 mL di DAM (diacetylmonoxime 0.5 g/10 mL), 0.15 mL TSC (thiosemicarbazide 0.2 g/100 mL), 0.15 mL $FeCl_3$, $6H_2O$ 40.56 mg/100 mL), 12 mL (5%, v/v) H_2SO_4 were added in sequence. The reaction mixtures were heated in a water bath at 80 °C for one hour.

The absorbance of the complex was measured at 520 nm in a spectrophotometer (Hitachi U-2000), previously calibrated with solutions of urea at given concentrations treated as described above.

2.4.2. Determination of Ammonia Nitrogen

A 10 mL aliquot of the supernatant was transferred to a Kjeldahl tube. After addition of 1 g MgO and 70 mL water, NH_3 was distilled in a Kjeldahl instrument and collected in a beaker containing 20 mL boric acid and 2 drops of methyl red and bromocresol green indicator.

2.4.3. Determination of Total Nitrogen

A 1 mL aliquot of the supernatant was suspended in 25 mL sulfuric acid and 0.5 g selenic mixture and heated to boiling point until complete mineralization. The solution was transferred to a 25 mL volumetric flask and brought to volume with water. A 10 mL aliquot was transferred to a Kjeldahl tube and treated as above after addition of 40 mL 40% NaOH instead of MgO.

2.4.4. Determination of Organic Nitrogen

Organic nitrogen was calculated by subtracting ammonia and urea N from total N.

2.5. Statistical Treatment of Data

The data were evaluated by one-way ANOVA ($p < 0.05$) followed by the Tukey's test for multiple comparison procedures.

3. Results

3.1. The Twin-Screw Extrusion Process to Manufacture the SPC-Based Composite Pellets

A process was patented [22], in which a material rich in protein, such as sunflower or rapeseed oil cake, can be rendered thermoplastic, capable of being injected at about 120 °C. This is achieved by previously destructuring the oil cake in a continuous twin-screw extruder, while adding an appropriate plasticizer (Figure 1). The most interesting part of the invention is that water can act as plasticizer, which is a neat environmental advantage. The extrusion process, followed by an injection process, allowed obtaining biodegradable planting pots [29].

A further improvement in the process of plasticizing sunflower proteins by thermo-mechano-chemical destructuring in a twin-screw extruder consists of adding a sodium sulphite salt to the water in such a way that the proportion of sodium sulphite is between 1 and 10 g per 100 g of protein [23]. In addition to the plasticizing effect of water, sodium sulfite allows a reduction of the disulphide covalent bonds linking the amino residues of cysteine together. Proteins are thus further destructured, and the melt rheology of the obtained agro-granulates is reduced. This makes it easier to produce thermoplastic injection-molded articles.

Another process was also patented for the production of water soluble SPC from industrial SOC reacted in alkaline water at pH 12 and 50 °C [30]. By this procedure, the SPC contains 90% protein, 1.7% lignin, no polysaccharide, and 2.4% ash against 34% protein, 32% polysaccharides, 5.2% lignin, and 7.6% ash in the pristine material. The SPC protein isolate could be relevant in the present work for two reasons. Its thermoplastic properties allow using it as matrix for the manufacture of extruded articles from blends with additives, which have no film forming properties or cannot be processed by extrusion. In addition to the matrix plastic properties, the SPC content of N [31], and of P and K [32], adds the capability to perform also as plant nutrient and/or biostimulant [33] release agent in CRF agriculture applications.

In the present work, twin-screw extrusion and injection-molding were carried out from a sunflower protein concentrate (SPC) having a 51% content in proteins. Figure 2 describes a schematic diagram of the process developed for the preparation of the fertilizers. First, the extrusion tool was used with success to conduct the thermo-mechano-chemical de-structuring of SPC, thus allowing the protein plasticization. The latter was effective thanks to the addition of the sodium sulfite solution, water acting as a plasticizer for proteins [8] on the one hand and the sodium sulfite salt as a reducing agent for the disulfide bonds, i.e., the cysteine bridges [23], on the other hand. In addition, thanks to the high versatility of the twin-screw extrusion technology [34], the simultaneous addition of BP and/or U in the same twin-screw extruder pass resulted in the unexpected synergic intensification of the developed process. All the obtained formulations (i.e., SPC, SPC-U, SPC-BP, and SPC-BP-U) were produced in a controlled way as shown by the stability of the current consumed by the motor of the twin-screw machine throughout the sampling process.

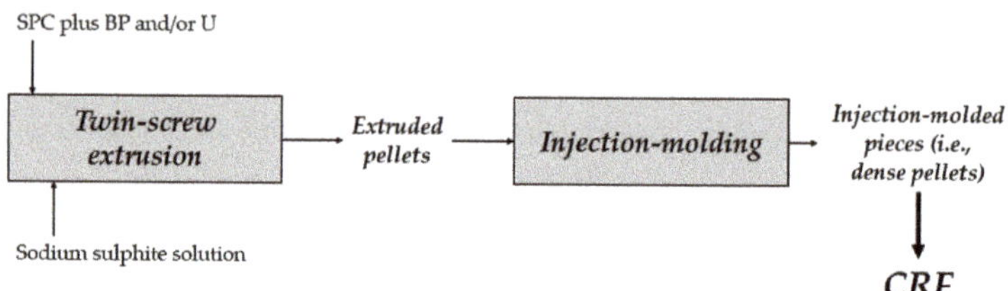

Figure 2. Schematic diagram of the process developed for the preparation of the fertilizers.

However, for the two formulations including BP (i.e., SPC-BP and SPC-BP-U), it has to be noted that a significant self-heating of the mixture was observed in the die (up to 113 °C instead of 96–104 °C without BP). This resulted in higher electric current (up to 66 A instead of 41–46 A), higher specific mechanical energy (up to 143 W·h/kg instead of 89–108 W·h/kg), and higher pressure at the die (up to 14.9 bar instead of 6.0–7.5 bar). The self-heating resulted also in lower moisture content of the SPC-BP and SPC-BP-U extruded pellets (up to 14.3% instead of 26.2–26.7%) at the twin-screw extruder outlet, being associated with a 60% water evaporation during the extrusion process instead of no more than 17% for the two other formulations (i.e., SPC and SPC-U). For future work, it will be therefore necessary to be aware of the risk of abrasive wear on the screw elements and the inner walls of the twin-screw extruder barrel during the production of large quantities of BP-based extruded pellets.

Due to the rehydration of the extruded pellets up to 20% moisture content before shaping, the injection-molding of the test specimens was possible for all the formulations tested. Indeed, the correct filling of the mold cavities was always effective, including for the BP-based products that generated self-heating in extrusion. Immediately after injection-molding, tensile and bending test specimens were positioned in a climatic chamber for conditioning. A three-week duration was required for complete equilibration (i.e., for obtaining a constant weight for the specimens).

3.2. Composition of the SPC-Based Composites

Table 1 reports the C, N, P, K content for the SPC-U, SPC-BP, and SPC-BP-U composite pellets fabricated in the present work and for the neat SPC, U, BP components used in the fabrication of the composite pellets. All three components contain N. Differently from synthetic urea, SPC and BP originating from natural renewable materials also contain the two additional P and K plant nutrient elements inherited from the pristine sources. It may be observed that the N content in the pellets decreases significantly in the order SPC-U > SPC-BP-U > SPC-BP = SPC. The K content of the SPC-BP pellets is significantly higher than the K content in the other three pellets, as result of the K content in the neat BP sample, being nearly four times higher than the K content in the neat SPC pellet. The data confirm that during the pellets manufacture no degradation of the neat components occurs.

Table 1. Total C, N, P, and K concentrations (%, w/w) in neat sunflower protein concentrate (SPC), biopolymer (BP), urea (U), and in the SPC pellets' samples [1].

Formulation	C	N	P_2O_5	K_2O
BP	39.6	6.59	1.14	5.5
U	20.0	46.6	-	-
SPC	38.7 ± 5.2 [a]	6.9 ± 0.9 [a]	2.5 ± 0.1 [a]	1.5 ± 0.1 [a]
SPC-U	37.7 ± 0.2 [a]	10.5 ± 0.1 [b]	2.3 ± 0.1 [a]	1.3 ± 0.1 [a]
SPC-BP	39.5 ± 2.6 [a]	7.0 ± 0.6 [a]	2.6 ± 0.3 [a]	1.8 ± 0.1 [b]
SPC-BP-U	38.6 ± 0.4 [a]	8.5 ± 0.0 [c]	2.1 ± 0.1 [a]	1.5 ± 0.1 [a]

[1] Values in the same column followed by different letters (a–c) are significantly different ($p < 0.05$).

3.3. Mechanical Properties of the SPC-Based Composites

Table 2 reports the density of the extruded pellets and the injection-molded pieces from formulations SPC, SPC-U, SPC-BP, and SPC-BP-U. For the extruded pellets, those containing U and/or BP appear denser than those related to the SPC formulation. Urea contributes more to the densification of the pellets than BP. A very clear densification is observed for all the formulations tested after the shaping step, with the density of the injection-molded pieces remaining relatively independent of the formulation (1362–1398 kg/m^3).

Table 2. Density (kg/m^3) of the extruded pellets and the resulting injection-molded pieces.

Formulation	Extruded Pellets	Injection-Molded Pieces
SPC	783.2 ± 4.1	1383.2 ± 10.0
SPC-U	1122.5 ± 29.4	1398.2 ± 4.5
SPC-BP	878.2 ± 29.9	1384.9 ± 2.7
SPC-BP-U	998.7 ± 26.9	1361.8 ± 19.2

Table 3 reports the results for the resistance to mechanical abrasion for both extruded pellets and injection-molded pieces from formulations SPC, SPC-U, SPC-BP, and SPC-BP-U. These results show a poor abrasion resistance of the extruded pellets, especially those containing neither U nor BP, which results in the generation of a high proportion of fines during the test. In the presence of one of these additives, the abrasion sensitivity of the pellets is reduced, especially for those containing urea. However, the generation of fines is still important in that case, i.e., at least 11.5% (case of the SPC-U formulation). On the opposite, the densification of the materials during the injection-molding step contributes to a very good resistance to mechanical abrasion, with the number of fines generated never exceeding 0.7% (w/w).

Table 3. Resistance to mechanical abrasion of the extruded pellets and the resulting injection-molded pieces (% of fines, w/w).

Formulation	Extruded Pellets	Injection-Molded Pieces
SPC	68.8 ± 1.6	0.6 ± 0.0
SPC-U	11.5 ± 3.0	0.4 ± 0.0
SPC-BP	55.9 ± 2.4	0.4 ± 0.0
SPC-BP-U	32.5 ± 2.1	0.7 ± 0.1

Tables 4 and 5 present the tensile and bending properties, respectively, of the injection-molded pieces made from the SPC, SPC-U, SPC-BP, and SPC-BP-U formulations. In parallel, Figure 3 represents the mean stress-strain curves obtained during the tensile tests, and Figure 4 the mean load-displacement ones obtained during the bending tests. These results reveal a reduction in the rigidity of the injection-molded pieces when U and/or BP are contained inside the formulation, especially for the SPC-U formulation. This is illustrated by the decreases in the elastic modules (Tables 4 and 5), which also results in a reduction in the gradient of the curves at their origin (Figures 3 and 4). Additionally, an increase in the elongation at break is observed during the tensile tests, especially for the SPC-U formulation. The same result is also evidenced for the displacement at rupture during the bending tests.

Table 4. Tensile properties [1] of the injection-molded pieces.

Formulation	E_y (MPa)	σ_{max} (MPa)	ε_{max} (%)	σ_r (MPa)	ε_r (%)
SPC	1490 ± 120	8.2 ± 1.7	0.9 ± 0.4	7.8 ± 2.6	0.9 ± 0.4
SPC-U	476 ± 210	6.4 ± 1.0	4.3 ± 3.2	6.4 ± 1.1	4.4 ± 3.3
SPC-BP	858 ± 321	6.0 ± 1.6	1.3 ± 1.0	5.8 ± 1.9	1.3 ± 1.0
SPC-BP-U	497 ± 77	4.6 ± 1.3	1.3 ± 0.5	4.6 ± 1.3	1.3 ± 0.5

[1] E_y, Young's modulus; σ_{max}, maximal tensile strength; ε_{max}, elongation at maximal tensile strength; σ_r, tensile strength at rupture; ε_r, elongation at rupture.

Table 5. Bending properties [1] of the injection-molded pieces.

Formulation	E_f (MPa)	d_r (mm)	σ_f (MPa)
SPC	1471 ± 85	1.2 ± 0.2	20.6 ± 3.9
SPC-U	295 ± 20	6.7 ± 1.4	14.0 ± 1.3
SPC-BP	434 ± 69	3.2 ± 0.2	13.3 ± 2.5
SPC-BP-U	716 ± 34	2.0 ± 0.1	14.9 ± 0.8

[1] E_f, bending modulus; d_r, displacement at rupture; σ_f, flexural strength at break.

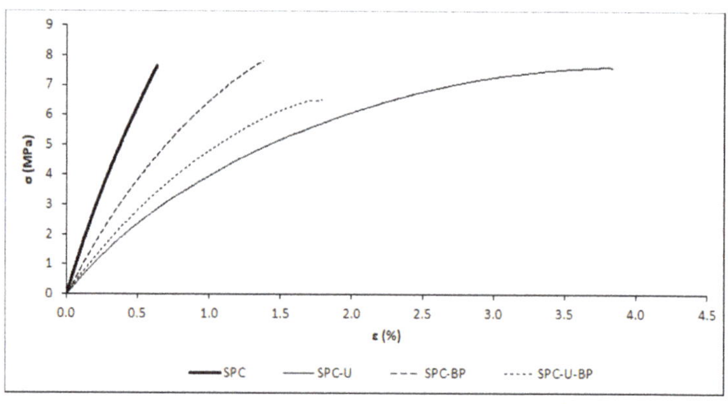

Figure 3. Mean stress-strain curves obtained during the tensile test conducted on the injection-molded pieces.

Table 6 reports the Shore D surface hardness results. Compared to the SPC injection-molded product, a slight reduction in the surface hardness is observed for the SPC-U formulation. On the contrary, it is clear that the addition of BP inside SPC contributes to much harder injection-molded pieces on their surface.

Table 6. Shore D surface hardness of the injection-molded pieces.

Formulation	Shore D (°)
SPC	52.1 ± 3.4
SPC-U	46.9 ± 2.4
SPC-BP	73.2 ± 2.3
SPC-BP-U	62.0 ± 3.5

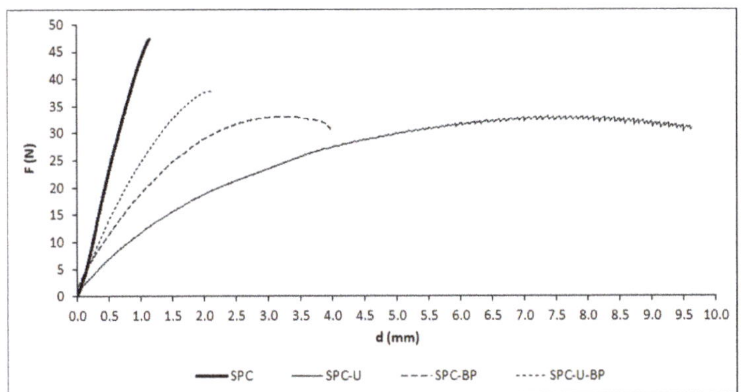

Figure 4. Mean load-displacement curves obtained during the bending test conducted on the injection-molded pieces.

To complete the mechanical characterization of the composites, the latter were also tested with respect to their water resistance after immersion in water during 1, 3, 6, and 24 h. TS and WA values were determined after immersion, and the results are mentioned in Figures 5 and 6, respectively. A progressive increase in the water absorption is observed with the increasing soaking duration for all the tested formulations, with the WA mean value ranging from 45% after 1 h immersion to 158% after 24 h. However, the SPC materials are those with the lowest WA values, regardless of the immersion duration. In parallel, after 1 h immersion in water, thickness swelling is much lower for the SPC and SPC-U formulations (30% in average for TS), compared to the two others (41% in average). TS continues to increase for longer soaking durations, and it remains in the same order of magnitude for all the tested formulations after 3 h just as 6 h (52% in average). After 24 h, SPC, SPC-U, and SPC-BP formulations are so impregnated with water and so fragile when handled that their TS is no longer measurable.

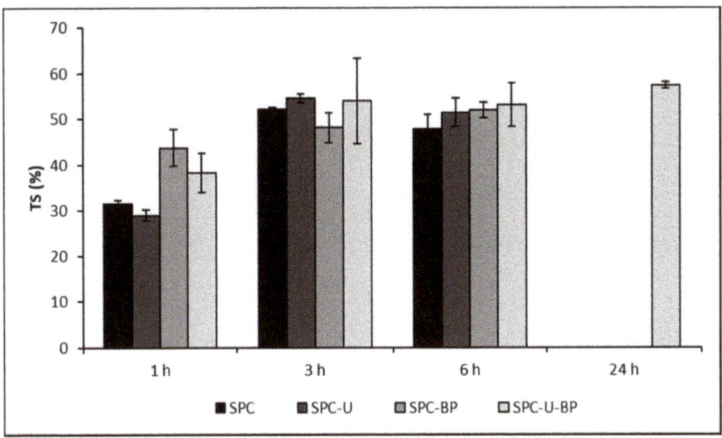

Figure 5. Thickness swelling (TS) of the injection-molded pieces after immersion in water.

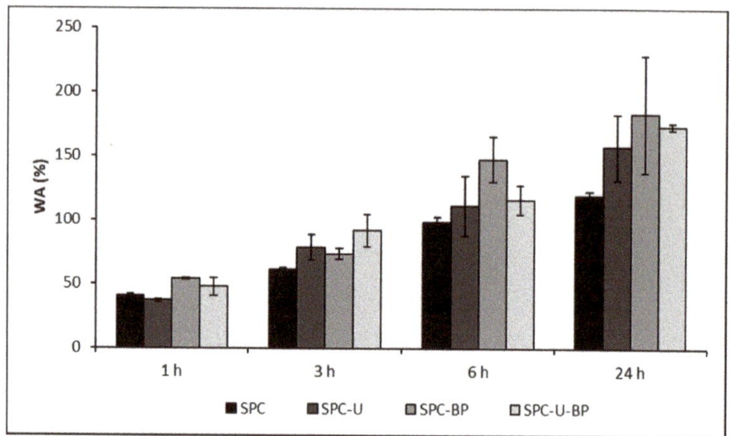

Figure 6. Water absorption (WA) of the injection-molded pieces after immersion in water.

3.4. Urea, Ammonia, and Total N Release from the Composite SPC-Based Pellets

The release of water solubles from the SPC injection-molded pieces is visually appreciated in Figure 7 by the color of the solution becoming darker upon increasing the water contact time.

Figure 7. Macroscopical image illustrating the change of the visual of the SPC injection-molded piece-water system upon increasing the solid–liquid contact time (from left to right, 1, 3, and 6 h).

Table 7 reports the total N released in solution from the pellets after 25 days' contact time with water, as % of the initial N. The data were calculated from the experimental N values measured in solution and from the experimental N values measured in the residual solid pellet. Considering the standard deviation values, for each pellet, the two methods give the same values. This validates the analytical procedure adopted to assess the performance of the pellets as CRF.

Table 7. N mass balance (comparison of % total N in solution measured by mineralization and Kjeldahl analysis, and % total N in the solid samples measured by elemental analysis) [1].

Formulation	% of Initial N, Measured in Solution	% of Initial N, Calculated on the Basis of the Initial and Final Weight and N Concentration of the Samples
SPC	46.1 ± 6.1 [a]	59.9 ± 10.0 [a]
SPC-U	79.4 ± 7.4 [b]	72.7 ± 2.1 [a]
SPC-BP	60.5 ± 9.9 [a]	68.7 ± 15.7 [a]
SPC-BP-U	82.0 ± 2.5 [b]	71.8 ± 11.9 [a]

[1] Values in the same column followed by different letters (a,b) are significantly different ($p < 0.05$).

The cumulative urea release in solution from the urea loaded pellets is illustrated in Figure 8. It is about 3% and 1.5% of the initial urea content of SPC-BP-U and SPC-U, respectively. The initial urea was 5% in the SPC-BP-U and 10% in the SPC-U sample. At the end of the release experiment, the concentration of urea was less than 1% of the total N content in both pellets. Thus, the loss of urea from the SPC-U pellet was about twice that from the SPC-BP-U sample. Yet, the cumulative urea released from the SPC-U sample was half that released from the SPC-BP-U sample. The two pellets were tested under the same conditions, i.e., 4 g pellet sample in 80 mL water. The apparent inconsistency of the release experimental data in Figure 8 suggests two plausible explanations. The relatively low and constant cumulative urea release suggests that, besides its slow release, urea is promptly hydrolyzed to ammonia [35,36]. The higher cumulative urea release found for the SPC-BP-U sample may suggest that BP acts as inhibitor of urea hydrolysis. Inhibition of urea hydrolysis implies reduction of ammonia formation. These effects are in agreement with a recent paper [37] reporting that BP reduces the emission of ammonia from animal urine by 30%.

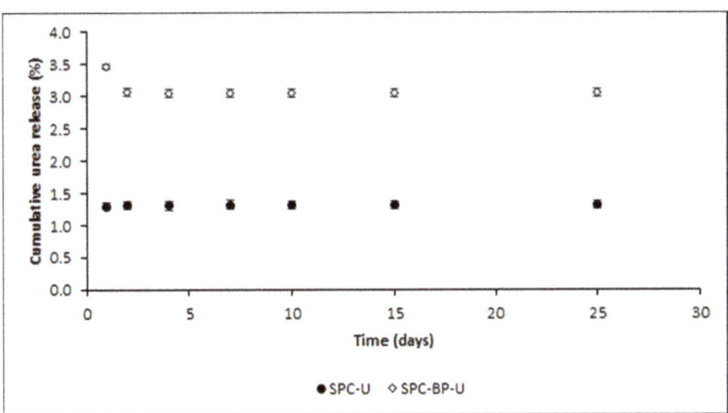

Figure 8. Cumulative urea N released expressed as percentage of the initial urea content.

Table S1 (see Supplementary Material) reports the calculated amount of N from each component (N_{SPC}, N_{BP} and N_U) in the blended SPC-U and SPC-BP-U pellets. Based on the residual amount of urea in the pellets at the end of the experiments, one can calculate that the urea release is 88% from SPC-U and 61% from SPC-BP-U. Based on Figure 8, these amounts of urea appear released within the first day. However, due to the rapid hydrolysis of urea, this cannot be assessed. Other workers did not report such rapid hydrolysis. For urea coated with 10–20% Kraft lignin [13], the increase of urea release upon increasing time was much less steep than that in Figure 8. The former was almost linear over the 15 days' duration of the experiment, during which time the investigated material released 80% of its initial urea content. Urea granules coated with 20% mix made of polyvinyl alcohol and other natural biodegradable polymers [36] showed a similar pattern as in Figure 8, but they released 100% urea within 15–120 min.

Under the above circumstances, total N, ammonia N, and organic N were found better indicators of the behavior of the SPC composite pellets as CRFs.

3.5. Release of N in Solution as Ammonia

The release of N as ammonia, expressed as percentage of the initial N amount, is illustrated in Figure 9. The plot pattern is much less steep than that in Figure 8. In the first days of the experiment, the samples containing urea, SPC-U and SPC-BP-U, released more ammonia than the other two. After seven days, the data were affected by a large variability therefore no significant differences between the samples could be appreciated. In all samples, after 25 days, about 10% of the initial N amount was released as ammonia.

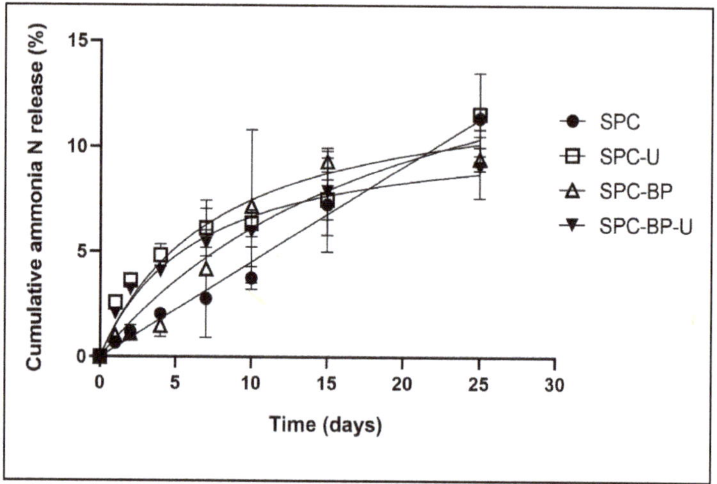

Figure 9. Percentage of initial N released in solution as ammonia.

The results indicate that in the first five days, urea was the main source of ammonia. In the same period of time, the release of ammonia from the SPC-BP-U sample is shown lower than that from the SPC-U sample. This fact supports the hypothesis of BP inhibiting the hydrolysis of urea. It is therefore consistent with the higher release of urea shown in Figure 8.

On the other hand, the lack of difference of the cumulative ammonia release among all samples at the end of the experiments suggests that hydrolysis of other N-containing compounds, likely the proteins of the sunflower cake, was the main source of ammonia. Indeed, while the release of ammonia from SPC was significantly the lowest until day 10, at longer contact time with water, SPC ends up releasing the same amount of ammonia as SPC-U. This confirms that part of the SPC proteins is converted into ammonia.

The data indicate that the rate of solubility and nitrogen mineralization of the three pellets components follows the order U > BP > SPC. The slowest rates exhibited by BP and SPC, compared to urea, is reasonably expected as these samples contain protein matter, which has first to dissolve in water, then be hydrolyzed to amino acids (AA), before the conversion of N from AA into ammonia.

3.6. Release of Total and Organic N in Solution

Figure 10 shows the plots of the cumulative total N release in solution. The total N data are far greater than the N release values as ammonia in Figure 9. This shows that most of the total N in solution is organic N. As the SPC content in the pellet samples is 90%, most of the organic N in solution must be contributed from solubilization of SPC. Indeed, based on the 4 g sample weight (see Materials and Methods), on the pellets' formulations, and on Table 1 and Figure 10 data, one can calculate that the total N release from the SPC-U, SPC-BP, and SPC-BP-U pellets was always much higher than the amount of N contributed by both BP and U in the starting composite pellets (Table S1).

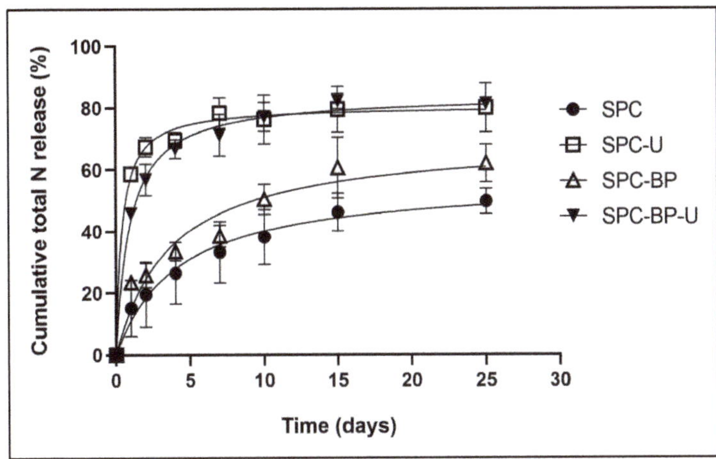

Figure 10. Percentage of initial N released in solution as total N.

The plots in Figure 10 reflect the above order of solubility/N mineralization for the three samples formulations, which is indicated by the cumulative ammonia release plots in Figure 9. The total N release data clearly point out the difference between the pellets containing urea and the pellets containing no urea. SPC sample released 13.4% N in 24 h and 49.7% N in 25 days. This sample meets the requirements [1] for a slow-release fertilizer: No more than 15% released in 24 h, no more than 75% released in 28 days at 25 °C. The SPC-BP sample is close to this behavior, except for the 24 h release, which was 23.4%. The urea-containing samples tended to release N more rapidly than the others all over the experiment, especially within the first day (45.8% and 58.7% for SPC-BP-U and SPC-U, respectively). After 25 days, the total N found in solution was about 80% of the initial amount. These results are close to those reported by other workers [38] in a similar experiment, conducted with urea-impregnated woodchips (U-WC) having the same C and N contents as the SPC composite pellets (Table 1). The cumulative N release pattern of the urea-impregnated woodchips was similar to that of the SPC composite pellets in Figure 10. The U-WC released 40% N in solution, relative to the initial N content in the solid material, after 24 h, nearly 60% N after 4 days, and 62–68% N after 32 days, depending on the nature of the woody material. Statistically, at the end of the experiment, all above SPC samples yield total N release after 24 days within the above highest 75% limit. Thus, all four sample formulations may be classified as controlled-release fertilizers.

Figure 11 shows that both BP and, much more so, urea enhance the organic N release, compared to the neat SPC pellet. The behavior of the SPC-BP sample is very interesting. In the first 10 days, the total organic release from SPC-BP is not much different from that of the neat SPC pellets. After 10 days, the total organic N release from SPC-BP becomes significantly higher, compared to the neat SPC. At day 25, the cumulative organic N release from SPC-BP approaches the value for the pellets containing urea. It seems, therefore, that BP allows a better more gradual organic N release from the SPC pellets. This may happen due to interaction of the two biopolymers yielding a product with different solubility properties than the pristine SPC and BP.

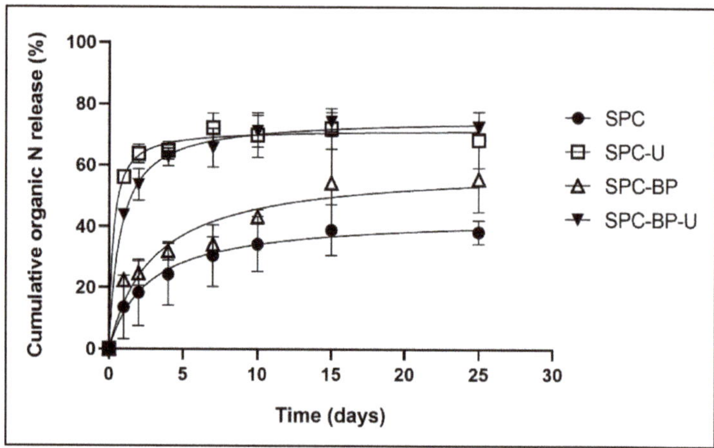

Figure 11. Percentage of initial N released in solution as total organic N.

3.7. Evaluation of the Total C and Nutrients' Release

At the end of the experiment, the solid residues of the samples were dried, weighed, and analyzed for the C, N, P, and K content. The comparison between the initial (Table 1) and final amounts of these elements, allows calculating the percentage of release of each one. The results are reported in Table 8. The loss of weight was about 70% in 25 days, without significant differences between the samples. The nutrient release results were affected by a high variability because they were derived from the elaboration of several parameters (weight, initial and final nutrient contents), therefore subject to the error propagation laws. The N release reflected the results obtained by direct analysis of the solution (Figure 9), but no significant difference between the samples could be pointed out for the reason explained above. The average C and P release were, respectively, 64.7% and 73.2%, more or less reflecting the loss of weight. In contrast, nearly all the initial K (>97%) was solubilized.

Table 8. Percentage of release in solution of total C and nutrients after 25 days experiment, and pellets mass decrease, relative to the initial values in the formulation [1].

Formulation	Mass Decrease	C Release	N Release	P Release	K Release
SPC	69.7 ± 5.0 [a]	63.3 ± 5.7 [a]	59.9 ± 10.0 [a]	70.0 ± 6.8 [a]	98.3 ± 0.1 [a]
SPC-U	71.9 ± 1.0 [a]	63.5 ± 1.5 [a]	72.7 ± 2.1 [a]	78.2 ± 21.2 [a]	98.6 ± 1.2 [a]
SPC-BP	72.8 ± 8.8 [a]	67.7 ± 10.3 [a]	68.7 ± 15.7 [a]	66.7 ± 19.6 [a]	97.8 ± 1.3 [a]
SPC-BP-U	71.6 ± 7.1 [a]	64.2 ± 8.3 [a]	71.8 ± 11.9 [a]	77.9 ± 5.8 [a]	99.0 ± 0.2 [a]

[1] As all values in the same column are always followed by the same letter (a), they are not significantly different ($p < 0.05$).

4. Discussion

4.1. Manufacture of the Pellets Through Twin-Screw Extrusion

In this study, in addition to the pellets produced using the twin-screw extrusion technology, it has also been envisaged to convert them into injection-molded pieces. Because it is carried out under pressure, the injection-molding technique allowed producing pieces that were much denser than the pellets from which they were obtained (Table 2). The higher density of the injected pieces was indicative of largely reduced internal porosity. Thanks to their densification, the injected pieces were much more mechanically resistant than the extruded pellets they came from. This was evidenced by their quite promising tensile (Table 4) and bending (Table 5) properties, as well as their high Shore D surface hardness (Table 6). The injection-molded pieces were also much more resistant to mechanical abrasion than the pellets (Table 3), enabling them to be suitable for their intended use, i.e., their spreading to the field, with the assurance that very few fines will be generated during this mechanized operation. This could be a real advantage for high-throughput applications. In parallel, because of their lower internal porosity, it is reasonable to assume that injection-molded pieces would be more resistant to water than the pellets, thus reducing the kinetics of urea release over time. The water immersion tests carried out on the injection-molded pieces nevertheless showed that the longer they were soaked in water, the more water they absorbed (Figure 6). The same trend was also evidenced in terms of thickness swelling (Figure 5).

In more detail, when looking at the results of resistance to mechanical abrasion, it should be acknowledged that when fines (%) are increasing, the material is less resistant to mechanical abrasion. The results in Table 3 show that the injected pieces are all really resistant to mechanical abrasion as the fines generated during the test are always very low (i.e., 0,7% (w/w) max). This can be explained by the densification of the injected pieces compared with the extruded pellets. However, when comparing the values obtained for the four injected formulations, there is no significant difference in terms of resistance to mechanical abrasion between SPC, SPC-U, SPC-BP, and SPC-BP-U. Indeed, the amount of fines generated ranged from 0.4% to 0.7% (w/w), and this was expected as all the injected formulations have quite the same density values (1362–1398 kg/m^3).

On the opposite, compared with the injected pieces, the resistance to mechanical abrasion of the extruded pellets is really bad. Fines range from 11.5% to 68.8% (w/w). This is due to the much higher internal porosity as related to the lower density values (783–1122 kg/m^3). The result is that the extruded pellets are more friable. Based on the fines generated, the extruded pellets rank in the following order of decreasing resistance to mechanical abrasion: SPC-U (11.5% fines) > SPC-BP-U (32.5% fines) > SPC-BP (55.9% fines) > SPC (68.8% fines). This is in line with the decreasing pellet density values. A second reason may be the unexpected important plasticizing effect of urea on proteins that will be discussed later, which gives to the SPC-U extruded pellets a better elasticity. Thus, they are better supporting deformation at impact, and fines generation from the SPC-U extruded pellets is unfavored. This same remark applies when comparing SPC-BP-U, SPC-BP, and SPC but in a lesser extent. In essence, BP exhibit a reduced plasticizing effect on SPC proteins, compared with urea. This is in perfect accordance with the tensile and bending results conducted from the injected pieces: The less the elastic modules, the more the plasticizing effect.

Differences in mechanical strength were also found by other authors investigating urea granules coated with 20% mix made of polyvinyl alcohol and other natural biodegradable polymers [36]. They reported 140% higher crushing strength for urea granules coated with 10% starch—5% polyvinyl alcohol mix, compared with uncoated urea granules, and 10% higher crushing strength, compared with urea granules coated with other biodegradable polymers such as molasses, gelatin, gum Arabic, and paraffin wax. The as defined "crushing strength" is in fact more comparable to the resistance to mechanical abrasion presented in this study, as it was used by the authors to estimate the tendency of granules to turn into fine powder. The results reported for the SPC-based composites in the present work show that all the injection-molded pieces are fully abrasion-resistant as fines generated are always negligible (Table 3). The differences in mechanical behavior between the SPC-based composite pellets obtained in the present work by extrusion (Table 3) and between the coated urea granules reported by other workers [36], as compared to the high resistance and no differences shown by the injection-molded SPC-based composites, evidences the great advantage of the injection-molding process. When conducted, it results in high densification, surface hardening, and negligible fines generation during the resistance to mechanical abrasion test. In fact, the injection-molding step levels out the differences in resistance to mechanical abrasion due to the different formulations of the SPC-based materials.

The surface hardness values of the injected pieces show a different order: SPC-BP > SPC-BP-U > SPC > SPC-U. It is evident that BP contributes more than urea to the hardening of the composite surface. The more BP is added, the higher the surface hardness.

Regarding the plasticizing effect of urea on proteins mentioned just above, it could be explained by the low steric dimension (i.e., low molecular weight) of the urea molecule on the one hand, and by its affinity of chemical structure with proteins on the other hand. As a reminder, proteins are co-polymers of amino acids linked together by peptide bonds, the latter being in the form of a CO–NH (i.e., type amide) function. This same functional group is also present in the urea molecule. Urea can thus easily position itself between the protein chains within the SPC matrix, leading to their plasticization. This plasticizing effect was confirmed thanks to the tensile tests conducted on the injection-molded pieces (Table 4 and Figure 3), as the presence of U inside the formulation resulted in a large decrease in the elastic modulus, simultaneously with a large increase in both elongations at maximal strength and at break. At the same time, maximal tensile strength and tensile strength at rupture were slightly reduced. All these observations are in perfect accordance with a plasticizing effect. The same observations were also made during the bending tests (Table 5 and Figure 4), with nevertheless a decrease in the flexural strengths at break in a greater proportion. In the same way, a plasticizing effect was also evidenced for BP and for the BP/U mixture at the 50/50 (*w*/*w*) proportion. However, it was less significant. As a result, the extruded pellets and especially the injection-molded pieces have more ductility when mixed with U, BP, or BP-U. These related materials thus become more flexible (i.e., less brittle), consistently with the decrease in the elastic modules.

4.2. The Pellets' Properties

BP does not have film forming properties. The chemical memory of the pristine lignocellulosic structure does not allow fabricating flexible films with adequate properties for use in the manufacture of plastic articles. However, in the case of the composite pellets of the present work, the mechanical strength of their lignocellulosic structural memory contributes the advantage of increasing the mechanical resistance of the composite pellets. Table 6 shows that the two SPC-BP and SPC-BP-U injection-molded pieces, respectively, exhibit 56% and 32% higher Shore D surface hardness than that of the SPC-U injection-molded pieces, which in turn is even lower than that of the SPC ones. The data indicate clearly that urea lowers the surface hardness of the neat SPC. On the contrary, BP not only increases the surface hardness of the neat SPC pieces, but also compensates the negative effect of urea. This specific property of BP is not new. Biagini and coworkers [39] fabricated and measured the mechanical resistance of animal feed pellets containing proteins and fibers, to which 5–10% BP was added. They found that, compared to the control pellets (no added BP), the pellets containing BP exhibited 18–20% higher resistance to shear and compression. Other authors reported an increase in the crushing strength of 59% for urea coated with 20% pine lignin [13].

Mechanical strength is not the only benefit contributed by BP. This biopolymer is also shown to inhibit the hydrolysis of urea and formation of ammonia. Thus, a further environmental benefit is expected from BP, i.e., lowering ammonia and derived N_2O gas emission and/or decreasing ammonia and derived nitrates leaching through soil into ground water. Because of the rapid hydrolysis of urea occurring in the release trials performed in the present work (Figure 8), urea is an unsuitable indicator of the behavior of the SPC-based composite materials as CRFs. Ammonia, as product of urea hydrolysis, is an indirect indicator of the urea release rate (Figure 9). However, all investigated SPC-based materials contain organic N, which can be hydrolyzed to ammonia. Moreover, all three components of the SPC-based composites have fertilizer power. Under these circumstances, total N appears the best indicator of the SPC composites' performance as CRFs.

Table 9 reports the results from the interpolation of Figure 10 data with Equation (1).

$$N_t = N_e \times t/(Kd + t) \tag{1}$$

where N_t is the total nitrogen released at time t, expressed as % of the total nitrogen in the material at time 0, N_e is the maximum total nitrogen released, expressed as % of the total nitrogen in the material at time 0, extrapolated to very high solid–water contact time upon reaching equilibrium, and Kd is the time t needed to reach 50% N_e. The first derivative of Equation (1) yields Equation (2).

$$dN_t/dt = N_e \times [(Kd + 2t)/(Kd + t)^2] \tag{2}$$

which describes the decrease of the total nitrogen release rate upon increasing the contact time between the solid and water.

Table 9. Results from the interpolation of Figure 10 data with Equation (1): Constants (N_e and Kd), and correlation coefficient (R).

Formulation	SPC	SPC-U	SPC-BP	SPC-BP-U
Total N in pellet at time 0 (%, w/w)	6.9	10.5	7.0	8.5
N_e	56.4	80.4	69.7	83.8
Kd	4.100	0.392	3.660	0.915
R	0.913	0.983	0.953	0.990
Total N in pellet at equilibrium (%, w/w)	3.01	2.06	2.12	1.38

The N_e values in Table 9 reflect the mass and nutrients decrease data in Table 8. This allows calculating the residual total nitrogen in the pellet at equilibrium. The Kd values evidence the different release rates of the pellets. The order of decreasing rate is SPC-U > SPC-BP-U > SPC-BP > SPC. The same is also evidenced in the plots of Equation (2) (Figure 12). It is clear that urea is the fastest component to be released, and SPC is the slowest one. BP allows modulating the nitrogen release of the other two components. For instance, BP in the SPC-BP-U pellet allows slowing down the release rate of urea, whereas BP in the SPC-BP one accelerates the release rate of SPC nitrogen. In fact, the four different formulations represent a range of materials with different nutrient release rates.

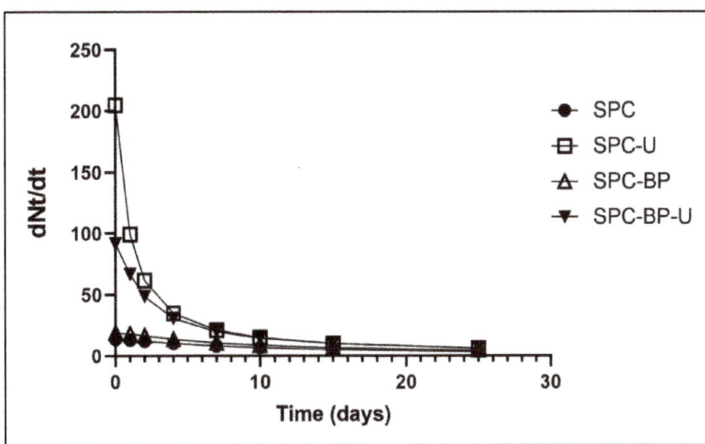

Figure 12. Change of total nitrogen release rate upon increasing the SPC-based composite pellets-water contact time.

The BP biopolymer is a multifunctional product. It contains several acid, basic, and chelating functional groups [9]. The BP effect to control the release of urea and organic nitrogen from the composite pellets is likely to be due to the interaction of its functional groups with urea and SPC. The BP is also a proven soil fertilizer and plant growth biostimulants [9]. All of the above BP properties may work synergically with the properties of urea and SPC, making SPC-BP-U a powerful eco-friendly CRF containing, and capable of releasing and providing soil and plants, organic C and the three major N, P, K nutrients (Table 8). The combination of the three components in different relative ratios may allow tailoring the release rate to the specific requirements in real agriculture practices for specific cultivations. This perspective offers scope for worthwhile testing the performance of the SPC-BP-U pellets in specific plants' cultivation trials.

5. Conclusions

An efficient process has been developed to manufacture a new composite biomaterial made from the two SPC and BP biopolymers, and urea. The composite material has been fabricated in the forms of extruded pellets and then injection-molded pieces, although it contains components having no film forming properties or not processable through extrusion. First of all, the twin-screw extrusion process comprises two main phases, carried out continuously in the extruder depicted in Figure 1. First, SPC is destructured and rendered thermo-plastic by adding aqueous sodium sulfite. In the second phase, BP and urea are added into the thermoplastic SPC matrix and homogeneized under pressure as the blend proceeds to the exit of the extruder. Additionally, the extruded pellets generated can be transformed into denser pieces through injection-molding.

The injection-molded composites, in the form of dense pellets and characterized for their mechanical and nitrogen release properties, have evidenced a number of specific benefits contributed by BP. These allow for concluding that the manufacture process and the formulation of the composite make possible producing new biomaterials in the form of dense pellets from the renewable SPC and BP biopolymers, and urea. These have all mechanical and nutrient release properties for being tested as eco-friendly CRFs for the cultivation of specific plants.

Supplementary Materials: The following are available online at https://www.mdpi.com/2079-6412/11/1/43/s1, Table S1: Calculated amount of N from each component (N_{SPC}, N_{BP}, N_{U}) in the blended SPC-U, SPC-BP, SPC-BP-U pellets compared to the total N amount released in solution from the pellets. Calculations based on the 4 g sample weight (see Materials and Methods), on the composite formulations, and on Table 1 data.

Author Contributions: C.V.-G. and E.M. developed the concept, acquired the funds for the work dedicated to the manufacture of the composite pellets and the characterization of their properties, and supervised it. P.E., L.L., and C.V.-G. developed the process, manufactured, and characterized the mechanical properties of the composite pellets. E.P., V.B., and M.N. performed the characterization, and developed the analytical procedure for the release properties of the composite pellets. E.M. developed the process to manufacture BP, and its analytical protocol. E.M. wrote the original draft of the paper. P.E. edited the final version of the paper. C.V.-G. and M.N. administered the funds at their institution, respectively, to carry out the work. All authors reviewed and completed the paper before submission and agreed to the published version of the manuscript.

Funding: This research was supported partly by endowed funds of the authors' institutions, and partly funded by the European Commission in order to support the implementation of the actions pursued in the LIFE16 ENV/IT/000179-LIFECAB and the LIFE19 ENV/IT/000004-LIFEEBP projects.

Data Availability Statement: Data is contained within the article or supplementary material.

Acknowledgments: The authors acknowledge the kind assistance of Claudio Spitaleri (Istituto Nazionale di Fisica Nucleare, LNS, Catania, Italy) for the mathematical elaboration of the kinetics data.

Conflicts of Interest: The authors declare no conflict of interest.

References

1. Trenkel, M.E. *Slow- and Controlled-Release and Stabilized Fertilizers: An Option for Enhancing Nutrient Use Efficiency in Agriculture*; International Fertilizer Industry Association (IFA): Paris, France, 2010; Available online: www.fertilizer.org (accessed on 4 December 2020).
2. Ramli, R.A. Slow release fertiliser hydrogels: A review. *Polymer* **2019**, *10*, 6073–6090.
3. Razza, F.; Briani, C.; Breton, T.; Marazza, D. Metrics for quantifying the circularity of bioplastics: The case of bio-based and biodegradable mulch films. *Resour. Conserv. Recycl.* **2020**, *159*, 104753. [CrossRef]
4. Majeed, Z.; Ramli, N.K.; Mansor, N.; Man, Z. A comprehensive review on biodegradable polymers and their blends used in controlled-release fertilizer processes. *Rev. Chem. Eng.* **2015**, *31*, 69–95. [CrossRef]
5. Bio Technologies LLC. Sunprotein 2018. Available online: http://www.biot.pro/eng/products/food/sunprotein (accessed on 4 December 2020).
6. Salgado, P.R.; Ortiz, S.E.M.; Petruccelli, S.; Mauri, A.N. Functional food ingredients based on sunflower protein concentrates naturally enriched with antioxidant phenolic compounds. *J. Am. Oil Chem. Soc.* **2012**, *89*, 825–836. [CrossRef]
7. Salgado, P.R.; López-Caballero, M.E.; Gómez-Guillén, C.; Mauri, A.N.; Montero, M.P. Exploration of the antioxidant and antimicrobial capacity of two sunflower protein concentrate films with naturally present phenolic compounds. *Food Hydrocoll.* **2012**, *29*, 374–381. [CrossRef]
8. Rouilly, A.; Mériaux, A.; Geneau, C.; Silvestre, F.; Rigal, L. Film extrusion of sunflower protein isolate. *Polym. Eng. Sci.* **2006**, *46*, 1635–1640. [CrossRef]
9. Montoneri, E. Municipal waste treatment, technological scale up and commercial exploitation: The case of bio-waste lignin to soluble lignin-like polymers. In *Food Waste Reduction and Valorisation*; Morone, P., Papendiek, F., Tartiu, V.E., Eds.; Springer: Berlin/Heidelberg, Germany, 2017; Chapter 6. [CrossRef]
10. Nisticò, R.; Evon, P.; Labonne, L.; Vaca-Medina, G.; Montoneri, E.; Francavilla, M.; Vaca-Garcia, C.; Magnacca, G.; Franzoso, F.; Negre, M. Extruded poly(ethylene-co-vinyl alcohol) composite films containing biopolymers isolated from municipal biowaste. *Chem. Sel.* **2016**, *1*, 2354–2365. [CrossRef]
11. Franzoso, F.; Causone, D.; Tabasso, S.; Antonioli, D.; Montoneri, E.; Persico, P.; Laus, M.; Mendichi, R.; Negre, M. Films made from polyethylene-co-acrylic acid and soluble biopolymers sourced from agricultural and municipal biowaste. *J. Appl. Polym. Sci.* **2015**, *132*, 5803. [CrossRef]
12. IFA. Consumption Urea World 2017. Available online: https://www.ifastat.org/databases/plant-nutrition (accessed on 4 December 2020).
13. Garcia, M.C.; Daez, J.A.; Vallejo, A.; Garcia, L.; Cartagena, M.C. Use of kraft pine lignin in controlled-release fertilizer formulations. *Ind. Eng. Chem. Res.* **1996**, *35*, 245–249. [CrossRef]
14. Mulder, W.J.; Gosselink, R.J.A.; Vingerhoeds, M.H.; Harmsen, P.F.H.; Eastham, D. Lignin based controlled release coatings. *Ind. Crops Prod.* **2011**, *34*, 915–920. [CrossRef]
15. Fernández-Pérez, M.; Garrido-Herrera, F.J.; González-Pradas, E.; Villafranca-Sánchez, M.; Flores-Céspedes, F. Lignin and ethylcellulose as polymers in controlled release formulations of urea. *J. Appl. Polym. Sci.* **2008**, *108*, 3796–3803. [CrossRef]
16. Ariyanti, S.; Man, Z.; Azmi, B.M. Improvement of hydrophobicity of urea modified tapioca starch film with lignin for slow release fertilizer. *Adv. Mater. Res.* **2012**, *626*, 350–354. [CrossRef]
17. ISO. *ISO 5983-1:2005, Animal Feeding Stuffs–Determination of Nitrogen Content and Calculation of Crude Protein Content—Part 1: Kjeldahl Method*; International Organization for Standardization: Geneva, Switzerland, 2005.
18. ISO. *ISO 749:1977, Oilseed Residues–Determination of Total Ash*; International Organization for Standardization: Geneva, Switzerland, 1977.
19. ISO. *ISO 659:2009, Oilseeds–Determination of Oil Content*; International Organization for Standardization: Geneva, Switzerland, 2009.
20. Van Soest, P.J.; Wine, R.H. Use of detergents in the analysis of fibrous feeds. Part IV. Determination of plant cell-wall constituents. *J. Assoc. Off. Agric. Chem.* **1967**, *50*, 50–55.
21. Van Soest, P.J.; Wine, R.H. Determination of lignin and cellulose in acid-detergent fiber with permanganate. *J. Assoc. Off. Agric. Chem.* **1968**, *51*, 780–785. [CrossRef]
22. Rigal, L.; Silvestre, F.; Doumeng, C.; Leyris, J.; Gaset, A. Method For Making Shaped Objects from a Vegetable Raw Material by Injection-Moulding. European Patent EP0988948, 29 March 2000.
23. Humbert, J.; Makoumbou, U.; Rigal, L.; Chelle, R.; Rouilly, A.; Geneau Sbartai, C. Matériau Plastique Composite Sous la Forme de Granulats Issus de Matières Protéiques Végétales et son Procédé de Fabrication. French Patent FR2940297, 25 June 2010.
24. ISO. *ISO 527-4:1997, Plastics–Determination of Tensile Properties–Part 4: Test Conditions for Isotropic and Orthotropic Fibre-Reinforced Plastic Composites*; International Organization for Standardization: Geneva, Switzerland, 1997.
25. ISO. *ISO 178:2010, Plastics–Determination of Flexural Properties*; International Organization for Standardization: Geneva, Switzerland, 2010.
26. ISO. *ISO 868:2003, Plastics and Ebonite–Determination of Indentation Hardness by Means of a Durometer (Shore Hardness)*; International Organization for Standardization: Geneva, Switzerland, 2003.
27. ISO. *ISO 16983:2003, Wood-Based Panels–Determination of Swelling in Thickness after Immersion in Water*; International Organization for Standardization: Geneva, Switzerland, 2003.

28. Chen, L.; Ma, J.; Huang, Y.; Dai, M.; Li, X. Optimization of a colorimetric method to determine trace urea in sea water. *Limnol. Oceanogr. Methods* **2015**, *13*, 303–311. [CrossRef]
29. Geneau-Sbartaï, C. Manufacturing Process of Agromaterial Natural Composite by Twin-Screw Extrusion and Injection Moulding from Sunflower Cake. Ph.D. Thesis, INP, Toulouse, France, 2006.
30. Silvestre, F.; Rigal, L.; Leyris, J.; Gaset, A. Aqueous Adhesive Based on a Vegetable Protein Extract and Process for the Preparation. French Patent FR2785288, 5 May 2000.
31. Ordonez, C.; Tejada, M.; Benitez, C.; Gonzalez, J.L. Characterization of a phosphorus-potassium solution obtained during a protein concentrate process from sunflower flour. Application on rye-grass. *Bioresour. Technol.* **2006**, *97*, 522–528. [CrossRef]
32. Ordonez, C.; Benitez, C.; Gonzalez, J.L. Amino acid production from a sunflower wholemeal protein concentrate. *Bioresour. Technol.* **2008**, *99*, 4749–4754. [CrossRef]
33. Ugolini, L.; Cintia, S.; Righetti, L.; Stefan, A.; Matteo, R.; D'Avino, L.; Lazzeria, L. Production of an enzymatic protein hydrolyzate from defattedsunflower seed meal for potential application as a plant biostimulant. *Ind. Crops Prod.* **2015**, *75*, 15–23. [CrossRef]
34. Evon, P.; Vandenbossche, V.; Candy, L.; Pontalier, P.Y.; Rouilly, A. Twin-screw extrusion: A key technology for the biorefinery. In *Biomass Extrusion and Reaction Technologies: Principles to Practices and Future Potential*; American Chemical Society, ACS Symposium Series, eBooks; American Chemical Society: Washington, DC, USA, 2018; Volume 1304, pp. 25–44.
35. Sigurdarson, J.J.; Svane, S.; Karring, H. The molecular processes of urea hydrolysis in relation to ammonia emissions from agriculture. *Rev. Environ. Sci. Biotechnol.* **2018**, *17*, 241–258. [CrossRef]
36. Beig, B.; Niazi, M.B.K.; Jahan, Z.; Kakar, S.J.; Shah, G.A.; Shahi, M.; Zia, M.; Haq, M.U.; Rashid, M.I. Biodegradable polymer coated granular urea slows down N release kinetics and improves spinach productivity. *Polymers* **2020**, *12*, 2623. [CrossRef]
37. Biagini, D.; Montoneri, E.; Rosato, R.; Lazzaroni, C.; Dinuccio, E. Reducing ammonia and GHG emissions from rabbit rearing through a feed additive produced from green urban residues. *Sustain. Prod. Consum.* **2021**, *27*, 1–9. [CrossRef]
38. Khalid, N.N.A.; Ashaari, Z.; Mohd, A.H.; Mohamed, H.A.; Lee, S.H. Nitrogen deposition and release pattern of slow release fertiliser made from urea-impregnated oil palm frond and rubberwood chips. *J. For. Res.* **2019**, *30*, 208762094. [CrossRef]
39. Biagini, D.; Gasco, L.; Rosato, R.; Peiretti, P.G.; Gai, F.; Lazzaroni, C.; Montoneri, C.; Ginepro, M. Compost-sourced substances (SBO) as feedstuff additives in rabbit production. *Anim. Feed Sci. Technol.* **2016**, *214*, 66–76. [CrossRef]

Review

Developments in Chemical Treatments, Manufacturing Techniques and Potential Applications of Natural-Fibers-Based Biodegradable Composites

Muhammad Yasir Khalid [1,*], Ramsha Imran [2], Zia Ullah Arif [1], Naveed Akram [3,4], Hassan Arshad [1], Ans Al Rashid [5] and Fausto Pedro García Márquez [6,*]

1 Department of Mechanical Engineering, University of Management & Technology, Lahore 51040, Pakistan; zia.arif@skt.umt.edu.pk (Z.U.A.); hassan.arshad@skt.umt.edu.pk (H.A.)
2 Department of Mechanical Engineering, Institute of Space Technology, Islamabad 44000, Pakistan; ramshaimran880@gmail.com
3 Department of Mechanical Engineering, Mirpur University of Science and Technology (MUST), New Mirpur City, Azad Jammu and Kashmir 10250, Pakistan; naveed.me@must.edu.pk
4 Department of Mechanical Engineering, Faculty of Mechanical Engineering, University of Malaya, Kuala Lumpur 50603, Malaysia
5 Division of Sustainable Development, College of Science and Engineering, Hamad Bin Khalifa University, Qatar Foundation, Doha 0000, Qatar; anrashid@hbku.edu.qa
6 Ingenium Research Group, University of Castilla-La Mancha, 13071 Ciudad Real, Spain
* Correspondence: yasirkhalid94@gmail.com (M.Y.K.); faustopedro.garcia@uclm.es (F.P.G.M.); Tel.: +34-926-95300 (F.P.G.M.)

Citation: Khalid, M.Y.; Imran, R.; Arif, Z.U.; Akram, N.; Arshad, H.; Al Rashid, A.; García Márquez, F.P. Developments in Chemical Treatments, Manufacturing Techniques and Potential Applications of Natural-Fibers-Based Biodegradable Composites. *Coatings* 2021, 11, 293. https://doi.org/10.3390/coatings11030293

Academic Editor: Philippe Evon

Received: 2 February 2021
Accepted: 26 February 2021
Published: 4 March 2021

Abstract: The utilization of synthetic materials stimulates environmental concerns, and researchers worldwide are effectively reacting to environmental concerns by transitioning towards biodegradable and sustainable materials. Natural fibers like jute and sisal have been being utilized for ages in several applications, such as ropes, building materials, particle boards, etc. The absence of essential information in preparing the natural-fiber-reinforced materials is still a challenge for future applications. Chemical treatments and surface modifications can improve the quality of the natural fibers. Natural-fiber-based composites are a potential candidate for many lightweight engineering applications with significant mechanical properties. In the view of the progressive literature reported in the field, this work aims to present the significance of natural fibers, their composites, and the main factors influencing these materials for various applications (automotive industry, for instance). Secondly, we aim to address different surface modifications and chemical treatments on natural fibers and finally provide an overview of natural fiber reinforced polymer composites' potential applications.

Keywords: natural-fiber-reinforced polymer composites; chemical treatments; natural fibers; manufacturing techniques; green composites

1. Introduction

The utilization of synthetic materials (Kevlar, carbon, and aramid fibers) stimulates environmental concerns. The production of composites reinforced with these synthetic fibers is costly [1]. Researchers around the globe are effectively reacting to environmental concerns and now shifting towards biodegradable and sustainable materials. The natural fiber-reinforced polymer composites (NFRPCs) are becoming a prime choice in several industrial applications [2,3]. For instance, considering the automotive sector, the overall car production rate is expanding every year. Restricted oil assets will elevate the oil-based items' costs soon [4]. It is assessed that a 25% decrease in vehicle weight would be equal to sparing 250 million barrels of unrefined petroleum. Utilizing the NFRPCs could prompt a weight decrease of 10–30%; thus, it is conceivable that producers will consider extending the utilization of NFRPCs in their new products [5]. Natural fibers are seen in many

321

applications, like in constructing sewage-system septic pipes, which triggering their use in many recyclable products [6]. The utilization of these natural fibers in the car business is now practiced. It can be seen in vehicles' interior, for example, seatbacks, bumpers, dashboards, truck liners, main events, decking, railing, windows, and casings.

Another crucial factor is the massive manufacturing of plastic which rapidly turns into waste and becomes an issue to the climate after its application. From 1950 to 2014, global plastic production has expanded from 1.7 million tons to more than 300 million tons. Plastics use around 8% of globally produced oil in their design as crude material and energy for their production. It was discovered that 95% of these synthetic plastic materials were situated on the seashore, seabed, and top of the ocean [7]. The fabrication of an enormous amount of synthetic-fiber-reinforced materials, such as glass/carbon-fiber polymer composites, is a severe threat to our environment, as reusing glass-fiber-reinforced composites is not straightforward [8]. Furthermore, the synthetic fiber-reinforced composite poses an estimated cost of for glass US$1200–$1800/ton for glass and 30 GJ/ton energy to create, and for carbon, it is US$12,500/ton and 130 GJ/ton energy to deliver, whereas NFRPCs have a lower cost of US$200–$1000/ton and 4 GJ/ton energy to produce [9,10].

Two major challenges are hindering the vast utilization of NFRPCs. The first one is the inferior mechanical properties, compared to synthetic fiber composites [11], which usually is a consequence of the inconsistency between the fiber and the matrix [12]. The fibers' wettability is significantly lower than synthetic fibers (like glass fiber) [13]. Lower wettability is still a challenge for their extensive use in industries [14], even though the strength of these natural fibers is comparable to synthetic fibers [15,16]. The water-absorbing capacity is another factor restricting the enormous use of NFRPCs [17,18]. This behavior hinders the utilization of NFRPCs, since their mechanical properties decline in wet conditions. The opportunities to implement these new materials in open-air applications need to be considered, especially in damp climates [19], as these fibers retain water from the air and on direct contact with the earth, resulting in voids creation within the material [20,21]. Another factor is the exceptionally polar surface of natural fibers; this actuates an interfacial dissimilarity with non-polar polymers like polypropylene and polyethylene [22,23]. Briefly, natural fibers possess inferior hydrophilicity, as well as surface and mechanical properties, and have a porous structure, restricting their use in commercial applications [24].

Besides the full replacement of synthetic fibers as reinforcement to polymers, researchers widely use another technique to reinforce the polymers with a combination of natural and synthetic fibers, limiting the use of synthetic fibers. This technique is often termed hybridization. Khalid et al. [25] studied the effect of hybridization of jute and carbon fibers. Results revealed that the jute/carbon-reinforced composites provide sufficient tensile strength compared to merely carbon-fiber-reinforced composites, broadening the scope to use hybridized fiber-reinforced composites for various applications [26].

A vast majority of the research data published in recent years on natural fiber composites provide limited information on the classification of natural fibers [27], the effect of chemical treatments, and their applications [28–30]. Thus, comprehensive work is needed for the recent accomplishments in this field. This work aims to present the significance of natural fibers, their composites, and the main factors influencing these materials' choices for various applications (automotive industry, for instance). Secondly, this work aims to address different surface modifications and chemical treatments on natural fibers and finally provide an overview of potential applications of NFRPCs.

2. Composition and Mechanical Properties of Natural Fibers

The polymers containing natural fibers as reinforcement are called green composites or natural-fiber-reinforced polymer composites (NFRPCs) [31,32]. Green composites provide several benefits over synthetic-strands/fibers-based composites [9], for example, ease of availability, cost adequacy, modulus, biodegradability, sustainability, and handling simplicity [33–35]. The research interest in natural fibers is developing due to their potential use in modern applications [36]. They are sustainable, cheaper, totally/partially recyclable, and biodegradable [37–39]. The natural fibers can be extracted from several sources, including animals, plants, and minerals [40]. The NFRPCs are also globally accepted sustainable materials [41]. Their brief classification, according to their source, is illustrated in Figure 1. Natural fibers are abundantly available in the South Asian region, specifically Pakistan, India, and Sri Lanka. Among all the natural fibers, jute, bamboo, coir, sisal, and hemp are most commonly reported in the literature.

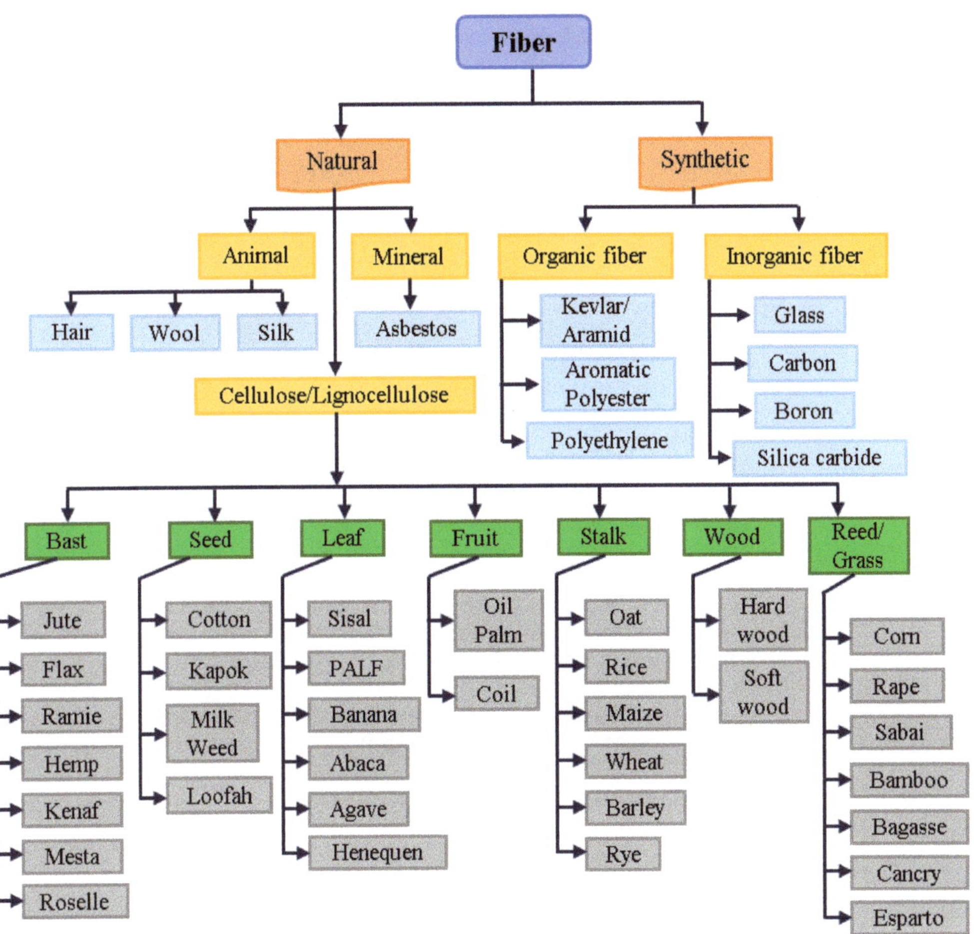

Figure 1. Classification of different types of natural fibers and their sources.

The fibers obtained from plant stems reveal better mechanical properties and yield strength as they contain higher cellulose content [42]. Cellulose content characterizes the strength and firmness of fiber [43], owing to the number of hydrogen bonds and different linkages that exist in the cellulose [44,45]. Generally, fibers' mechanical properties are highly dependent on the production process, chemical processing, and final composite manufacturing techniques [46]. Pretty much all the fibers, barring cotton, are predominantly made out of cellulose, hemicelluloses, lignin, waxes, and a few water-solvent compounds [47–49]. Table 1 reports the most significant mechanical properties and composition of the most common natural fibers.

Coir fiber is an important class of natural fibers with attractive mechanical properties [22]. Coir fiber is cheap and widely utilized in several applications. Coir fibers are obtained from coconut trees, which are found in tropical regions. Moreover, the coir contains flexible lignocellulosic fibers with up to 42% cellulose content, 0.2% hemicellulose content, and 40–45% lignin content [50].

Jute fiber has a high perspective as it reveals high strength to weight proportion and excellent impact strength [51]. The largest producers of jute are India, China, and Bangladesh. Jute fiber reinforced polymer composites has many reported applications, e.g., for windows, furniture, I-shaped beam, skateboards, underground channel drains, water pipes, floor tiles, etc. [52].

However, bamboo fibers exist abundantly in Asia and South America but have not been investigated entirely to their potential, even though they are considered characteristic engineering material [53]. Generally, bamboo has been utilized in living spaces, offices, and instruments, owing to its high strength to its weight due to the longitudinal arrangement of fibers. The bamboo fibers generally have better mechanical properties yet are weak compared to other natural fibers because of the additional lignin content covering the bamboo strands. Furthermore, bamboo fiber has higher elasticity than most natural fibers, such as jute [54] and sisal [53].

Hemp has been cultivated annually in Central Asia for the last 12,000 years and reveals extraordinary tensile strength and modulus, making it useful for reinforcement in composite materials [55,56]. The applications of hemp fibers are in geotextiles, furniture, and manufacturing pipes [57].

Like other natural fibers, kenaf can replace wood, which is gathered once in 20–25 years, while the kenaf plant is collected two or three times a year. kenaf plants can grow 3–4 m high within four to five months and consists of three layers; bast, pith, and core [10,58]. The kenaf plant consists of 33% of the bast, and the remaining part accounts for pith and core. Kenaf bast fiber has been determined for to have predominant mechanical properties than other fibers [59,60].

Banana fiber, at present, is a byproduct of banana development [61]. Subsequently, banana fiber can be utilized for several applications [62]. Banana fiber is found to be a reasonable reinforcement to polyester tar. The properties of the composites are unequivocally impacted by the fiber length [63].

Table 1. Mechanical properties and composition of the most common natural fibers [50,57,64–70].

Properties	Natural Fibers								
	Cotton	Jute	Hemp	Coir	Date Palm	Flax	Ramie	Sisal	Pineapple
Elongation (%)	3–10	1.5–1.8	1.6	15–30	2–19	1.2–3.0	2.0–9.0	2–14	14
Density (g/cm^3)	1.5–1.6	1.3–1.4	1.4	1.2	0.9–12	1.4–1.5	1.5	1.3–1.5	1.4
Tensile Strength (MPa)	280–580	400–800	550–900	175–220	300–800	400–1500	220–938	400–700	400–1600
Cellulose (%)	82–91	60–70	71–75	32–42	46	71	68–76	67–78	70–82
Lignin (%)	–	12–13	3.7–5.7	40–45	20	2.5	0.5–0.7	8–11	5–12
Tensile Modulus (GPa)	6–13	10–30	70	4–6	7	28–80	44–128	9–38	34–82

Likewise, sisal is one of the natural fibers which is abundantly available, and its synthesis is mostly in a regularly developed wasteland, which helps in soil protection [71]. Tanzania and Brazil are the largest producers of sisal in the world [72]. The use of sisal fibers has been seen in many applications like in railway industries as the gear case, main doors, luggage racks, berths, interior panels and partitions, and modular toilets [73,74].

Figure 2 presents the comparison between the unit price of the most widely used natural fibers. The unit cost of natural fibers is significantly lower than synthetic fibers like glass and carbon [75]. Henceforth, taking the mechanical properties, price, and environmental impact into account, flax [76,77], hemp [78], and jute [2] fibers can be considered as potential reinforcement to polymers [79].

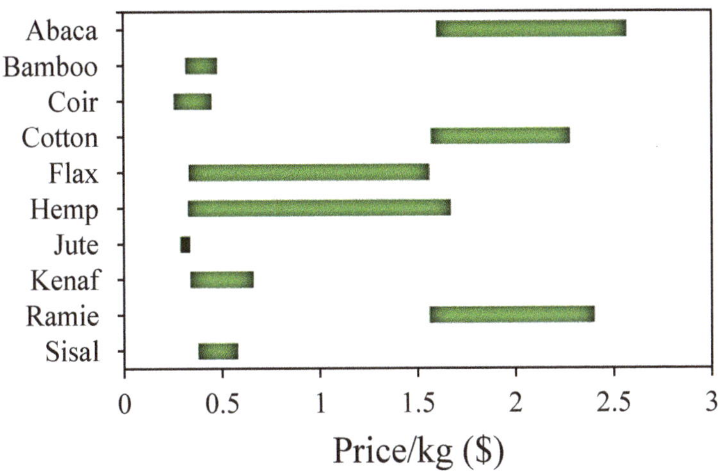

Figure 2. Comparison of cost/weight of different fibers.

3. Different Treatments for Natural Fibers

3.1. Physical Treatments

It is essential to understand the properties of the cell wall segments and their dependencies on fiber properties, to acknowledge how lignocellulosic fiber can be utilized in high-performance industrial applications [80–82]. Natural fibers mostly lag in the hydrophilic property, causing poor chemical resistance, inferior mechanical properties, and porous structure, which limit their engineering applications [83,84]. Hydrophilicity also reduces the applicability of textile products, especially in transport and packaging [85]. Another significant factor in NFRPCs production that virtually affects their properties and interfacial behavior is the different treatments. Many physical treatments [30] are performed on natural fibers, prior to chemical treatments [86]. These treatments [87] include Corona Discharge, Plasma Treatment, Ultraviolet (UV) Treatment, Fiber Beating, and Heat Treatment.

Corona treatment is possibly the most intriguing strategy for surface oxidation actuation. This treatment changes the cellulose strands' surface energy and is responsible for the improved compatibility between the hydrophilic matrix and fibers [88].

Plasma treatment has been effectively used to eliminate the pollutions/dust particles on the fibers, thus providing an improved fibers' surface. The gas type, pressing factor, and concentration need to be controlled precisely, for effective processing [87].

UV is a generally new approach, acknowledged for eliminating dust particles from the plant fiber surface. In UV treatment, some factors (e.g., stream, gas type, etc.) are uncontrolled [83,89]. During the treatment, the strands are set in a chamber for the surface oxidation of fibers. Besides, the UV treatment builds the polarity on the fiber surface, which prompts better wettability of the fibers, leading to the higher strength of the NFRPCs [87,88].

3.2. Chemical Treatments

Chemical treatments have a considerable effect on NFRPC's mechanical properties [90–92], owing to hydroxyl groups from cellulose and lignin [93]. Chemical-treatment strategies commonly rely on reagent functional groups/active groups capable of responding superiorly with natural-fiber structures, just as adequately removing non-cellulosic materials from the fibers [94]. Besides, hydroxyl groups resulting from chemical treatments might be engaged with the hydrogen bonding inside the cellulose atoms, limiting the movement towards the matrix [6,90]. Thus, chemical modifications enact these groups or present new moieties that can viably interlock with the matrix, resulting in good bonding [95,96].

The poisonous and reactive synthetic chemicals are sometimes added to the composite to stop an organic attack [97], which is the basis for the wood-preservation industry [11]. Many surface treatments, for example, alkali, acetylation, silane, and peroxide treatments, have been conducted to improve fiber properties. Different chemical treatments and their particular significance on various natural fibers are reported in Table 2. Table 3 presents optimum concentrations of the solution in different chemical treatment procedures.

Table 2. Different chemical treatments and their effect on natural fibers.

Chemical Treatments Name	Coconut Fiber	Sisal Fiber	Jute Fiber	Banana Fiber	Hemp Fiber	Kenaf Fiber	Hemp Fiber	Flax Fiber	Oil Palm Fiber	Cotton Fiber	Significance or Improvement	References
Alkaline treatment	yes	yes	yes	yes	–	yes	–	–	yes	–	Adhesion	[88,98,99]
Silane treatment	–	yes	yes	yes	–	–	–	–	–	–	Control Fiber Swelling	[86,99,100]
Acetylation treatment	–	-	-	–	–	–	yes	yes	–	–	Moisture absorption	[20,101]
Benzoylation treatment	–	-	yes	–	–	–	–	yes	–	–	Thermal stability	[102,103]
Peroxide treatment	–	-	-	–	–	–	–	yes	–	–	Adhesion	[103,104]
Maleated coupling agents	–	yes	yes	–	yes	yes	–	yes	–	–	Bonding between fibers and matrix	[105]
Sodium chlorite treatment	–	yes	–	–	yes	yes	yes	yes	–	–	Moisture absorption	[105]
Acrylation and acrylonitrile grafting	–	yes	yes	–	–	–	–	yes	–	yes	coupling	[6,106]
Isocyanate treatment	–	yes	yes	–	–	yes	–	yes	–	–	Bonding	[94,106]
Oleoyl chloride treatment	–	yes	yes	–	–	–	–	–	–	–	Wettability	[85,93,106]
Stearic acid treatment	yes	–	yes	–	–	–	–	yes	–	yes	Water resistance	[99,105,107]
Permanganate treatment	–	yes	yes	–	–	–	–	yes	yes	–	Adhesion	[20]
Fungal treatment	–	–	–	–	yes	–	yes	yes	–	–	Remove lignin	[78,108]
Triazine treatment	–	–	–	–	–	–	–	–	–	yes	Adhesion	[109]

Table 3. Strength of chemicals and soaking time of different chemical treatments.

Treatment Name	Strength of Chemical (of Weight)	Soaking Time	Curing Temperature	References
Alkaline	5%	2 h	Room temperature	[76,110,111]
Bleaching	5%	1 h	60 °C	[7,112]
Benzoyl Chloride	30%	30 min	80 °C	[72]
Potassium Permanganate	0.125%	3 min	Room temperature	[7,103]
Maleated Coupling Agents	20%	5–10 min	Room temperature	[105]
Acetylation	2%	2 h	Room temperature	[113]
Isocyanates	5%	2 h	40–45 °C	[114]

3.2.1. Alkali Treatment

Among different methods/techniques, alkali treatment is considered to be cost-effective and most efficient strategy, resulting in changes in the fiber surface, just as removing amorphous hemicelluloses and lignin [94]. Alkali treatment is a typical method to clean and adjust the fiber surface to lower surface strain and improve the interfacial attachment between natural fiber and a polymer matrix [115].

To this end, a few endeavors have been made to create effective natural fibers for reinforcement. A past report showed that the cellulosic fibers removed from leaves of the Saharan aloe vera cactus plant utilizing an antacid treatment could potentially be used as strengthening material [83].

3.2.2. Silane Treatment

Silanes are productive coupling specialists widely used in fiber-reinforced composites and have an equivalent effect on synthetic-fiber-reinforced composite. Silanes also have the potential to act as an adhesion promoter in numerous glue formulations or utilized as substrate primers, providing strong adhesion. Various studies indicated that a layer was formulated on the fiber surface by silane adsorption. The surface morphology of sisal fibers changed by developing compound connections between the silane and the sisal fiber surface. Therefore, they are now extensively used in many studies [116]. Different steps involved in the silane treatment process are as follows [116]:

- Hydrolysis,
- Self-condensation,
- Adsorption,
- Chemical grafting.

3.2.3. Acetylation Treatment

The acetylation treatment reduces plant fibers' hydrophilicity and improves the fibers and matrix's interfacial bonding. It works on acetyl groups (CH_3CO), which removes the hydrophilic hydroxyl gatherings (OH) of the fiber. Accordingly, the fiber's hydrophilic nature diminishes and improves the composites' dimensional stability [36]. This treatment also results in a rough surface topology with fewer void substances, thus providing better mechanical interlocking with the matrix. Senthilraja et al. [113] reported that the mechanical properties of NFRPCs could be improved through acetylation treatment. These NFRPCs were utilized later for applications like entryway boards, vehicle parts, etc. Acetylation treatment reduces the dampness retention furthermore, it improves their life [117].

3.2.4. Benzoylation Treatment

Benzoylation produces a significant change in the natural blend [43]. Benzoyl chloride is regularly utilized in fiber treatment. Benzoyl chloride incorporates benzoyl ($C_6H_5C=O$), ascribed to the treated fiber's diminished hydrophilic nature and improved epoxy adhesion [20]. Kalia et al. [118] explained benzoylation treatment on the sisal fibers through a chemical reaction described in Figure 3. Results revealed that the outside surface of sisal strands got coarse, contrary to the sisal fibers' clear and smooth texture.

3.2.5. Maleated Coupling Agents

The maleated coupling technique is commonly used to strengthen the composites containing fillers and fiber. The setup function of this technique results from two principal factors, cost-effective production and the practical interaction of maleic anhydride with the fibers. Coupling specialists, for example, Polypropylene (PP) joined with maleic anhydride (PP-MAH) and PP united with acrylic corrosive (PP-AA), are generally utilized to improve interfacial properties [119]. Esterification response and H-bond collaborations may occur at the cellulosic filler interface, the PP-MAH, proposed in the current writing and portrayed in Figure 4.

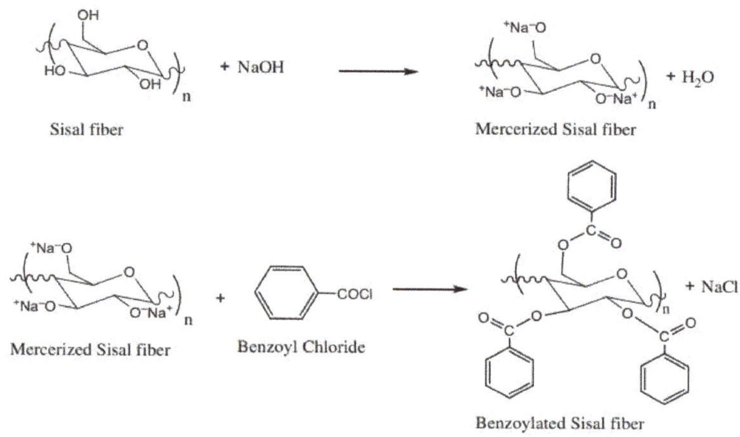

Figure 3. Reactions involved in benzoylation treatment with sisal fibers (Reprinted from permission from [72]. Copyright 2021 Copyright Taylor & Francis).

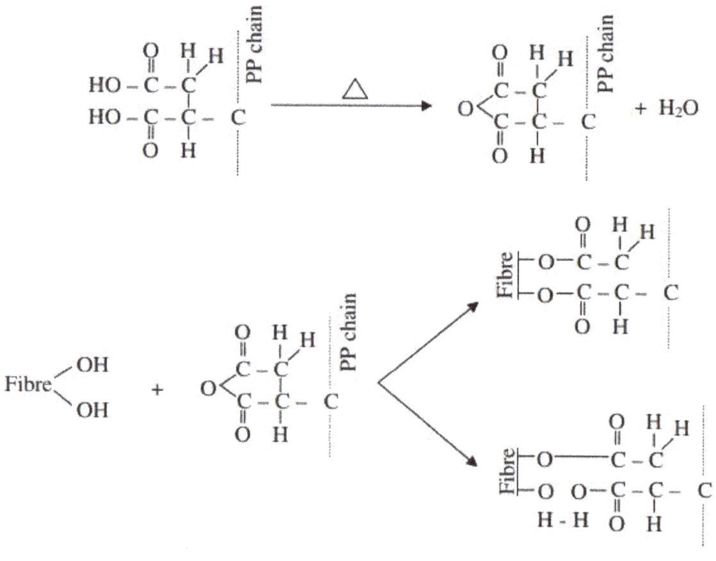

Figure 4. Reactions involved in maleated coupling agent technique (Reprinted from permission from [105]. Copyright 2021 Copyright Elsevier).

Other minor medications, for example, potassium permanganate [120,121], and acrylic acid can also be utilized. The effects of different treatments on a single flax fiber are presented in Figure 5. For example Figure 5a,b shows the comparison of silane treatment on the flax fiber surface. This treatment improves the good adhesion of flax fiber with matrix, resulting in less pull out of fibers from matrix surfaces [122]. Furthermore, it improves the flexibility of flax fiber. Figure 5c,d presents the effect of stearic acid treatment on the flax fiber surface. This technique removes the non-crystalline constituents of the fiber structure. Consequently, the strands are scattered better in the matrix by separating the fiber groups with more fibrillation [123].

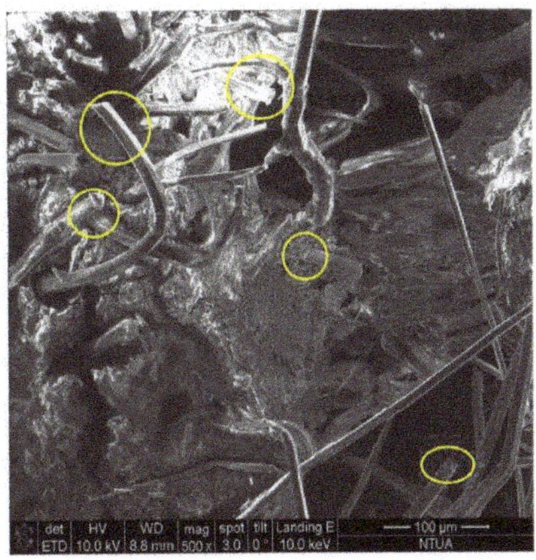

(**a**) Fractured surfaces of Untreated PLAflax

(**b**) Fractured surfaces of PLAflax Treated with Silane content 2wt.%

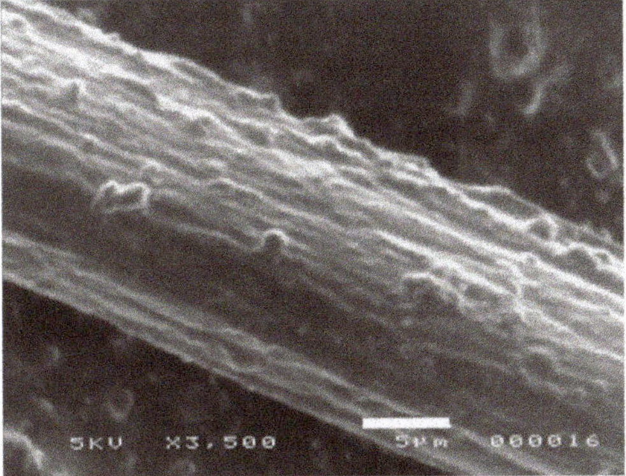

(**c**) Dew Retted (DR) Flax

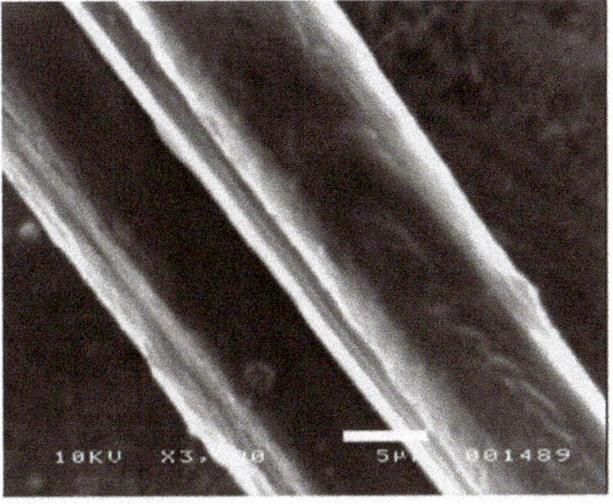

(**d**) Stearic acid-treated Dew Retted (DR) Flax

Figure 5. SEM Images of different chemical treatments on flax fiber ((**a**,**b**) Reprinted from permission from [122]. Copyright 2021 Copyright Elsevier and (**c**,**d**) Reprinted from permission from [123]. Copyright 2021 Copyright Elsevier).

3.3. Effect of Treatments on Mechanical Properties of Composites

Table 4 reports the effect of different treatments on mechanical properties of natural fiber-reinforced (specifically sisal fiber) composites.

Table 4. Effect of different chemical treatments on sisal-fiber-reinforced composites [100,124,125].

Chemical Treatments	Concentration	Tensile Strength (MPa)	Elongation at Break (%)	Young's Modulus (GPa)
Alkali	5% and 10%	34	1	3.3
Benzoylation	Different weight %	45	7	1.01
Isocyanate	Different weight %	42	4	4.1
Permanganate	0.06%	33	5	1.1
Stearic	4%	32	5	1

4. Synthesis of Natural Fiber Reinforced Polymer Composites

NFRPCs composites are fabricated by different manufacturing techniques [87]. These manufacturing techniques include; automated fiber placement, extrusion, compression molding, Resin Transfer Molding (RTM), long fiber thermoplastic-direct (LFT-D) method, sheet molding compound (SMC), pultrusion, and Thermoset Compression Molding. These techniques have been evolved progressively and used by many researchers, revealing the potential of these techniques in composites manufacturing with good quality [74].

Innovative advancements and process solutions should focus on achieving high-quality NFRPCs identified with their new application zones [65]. However, sometimes the following factors affect the manufacturing process:

- Moisture;
- Constituents like cellulose and lignin in natural fiber;
- Final composite structure.

In any case, new downstream and supplementary equipment have been planned. Single or double venting frameworks, for example, are mostly used for heating for in-line drying, high-power splash cooling tanks [126]. An assortment of new feeding systems' arrangements through any gravimetric or vertical crammer, blends the expulsion infusion shaping expulsion pressure forming just as screw, bite the dust, and shape plan [88]. Most of the NFRPCs fabricated to date by these procedures are presented in Figure 6. The manufacturers aim to improve the possibility of utilizing these techniques [64,127].

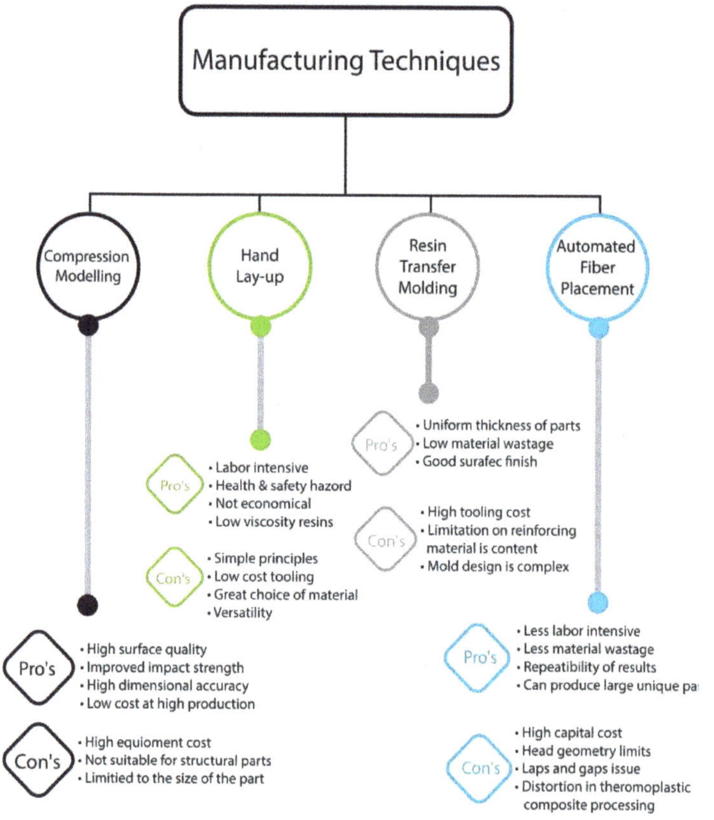

Figure 6. Different Manufacturing techniques for producing natural fiber composites (NFCs).

5. Potential Scope/Application of Natural of Fibers Reinforced Composites

The interest in the use of NFRPCs is developing rapidly in various engineering fields [128]. Figure 7 illustrates the applications of NFRPCs in different industrial areas. Natural fibers, like kenaf, hemp, flax, jute, and sisal, are providing several benefits, such as decreases in weight, cost, and CO_2 footprints; less dependence on oil resources; and recyclability [16,31,129]. Therefore, vehicle manufacturers have been keen on utilizing naturals fiber composites both inside and outside of automobiles. For example, the various types of characteristic NFRPCs have gained incredible attention across multiple car applications and by numerous car manufacturers, for example, German auto organizations (BMW, Audi Group, Ford, Opel, Volkswagen, Daimler Chrysler, and Mercedes), Proton organization (Malaysian national carmaker), and Cambridge industry (a car industry in the USA) [48,68]. The use of NFRPCs serves multiple objectives to these organizations: to bring down the vehicle's overall load along these lines expanding eco-friendliness, and build the supportability of their assembling procedure [130].

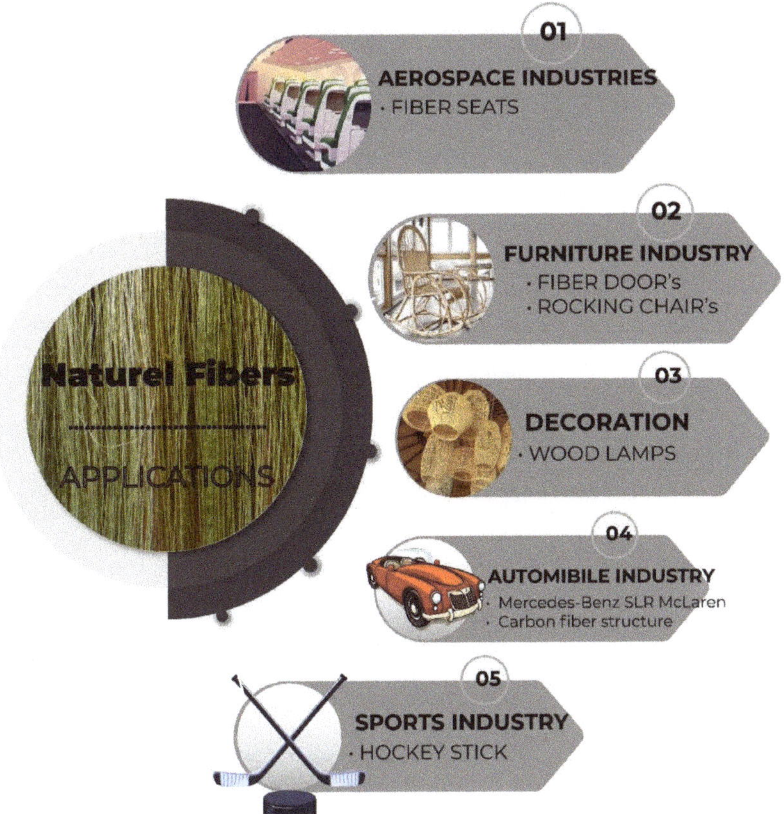

Figure 7. General applications of natural fibers.

Adjacent to the car business, the use of NFRPCs can be seen in the building and development industry, sports, aviation, and others, for instance, boards, window outline, decking, and bike outline frame [77,131]. The specific applications of different natural fibers in industrial products are discussed in Table 5.

Table 5. Applications of different natural fibres [38,48,132–136].

Natural Fibers	Applications
Jute Fiber	Skateboards, Tensil, Hockey, wind-turbines blades, door knobs, automobile interior and exterior parts
Sisal Fiber	In construction industry and also flax and sisal used in interior door-lining panels, MercedesBenz E-Class model
Oil Palm Fiber	Structural insulated panel, fencing and decking
Wood Fiber	In decks, doors, and back seat of automobile as well as molded panel components
Coconut Fiber	Soundproofing, cotton with PP/PET fibers used in trunk
Flax Fiber	Snowboarding laptop cases, tennis racket, automobile parts such as rearview mirror, and visor in two-wheeler; geopolymers panels and Chevrolet Impala automobile trim panels
Kenaf Fiber	Clothing-grade cloth passenger car bumper beam, animal bedding, material that absorbs oil and liquids, packing material and mobile cases, and bags
Coir Fiber	Storage tank, packing material, and engine transmission cover; Brazilian trucks' trim parts of seat cushions
Bagasse Fiber	Window frames and panels

6. Conclusions and Future Perspective

Alarming atmospheric conditions, environmental concerns, and the need for innovation have motivated many researchers to create novel materials. Researchers around the globe are effectively reacting to environmental concerns and now shifting towards biodegradable and sustainable materials. Subsequently, the natural-fiber-reinforced polymer composites (NFRPCs) are becoming a prime choice in several industrial applications. In view of growing interest in NFRPCs, this review aimed to present up-to-date knowledge about different natural fibers, the manufacturing and chemical treatments of natural fibers, and their applications in various fields. Chemical treatments were also highlighted in detail for getting better surface attributes, which permits us to lessen the hydrophilic capacity and improve the adhesion between the fibers and polymers [118]. Natural fibers' choice depends upon their availability, weight/cost, and mechanical properties [137]. The mechanical properties of NFRPCs are primarily affected by the adhesion property of the fibers. This property can be improved by pretreating the fibers. Alkali treatment can be utilized, as it is reported as the most effective technique for all NFRPCs. A combination of other treatments like (alkali + silane) can also be used. The higher concentration of chemicals and the longer soaking time of fibers result in significant mechanical properties improvement. It can be concluded that the development of NFRPCs is rapidly expanding and is visualized as potential future material for several applications [131].

Interfacial adhesion of natural fibers with matrix remains the critical issue in overall performance [99]. Among every chemical treatment studied, alkali treatment is a simple, traditional, and effective procedure for treating many fibers, which can be done on almost all-natural fibers to improve interfacial adhesion [138,139]. Biodegradation phenomena or potential ignition for NFRPCs is moderately new. Therefore, the research should also focus on evaluating the decomposition behavior of NFRPCs. It is necessary to get trustworthy information on the level of biodegradation and the total time for deterioration.

In the future, these NFRPCs will probably achieve complete biodegradability and comparable mechanical properties to those of synthetic-fibers-reinforced composites. Future improvement areas in NFRPCs composites are; mass production of NFRPCs, and thinking about their uses and popularity in the large-scale market are likely to be the primary goals of many researchers. Moreover, the consolidation of nano-cellulose (potentially nanoclay) into NFRPC, can enhance different useful properties, and finally, the examination of the tribological properties of NFRPCs should be one of the prime areas of interest in future research-based studies.

Author Contributions: Conceptualization, A.A.R.; methodology, Z.U.A., M.Y.K., and N.A.; software, H.A. and Z.U.A.; formal analysis, M.Y.K., H.A., and N.A.; investigation, M.Y.K., A.A.R., Z.U.A., and H.A.; writing—original draft preparation, M.Y.K.; writing—review and editing, A.A.R., M.Y.K., R.I., and H.A.; supervision, F.P.G.M.; project administration, F.P.G.M.; funding acquisition, F.P.G.M. and R.I. All authors have read and agreed to the published version of the manuscript.

Funding: The work reported herewith has been financially by the Dirección General de Universidades, Investigación e Innovación of Castilla-La Mancha, under Research Grant ProSeaWind project (Ref.: SBPLY/19/180501/000102).

Institutional Review Board Statement: Not applicable.

Informed Consent Statement: Not applicable.

Conflicts of Interest: The authors declare no conflict of interest.

References

1. Badie, M.A.; Mahdi, E.; Hamouda, A.M.S. An investigation into hybrid carbon/glass fiber reinforced epoxy composite automotive drive shaft. *Mater. Des.* **2011**, *32*, 1485–1500. [CrossRef]
2. Khalid, M.; Al Rashid, A.; Sheikh, M. Correction to: Effect of anodizing process on inter laminar shear strength of GLARE composite through T-peel test: Experimental and numerical approach. *Exp. Tech.* **2021**, 1. [CrossRef]
3. Biagiotti, J.; Puglia, D.; Kenny, J.M. A Review on Natural Fibre-Based Composites—Part II. *J. Nat. Fibers* **2004**, *1*, 37–68. [CrossRef]
4. Al Rashid, A.; Imran, R.; Khalid, M.Y. Determination of opening stresses for railway steel under low cycle fatigue using digital image correlation. *Theor. Appl. Fract. Mech.* **2020**, *108*, 102601. [CrossRef]
5. Davoodi, M.M.; Sapuan, S.M.; Ahmad, D.; Ali, A.; Khalina, A.; Jonoobi, M. Mechanical properties of hybrid kenaf/glass reinforced epoxy composite for passenger car bumper beam. *Mater. Des.* **2010**, *31*, 4927–4932. [CrossRef]
6. Singha, S.; Rana, R.K. Functionalization of cellulosic fibers by graft copolymerization of acrylonitrile and ethyl acrylate from their binary mixtures. *Carbohydr. Polym.* **2012**, *87*, 500–511. [CrossRef]
7. Zie, F.W.W.; Debnath, S.; Anwar, M.; Abdullah, A.H. Enhancing mechanical performance of bagasse fiber-epoxy composite by surface treatment. *Solid State Phenom.* **2020**, *305*, 8–17. [CrossRef]
8. Zahid, S.; Nasir, M.A.; Nauman, S.; Karahan, M.; Nawab, Y.; Ali, H.M.; Khalid, Y.; Nabeel, M.; Ullah, M. Experimental analysis of ILSS of glass fibre reinforced thermoplastic and thermoset textile composites enhanced with multiwalled carbon nanotubes. *J. Mech. Sci. Technol.* **2019**, *33*, 197–204. [CrossRef]
9. Shalwan, A.; Yousif, B. In state of art: Mechanical and tribological behaviour of polymeric composites based on natural fibres. *Mater. Des.* **2013**, *48*, 14–24. [CrossRef]
10. Yousif, F.; Shalwan, A.; Chin, C.W.; Ming, K.C. Flexural properties of treated and untreated kenaf/epoxy composites. *Mater. Des.* **2012**, *40*, 378–385. [CrossRef]
11. Gon, K.; Das, P.; Maity, S. Jute Composites as Wood Substitute. *Int. J. Text. Sci.* **2013**, *1*, 84–93. [CrossRef]
12. Parre, B.; Karthikeyan, A.B.; Udhayasankar, R. Investigation of chemical, thermal and morphological properties of untreated and NaOH treated banana fiber. *Mater. Today Proc.* **2020**, *22*, 347–352. [CrossRef]
13. Valadez-Gonzalez, A.; Cervantes-Uc, J.; Olayo, R.; Herrera-Franco, P. Effect of fiber surface treatment on the fiber–matrix bond strength of natural fiber reinforced composites. *Compos. Part. B Eng.* **1999**, *30*, 309–320. [CrossRef]
14. Sanjay, M.R.; Siengchin, S.; Parameswaranpillai, J.; Jawaid, M.; Pruncu, C.I.; Khan, A.A. comprehensive review of techniques for natural fibers as reinforcement in composites: Preparation, processing and characterization. *Carbohydr. Polym.* **2019**, *207*, 108–121. [CrossRef]
15. Mohanty, A.; Wibowo, M.; Misra, M.; Drzal, L. Effect of process engineering on the performance of natural fiber reinforced cellulose acetate biocomposites. *Compos. Part. A Appl. Sci. Manuf.* **2004**, *35*, 363–370. [CrossRef]
16. Siakeng, R.; Jawaid, M.; Ariffin, H.; Sapuan, S.M.; Asim, M.; Saba, N. Natural fiber reinforced polylactic acid composites: A review. *Polym. Compos.* **2019**, *40*, 446–463. [CrossRef]
17. Venkateshwaran, N.; ElayaPerumal, A.; Alavudeen, A.; Thiruchitrambalam, M. Mechanical and water absorption behaviour of banana/sisal reinforced hybrid composites. *Mater. Des.* **2011**, *32*, 4017–4021. [CrossRef]
18. Saheb, D.N.; Jog, J.P. Natural fiber polymer composites: A review. *Adv. Polym. Technol.* **1999**, *18*, 351–363. [CrossRef]
19. Assarar, M.; Scida, D.; el Mahi, A.; Poilâne, C.; Ayad, R. Influence of water ageing on mechanical properties and damage events of two reinforced composite materials: Flax-fibres and glass-fibres. *Mater. Des.* **2011**, *32*, 788–795. [CrossRef]
20. Li, X.; Tabil, L.G.; Panigrahi, S. Chemical treatments of natural fiber for use in natural fiber-reinforced composites: A review. *J. Polym. Environ.* **2007**, *15*, 25–33. [CrossRef]
21. Jena, P.K.; Mohanty, J.R.; Nayak, S. Effect of Surface Modification of Vetiver Fibers on Their Physical and Thermal Properties. *J. Nat. Fibers* **2020**, 1–12. [CrossRef]

22. Arrakhiz, F.Z.; Elachaby, M.; Bouhfid, R.; Vaudreuil, S.; Essassi, M.; Qaiss, A. Mechanical and thermal properties of polypropylene reinforced with Alfa fiber under different chemical treatment. *Mater. Des.* **2012**, *35*, 318–322. [CrossRef]
23. Arrakhiz, F.; El Achaby, M.; Malha, M.; Bensalah, M.; Fassi-Fehri, O.; Bouhfid, R.; Benmoussa, K.; Qaiss, A. Mechanical and thermal properties of natural fibers reinforced polymer composites: Doum/low density polyethylene. *Mater. Des.* **2013**, *43*, 200–205. [CrossRef]
24. Arib, R.M.N.; Sapuan, S.M.; Ahmad, M.M.H.M.; Paridah, M.T.; Zaman, H.M.D.K. Mechanical properties of pineapple leaf fibre reinforced polypropylene composites. *Mater. Des.* **2006**, *27*, 391–396. [CrossRef]
25. Khalid, M.Y.; Nasir, M.A.; Ali, A.; Al Rashid, A.; Khan, M.R. Experimental and numerical characterization of tensile property of jute/carbon fabric reinforced epoxy hybrid composites. *SN Appl. Sci.* **2019**, *2020*, 1–10. [CrossRef]
26. Jeyapragash, R.; Srinivasan, V.; Sathiyamurthy, S. Mechanical properties of natural fiber/particulate reinforced epoxy composites—A review of the literature. *Mater. Today Proc.* **2020**, *22*, 1223–1227. [CrossRef]
27. Sobczak, L.; Lang, R.W.; Haider, A. Polypropylene composites with natural fibers and wood—General mechanical property profiles. *Compos. Sci. Technol.* **2012**, *72*, 550–557. [CrossRef]
28. Prasad, V.; Joy, A.; Venkatachalam, G.; Narayanan, S.; Rajakumar, S. Finite Element Analysis of Jute and Banana Fibre Reinforced Hybrid Polymer Matrix Composite and Optimization of Design Parameters Using ANOVA Technique. *Proc. Eng.* **2014**, *97*, 1116–1125. [CrossRef]
29. Li, Z.; Wang, L.; Wang, X. Flexural characteristics of coir fiber reinforced cementitious composites. *Fibers Polym.* **2006**, *7*, 286–294. [CrossRef]
30. Ghani, M.A.A.; Salleh, Z.; Hyie, K.M.; Berhan, M.N.; Taib, Y.M.D.; Bakri, M.A.I. Mechanical properties of kenaf/fiberglass polyester hybrid composite. *Proc. Eng.* **2012**, *41*, 1654–1659. [CrossRef]
31. Parikh, D.; Calamari, T.; Sawhney, A.; Blanchard, E.; Screen, F.; Myatt, J.; Stryjewski, D.; Muller, D. Thermoformable Automotive Composites Containing Kenaf and Other Cellulosic Fibers. *Text. Res. J.* **2002**, *72*, 668–672. [CrossRef]
32. Jamshaid, H.; Mishra, R. A green material from rock: Basalt fiber—A review. *J. Text. Inst.* **2016**, *107*, 923–937. [CrossRef]
33. Dawit, J.B.; Regassa, Y.; Lemu, H.G. Property characterization of acacia tortilis for natural fiber reinforced polymer composite. *Results Mater.* **2020**, *5*, 100054. [CrossRef]
34. Khan, Z.; Yousif, B.F.; Islam, M. Fracture behaviour of bamboo fiber reinforced epoxy composites. *Compos. Part. B Eng.* **2017**, *116*, 186–199. [CrossRef]
35. Rahman, N.; Shing, L.W.; Simon, L.; Philipp, M.; Alireza, J.; Hebel, D.E.; Ling, C.S.; Wuan, L.H.; Valavan, S.; Nee, S.S. Enhanced bamboo composite with protective coating for structural concrete application. *Energy Proc.* **2017**, *143*, 167–172. [CrossRef]
36. Tavares, T.D.; Antunes, J.C.; Ferreira, F.; Felgueiras, H.P. Biofunctionalization of natural fiber-reinforced biocomposites for biomedical applications. *Biomolecules* **2020**, *10*, 148. [CrossRef]
37. Reddy, M.; Reddy, Y.V.M.; Reddy, B.C.M.; Reddy, R.M. Mechanical, morphological, and thermogravimetric analysis of alkali-treated Cordia-Dichotoma natural fiber composites. *J. Nat. Fibers* **2020**, *17*, 759–768. [CrossRef]
38. Kumar, S.S. Dataset on mechanical properties of natural fiber reinforced polyester composites for engineering applications. *Data Br.* **2020**, *28*, 105054. [CrossRef] [PubMed]
39. Alsaeed, T.; Yousif, B.F.; Ku, H. The potential of using date palm fibres as reinforcement for polymeric composites. *Mater. Des.* **2013**, *43*, 177–184. [CrossRef]
40. Sumesh, K.R.; Kanthavel, K.; Kavimani, V. Peanut oil cake-derived cellulose fiber: Extraction, application of mechanical and thermal properties in pineapple/flax natural fiber composites. *Int. J. Biol. Macromol.* **2020**, *150*, 775–785. [CrossRef]
41. Ljungberg, L.Y. Materials selection and design for development of sustainable products. *Mater. Des.* **2007**, *28*, 466–479. [CrossRef]
42. Silva, G.; Kim, S.; Aguilar, R.; Nakamatsu, J. Natural fibers as reinforcement additives for geopolymers—A review of potential eco-friendly applications to the construction industry. *Sustain. Mater. Technol.* **2020**, *23*, e00132. [CrossRef]
43. Kumar, A.; Sharma, K.; Dixit, A.R. A review of the mechanical and thermal properties of graphene and its hybrid polymer nanocomposites for structural applications. *J. Mater. Sci.* **2019**, *54*, 5992–6026. [CrossRef]
44. Rokbi, M.; Osmani, H.; Imad, A.; Benseddiq, N. Effect of chemical treatment on flexure properties of natural fiber-reinforced polyester composite. *Proc. Eng.* **2011**, *10*, 2092–2097. [CrossRef]
45. Reddy, N.; Yang, Y. Properties and potential applications of natural cellulose fibers from the bark of cotton stalks. *Bioresour. Technol.* **2009**, *100*, 3563–3569. [CrossRef]
46. Shrivastava, R.; Christy, A.; Telang, A.; Rana, R.S. Effect of Chemical Treatment and Curing Parameters on Mechanical Properties of Natural Fiber Reinforced Polymer Matrix Composites-A Review. *J. Eng. Res. Appl.* **2015**, *5*, 70–75.
47. Parikh, V.; Chen, Y.; Sun, L. Reducing Automotive Interior Noise with Natural Fiber Nonwoven Floor Covering Systems. *Text. Res. J.* **2006**, *76*, 813–820. [CrossRef]
48. Saxena, M.; Pappu, A.; Haque, R.; Sharma, A. *Cellulose Fibers: Bio- and Nano-Polymer Composites*; Springer: Berlin/Heidelberg, Germany, 2011.
49. He, J.; He, L.; Yang, B. Analysis on the impact response of fiber-reinforced composite laminates: An emphasis on the FEM simulation. *Sci. Eng. Compos. Mater.* **2019**, *26*, 1–11. [CrossRef]
50. Ramamoorthy, S.K.; Skrifvars, M.; Persson, A. A review of natural fibers used in biocomposites: Plant, animal and regenerated cellulose fibers. *Polym. Rev.* **2015**, *55*, 107–162. [CrossRef]

51. Ali, A.; Nasir, M.A.; Khalid, M.Y.; Nauman, S.; Shaker, K.; Khushnood, S.; Altaf, K.; Zeeshan, M.; Hussain, A. Experimental and numerical characterization of mechanical properties of carbon/jute fabric reinforced epoxy hybrid composites. *J. Mech. Sci. Technol.* **2019**, *33*, 4217–4226. [CrossRef]

52. Chandekar, H.; Chaudhari, V.; Waigaonkar, S. A review of jute fiber reinforced polymer composites. *Mater. Today Proc.* **2020**, *26*, 2079–2082. [CrossRef]

53. Khalil, H.P.S.A.; Bhat, I.U.H.; Jawaid, M.; Zaidon, A.; Hermawan, D.; Hadi, Y.S. Bamboo fibre reinforced biocomposites: A review. *Mater. Des.* **2012**, *42*, 353–368. [CrossRef]

54. Khalid, M.Y.; Al Rashid, A.; Arif, Z.U.; Akram, N.; Arshad, H.; Márquez, F.P.G. Characterization of Failure Strain In Fiber Reinforced Composites: Under On-Axis and Off-Axis Loading. *Crystals* **2021**, *11*, 216. [CrossRef]

55. Elkhaoulani, A.; Arrakhiz, F.; Benmoussa, K.; Bouhfid, R.; Qaiss, A. Mechanical and thermal properties of polymer composite based on natural fibers: Moroccan hemp fibers/polypropylene. *Mater. Des.* **2013**, *49*, 203–208. [CrossRef]

56. Asim, M.; Abdan, K.; Jawaid, M.; Nasir, M.; Dashtizadeh, Z.; Ishak, M.R.; Hoque, M.E. A review on pineapple leaves fibre and its composites. *Int. J. Polym. Sci.* **2015**, *2015*. [CrossRef]

57. Mohammed, L.; Ansari, M.N.M.; Pua, G.; Jawaid, M.; Islam, M.S. A Review on Natural Fiber Reinforced Polymer Composite and Its Applications. *Int. J. Polym. Sci.* **2015**, *2015*. [CrossRef]

58. Yahaya, R.; Sapuan, S.M.; Jawaid, M.; Leman, Z.; Zainudin, E.S. Effect of layering sequence and chemical treatment on the mechanical properties of woven kenaf-aramid hybrid laminated composites. *Mater. Des.* **2015**, *67*, 173–179. [CrossRef]

59. El-Shekeil, Y.A.; Sapuan, S.M.; Abdan, K.; Zainudin, E.S. Influence of fiber content on the mechanical and thermal properties of Kenaf fiber reinforced thermoplastic polyurethane composites. *Mater. Des.* **2012**, *40*, 299–303. [CrossRef]

60. Akil, H.M.; Omar, M.F.; Mazuki, A.A.M.; Safiee, S.; Ishak, Z.A.M.; Bakar, A.A. Kenaf fiber reinforced composites: A review. *Mater. Des.* **2011**, *32*, 4107–4121. [CrossRef]

61. Al Rashid, A.; Khalid, M.Y.; Imran, R.; Ali, U.; Koc, M. Utilization of Banana Fiber-Reinforced Hybrid Composites in the Sports Industry. *Materials* **2020**, *13*, 3167. [CrossRef]

62. Venkateshwaran, N.; Perumal, A.E.; Arunsundaranayagam, D. Fiber surface treatment and its effect on mechanical and visco-elastic behaviour of banana/epoxy composite. *Mater. Des.* **2013**, *47*, 151–159. [CrossRef]

63. Sapuan, S.M.; Leenie, A.; Harimi, M.; Beng, Y.K. Mechanical properties of woven banana fibre reinforced epoxy composites. *Mater. Des.* **2006**, *27*, 689–693. [CrossRef]

64. Balla, V.K.; Kate, K.H.; Satyavolu, J.; Singh, P.; Tadimeti, J.G.D. Additive manufacturing of natural fiber reinforced polymer composites: Processing and prospects. *Compos. Part. B Eng.* **2019**, *174*, 106956. [CrossRef]

65. Bourmaud, A.; Beaugrand, J.; Shah, D.U.; Placet, V.; Baley, C. Towards the design of high-performance plant fibre composites. *Prog. Mater. Sci.* **2018**, *97*, 347–408. [CrossRef]

66. Hassanzadeh, S.; Hasani, H. A review on milkweed fiber properties as a high-potential raw material in textile applications. *J. Ind. Text.* **2017**, *46*, 1412–1436. [CrossRef]

67. Kumar, R.; Haq, M.I.U.; Raina, A.; Anand, A. Industrial applications of natural fibre-reinforced polymer composites–challenges and opportunities. *Int. J. Sustain. Eng.* **2019**, *12*, 212–220. [CrossRef]

68. Onuaguluchi, O.; Banthia, N. Plant-based natural fibre reinforced cement composites: A review. *Cem. Concr. Compos.* **2016**, *68*, 96–108. [CrossRef]

69. Al-Oqla, F.M.; Sapuan, S.M. Natural fiber reinforced polymer composites in industrial applications: Feasibility of date palm fibers for sustainable automotive industry. *J. Clean. Prod.* **2014**, *66*, 347–354. [CrossRef]

70. Venkateshwaran, N.; Elayaperumal, A. Banana fiber reinforced polymer composites—A review. *J. Reinf. Plast. Compos.* **2010**, *29*, 2387–2396. [CrossRef]

71. Prasad, V.R.; Rao, K.M. Mechanical properties of natural fibre reinforced polyester composites: Jowar, sisal and bamboo. *Mater. Des.* **2011**, *32*, 4658–4663. [CrossRef]

72. Kalia, S.; Kaushik, V.K.; Sharma, R.K. Effect of benzoylation and graft copolymerization on morphology, thermal stability, and crystallinity of sisal fibers. *J. Nat. Fibers* **2011**, *8*, 27–38. [CrossRef]

73. Kumre, A.; Rana, R.; Purohit, R. A Review on mechanical property of sisal glass fiber reinforced polymer composites. *Mater. Today Proc.* **2017**, *4*, 3466–3476. [CrossRef]

74. Lotfi, A.; Li, H.; Dao, D.V.; Prusty, G. Natural fiber–reinforced composites: A review on material, manufacturing, and machinability. *J. Thermoplast. Compos. Mater.* **2019**. [CrossRef]

75. Zaini, S.; Azaman, M.D.; Jamali, M.S.; Ismail, K.A. Synthesis and characterization of natural fiber reinforced polymer composites as core for honeycomb core structure: A review. *J. Sandw. Struct. Mater.* **2020**, *22*, 525–550. [CrossRef]

76. Flynn, J.; Amiri, A.; Ulven, C. Hybridized carbon and flax fiber composites for tailored performance. *Mater. Des.* **2016**, *102*, 21–29. [CrossRef]

77. Dahlke, B.; Larbig, H.; Scherzer, H.D.; Poltrock, R. Natural fiber reinforced foams based on renewable resources for automotive interior applications. *J. Cell. Plast.* **1998**, *34*, 361–378. [CrossRef]

78. Gulati, D.; Sain, M. Fungal-modification of natural fibers: A novel method of treating natural fibers for composite reinforcement. *J. Polym. Environ.* **2006**, *14*, 347–352. [CrossRef]

79. Ali, A.; Shaker, K.; Nawab, Y.; Jabbar, M.; Hussain, T.; Militky, J.; Baheti, V. Hydrophobic treatment of natural fibers and their composites—A review. *J. Ind. Text.* **2018**, *47*, 2153–2183. [CrossRef]

80. Basak, M.K.; Chanda, S.; Bhaduri, S.K.; Mondal, S.B.; Nandi, R. Recycling of jute waste for edible mushroom production. *Ind. Crop. Prod.* **1996**, *5*, 173–176. [CrossRef]

81. Monteiro, S.N.; Lopes, F.P.D.; Ferreira, A.S.; Nascimento, D.C.O. Natural-fiber polymer-matrix composites: Cheaper, tougher, and environmentally friendly. *JOM* **2009**, *61*, 17–22. [CrossRef]

82. Camargo, P.H.C.; Satyanarayana, K.G.; Wypych, F. Nanocomposites: Synthesis, structure, properties and new application opportunities. *Mater. Res.* **2009**, *12*, 1–39. [CrossRef]

83. Sonawane, G.H.; Patil, S.P.; Sonawane, S.H. *Nanocomposites and Its Applications*; Elsevier Ltd.: Amsterdam, The Netherlands, 2018.

84. Meneghetti, P.; Qutubuddin, S. Synthesis, thermal properties and applications of polymer-clay nanocomposites. *Thermochim. Acta* **2006**, *442*, 74–77. [CrossRef]

85. Vilaseca, F.; Corrales, F.; Llop, M.F.; Pèlach, M.À.; Mutjé, P. Chemical treatment for improving wettability of biofibres into thermoplastic matrices. *Compos. Interfaces* **2005**, *12*, 725–738. [CrossRef]

86. Pothan, L.A.; Thomas, S.; Groeninckx, G. The role of fibre/matrix interactions on the dynamic mechanical properties of chemically modified banana fibre/polyester composites. *Compos. Part. A Appl. Sci. Manuf.* **2006**, *37*, 1260–1269. [CrossRef]

87. Gholampour, A.; Ozbakkaloglu, T. *A Review of Natural Fiber Composites: Properties, Modification and Processing Techniques, Characterization, Applications*; Springer: New York, NY, USA, 2020; Volume 55, p. 3.

88. Faruk, O.; Bledzki, A.K.; Fink, H.-P.; Sain, M. Biocomposites reinforced with natural fibers: 2000–2010. *Prog. Polym. Sci.* **2012**, *37*, 1552–1596. [CrossRef]

89. Rahman, S.; Mathur, V.; Asmatulu, R. Effect of nanoclay and graphene inclusions on the low-velocity impact resistance of Kevlar-epoxy laminated composites. *Compos. Struct.* **2018**, *187*, 481–488. [CrossRef]

90. Amiandamhen, S.O.; Meincken, M.; Tyhoda, L. Natural Fibre Modification and Its Influence on Fibre-matrix Interfacial Properties in Biocomposite Materials. *Fibers Polym.* **2020**, *21*, 677–689. [CrossRef]

91. Rippon, J.A.; Evans, D.J. *Improving the Properties of Natural Fibres by Chemical Treatments*; Elsevier Ltd.: Amsterdam, The Netherlands, 2020.

92. Khalid, M.Y.; Rashid, A.A.; Sheikh, M.F. Effect of anodizing process on inter laminar shear strength of GLARE composite through T-peel test: Experimental and numerical approach. *Exp. Tech.* **2021**. [CrossRef]

93. Corrales, F.; Vilaseca, F.; Llop, M.; Gironès, J.; Méndez, J.; Mutjè, P. Chemical modification of jute fibers for the production of green-composites. *J. Hazard. Mater.* **2007**, *144*, 730–735. [CrossRef]

94. Gallego, R.; Piras, C.C.; Rutgeerts, L.A.J.; Fernandez-Prieto, S.; De Borggraeve, W.M.; Franco, J.M.; Smets, J. Green approach for the activation and functionalization of jute fibers through ball milling. *Cellulose* **2020**, *27*, 643–656. [CrossRef]

95. Haneefa, A.; Bindu, P.; Aravind, I.; Thomas, S. Studies on tensile and flexural properties of short banana/glass hybrid fiber reinforced polystyrene composites. *J. Compos. Mater.* **2008**, *42*, 1471–1489. [CrossRef]

96. Arbelaiz, A.; Txueka, U.; Mezo, I.; Orue, A. Biocomposites Based on Poly(Lactic Acid) Matrix and Reinforced with Lignocellulosic Fibers: The Effect of Fiber Type and Matrix Modification. *J. Nat. Fibers* **2020**, 1–14. [CrossRef]

97. Arju, S.N.; Afsar, A.M.; Das, D.K.; Khan, M.A. Role of reactive dye and chemicals on mechanical properties of jute fabrics polypropylene composites. *Proc. Eng.* **2014**, *90*, 199–205. [CrossRef]

98. Van Voorn, B.; Smit, H.; Sinke, R.; De Klerk, B. Natural fibre reinforced sheet moulding compound. *Compos. Part. A Appl. Sci. Manuf.* **2001**, *32*, 1271–1279. [CrossRef]

99. Roy, K.; Debnath, S.C.; Tzounis, L.; Pongwisuthiruchte, A.; Potiyaraj, P. Effect of various surface treatments on the performance of jute fibers filled natural rubber (NR) composites. *Polymers* **2020**, *12*, 369. [CrossRef] [PubMed]

100. Zhou, F.; Cheng, G.; Jiang, B. Effect of silane treatment on microstructure of sisal fibers. *Appl. Surf. Sci.* **2014**, *292*, 806–812. [CrossRef]

101. De Klerk, M.D.; Kayondo, M.; Moelich, G.M.; de Villiers, W.I.; Combrinck, R.; Boshoff, W.P. Durability of chemically modified sisal fibre in cement-based composites. *Constr. Build. Mater.* **2020**, *241*, 117835. [CrossRef]

102. Khan, M.A.; Terano, M.; Gafur, M.A.; Alam, M.S. Studies on the mechanical properties of woven jute fabric reinforced poly(L-lactic acid) composites. *J. King Saud Univ. Eng. Sci.* **2016**, *28*, 69–74. [CrossRef]

103. Wang, B.; Panigrahi, S.; Tabil, L.; Crerar, W. Pre-treatment of flax fibers for use in rotationally molded biocomposites. *J. Reinf. Plast. Compos.* **2007**, *26*, 447–463. [CrossRef]

104. Paul, A.; Joseph, K.; Thomas, S. Effect of surface treatments on the electrical properties of low-density polyethylene composites reinforced with short sisal fibers. *Compos. Sci. Technol.* **1997**, *57*, 67–79. [CrossRef]

105. Kabir, M.M.; Wang, H.; Lau, K.T.; Cardona, F. Chemical treatments on plant-based natural fibre reinforced polymer composites: An overview. *Compos. Part. B Eng.* **2012**, *43*, 2883–2892. [CrossRef]

106. Ali, K.I.; Khan, M.A.; Ali, M.A. Study on Jute Material with Urethane Acrylate by UV Curing. *Radiat. Phys. Chem.* **1997**, *49*, 383–388.

107. Khattab, T.A.; Mohamed, A.L.; Hassabo, A.G. Development of durable superhydrophobic cotton fabrics coated with silicone/stearic acid using different cross-linkers. *Mater. Chem. Phys.* **2020**, *249*, 122981. [CrossRef]

108. Onwulata, C.I.; Cooke, P.H.; Harden, J.; Liu, Z.; Erhan, S.Z.; Akin, D.E.; Barton, F.E. Farrell Interfacial Modification of Hemp Fiber Reinforced Composites Using Fungal and Alkali Treatment. *J. Biobased Mater. Bioenergy* **2007**, *1*, 109–117. [CrossRef]

109. Sun, L.; Wang, S.; Zhang, J.; Li, W.; Lu, Z.; Zhang, Z.; Zhu, P.; Dong, Z. Preparation of a novel flame retardant containing triazine groups and its application on cotton fabrics. *New J. Chem.* **2020**. [CrossRef]

110. Mylsamy, K.; Rajendran, I. Influence of alkali treatment and fibre length on mechanical properties of short Agave fibre reinforced epoxy composites. *Mater. Des.* **2011**, *32*, 4629–4640. [CrossRef]

111. Wei, B.; Cao, H.; Song, S. Tensile behavior contrast of basalt and glass fibers after chemical treatment. *Mater. Des.* **2010**, *31*, 4244–4250. [CrossRef]

112. Khondker, A.; Ishiaku, U.S.; Nakai, A.; Hamada, H. Tensile, flexural and impact properties of jute fibre-based thermosetting composites. *Plast. Rubber Compos.* **2005**, *34*, 450–462. [CrossRef]

113. Senthilraja, R.; Sarala, R.; Antony, A.G. Effect of acetylation technique on mechanical behavior and durability of palm fibre vinyl-ester composites. *Mater. Today Proc.* **2020**, *21*, 634–637. [CrossRef]

114. Gironès, J.; Pimenta, M.T.B.; Vilaseca, F.; de Carvalho, A.J.F.; Mutjé, P.; Curvelo, A.A.S. Blocked isocyanates as coupling agents for cellulose-based composites. *Carbohydr. Polym.* **2007**, *68*, 537–543. [CrossRef]

115. Bachtiar, D.; Sapuan, S.; Hamdan, M. The effect of alkaline treatment on tensile properties of sugar palm fibre reinforced epoxy composites. *Mater. Des.* **2008**, *29*, 1285–1290. [CrossRef]

116. Vijay, R.; Manoharan, S.; Arjun, S.; Vinod, A.; Singaravelu, D.L. Characterization of Silane-Treated and Untreated Natural Fibers from Stem of Leucas Aspera. *J. Nat. Fibers* **2020**, 1–17. [CrossRef]

117. Bledzki, A.; Gassan, J. Composites reinforced with cellulose based fibres. *Prog. Polym. Sci.* **1999**, *24*, 221–274. [CrossRef]

118. Kaushik, V.K.; Kumar, A.; Kalia, S. Effect of Mercerization and Benzoyl Peroxide Treatment on Morphology, Thermal Stability and Crystallinity of Sisal Fibers. *Int. J. Text. Sci.* **2013**, *1*, 101–105. [CrossRef]

119. Keener, T.J.; Stuart, R.K.; Brown, T.K. Maleated coupling agents for natural fibre composites. *Compos. Part. A Appl. Sci. Manuf.* **2004**, *35*, 357–362. [CrossRef]

120. Madhu, P.; Sanjay, M.R.; Jawaid, M.; Siengchin, S.; Khan, A.; Pruncu, C.I. A new study on effect of various chemical treatments on Agave Americana fiber for composite reinforcement: Physico-chemical, thermal, mechanical and morphological properties. *Polym. Test.* **2020**, *85*, 106437. [CrossRef]

121. Knežević, M.; Kramar, A.; Hajnrih, T.; Korica, M.; Nikolić, T.; Žekić, A.; Kostić, M. Influence of Potassium Permanganate Oxidation on Structure and Properties of Cotton. *J. Nat. Fibers* **2020**, 1–13. [CrossRef]

122. Georgiopoulos, P.; Kontou, E.; Georgousis, G. Effect of silane treatment loading on the flexural properties of PLA/flax unidirectional composites. *Compos. Commun.* **2018**, *10*, 6–10. [CrossRef]

123. Zafeiropoulos, N.E.; Baillie, C.A.; Hodgkinson, J.M. Engineering and characterisation of the interface in flax fibre/polypropylene composite materials. Part II. The effect of surface treatments on the interface. *Compos. Part. A Appl. Sci. Manuf.* **2002**, *33*, 1185–1190. [CrossRef]

124. Joseph, K.; Varghese, S.; Kalaprasad, G.; Thomas, S.; Prasannakumari, L.; Koshy, P.; Pavithran, C. Influence of interfacial adhesion on the mechanical properties and fracture behaviour of short sisal fibre reinforced polymer composites. *Eur. Polym. J.* **1996**, *32*, 1243–1250. [CrossRef]

125. Ray, K.; Patra, H.; Swain, A.K.; Parida, B.; Mahapatra, S.; Sahu, A.; Rana, S. Glass/jute/sisal fiber reinforced hybrid polypropylene polymer composites: Fabrication and analysis of mechanical and water absorption properties. *Mater. Today Proc.* **2020**. [CrossRef]

126. Parameshwaran, R.; Kalaiselvam, S. Energy efficient hybrid nanocomposite-based cool thermal storage air conditioning system for sustainable buildings. *Energy* **2013**, *59*, 194–214. [CrossRef]

127. Zhang, L.; Liu, W.; Wang, L.; Ling, Z. On-axis and off -axis compressive behavior of pultruded GFRP composites at elevated temperatures. *Compos. Struct.* **2020**, *236*, 111891. [CrossRef]

128. Sanjay, M.R.; Arpitha, G.R.; Naik, L.L.; Gopalakrishna, K.; Yogesha, B. Applications of Natural Fibers and Its Composites: An Overview. *Nat. Resour.* **2016**, *7*, 108–114. [CrossRef]

129. Viscusi, G.; Barra, G.; Verdolotti, L.; Galzerano, B.; Viscardi, M.; Gorrasi, G. Natural fiber reinforced inorganic foam composites from short hemp bast fibers obtained by mechanical decortation of unretted stems from the wastes of hemp cultivations. *Mater. Today Proc.* **2020**, *34*, 176–179. [CrossRef]

130. Westman, M.P.; Fifield, L.S.; Simmons, K.L.; Laddha, S.; Kafentzis, T.A. *Natural Fiber Composites: A Review*; Pacific Northwest National Laboratory: Richland, WA, USA, 2010.

131. Holbery, J.; Houston, D. Natural-fiber-reinforced polymer composites in automotive applications. *JOM* **2006**, *58*, 80–86. [CrossRef]

132. Lertwattanaruk, P.; Suntijitto, A. Properties of natural fiber cement materials containing coconut coir and oil palm fibers for residential building applications. *Constr. Build. Mater.* **2015**, *94*, 664–669. [CrossRef]

133. Leao, A.L.; Rowell, R.; Tavares, N. Applications of natural fibers in automotive industry in Brazil—Thermoforming process. In *Science and Technology of Polymers and Advanced Materials*; Prasad, P.N., Mark, J.E., Kandil, S.H., Kafafi, Z.H., Eds.; Springer: Boston, MA, USA, 1998; pp. 755–761. [CrossRef]

134. Jothibasu, S.; Mohanamurugan, S.; Vijay, R.; Singaravelu, D.L.; Vinod, A.; Sanjay, R.M. Investigation on the mechanical behavior of areca sheath fibers/jute fibers/glass fabrics reinforced hybrid composite for light weight applications. *J. Ind. Text.* **2018**. [CrossRef]

135. Ramesh, M.; Palanikumar, K.; Reddy, K.H. Comparative evaluation on properties of hybrid glass fiber-sisal/jute reinforced epoxy composites. *Proc. Eng.* **2013**, *51*, 745–750. [CrossRef]

136. Satyanarayana, K.G.; Sukumaran, K.; Mukherjee, R.S.; Pavithran, C.; Piuai, S.G.K. Natural Fibre-Polymer Composites. *Cement Concrete Compos.* **1990**, *12*, 117–136. [CrossRef]

137. Tholibon, D.; Tharazi, I.; Sulong, A.B.; Muhamad, N.; Ismial, N.F.; Radzi, M.K.F.M.; Radzuan, N.M.; Hui, D. Kenaf Fiber Composites: A Review on Synthetic and Biodegradable Polymer Matrix. *J. Kejuruter.* **2019**, *31*, 65–76.
138. Sanjeevi, S.; Ayyanar, A.; Kalimuthu, R.; Susaiyappan, S. Effects of chemical modification on the mechanical properties of Calotropis Gigantea fiber-reinforced phenol formaldehyde biocomposites. *Medziagotyra* **2019**, *26*, 295–299. [CrossRef]
139. Jaafar, J.; Siregar, J.P.; Salleh, S.M.; Hamdan, M.H.M.; Cionita, T.; Rihayat, T. Important Considerations in Manufacturing of Natural Fiber Composites: A Review. *Int. J. Precis. Eng. Manuf. Green Technol.* **2019**, *6*, 647–664. [CrossRef]

MDPI

St. Alban-Anlage 66

4052 Basel

Switzerland

Tel. +41 61 683 77 34

Fax +41 61 302 89 18

www.mdpi.com

Coatings Editorial Office

E-mail: coatings@mdpi.com

www.mdpi.com/journal/coatings